Porter and Chester Institute

Wethersfield CT 06109

FOURTH EDITION

# Digital Principles and Applications

**ALBERT PAUL MALVINO, PH.D.**
President
Malvino, Inc.

**DONALD P. LEACH, PH.D.**
Associate Professor
Santa Clara University

**McGRAW-HILL BOOK COMPANY**

New York   Atlanta   Dallas   St. Louis   San Francisco
Auckland   Bogotá   Guatemala   Hamburg   Johannesburg
Lisbon   London   Madrid   Mexico   Montreal   New Delhi
Panama   Paris   San Juan   São Paulo   Singapore
Sydney   Tokyo   Toronto

**Sponsoring Editor:** Paul Berk
**Editing Supervisor:** Mitsy Kovacs
**Design and Art Supervisor:** Caryl Valerie Spinka
**Production Supervisor:** Albert Rihner

**Text Designer:** Lilly Caporlingua
**Cover Designer:** Renée Kilbride Edelman


**Library of Congress Cataloging-in-Publication Data**

Malvino, Albert Paul.
  Digital principles and applications.

  Includes index.
  1. Digital electronics.   2. Electronic digital
computers—Circuits.   I. Leach, Donald P.   II. Title.
TK7868.D5M3   1986        621.395        85-23979
ISBN 0-07-039883-6

**Digital Principles and Applications,** Fourth Edition

1 2 3 4 5 6 7 8 9 0 DOCDOC 8 9 2 1 0 9 8 7 6 5

ISBN 0-07-039883-6

# CONTENTS

# PREFACE

The fourth edition of *Digital Principles and Applications* maintains the approach of earlier editions. Rather than focusing on computers, architecture, and programming, the emphasis is placed on those principles that apply not only to computers but also to applications for automobiles, communications, industrial automation, process control, and so on. This general introduction to digital electronics provides a broader base for study in specialized areas.

This edition contains many improvements, including updates and expansions of areas that were either weak or nonexistent in the previous edition. Actual IC chips and pin numbers now are used in illustrative examples, and there are new sections and problems on digital troubleshooting. Overall, this fourth edition is more accurately aimed at developing the skills needed to work effectively in digital electronics.

Some other improvements are:

1. Timing diagrams as well as truth tables are used to analyze logic circuits.
2. The logic clip, probe, and pulser are discussed in the text, then used in examples and problems.
3. New material on ROMs, PROMs, and PALs is included in Chap. 3.
4. 2's complement arithmetic is stressed in Chap. 5.
5. TTL input/output voltages and currents, driver/load models, internal TTL structures, outside-world interfacing, and assertion-level logic with signal labeling are all described in detail.
6. A comfortable introduction to the details of CMOS chips, including the interfacing to TTL devices, is included.
7. There is a modern presentation of flip-flops, timers, shift registers, and counters with the emphasis on commercially available MSI and LSI chips.
8. An entirely new chapter on semiconductor memories has been added that includes widely used RAMs, ROMs, and PROMs.
9. There is a comprehensive discussion of seven-seg-

ment indicators and display circuits, including the logic and timing necessary for display multiplexing.

10. DACs and ADCs are fully covered, including single and dual-slope conversion techniques.
11. Numerous practical examples are presented throughout the text, and the final chapter is a sampling of representative applications that demonstrate many of the basic principles of digital circuits.

The main prerequisite for this book is an understanding of semiconductor diodes and transistors. The length and level make the book suitable for a beginning course in digital electronics. Study aids include end-of-chapter summary, glossary, and problems.

## Acknowledgments

The authors and publisher wish to give grateful acknowledgment to the following individuals, whose experience in digital electronics instruction and industry practice allowed them to make invaluable suggestions for the improvement of this fourth edition: Gary Boyington, Chemeketa Community College, Salem, Oregon; Charles E. Center, Central Kentucky State Vo-Tech, Lexington, Kentucky; Harold L. Coomes, Patterson State Technical College, Montgomery, Alabama; Larry Flicker, New York City Technical College, Brooklyn, New York; Robert D. King, Texarkana College, Texarkana, Texas; Michael Mason, Cabrillo College, Aptos, California; Richard A. Shannon, South Plains College, Lubbock, Texas; Tom Strand, Foothill College, Los Altos Hills, California; Wayne Vyrostek, Westark Community College, Fort Smith, Arkansas; Willard Waters, RETS Electronic Training Center, Birmingham, Alabama; and Ronald L. Way, Muskegon Community College, Grant, Michigan.

**A. P. Malvino**
**D. P. Leach**

# 1

## L O G I C
## C I R C U I T S

The main difference between *analog* and *digital* operation is the way the load line is used. With analog circuits, adjacent points on the load line may be used, so that the output voltage is continuous. Because of this, the output voltage can have an infinite number of values. One way to get analog operation is with a sinusoidal input. The continuously changing input voltage produces a continuously changing output voltage.

Digital circuits are different. Almost all digital circuits are designed for *two-state* operation. This means using only two nonadjacent points on the load line, typically saturation and cutoff. As a result, the output voltage has only two states (values), either low or high. One way to get digital operation is with a square-wave input. If large enough, this type of input drives the transistors into saturation and cutoff, producing a two-state output.

This chapter begins with the binary number system because it is used with digital circuits. Next, we examine the *gate*, a digital circuit with one or more input voltages but only one output voltage. The most basic gates are called the NOT gate, the OR gate, and the AND gate. By connecting these gates in different ways, we can build circuits that perform arithmetic and other functions associated with the human brain. Because they simulate mental processes, gates are often called *logic circuits*.

## 1-1 BINARY NUMBER SYSTEM

The *binary number system* is a system that uses only the digits 0 and 1 as codes. All other digits (2 to 9) are thrown away. To represent decimal numbers and letters of the alphabet with the binary code, you have to use different strings of binary digits for each number or letter. The idea is similar to the Morse code, where strings of dots and dashes are used to code all numbers and letters. What follows is a discussion of decimal and binary counting.

### Decimal Odometer

To understand how to count with binary numbers, it helps to review how an *odometer* (miles indicator of a car) counts with decimal numbers. When a car is new, its odometer starts with

$$00000$$

After 1 mile the reading becomes

$$00001$$

Successive miles produce 00002, 00003, and so on, up to

$$00009$$

A familiar thing happens at the end of the tenth mile. When the units wheel turns from 9 back to 0, a tab on this wheel forces the tens wheel to advance by 1. This is why the numbers change to

$$00010$$

### Reset-and-Carry

The units wheel has reset to 0 and sent a carry to the tens wheel. Let's call this familiar action *reset and carry*. The other wheels of an odometer also reset and carry. For instance, after 999 miles the odometer shows

$$00999$$

What does the next mile do? The units wheel resets and carries, the tens wheel resets and carries, the hundreds wheel resets and carries, and the thousands wheel advances by 1, to get

$$01000$$

### Binary Odometer

Visualize a binary odometer, a device whose wheels have only two digits, 0 and 1. When each wheel turns, it displays 0, then 1, then back to 0, and the cycle repeats. A four-digit binary odometer starts with

$$0000 \qquad \text{(zero)}$$

After 1 mile, it indicates

$$0001 \qquad \text{(one)}$$

The next mile forces the units wheel to reset and carry, so the numbers change to

$$0010 \qquad \text{(two)}$$

The third mile results in

$$0011 \qquad \text{(three)}$$

After 4 miles, the units wheel resets and carries, the second wheel resets and carries, and the third wheel advances by 1:

$$0100 \qquad \text{(four)}$$

Table 1–1 shows all the binary numbers from 0000 to 1111, equivalent to decimal 0 to 15. Study this table carefully and practice counting from 0000 to 1111 until you can do it easily. Why? Because all kinds of logic circuits are based on counting from 0000 to 1111.

Incidentally, the word *bit* is used as an abbreviation for *binary digit*. Table 1–1 is a list of 4-bit numbers from 0000 to 1111. When a binary number has 4 bits, it is sometimes called a *nibble*. Table 1–1 shows 16 nibbles (0000 to 1111). A binary number with 8 bits is known as a *byte*; this has become the basic unit of data used in computers. You will learn more about bits, nibbles, and bytes in later chapters. For now, memorize these definitions:

$$\text{bit} = X$$
$$\text{nibble} = XXXX$$
$$\text{byte} = XXXXXXXX$$

where the X may be a 0 or a 1.

**TABLE 1–1** *4-Digit Binary Numbers*

| Binary | Decimal |
|--------|---------|
| 0000 | 0 |
| 0001 | 1 |
| 0010 | 2 |
| 0011 | 3 |
| 0100 | 4 |
| 0101 | 5 |
| 0110 | 6 |
| 0111 | 7 |
| 1000 | 8 |
| 1001 | 9 |
| 1010 | 10 |
| 1011 | 11 |
| 1100 | 12 |
| 1101 | 13 |
| 1110 | 14 |
| 1111 | 15 |

## 1-2 INVERTERS

The simplest way to use a transistor is as a *switch,* meaning that we operate it at either saturation or cutoff but nowhere else along the load line. When saturated, a transistor is like a closed switch. When cut off, it's like an open switch. This is two-state operation because only two distinct points on the load line are used.

### Common-Emitter Circuit

For instance, Fig. 1–1a shows a common-emitter circuit, and Fig. 1-1b is a simplified drawing of the same circuit. The base current is given by

$$I_B = \frac{V_{\text{in}} - V_{BE}}{R_B} \qquad (1\text{--}1)$$

Because of the transistor's current gain, the collector current is

$$I_C = \beta_{dc} I_B \qquad (1\text{--}2)$$

and the output voltage is

$$V_{\text{out}} = V_{CC} - I_C R_C \qquad (1\text{--}3)$$

### Load Line

Figure 1–2a shows the dc and ac load line. When $V_{\text{in}}$ is zero in Fig. 1–1, the transistor goes into cutoff and the operating point is at the lower end of the load line. To a first approximation, the transistor is like an open switch between the collector and the emitter (see Fig. 1–2b). On the other hand, when $V_{\text{in}}$ is large, the transistor goes into saturation and the operating point is at the upper end of the load line. Ideally, the transistor is like a closed switch (Fig. 1–2c).

### Hard Saturation

*Soft saturation* means the transistor is barely saturated; the base current is just enough to operate the transistor at the upper end of the load line. Soft saturation is not reliable in

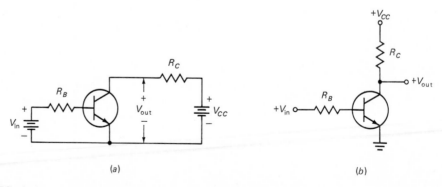

(a)

(b)

**FIG. 1–1.** (a) Common-emitter circuit (b) simplified drawing

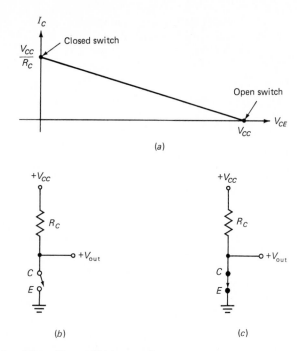

(a)

(b)                                    (c)

**FIG. 1-2.** (a) Load line (b) cutoff (c) saturation

mass production because of the variation in $\beta_{dc}$ (same as $h_{FE}$). A circuit using soft saturation can easily come out of saturation with temperature change or transistor replacement.

*Hard saturation* means the transistor has sufficient base current to be saturated under all operating conditions. To get hard saturation, a designer makes $I_{C,\text{sat}}$ approximately 10 times the value of $I_{B,\text{sat}}$. A ratio of 10:1 is low enough for almost any transistor to remain saturated, despite temperature extremes, transistor replacement, supply-voltage changes, etc.

For instance, the input voltage of Fig. 1-3 may be either 0 or +5 V. When $V_{\text{in}}$ is zero, the transistor cuts off and $V_{\text{out}}$ equals +5 V. When $V_{\text{in}}$ is +5 V, the transistor goes into hard saturation and $V_{\text{out}}$ is approximately zero. In this case, $I_{C,\text{sat}}$ is ideally 5 mA. Ignoring

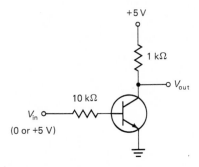

**FIG. 1-3.** Example

the $V_{BE}$ of the transistor, $I_{B,\text{sat}}$ is approximately 0.5 mA. Therefore, the ratio of collector to base current is approximately 10 to 1.

When the maximum input voltage equals the supply voltage, you can get hard saturation by using a ratio of approximately 10:1 for $R_B/R_C$. For example, $R_B = 10$ k$\Omega$ and $R_C = 1$ k$\Omega$ in Fig. 1–3; the ratio of $R_B/R_C$ is 10, so the transistor is in hard saturation. Other values like $R_B = 47$ k$\Omega$ and $R_C = 4.7$ k$\Omega$ also produce hard saturation because $R_B/R_C$ still equals 10.

## Transistor Inverter

An *inverter* is a gate with only one input and one output. It is called an inverter because the output state is always opposite the input state. Specifically, when the input voltage is high, the output is low. On the other hand, when the input voltage is low, the output is high.

Figure 1–3 is an example of an inverter. Since the input voltage can have only two values (0 or +5 V), the transistor operates only at cutoff or saturation. As a result, we get two-state operation, which means the output voltage can have only two steady-state values (low or high).

Table 1–2 summarizes the operation of an inverter. A low input produces a high output, and a high input produces a low output. An inverter is also called a NOT gate because the output is not the same as the input. The output is also called the *complement* (opposite) of the input.

## Why Two-State Operation Is Used

Two-state operation is widespread in digital electronics because it is the most reliable way to operate transistors and other switching devices. With two-state operation all signals are easily recognized as either low or high. This simplicity is the key to building reliable computer circuits. Since the active devices need only appear open or closed, designers can overcome the wide tolerances of $\beta_{dc}$, $r'_e$, etc., and produce the highly reliable circuits found in modern computers.

## Inverter Symbol

The transistor inverter of Fig. 1–3 is the simplest way to build an inverter. Many other designs are possible to improve the input impedance, output impedance, and so on. Figure 1–4a is the symbol for an inverter of any design. Sometimes a schematic diagram will use the alternative symbol of Fig. 1–4b. The bubble (small circle) is on the input side. Whenever you see either of these symbols, remember that the output is the complement of the input.

**TABLE 1–2** *Inverter*

| $V_{\text{in}}$ | $V_{\text{out}}$ |
| --- | --- |
| Low | High |
| High | Low |

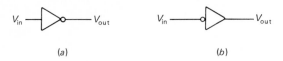

FIG. 1–4.  (a) Inverter symbol (b) another symbol

## TTL Circuits

In 1964 Texas Instruments introduced *transistor-transistor logic* (TTL), a widely used family of digital devices. TTL is fast, inexpensive, and easy to use. Figure 1–5 shows the pinout diagram of a 7404, a TTL hex inverter. This integrated circuit contains six inverters. After applying +5 V (the supply voltage for all TTL devices) to pin 14 and grounding pin 7, you can connect any or all of the inverters to other TTL devices. For instance, if you only need one inverter, you can connect an input signal to pin 1 and take the output signal from pin 2; the other inverters can be left unconnected.

☐ EXAMPLE 1—1

A 1-kHz square wave drives pin 1 of a 7404 (see Fig. 1–5). What does the voltage waveform at pin 2 look like?

☐ SOLUTION

Figure 1-6a shows what you will see on a dual-trace oscilloscope. Assuming you have set the sweep timing to get the upper waveform (pin 1), then you would see an inverted square wave on pin 2.

☐ EXAMPLE 1—2

If a 500-Hz square wave drives pin 3 of a 7404, what is the waveform on pin 4?

☐ SOLUTION

Pins 3 and 4 are the input and output pins of an inverter (see Fig. 1–5). A glance at Fig. 1–6b shows the typical waveforms on the input (pin 3) and output (pin 4) of a 7404. Again, the output waveform is the complement of the input waveform. Because of two-state operation, rectangular waveforms like this are the normal shape of digital signals.

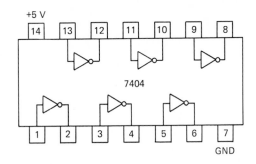

FIG. 1–5.  Pinout diagram of a 7404

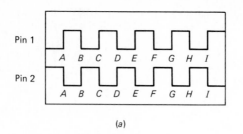 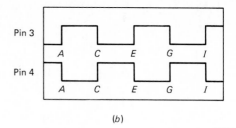

Pin 1 ... A B C D E F G H I

Pin 2 ... A B C D E F G H I

(a)

Pin 3 ... A C E G I

Pin 4 ... A C E G I

(b)

**FIG. 1–6.**

Incidentally, a *timing diagram* is a picture of the input and output waveforms of a digital circuit. Figures 1–6*a* and *b* are examples of timing diagrams.

## 1·3 OR GATES

An OR gate has two or more input signals but only one output signal. It is called an OR gate because the output voltage is high if any or all of the input voltages are high. For instance, the output of a 2-input OR gate is high if either or both inputs are high.

Figure 1–7*a* shows one way to build a 2-input OR gate, and Fig. 1–7*b* is a simplified drawing of the same circuit. The input voltages are labeled $A$ and $B$, while the output voltage is $Y$. Let us assume the input voltages are either 0 V (low state) or +5 V (high state). There are only four possible cases:

**CASE 1** $A$ is low and $B$ is low. With both input voltages low, the output voltage is low because both diodes are nonconducting. Therefore, $Y$ is low.

**CASE 2** $A$ is low and $B$ is high. The high $B$ input voltage (+5 V) forward-biases the lower diode, producing an output voltage that is ideally +5 V, or approximately +4.3 V including the diode voltage drop. Whether you use the ideal output (+5 V) or second approximation (+4.3 V), $Y$ is high. Note that the upper diode is reverse-biased.

**CASE 3** $A$ is high and $B$ is low. Because of the symmetry of the circuit, the circuit operation is similar to case 2. The upper diode is on, the lower diode is off, and $Y$ is high.

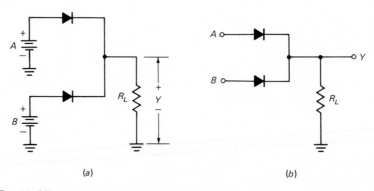

(a)          (b)

**FIG. 1–7.**   (a) OR gate (b) simplified drawing

| TABLE 1–3 *Two-Input OR Gate* | | | | TABLE 1–4 *Binary Equivalent* | | |
|---|---|---|---|---|---|---|
| *A* | *B* | *Y* | | *A* | *B* | *Y* |
| Low | Low | Low | | 0 | 0 | 0 |
| Low | High | High | | 0 | 1 | 1 |
| High | Low | High | | 1 | 0 | 1 |
| High | High | High | | 1 | 1 | 1 |

**CASE 4** *A* is high and *B* is high. With both inputs at +5 V, both diodes are forward-biased. Since the input voltages are in parallel, the output voltage is +5 V ideally (+4.3 V to a second approximation). Therefore, *Y* is high.

## Truth Table

A *truth table* is a table that shows all the input-output possibilities of a logic circuit. Table 1–3 is the truth table for an OR gate. Examine this table carefully and note the following: the OR gate has a high output when either *A* or *B* or both are high. In other words, the OR gate is an any-or-all gate; an output occurs when any or all of the inputs are high.

Because virtually all digital circuits are based on two-state operation, it is convenient to use *binary numbers* when troubleshooting, analyzing, and designing digital systems. In one approach, the binary digit 0 is the code for low voltage, and binary digit 1 is the code for high voltage. This is why you often see the OR-gate truth table as shown in Table 1–4.

## Three Inputs

Figure 1–8 shows a 3-input OR gate. The inputs are *A, B,* and *C*. When all inputs are low, none of the diodes is conducting; therefore, *Y* is low. If *A* or *B* or *C* is high, *Y* will be high because the diode connected to the high input will turn on. Table 1–5 summarizes all input possibilities. Table 1–6 shows the same truth table in binary form.

Table 1–6 allows us to check that all input possibilities are included. Why? Because every possibility is included when the input entries follow a binary sequence. In Table 1–6, for example, the first *ABC* entry is 000, the next is 001, then 010, and so on, up to the final entry of 111. Since all binary numbers are present, all input possibilities are included.

Incidentally, the number of rows in a truth table equals $2^n$, where *n* is the number of inputs. For a 2-input OR gate, the truth table has $2^2$, or 4 rows. A 3-input OR gate has a

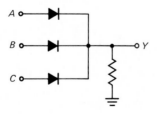

**FIG. 1–8.** Three-input OR gate

| TABLE 1–5 *Three-Input OR Gate* | | | |
|---|---|---|---|
| A | B | C | Y |
| Low | Low | Low | Low |
| Low | Low | High | High |
| Low | High | Low | High |
| Low | High | High | High |
| High | Low | Low | High |
| High | Low | High | High |
| High | High | Low | High |
| High | High | High | High |

| TABLE 1–6 *Binary Equivalent* | | | |
|---|---|---|---|
| A | B | C | Y |
| 0 | 0 | 0 | 0 |
| 0 | 0 | 1 | 1 |
| 0 | 1 | 0 | 1 |
| 0 | 1 | 1 | 1 |
| 1 | 0 | 0 | 1 |
| 1 | 0 | 1 | 1 |
| 1 | 1 | 0 | 1 |
| 1 | 1 | 1 | 1 |

truth table with $2^3$, or 8 rows, while a 4-input OR gate results in $2^4$, or 16 rows, and so on.

An OR gate can have as many inputs as desired; add one diode for each additional input. Six diodes result in a 6-input OR gate, nine diodes in a 9-input OR gate. No matter how many inputs, the action of any OR gate is summarized like this: One or more high inputs produce a high output.

## Logic Symbols

Figure 1–9a shows the symbol for a 2-input OR gate of any design. Whenever you see this symbol, remember the output is high if either input is high.

Shown in Fig. 1–9b is the logic symbol for a 3-input OR gate. Figure 1–9c is the symbol for a 4-input OR gate. For these gates, the output is high when any input is high. The only way to get a low output is by having all inputs low.

When there are many input signals, it's common drafting practice to extend the input side as needed to allow sufficient space between the input lines. For instance, Fig. 1–9d is the symbol for a 12-input OR gate. The same idea applies to any type of gate; extend the input side when necessary to accommodate a large number of input signals.

## TTL OR Gates

Figure 1–10 shows the pinout diagram of a 7432, a TTL quad 2-input OR gate. This digital integrated circuit (IC) contains four 2-input OR gates inside a 14-pin dual-in-line package (DIP). After connecting a supply voltage of +5 V to pin 14 and a ground to pin 7, you can connect one or more of the OR gates to other TTL devices.

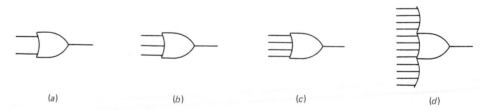

| (a) | (b) | (c) | (d) |

**FIG. 1–9.** OR gate symbols (a) two-input (b) three-input (c) four-input (d) twelve-input

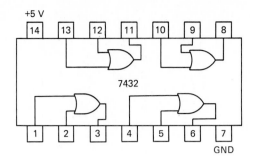

**FIG. 1–10.** Pinout diagram of a 7432

## Timing Diagram

Figure 1–11 shows an example of a timing diagram for a 2-input OR gate. The input voltages drive pins 1 and 2 of a 7432. Notice that the output (pin 3) is low only when both inputs are low. The output is high the rest of the time because one or more input pins are high.

☐ EXAMPLE 1—3

Work out the truth table for Fig. 1–12a.

☐ SOLUTION

With two input signals (A and B), four input cases are possible: low–low, low–high, high–low, and high–high. For convenience, let L stand for low and H for high. Then, the input possibilities are LL, LH, HL, and HH as listed in Table 1–7. Here is what happens for each input possibility.

**CASE 1** A is low and B is low. With both input voltages in the low state, each inverter has a high output. This means the OR gate has a high output, the first entry of Table 1–7.

**CASE 2** A is low and B is high. With these inputs the upper inverter has a high output, while the lower inverter has a low output. Since the OR gate still has a high input, the output Y is high.

**CASE 3** A is high and B is low. Now, the upper inverter has a low output and the lower inverter has a high output. Again, the OR gate produces a high output, so that Y is high.

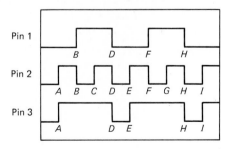

**FIG. 1–11.** Timing diagram

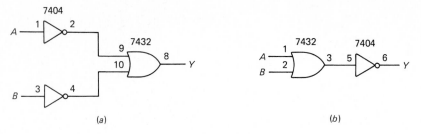

FIG. 1–12.

CASE 4 *A* is high and *B* is high. With both inputs high, each inverter has a low output. This time, the OR gate has all inputs in the low state, so that *Y* is low, as shown by the final entry of Table 1–7.

Incidentally, the circuit of Fig. 1–12*a* uses only one-third of a 7404 and one-fourth of a 7432. The other gates in these digital ICs are not connected, which is all right because you don't have to use all of the available gates.

☐ EXAMPLE 1—4

Work out the truth table for Fig. 1–12*b*.

☐ SOLUTION

Again, two input signals imply four possible input cases: low–low, low–high, high–low, and high–high. Here is what happens for each input possibility.

CASE 1 *A* is low and *B* is low. With both input voltages in the low state, the OR gate has a low output. Therefore, the inverter has a high output, as shown by the first entry of Table 1–8.

CASE 2 *A* is low and *B* is high. It only takes one high input to an OR gate to produce a high output. The inverter then complements this high voltage to get a low output, so that *Y* is low.

CASE 3 *A* is high and *B* is low. Because the circuit is symmetrical about the horizontal axis, the reasoning is identical to case 2, which means that *Y* is low.

CASE 4 *A* is high and *B* is high. With both inputs high, the OR gate has a high output, the inverter has a low output, and *Y* is low, as shown by the final entry of Table 1–8.

<table>
<tr><td colspan="3" align="center">TABLE 1–7</td></tr>
<tr><td align="center">A</td><td align="center">B</td><td align="center">Y</td></tr>
<tr><td align="center">L</td><td align="center">L</td><td align="center">H</td></tr>
<tr><td align="center">L</td><td align="center">H</td><td align="center">H</td></tr>
<tr><td align="center">H</td><td align="center">L</td><td align="center">H</td></tr>
<tr><td align="center">H</td><td align="center">H</td><td align="center">L</td></tr>
</table>

<table>
<tr><td colspan="3" align="center">TABLE 1–8</td></tr>
<tr><td align="center">A</td><td align="center">B</td><td align="center">Y</td></tr>
<tr><td align="center">L</td><td align="center">L</td><td align="center">H</td></tr>
<tr><td align="center">L</td><td align="center">H</td><td align="center">L</td></tr>
<tr><td align="center">H</td><td align="center">L</td><td align="center">L</td></tr>
<tr><td align="center">H</td><td align="center">H</td><td align="center">L</td></tr>
</table>

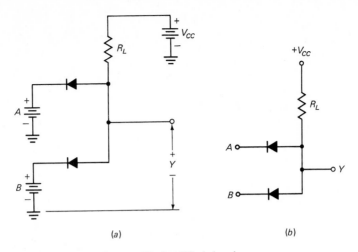

**FIG. 1–13.** (a) Two-input AND gate (b) simplified drawing

## 1-4 AND GATES

The AND gate has a high output only when all inputs are high. Figure 1–13a shows one way to build a 2-input AND gate, and Fig. 1–13b is a simplified drawing of the same circuit. The input voltages are labeled $A$ and $B$, while the output voltage is $Y$. Let us assume a supply voltage $V_{CC}$ of +5 V. Also, we will assume the input voltages are either 0 V (low state) or +5 V (high state). There are only four possible input cases:

**CASE 1** $A$ is low and $B$ is low. With both input voltages low, the cathode of each diode is grounded. Therefore, the positive supply forward-biases both diodes in parallel. Because of this, the output voltage is ideally zero (to a second approximation, 0.7 V). This means $Y$ is low.

**CASE 2** $A$ is low and $B$ is high. Since A is low, the upper diode is forward-biased, pulling the output down to a low voltage. With the $B$ input high, the lower diode goes into reverse bias. Whether you use the ideal output (0 V) or the second approximation (+0.7 V), $Y$ is low.

**CASE 3** $A$ is high and $B$ is low. Because of the symmetry of the circuit, the circuit operation is similar to case 2. The upper diode is off, the lower diode is on, and $Y$ is low.

**CASE 4** $A$ is high and $B$ is high. With both inputs at +5 V, both diodes are nonconducting because the voltage across each is zero. Since the diodes are off, there is no current through $R_L$, and the output is pulled up to the supply voltage (+5 V). Therefore, $Y$ is high.

### Truth Table

Table 1–9 summarizes all input-output possibilities for a 2-input AND gate. Examine this table carefully and remember the following: the AND gate has a high output only when $A$ and $B$ are high. In other words, the AND gate is an all-or-nothing gate; a high output occurs only when all inputs are high.

| TABLE 1–9 2-Input AND Gate | | | TABLE 1–10 Binary Equivalent | | |
|---|---|---|---|---|---|
| A | B | Y | A | B | Y |
| L | L | L | 0 | 0 | 0 |
| L | H | L | 0 | 1 | 0 |
| H | L | L | 1 | 0 | 0 |
| H | H | H | 1 | 1 | 1 |

Table 1–10 is the binary equivalent of Table 1–9, where a binary 0 represents the low state, while a binary 1 stands for the high state.

## Three Inputs

Figure 1–14 shows a 3-input AND gate. The inputs are *A, B,* and *C*. When all inputs are low, all diodes are conducting; therefore, *Y* is low. If even one input is low, *Y* is in the low state because the diode connected to the low input will remain on. The only way to get a high output is to raise all inputs to the high state (+5 V). Then, all diodes stop conducting and the supply voltage (+5 V) pulls the output up to the high state. Table 1–11 summarizes all input-output possibilities. Table 1–12 gives the same truth table in binary code.

## Logic Symbols

Figure 1–15*a* shows the symbol for a 2-input AND gate of any design. Shown in Fig. 1–15*b* is the logic symbol for a 3-input AND gate. Figure 1–15*c* is the symbol for a 4-input AND gate. Remember: For any of these gates, the output is high only if all inputs are high. As before, it's common drafting practice to extend the input sides when there are many input signals. For instance, Fig. 1–15*d* is the symbol for a 12-input AND gate.

## TTL AND Gates

Figure 1–16 shows the pinout diagram of a 7408, a TTL quad 2-input AND gate. This digital IC contains four 2-input AND gates. After connecting a supply voltage of +5 V to pin 14 and a ground to pin 7, you can connect one or more of the AND gates to other TTL devices. (TTL AND gates are also available in triple 3-input and dual 4-input packages. See Appendix 3 for pinout diagrams.)

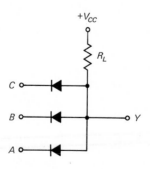

**FIG. 1–14.**   Three-input AND gate

| TABLE 1-11 | 3-Input AND Gate | | | | TABLE 1-12 | Binary Equivalent | | |
|---|---|---|---|---|---|---|---|---|
| A | B | C | Y | | A | B | C | Y |
| L | L | L | L | | 0 | 0 | 0 | 0 |
| L | L | H | L | | 0 | 0 | 1 | 0 |
| L | H | L | L | | 0 | 1 | 0 | 0 |
| L | H | H | L | | 0 | 1 | 1 | 0 |
| H | L | L | L | | 1 | 0 | 0 | 0 |
| H | L | H | L | | 1 | 0 | 1 | 0 |
| H | H | L | L | | 1 | 1 | 0 | 0 |
| H | H | H | H | | 1 | 1 | 1 | 1 |

## Timing Diagram

Figure 1–17 shows an example of a timing diagram for a 2-input AND gate. The input voltages drive pins 1 and 2 of a 7408. Notice that the output (pin 3) is high only when both inputs are high (between C and D, G and H, etc.). The output is low the rest of the time.

☐ EXAMPLE 1—5

Work out the truth table for Fig. 1–18*a*.

☐ SOLUTION

With two input signals, four input cases are possible: low–low, low–high, high–low, and high–high. Again, let L stand for low and H for high. Then, the input possibilities are LL, LH, HL, and HH, as listed in Table 1–13. Here is what happens for each input possibility.

**CASE 1** *A* is low and *B* is low. With both input voltages in the low state, each inverter has a high output. This means the AND gate has a high output, the first entry of Table 1–13.

**CASE 2** *A* is low and *B* is high. With these inputs the upper inverter has a high output, while the lower inverter has a low output. Since the AND gate produces a low output, *Y* is low.

**CASE 3** *A* is high and *B* is low. Now, the upper inverter has a low output and the lower inverter has a high output. Again, the AND gate produces a low output, so *Y* is low.

**CASE 4** *A* is high and *B* is high. With both inputs high, each inverter has a low output. Again, the AND gate has a low output, as shown by the final entry of Table 1–13.

☐ EXAMPLE 1—6

Work out the truth table for Fig. 1–18*b*.

☐ SOLUTION

Two input signals imply four possible input cases: low–low, low–high, high–low, and high–high. Here is what happens for each input possibility.

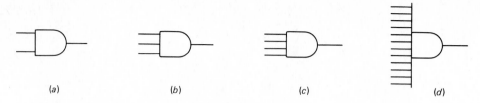

**FIG. 1–15.** AND gate symbols (a) two-input (b) three-input (c) four-input and (d) twelve-input

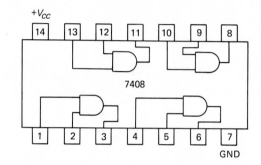

**FIG. 1–16.** Pinout diagram of a 7408

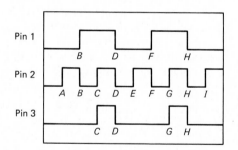

**FIG. 1–17.** Timing diagram

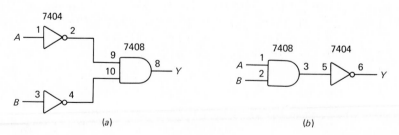

**FIG. 1–18.**

| TABLE 1–13 | | | | TABLE 1–14 | | |
|:---:|:---:|:---:|:---:|:---:|:---:|:---:|
| A | B | Y | | A | B | Y |
| L | L | H | | L | L | H |
| L | H | L | | L | H | H |
| H | L | L | | H | L | H |
| H | H | L | | H | H | L |

**CASE 1** *A* is low and *B* is low. With both input voltages in the low state, the AND gate has a low output. Therefore, the inverter has a high output, as shown by the first entry of Table 1–14.

**CASE 2** *A* is low and *B* is high. Because one of the inputs is low, the AND gate has a low output. The inverter then complements this low voltage to get a high output, so *Y* is high.

**CASE 3** *A* is high and *B* is low. Because the circuit is symmetrical about the horizontal axis, the reasoning is identical to case 2, which means that *Y* is high.

**CASE 4** *A* is high and *B* is high. With both inputs high, the AND gate has a high output, the inverter has a low output, and *Y* is low, as shown by the final entry of Table 1–14.

# 1-5 BOOLEAN ALGEBRA

Is an action right or wrong? A motive good or bad? A conclusion true or false? Much of our thinking involves trying to find the answer to two-valued questions like these. Two-state logic had a major influence on Aristotle, who worked out precise methods for getting to the truth. Logic next attracted mathematicians, who intuitively sensed some kind of algebraic process running through all thought.

Augustus De Morgan came close to finding the link between logic and mathematics. But it was George Boole (1854) who put it all together. He invented a new kind of algebra that replaced Aristotle's verbal methods. Boolean algebra did not have an impact on technology, however, until almost a century later. In 1938, Shannon applied the new algebra to telephone switching circuits. Because of Shannon's work, engineers soon realized that Boolean algebra could be used to analyze and design computer circuits.

## NOT Operation

In Boolean algebra a variable can be either a 0 or a 1. For digital circuits this means that a signal can be either low or high. Figure 1–19 is an example of a digital circuit because the input and output voltages are either low or high. Furthermore, the output *Y* is always the complement of the input *A*. In equation form,

$$Y = \text{NOT } A \qquad (1\text{–}4)$$

If *A* is 0,

$$Y = \text{NOT } 0 = 1$$

On the other hand, if *A* is 1,

**FIG. 1–19.** NOT gate

$$Y = \text{NOT } 1 = 0$$

In Boolean algebra, the overbar stands for the NOT operation. This means that Eq. (1–4) may be written as

$$y = \overline{A} \qquad\qquad (1\text{–}5)$$

Read this as "$Y$ equals NOT $A$" or "$Y$ equals the complement of $A$." Equation (1–5) is the standard way to write the output of an inverter.

Using Eq. (1–5) is easy. Given the value of $A$, substitute and solve for $Y$. For instance, if $A$ is 0,

$$Y = \overline{0} = 1$$

because NOT 0 is 1. On the other hand, if $A$ is 1,

$$Y = \overline{1} = 0$$

because NOT 1 is 0.

## OR Operation

One way to symbolize the action of an OR gate (Fig. 1–20) is by writing the following word equation:

$$Y = A \text{ OR } B \qquad\qquad (1\text{–}6)$$

This equation says the output is found by ORing the inputs. For example, if $A = 0$ and $B = 0$,

$$Y = A \text{ OR } B = 0 \text{ OR } 0 = 0$$

because 0 comes out of an OR gate when both inputs are 0s.
As another example, if $A = 0$ and $B = 1$,

$$Y = A \text{ OR } B = 0 \text{ OR } 1 = 1$$

because 1 is the output of an OR gate when either input is 1. Similarly, if $A = 1$ and $B = 0$,

$$Y = A \text{ OR } B = 1 \text{ OR } 0 = 1$$

If $A = 1$ and $B = 1$,

$$Y = A \text{ OR } B = 1 \text{ OR } 1 = 1$$

**FIG. 1–20.** Two-input OR gate

In Boolean algebra the + sign stands for the OR operation. In other words, Eq. (1–6) can be written as

$$Y = A + B \qquad\qquad (1–7)$$

Read this as ''$Y$ equals $A$ or $B$.'' Equation (1–7) is the standard way to write the output of an OR gate.

Given the inputs, you can substitute and solve for the output. For instance, if $A = 0$ and $B = 0$,

$$Y = A + B = 0 + 0 = 0$$

If $A = 0$ and $B = 1$,

$$Y = A + B = 0 + 1 = 1$$

because 0 ORed with 1 results in 1. If $A = 1$ and $B = 0$,

$$Y = A + B = 1 + 0 = 1$$

If both inputs are high,

$$Y = A + B = 1 + 1 = 1$$

because 1 ORed with 1 produces 1.

The last result, $1 + 1 = 1$, may disturb you. If so, what is bothering you is that you have always used the + sign for decimal addition. In Boolean algebra the + sign has a distinctly different meaning because it represents OR addition, the action of an OR gate combining the inputs. Put another way, the + sign is a code symbol for two totally different operations:

Decimal addition: $1 + 1 = 2$
OR addition: $1 + 1 = 1$

The way you read these equations is important in keeping things straight in your head. Read the first equation as one plus one equals two, and the second equation as 1 OR 1 equals 1.

## AND Operation

A word equation for Fig. 1–21 is

$$Y = A \text{ AND } B \qquad\qquad (1–8)$$

In Boolean algebra the multiplication sign stands for the AND operation. Therefore, Eq. (1–8) is usually written as

$$Y = A \cdot B$$

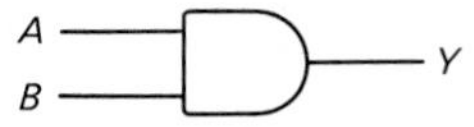

**FIG. 1–21.** Two-input AND gate

or simply

$$Y = AB \qquad\qquad (1-9)$$

Read this as "$Y$ equals $A$ AND $B$." Equation (1–9) is the standard way to write the output of an AND gate.

Given the inputs, you can substitute and solve for the output. For instance, if both inputs are low,

$$y = AB = 0 \cdot 0 = 0$$

because 0 ANDed with 0 gives 0. If $A$ is low and $B$ is high,

$$y = AB = 0 \cdot 1 = 0$$

because 0 comes out of an AND gate if any input is 0. If $A$ is 1 and $B$ is 0,

$$y = AB = 1 \cdot 0 = 0$$

When both inputs are high,

$$y = AB = 1 \cdot 1 = 1$$

because 1 ANDed with 1 gives 1.

## Boolean Equations of Logic Circuits

Think of the *Boolean operators* (plus, times, and overbar) as codes for the basic gates. Then you can write equations for logic circuits using the plus sign for an OR gate, the times sign for an AND gate, and the overbar for an inverter. In other words, you can use Boolean algebra as a shorthand notation for digital circuits.

For example, Fig. 1–22$a$ shows three cascaded OR gates in a 7432. The output of the first gate is

$$Y_3 = A + B$$

where the subscript of $Y$ is the pin number. The output of the second gate is

$$Y_6 = Y_3 + C = A + B + C$$

The final output is

$$Y_8 = Y_6 + D = A + B + C + D$$

This tells us that the logic circuit of Fig. 1–22$a$ is equivalent to a 4-input OR gate because 4 inputs are being ORed to produce the final output.

A 7411 is a triple 3-input AND gate, a TTL circuit with three 3-input AND gates in a 14-pin DIP. As usual, a positive supply voltage ($+5$ V) is connected to pin 14 and a ground to pin 7. Figure 1–22$b$ shows a cascaded connection of two of the 7411 AND gates. Here's how you can derive the Boolean equation for the final output. Begin with the first gate which has an output of

$$Y_{12} = ABC$$

This is 1 input to the second gate, so the final output is

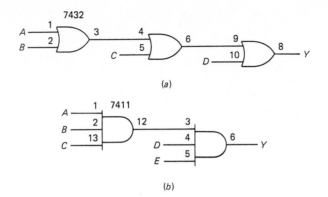

(a)

(b)

**FIG. 1–22.**

$$Y_6 = Y_{12}DE = ABCDE$$

Therefore, the circuit is equivalent to a 5-input AND gate.

## ☐ EXAMPLE 1—7

What is the Boolean equation for the logic circuit of Fig. 1–23?

## ☐ SOLUTION

This circuit is called an AND–OR network because input AND gates drive an output OR gate. The intermediate outputs are

$$Y_3 = AB$$
$$Y_6 = CD$$

The final output is

$$Y_8 = Y_3 + Y_6$$
or
$$Y = AB + CD$$

An equation in this form is referred to as a *sum-of-products* equation. AND–OR networks always produce sum-of-products equations.

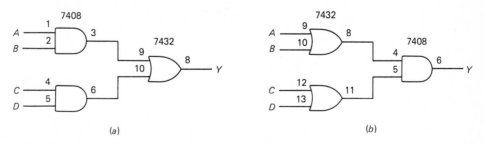

(a)          (b)

**FIG. 1–23.**

□ EXAMPLE 1—8

Write the Boolean equation for Fig. 1–23*b*.

□ SOLUTION

This logic circuit is called an OR–AND network because input OR gates drive an output AND gate. The intermediate outputs are

$$Y_8 = A + B$$
$$Y_{11} = C + D$$

The final output is

or
$$Y_6 = Y_8 Y_{11}$$
$$Y = (A + B)(C + D)$$

As shown in this equation, parentheses may be used to indicate a logical product (ANDing). Also notice that the final answer is a product of sums. OR–AND networks always produce *product-of-sums* equations.

□ EXAMPLE 1—9

What is the Boolean equation for the output of Fig. 1–24?

□ SOLUTION

Input *A* is inverted before it reaches the upper AND gate. Input *C* is inverted before it reaches the lower AND gate. Therefore, the intermediate outputs are

$$Y_{12} = \overline{A}BC$$
$$Y_6 = AB\overline{C}$$

The final output is

or
$$Y_8 = Y_{12} + Y_6$$
$$Y = \overline{A}BC + AB\overline{C}$$

□ EXAMPLE 1—10

What is the logic circuit whose Boolean equation is

$$Y = \overline{A}BC + A\overline{B}C$$

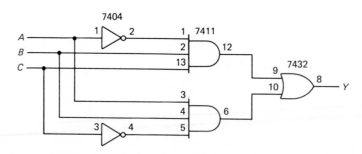

FIG. 1–24.

□ SOLUTION

This is a sum-of-products equation with some of the inputs in complemented form. Figure 1–25a shows an AND–OR circuit with the foregoing Boolean equation. The upper AND gate produces a logical product of

$$Y_{12} = \overline{A}BC$$

The lower AND gate produces

$$Y_6 = A\overline{B}C$$

The final output therefore equals the sum of the $Y_{12}$ and $Y_6$ products:

$$Y = \overline{A}BC + A\overline{B}C$$

The complemented inputs $A$ and $B$ may be produced by other circuits (discussed later). Alternatively, inverters on the $A$ and $B$ input lines may produce the complemented variables, as shown in Fig. 1–25b.

This example illustrates one method of logic design. Whenever you are given a sum-of-products equation, you can draw the corresponding AND–OR network using AND gates to produce the logical products and an OR gate to produce the sum.

## 1-6  NOR GATES

At one time, the logic circuit of Fig. 1–26a was called a NOT–OR gate because the output is

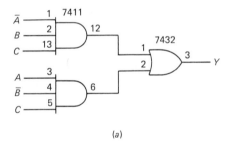

(a)

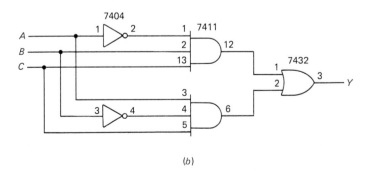

(b)

FIG. 1–25.

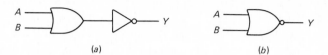

(a)  (b)

**FIG. 1–26.** (a) NOR gate (b) equivalent symbol

$$Y = \overline{A + B}$$

Read this as "$Y$ equals NOT $A$ OR $B$" or "$Y$ equals the complement of $A$ OR $B$." Because the circuit is an OR gate followed by an inverter, the only way to get a high output is to have both inputs low, as shown in the truth table of Table 1–15.

## NOR Gate Symbol

The logic circuit of Fig. 1–26*a* has become so popular that the abbreviated symbol of Fig. 1–26*b* is used for it. The bubble (small circle) on the output is a reminder of the inversion that takes place after the ORing. Furthermore, the words NOT–OR are contracted to the word NOR. So from now, we will call the circuit a NOR gate and will use the symbol of Fig. 1–26*b*. Whenever you see this symbol, remember that the output is NOT the OR of the inputs. With a NOR gate, all inputs must be low to get a high output. If any input is high, the output is low.

## Bubbled AND Gate

Figure 1–27*a* shows inverters on the input lines of an AND gate. This logic circuit is often drawn in the abbreviated form shown in Fig. 1–27*b*. The bubbles on the inputs are a reminder of the inversion that takes place before the AND operation. We will refer to the abbreviated drawing of Fig. 1–27*b* as a *bubbled* AND gate. Let us analyze this logic circuit for all input possibilities.

**CASE 1** If $A = 0$ and $B = 0$, the inverters produce high inputs at the AND gate. Therefore, $Y = 1$.

**CASE 2** If $A = 0$ and $B = 1$, the bottom inverter produces a low input to the AND gate, so $Y = 0$.

**CASE 3** If $A = 1$ and $B = 0$, the top inverter produces a low input to the AND gate, resulting in $Y = 0$.

**CASE 4** If $A = 1$ and $B = 1$, the inverters produce low inputs to the AND gate. Therefore, $Y = 0$.

**TABLE 1–15** *NOR Gate*

| A | B | Y |
|---|---|---|
| 0 | 0 | 1 |
| 0 | 1 | 0 |
| 1 | 0 | 0 |
| 1 | 1 | 0 |

**TABLE 1–16** *Bubbled AND Gate*

| A | B | Y |
|---|---|---|
| 0 | 0 | 1 |
| 0 | 1 | 0 |
| 1 | 0 | 0 |
| 1 | 1 | 0 |

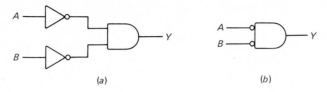

FIG. 1–27. (a) AND gate with inverted inputs (b) equivalent symbol

## Interchangeability

Table 1–16 shows the truth table for Fig. 1–27b. Compare this truth table to Table 1–15. What does this tell you? Right away, you can see that the two truth tables are identical. What does it mean? It means a bubbled AND gate produces the same output as a NOR gate. In other words, the circuits are *interchangeable*. Given any logic circuit with NOR gates, you can replace each NOR gate by a bubbled AND gate. Conversely, if a logic circuit has bubbled AND gates, you can replace each one by a NOR gate. Later examples will show you the practical application for the equivalence of NOR gates and bubbled AND gates.

## De Morgan's First Theorem

The Boolean equation for Fig. 1–26b is

$$Y = \overline{A + B}$$

The Boolean equation for Fig. 1–27b is

$$Y = \overline{A}\,\overline{B}$$

The first equation describes a NOR gate, and the second equation a bubbled AND gate. Since the outputs are equal for the same inputs, we can equate the right-hand members to get

$$\overline{A + B} = \overline{A}\,\overline{B} \qquad\qquad (1\text{–}10)$$

This identity is known as De Morgan's first theorem. In words, it says the complement of a sum equals the product of the complements. Figure 1–28 is the graphical meaning of De Morgan's first theorem. Visualize the left member of Eq. (1–10) as the output of a NOR gate. Visualize the right member of Eq. (1–10) as the output of a bubbled AND gate. Equation (1–10) is telling us something we already know; namely, a NOR gate is equivalent to a bubbled AND gate. In fact, many people refer to a bubbled AND gate as a NOR gate.

## Eye of the Beholder

Which brings us to a principle. Truth tables, logic circuits, and Boolean equations are different ways of looking at the same thing. Whatever we learn from one viewpoint

FIG. 1–28. DeMorgan's first theorem

applies to the other two. If we prove that truth tables are identical, this immediately tells us the corresponding logic circuits are interchangeable, and their Boolean equations are equivalent. When analyzing, we generally start with a logic circuit, construct its truth table, and summarize with the Boolean equation. When designing, we often start with a truth table, generate a Boolean equation, and arrive at a logic circuit.

## ☐ EXAMPLE 1—11

A 7402 is a quad 2-input NOR gate. This TTL IC has four 2-input NOR gates in a 14-pin DIP as shown in Appendix 3. What is the Boolean equation for the output of Fig. 1–29a?

## ☐ SOLUTION

The AND gates produce $AB$ and $CD$. These are ORed to get $AB + CD$. The final inversion gives

$$Y = \overline{AB + CD}$$

The circuit of Fig. 1–29a is known as an AND–OR–INVERT network because it starts with ANDing, follows with ORing, and ends with INVERTing.

The AND–OR–INVERT network is available as a separate TTL gate. For instance, the 7451 is a dual 2-input 2-wide AND–OR–INVERT gate, meaning two networks like Fig. 1–29a in a single 14-pin TTL package. Appendix 3 shows the pinout diagram. Figure 1–29b shows how we can use half of a 7451 to produce the same output as the circuit of Fig. 1–29a.

## ☐ EXAMPLE 1—12

Prove that Fig. 1–30c is logically equivalent to Fig. 1–30a.

## ☐ SOLUTION

De Morgan's first theorem says we can replace the final NOR gate of Fig. 1–30a by a bubbled AND gate to get the equivalent circuit of Fig. 1–30b. If you invert a signal twice, you get the original signal back again. Put another way, double inversion has no effect on the logic state; double invert a low and you still have a low; double-invert a high and you still have a high. Therefore, each double inversion in Fig. 1–30b (a pair of bubbles on the same signal line) cancels out, leaving the simplified circuit of Fig. 1–30c. Therefore, Fig. 1–30a and c are equivalent or interchangeable.

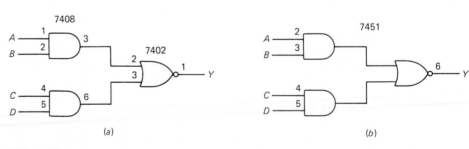

(a)                                        (b)

**FIG. 1–29.**

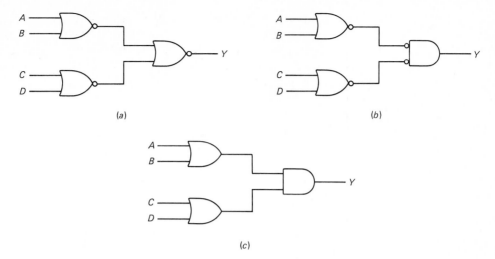

FIG. 1-30.

Why would anyone want to replace Fig. 1–30a by 1–30c? Suppose your shelves are full of AND gates and OR gates. If you have just run out of NOR gates and you are trying to build a NOR–NOR network like Fig. 1–30a, you can connect the OR–AND circuit of Fig. 1–30c because it produces the same output as the original circuit. In general, this idea applies to any circuit that you can rearrange with De Morgan's theorem. You can build whichever equivalent circuit is convenient.

## ☐ EXAMPLE 1–13

What is the truth table for the NOR–NOR circuit of Fig. 1–30a?

## ☐ SOLUTION

First, realize that the truth table of Fig. 1–30a is the same for 1–30b and 1–30c because all circuits are equivalent. Therefore, we can analyze whichever circuit we want to. Figure 1–30c is the easiest for most people to analyze, so let us use it in the following discussion.

Table 1–17 lists every possibility starting with all inputs low and progressing to all inputs high. Notice that the total number of entries equals $2^4$, or 16. By analyzing each input possibility, we can work out the corresponding output. For instance, when all inputs are low in Fig. 1–30c, both OR gates have low outputs, so the AND gate produces a low output. This is the first entry of Table 1–17. Proceeding like this, we can determine the output for the remaining possibilities and arrive at all the entries shown in Table 1–17.

## ☐ EXAMPLE 1–14

Convert the Table 1–17 into a timing diagram.

## ☐ SOLUTION

In Table 1–17, input $D$ changes states for each entry, input $C$ changes states every other entry, input $B$ every fourth entry, and input $A$ every eighth entry. Figure 1–31 shows how

**TABLE 1–17** *NOR–NOR Circuit*

| A | B | C | D | Y |
|---|---|---|---|---|
| 0 | 0 | 0 | 0 | 0 |
| 0 | 0 | 0 | 1 | 0 |
| 0 | 0 | 1 | 0 | 0 |
| 0 | 0 | 1 | 1 | 0 |
| 0 | 1 | 0 | 0 | 0 |
| 0 | 1 | 0 | 1 | 1 |
| 0 | 1 | 1 | 0 | 1 |
| 0 | 1 | 1 | 1 | 1 |
| 1 | 0 | 0 | 0 | 0 |
| 1 | 0 | 0 | 1 | 1 |
| 1 | 0 | 1 | 0 | 1 |
| 1 | 0 | 1 | 1 | 1 |
| 1 | 1 | 0 | 0 | 0 |
| 1 | 1 | 0 | 1 | 1 |
| 1 | 1 | 1 | 0 | 1 |
| 1 | 1 | 1 | 1 | 1 |

to draw the truth table in the form of a timing diagram. First, notice that the transitions on input $D$ are 1, 2, 3, and so on. Notice that input $D$ changes states each transition, input $C$ every other transition, input $B$ every fourth transition, and input $A$ every eighth transition. To agree with the truth table, output $Y$ is low up to transition 5, high between 5 and 8, low between 8 and 9, and so forth.

## 1-7 NAND GATES

Originally, the logic circuit of Fig. 1–32a was called NOT–AND gate because the output is

$$Y = \overline{AB}$$

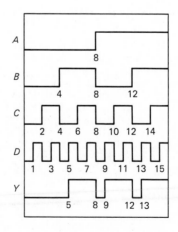

**FIG. 1–31.** Timing diagram

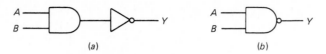

**FIG. 1–32.** (a) NAND gate (b) equivalent symbol

Read this as "*Y* equals NOT *A* AND *B*" or "*Y* equals the complement of *A* AND *B*." Because the circuit is an AND gate followed by an inverter, the only way to get a low output is for both inputs to be high, as shown in the truth table of Table 1–18.

## NAND-Gate Symbol

The logic circuit of Fig. 1–32*a* has become so popular that the abbreviated symbol of Fig. 1–32*b* is used for it. The bubble on the output reminds us of the inversion after the ANDing. Also, the words NOT–AND are contracted to NAND. Whenever you see this symbol, remember that the output is NOT the AND of the inputs. With a NAND gate, all inputs must be high to get a low output. If any input is low, the output is high.

## Bubbled OR Gate

Figure 1–33*a* shows inverters on the input lines of an OR gate. The circuit is often drawn in the abbreviated form shown in Fig. 1–33*b*, where the bubbles represent inversion. We will refer to the abbreviated drawing of Fig. 1–33*b* as a *bubbled* OR gate. Let us analyze this logic circuit for all input possibilities.

**CASE 1** If $A = 0$ and $B = 0$, the inverters produce high inputs at the OR gate. Therefore, $Y = 1$.

**CASE 2** If $A = 0$ and $B = 1$, the upper inverter produces a high input to the OR gate, so $Y = 1$.

**CASE 3** If $A = 1$ and $B = 0$, the lower inverter produces a high input to the OR gate, resulting in $Y = 1$.

**CASE 4** If $A = 1$ and $B = 1$, the inverters produce low inputs to the OR gate. Therefore, $Y = 0$.

## Interchangeability

Table 1–19 shows the truth table for Fig. 1–33*b*. Compare this truth table to Table 1–18. You can see the two truth tables are identical, which means the bubbled OR gate of Fig. 1–33*b* produces the same output signals as the NAND gate of Fig. 1–32*b*. Given any

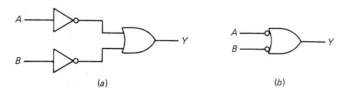

**FIG. 1–33.** (a) OR gate with inverted inputs (b) equivalent symbol

| **TABLE 1–18** *NAND Gate* | | | **TABLE 1–19** *Bubbled OR Gate* | | |
|:---:|:---:|:---:|:---:|:---:|:---:|
| *A* | *B* | *Y* | *A* | *B* | *Y* |
| 0 | 0 | 1 | 0 | 0 | 1 |
| 0 | 1 | 1 | 0 | 1 | 1 |
| 1 | 0 | 1 | 1 | 0 | 1 |
| 1 | 1 | 0 | 1 | 1 | 0 |

logic circuit with NAND gates, you can replace each NAND gate by a bubbled OR gate. Conversely, if a logic circuit has bubbled OR gates, you can replace each one by a NAND gate.

## De Morgan's Second Theorem

The Boolean equation for Fig. 1–32*b* is

$$Y = \overline{AB}$$

The Boolean equation for Fig. 1–33*b* is

$$Y = \overline{A} + \overline{B}$$

The first equation describes a NAND gate, and the second equation a bubbled OR gate. Since the outputs are equal for the same inputs, we can equate the right-hand members to get

$$\overline{AB} = \overline{A} + \overline{B} \qquad (1–11)$$

This identity is known as De Morgan's second theorem. It says the complement of a product equals the sum of the complements. Figure 1–34 is the graphical meaning of De Morgan's second theorem. Visualize the left member of Eq. (1–11) as the output of a NAND gate. Visualize the right member of Eq. (1–11) as the output of a bubbled OR gate. Equation (1–11) tells us a NAND gate is equivalent to a bubbled OR gate. In fact, many people refer to either symbol as a NAND gate because either produces the same output.

## TTL NAND Gates

The NAND gate is the backbone of the 7400 TTL series because most devices in this family are derived from it. Chapter 6 discusses the schematic diagram of a NAND gate and will show you how other gates are a redesign of the NAND gate. Because of its central role in TTL technology, the NAND gate has become the least expensive and most widely used TTL gate. Furthermore, the NAND gate is available in more configurations than other gates, as shown in Table 1–20. Notice that the NAND gate is available as a 2-, 3-, 4-, or 8-input gate. The other gates have fewer configurations, with the OR gate available only in 2-input form.

**FIG. 1–34.** DeMorgan's second theorem

TABLE 1-20 *Standard TTL Gates*

| Type | Quad 2-Input | Triple 3-Input | Dual 4-Input | Single 8-Input |
|------|--------------|----------------|--------------|----------------|
| NAND | 7400 | 7410 | 7420 | 7430 |
| NOR | 7402 | 7427 | 7425 | |
| AND | 7408 | 7411 | 7421 | |
| OR | 7432 | | | |

Incidentally, the NAND gate is sometimes called a *universal gate* because it can be connected to other NAND gates to generate any logic function. Some of the following examples will show you how to connect NAND gates to produce OR gates, AND gates, etc. (The NOR gate also is a universal gate because it too can be connected to other NOR gates to generate any logic function.)

◻ EXAMPLE 1—15

How can you connect a NAND gate to get an inverter?

◻ SOLUTION

Figure 1–35a shows how to connect one quarter of a 7400 to get an inverter. Because pins 1 and 2 are tied together, both have the same input voltage. When input A is low, output Y is high. When input A is high, output Y is low. Therefore, the circuit is equivalent to an inverter.

If an application requires three NAND gates and an inverter, a single 7400 does the job. You can use three of the NAND gates as NAND gates, and the fourth as an inverter. This way, you don't have to use a separate 7404 for an inverter.

◻ EXAMPLE 1—16

Show how to connect NAND gates to get an AND gate.

◻ SOLUTION

Figure 1–35b shows how it's done. Because the second NAND gate acts like an inverter, we have double inversion and the overall circuit is equivalent to an AND gate.

◻ EXAMPLE 1—17

How can you connect NAND gates to get an OR gate?

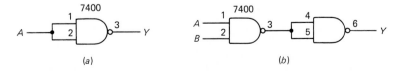

(a)  (b)

FIG. 1–35.

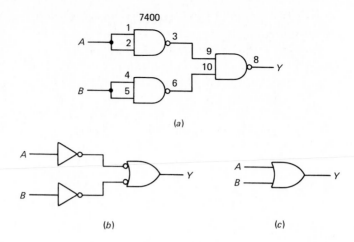

(a)

(b)                                        (c)

**FIG. 1–36.**

☐ SOLUTION

Figure 1–36a is the answer. The input NAND gates act like inverters and the output NAND gate like a bubbled OR gate, as shown in Fig. 1–36b. Since the double inversions cancel, the circuit is equivalent to the OR gate of Fig. 1–36c.

☐ EXAMPLE 1–18

Prove that Fig. 1–37c is logically equivalent to Fig. 1–37a.

☐ SOLUTION

De Morgan's second theorem says we can replace the final NAND gate of Fig. 1–37a by a bubbled OR gate to get the equivalent circuit of Fig. 1–37b. Each double inversion in

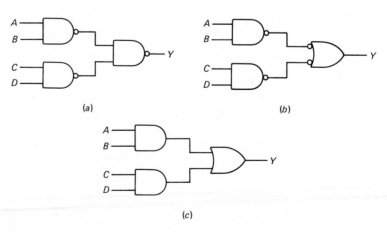

(a)                                        (b)

(c)

**FIG. 1–37.**

Fig. 1–37b cancels out, leaving the simplified circuit of Fig. 1–37c. Therefore, Figs. 1–37a and c are equivalent.

Incidentally, most people find Fig. 1–37b easy to analyze because they learn to ignore the double inversions and see only the simplified AND–OR circuit of Fig. 1–37c. For this reason, if you build a NAND–NAND circuit like Fig. 1–37a, you can draw it like Fig. 1–37b. Anyone who sees Fig. 1–37b on a schematic diagram will know it is two input NAND gates driving an output NAND gate. Furthermore, when troubleshooting the circuit, they can ignore the bubbles and visualize the easy-to-analyze AND–OR circuit of Fig. 1–37c.

☐ EXAMPLE 1—19

What is the truth table for the NAND–NAND circuit of Fig. 1–37a?

☐ SOLUTION

Let us analyze the equivalent circuit of Fig. 1–37c because it is simpler to work with. Table 1–21 lists every possibility starting with all inputs low and progressing to all inputs high. By analyzing each input possibility, we can determine the resulting output. For instance, when all inputs are low in Fig. 1–37c, both AND gates have low outputs, so the OR gate produces a low output. This is the first entry of Table 1–21. Proceeding like this, we can arrive at the output for the remaining possibilities of Table 1–21.

☐ EXAMPLE 1—20

Show a timing diagram for the NAND–NAND circuit of Fig. 1–37a.

☐ SOLUTION

All you have to do is convert the low–high states of Table 1–21 into low–high waveforms like Fig. 1–38. First, notice that the transitions on input $D$ are numbered 1, 2, 3, and so

**TABLE 1–21** *NAND–NAND Circuit*

| A | B | C | D | Y |
|---|---|---|---|---|
| 0 | 0 | 0 | 0 | 0 |
| 0 | 0 | 0 | 1 | 0 |
| 0 | 0 | 1 | 0 | 0 |
| 0 | 0 | 1 | 1 | 1 |
| 0 | 1 | 0 | 0 | 0 |
| 0 | 1 | 0 | 1 | 0 |
| 0 | 1 | 1 | 0 | 0 |
| 0 | 1 | 1 | 1 | 1 |
| 1 | 0 | 0 | 0 | 0 |
| 1 | 0 | 0 | 1 | 0 |
| 1 | 0 | 1 | 0 | 0 |
| 1 | 0 | 1 | 1 | 1 |
| 1 | 1 | 0 | 0 | 1 |
| 1 | 1 | 0 | 1 | 1 |
| 1 | 1 | 1 | 0 | 1 |
| 1 | 1 | 1 | 1 | 1 |

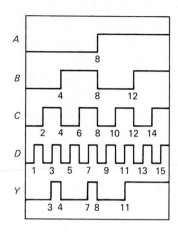

**FIG. 1–38.** Timing diagram

on. Input $D$ changes states each transition, input $C$ every other transition, input $B$ every fourth transition, and input $A$ every eighth transition. To agree with the truth table, output $Y$ is low up to transition 3, high between 3 and 4, low between 4 and 7, and so forth.

## S U M M A R Y

Binary numbers are strings of 0s and 1s. A binary odometer shows us how to count in binary. As each wheel turns from a 1 back to a 0, it advances the next higher wheel by 1. A bit is an abbreviation for binary digit. A nibble is a binary number with 4 bits. A byte is a binary number with 8 bits. The byte is the basic unit of data used in computers.

Almost all digital circuits are designed for two-state operation, which means the signal voltages are either at a low level or a high level. Because they duplicate mental processes, digital circuits are often called logic circuits. A gate is a digital circuit with 1 or more inputs, but only 1 output. The output is high only for certain combinations of the input signals.

An inverter is one type of logic circuit; it produces an output that is the complement of the input. An OR gate has 2 or more input signals; it produces a high output if any input is high. An AND gate has 2 or more input signals; it produces a high output only when all inputs are high. Truth tables often use binary 0 for the low state and binary 1 for the high state. The number of entries in a truth table equals $2^n$, where $n$ is the number of input signals.

The overbar is the algebraic symbol for the NOT operation, the plus sign is the symbol for the OR operation, and the times sign is the symbol for the AND operation. Since the Boolean operators are codes for the OR gate, AND gate, and inverter, we can use Boolean algebra to analyze digital circuits. An AND–OR circuit always produces a sum-of-products equation, while the OR–AND circuit results in a product-of-sums equation.

The NOR gate is equivalent to an OR gate followed by an inverter. De Morgan's first theorem tells us that a NOR gate is equivalent to a bubbled AND gate. Because of the De Morgan's first theorem, a NOR–NOR circuit is equivalent to an OR–AND circuit.

The NAND gate represents an AND gate followed by an inverter. De Morgan's second theorem says the NAND gate is equivalent to a bubbled OR gate. Furthermore, a NAND–NAND circuit is equivalent to an AND–OR circuit. The NAND gate is the backbone of the 7400 TTL series because most devices in this family are derived from the NAND-gate design. The NAND gate is a universal gate since any logic circuit can be built with NAND gates only.

# GLOSSARY

*AND Gate* A gate with 2 or more inputs. The output is high only when all inputs are high.

*Binary Numbers* A number code that uses only the digits 0 and 1 to represent quantities.

*Bit* An abbreviation for binary digit. It combines the first letter in *bi*nary with the last two letters in dig*it*.

*Byte* A binary number with 8 bits.

*De Morgan's First Theorem* In words, the complement of a logical sum equals the logical product of the complements. In terms of circuits, a NOR gate is equivalent to a bubbled AND gate.

*De Morgan's Second Theorem* In words, the complement of a logical product equals the logical sum of the complements. In terms of circuits, a NAND gate is equivalent to a bubbled OR gate.

*Gate* A digital circuit with one or more input voltages but only one output voltage.

*Hard Saturation* Using enough base current to ensure saturation under all operating conditions. A guideline for hard saturation is to design for a base current of one-tenth the collector saturation current.

*Inverter* A gate with only one input and a complemented output.

*Logic Circuit* A digital circuit, a switching circuit, or any kind of two-state circuit that duplicates mental processes.

*Nibble* A binary number with 4 bits.

*OR Gate* A gate with two or more inputs. The output is high when any input is high.

*Product-of-Sums Equation* A Boolean equation that is the logical product of logical sums. This type of equation applies to an OR–AND circuit.

*Soft Saturation* Driving a transistor with the minimum base current needed to produce saturation.

*Sum-of-Products Equation* A Boolean equation that is the logical sum of logical products. This type of equation applies to an AND–OR circuit.

*Timing Diagram* A picture that shows the input-output waveforms of a logic circuit.

*Transistor-Transistor Logic (TTL)* A family of digital devices which is produced as integrated circuits (ICs) in 14-, 16-, 20-, and 24-pin dual-in-line packages (DIPs).

*Truth Table* A table that shows all of the input-output possibilities of a logic circuit.

*Two-State Operation* The use of only two points on the load line of a device, resulting in all voltages being either low or high.

# PROBLEMS

*SECTION 1-1:*

**1-1.** What is the binary number that follows 01101111?

**1-2.** How many bits do each of the following contain:
    **a.** 001        **c.** 110011
    **b.** 01010    **d.** 10010011

**1-3.** How many nibbles are there in each of these:
    **a.** 1001        **c.** 110011110000
    **b.** 11110000    **d.** 1111000011001001

**1-4.** How many bits are there in 2 bytes?

*SECTION 1-2:*

**1-5.** The transistor of Fig. 1–1 has a collector saturation current of 20 mA. If $h_{FE} = 200$, how much base current do we need to produce hard saturation? Suppose $h_{FE} = 50$, how much base current do we need to get hard saturation? What effect does a change in $h_{FE}$ have when a transistor is in hard saturation?

**1-6.** A transistor with a $V_{BE}$ of 0.7 V and an $h_{FE}$ of 300 is used in Fig. 1–1. If $V_{CC} = +12$ V and $R_C = 470\ \Omega$, what is the value of $R_B$ that produces soft saturation for a $V_{\text{in}}$ of $+5$ V? Hard saturation?

**1-7.** What is the output in Fig. 1–39a if the input is low? The output if the input is high?

**1-8.** The input is low in Fig. 1–39b. Is the output low or high? Is the circuit equivalent to an inverter? If you cascade an odd number of inverters, what kind of gate is the overall circuit equivalent to?

*SECTION 1-3:*

**1-9.** Construct the truth table for Fig. 1–40a. After you are finished, discuss the relation between the circuit of Fig. 1–40a and a 3-input OR gate.

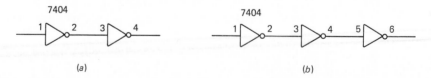

(a)                (b)

**FIG. 1–39.**

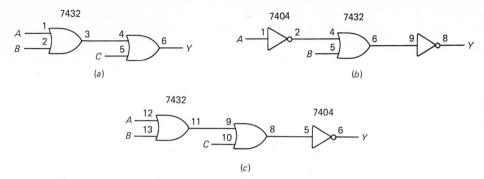

FIG. 1–40.

**1–10.** Construct the truth table for Fig. 1–40*b*.

**1–11.** Construct the truth table for Fig. 1–40*c*.

**1–12.** The circuit of Fig. 1–40*b* has trouble. Fig. 1–41 shows its timing diagram. Which of the following is the trouble:
   **a.** Input inverter acts like OR gate.
   **b.** Pin 6 is shorted to ground.
   **c.** AND gate is used instead of OR gate.
   **d.** Output inverter is faulty.

*SECTION 1-4:*

**1–13.** Construct the truth table for Fig. 1–42*a*. Then discuss the relation between the circuit of Fig. 1–42*a* and a 3-input AND gate.

**1–14.** Construct the truth table of Fig. 1–42*b*.

**1–15.** Construct the truth table for Fig. 1–42*c*.

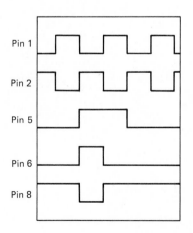

FIG. 1–41.

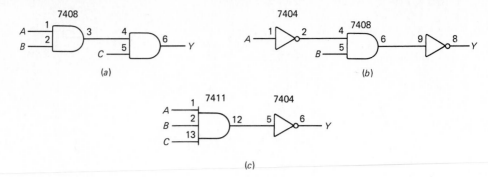

FIG. 1–42.

**1–16.** Assume the circuit of Fig. 1–42*b* has trouble. If Fig. 1–43 is the timing diagram, which of the following is the trouble:
  **a.** Input inverter is shorted.
  **b.** OR gate is used instead of AND gate.
  **c.** Pin 6 is shorted to ground.
  **d.** Pin 8 is shorted to +5 V.

*SECTION 1-5:*

**1–17.** What is the Boolean equation for the output of Fig. 1–40*a*? For Fig. 1–40*b*? For Fig. 1–40*c*?

**1–18.** Draw the logic circuit whose Boolean equation is

$$Y = \overline{A + B} + \overline{C}$$

Use the 7404 and the 7432 with pin numbers.

**1–19.** What is the Boolean equation for the output of Fig. 1–42*a*? For Fig. 1–42*b*? For Fig. 1–42*c*?

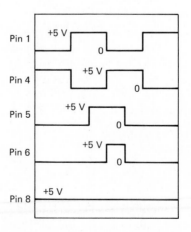

FIG. 1–43.

CHAPTER 1

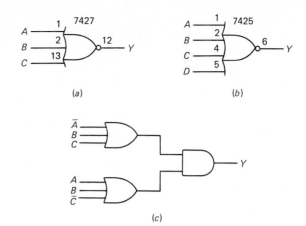

(a)

(b)

(c)

**FIG. 1–44.**

**1–20.** Draw the logic circuit described by

$$Y = (\overline{ABC})\overline{D}$$

Use a 7404, 7408, and 7411 with pin numbers.

**1–21.** Draw the logic circuit given by this Boolean equation:

$$Y = \overline{A}BC + A\overline{B}C + AB\overline{C} + \overline{A}B\overline{C}$$

Use the following devices with pin numbers: 7404, 7411, and 7432.

## SECTION 1-6:

**1–22.** Construct the truth table for the 3-input NOR gate of Fig. 1–44*a*.

**1–23.** Construct the truth table for the 4-input NOR gate of Fig. 1–44*b*.

**1–24.** Show an equivalent NOR–NOR circuit for Fig. 1–44*c*. Use the 7402 and the 7427 with pin numbers.

**1–25.** The circuit of Fig. 1–44*c* has trouble. If output *Y* is stuck in the high state, which of the following is the trouble:

**a.** Either input pin of the AND gate is shorted to ground.

**b.** Any input of either OR gate is shorted to ground.

**c.** Any input of either OR gate is shorted to a high voltage.

**d.** AND gate is defective.

## SECTION 1-7:

**1–26.** Construct the truth table for the 4-input NAND gate of Fig. 1–45*a*.

**1–27.** The inputs are $A_0$, $A_1$, $A_2$, . . . , $A_7$ in Fig. 1–45*b*. What is the Boolean equation for the input of the NAND gate?

**1–28.** Draw an equivalent NAND–NAND circuit for Fig. 1–45*c*. Use the 7420 and include pin numbers.

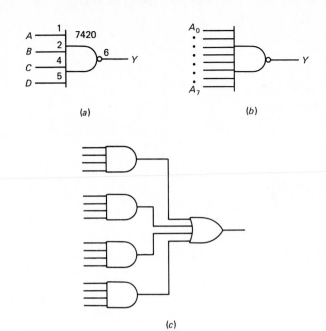

(a)

(b)

(c)

**FIG. 1–45.**

**1–29.** Suppose the final output of Fig. 1–45c is stuck in the high state. Which of the following is the trouble:
   **a.** Any input to the OR gate is shorted to a high voltage.
   **b.** Any AND-gate input is shorted to ground.
   **c.** Any AND-gate input is shorted to a high voltage.
   **d.** One of the AND gates is defective because its output is always low.

# 2

# CIRCUIT
# ANALYSIS
# AND DESIGN

This chapter discusses *Boolean algebra* and *Karnaugh maps*. After you learn the laws and theorems of Boolean algebra, you can rearrange Boolean equations to arrive at simpler logic circuits. An alternative method of simplification is based on the Karnaugh map. In this approach, geometric rather than algebraic techniques are used to simplify logic circuits.

There are two fundamental approaches in logic design: the sum-of-products method and the product-of-sums method. Either method produces a logic circuit corresponding to a given truth table. The sum-of-products solution results in a NAND–NAND circuit, while the product-of-sums solution results in a NOR–NOR circuit. Either can be used, although a designer usually selects the simpler circuit because it costs less and is more reliable.

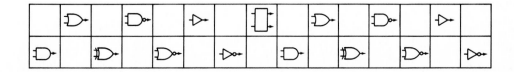

## 2-1 BOOLEAN LAWS AND THEOREM

You should know enough Boolean algebra to make obvious simplifications. What follows is a discussion of the basic laws and theorems of Boolean algebra. Some of them will look familiar from ordinary algebra, but others will be distinctly new.

### Basic Laws

The commutative laws are

$$A + B = B + A \tag{2-1}$$

$$AB = BA \tag{2-2}$$

These two equations indicate that the order of a logical operation is unimportant because the same answer is arrived at either way. As far as logic circuits are concerned, Fig. 2–1a shows how to visualize Eq. (2–1). All it amounts to is realizing that the inputs to an OR gate can be transposed without changing the output. Likewise, Fig. 2–1b is a graphical equivalent for Eq. (2–2).

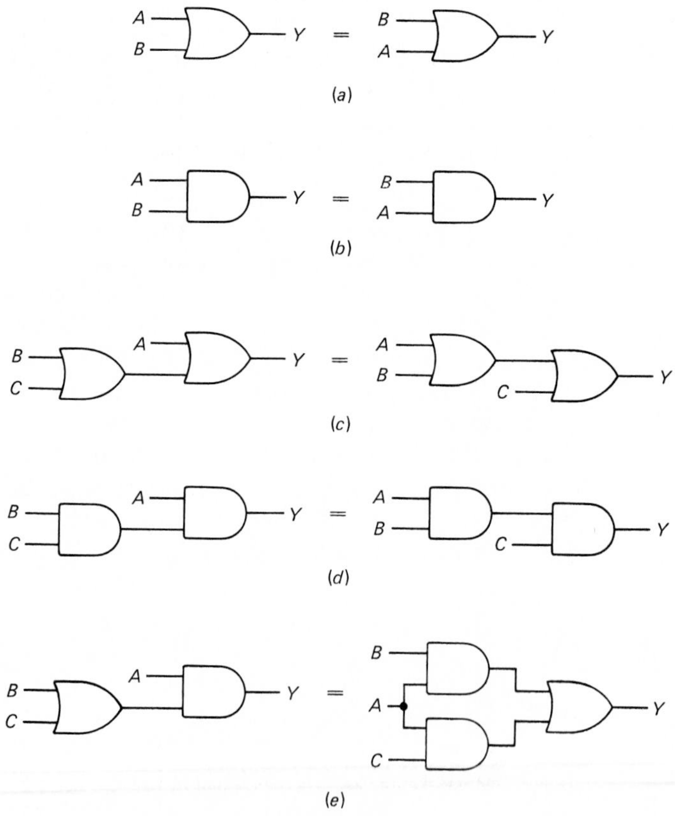

**FIG. 2-1.** Commutative, associative, and distributive laws

The associative laws are

$$A + (B + C) = (A + B) + C \qquad (2-3)$$

$$A(BC) = (AB)C \qquad (2-4)$$

These laws show that the order of combining variables has no effect on the final answer. In terms of logic circuits, Fig. 2–1c illustrates Eq. (2–3), while Fig. 2–1d represents Eq. (2–4).

The distributive law is

$$A(B + C) = AB + AC \qquad (2-5)$$

This law is easy to remember because it is identical to ordinary algebra. Figure 2–1e shows the corresponding logic equivalence. The distributive law gives you a hint about the value of Boolean algebra. If you can rearrange a Boolean expression, the corresponding logic circuit may be simpler.

The first five laws present no difficulties because they are identical to ordinary algebra. You can use these laws to simplify complicated Boolean expressions and arrive at simpler logic circuits. But before you begin, you have to learn other Boolean laws and theorems.

## OR Operations

The next four Boolean relations are about OR operations. Here is the first:

$$A + 0 = A \qquad (2-6)$$

This says that a variable ORed with 0 equals the variable. If you think about it, it makes perfect sense. When $A$ is 0,

$$0 + 0 = 0$$

And when $A$ is 1,

$$1 + 0 = 1$$

In either case, Eq. (2–6) is true.

Another Boolean relation is

$$A + A = A \qquad (2-7)$$

Again, you can see right through this by substituting the two possible values of $A$. First, when $A = 0$, Eq. (2–7) gives

$$0 + 0 = 0$$

which is true. Next, $A = 1$ results in

$$1 + 1 = 1$$

which is also true because 1 ORed with 1 produces 1. Therefore, any variable ORed with itself equals the variable.

Another Boolean rule worth knowing is

$$A + 1 = 1 \qquad (2-8)$$

Why is this valid? When $A = 0$, Eq. (2–8) gives

$$0 + 1 = 1$$

which is true. Also, $A = 1$ gives

$$1 + 1 = 1$$

This is correct because the plus sign implies OR addition, not ordinary addition. In summary, Eq. (2–8) says this: if one input to an OR gate is high, the output is high no matter what the other input.

Finally, we have

$$A + \overline{A} = 1 \qquad\qquad (2–9)$$

You should see this in a flash. If $A$ is 0, $\overline{A}$ is 1 and the equation is true. Conversely, if $A$ is 1, $\overline{A}$ is 0 and the equation still agrees. In short, a variable ORed with its complement always equals 1.

## AND Operations

Here are three AND relations:

$$A \cdot 1 = A \qquad\qquad (2–10)$$

$$A \cdot A = A \qquad\qquad (2–11)$$

$$A \cdot 0 = 0 \qquad\qquad (2–12)$$

When $A$ is 0, all the foregoing are true. Likewise, when $A$ is 1, each is true. Therefore, the three equations are valid and can be used to simplify Boolean equations.

One more AND formula is

$$A \cdot \overline{A} = 0 \qquad\qquad (2–13)$$

This one is easy to understand because you get either

$$0 \cdot 1 = 0$$

or

$$1 \cdot 0 = 0$$

for the two possible values of $A$. In words, Eq. (2–13) indicates that a variable ANDed with its complement always equals zero.

## Double Inversion and De Morgan's Theorems

The *double-inversion rule* is

$$\overline{\overline{A}} = A \qquad\qquad (2–14)$$

which shows that the double complement of a variable equals the variable. Finally, there are the De Morgan theorems discussed in Chap. 1:

$$\overline{A + B} = \overline{A} \ \overline{B} \qquad\qquad (2–15)$$

$$\overline{AB} = \overline{A} + \overline{B} \qquad\qquad (2–16)$$

You already know how important these are. The first says a NOR gate and a bubbled AND gate are equivalent. The second says a NAND gate and a bubbled OR gate are equivalent.

## Duality Theorem

The *duality theorem* is one of those elegant theorems proved in advanced mathematics. We will state the theorem without proof. Here is what the duality theorem says. Starting with a Boolean relation, you can derive another Boolean relation by

1. Changing each OR sign to an AND sign
2. Changing each AND sign to an OR sign
3. Complementing any 0 or 1 appearing in the expression

For instance, Eq. (2–6) says that

$$A + 0 = A$$

The dual relation is

$$A \cdot 1 = A$$

This dual property is obtained by changing the OR sign to an AND sign, and by complementing the 0 to get a 1.

The duality theorem is useful because it sometimes produces a new Boolean relation. For example, Eq. (2–5) states that

$$A(B + C) = AB + AC$$

By changing each OR and AND operation, we get the dual relation

$$A + BC = (A + B)(A + C)$$

This is new, not previously discussed. (If you want to prove it, construct the truth table for each side of the equation. The truth tables will be identical, which means the Boolean relation is true.)

## SUMMARY

For future reference, here are some Boolean relations and their duals:

$$A + B = B + A \qquad\qquad AB = BA$$
$$A + (B + C) = (A + B) + C \qquad A(BC) = (AB)C$$
$$A(B + C) = AB + AC \qquad A + BC = (A + B)(A + C)$$
$$A + 0 = A \qquad\qquad A \cdot 1 = A$$
$$A + 1 = 1 \qquad\qquad A \cdot 0 = 0$$
$$A + A = A \qquad\qquad A \cdot A = A$$
$$A + \overline{A} = 1 \qquad\qquad A \cdot \overline{A} = 0$$
$$\overline{\overline{A}} = A \qquad\qquad \overline{\overline{A}} = A$$
$$\overline{A + B} = \overline{A}\,\overline{B} \qquad\qquad \overline{AB} = \overline{A} + \overline{B}$$
$$A + AB = A \qquad\qquad A(A + B) = A$$
$$A + \overline{A}B = A + B \qquad A(\overline{A} + B) = AB$$

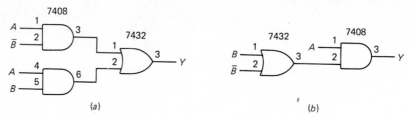

FIG. 2–2.

☐ EXAMPLE 2–1

Show the logic circuit for

$$Y = A\overline{B} + AB$$

Next, simplify this Boolean equation and the corresponding logic circuit.

☐ SOLUTION

We have a sum-of-products equation. This implies two AND gates driving an OR gate, as shown in Fig. 2–2a. Here is how to simplify the logic circuit. Factor the equation as follows:

$$Y = A\overline{B} + AB = A(\overline{B} + B)$$

Figure 2–2b shows the corresponding logic circuit. This is simpler because it uses only one AND gate.

As a matter of fact, a variable ORed with its complement always equals 1. Therefore,

$$Y = A(\overline{B} + B) = A(1) = A$$

This says output $Y$ equals input $A$. In other words, we don't even need a logic circuit. All we have to do is connect a wire between input $A$ and output $Y$.

☐ EXAMPLE 2–2

Show the logic circuit for this Boolean equation:

$$Y = (\overline{A} + B)(A + B)$$

Then, simplify the circuit as much as possible using algebra.

☐ SOLUTION

Figure 2–3a shows the logic circuit. Next, multiply the factors of the foregoing equation to get

$$Y = \overline{A}A + \overline{A}B + BA + BB$$

A variable ANDed with its complement equals zero, so the first term drops out. A variable ORed with itself equals itself, so the last term reduces to $B$. Because of the commutative law, $AB = BA$. The foregoing simplifications give us the following equation:

$$Y = \overline{A}B + AB + B$$

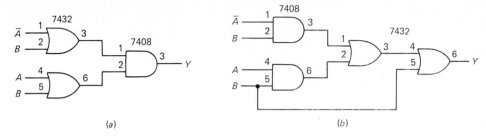

(a)                                   (b)

FIG. 2–3.

Figure 2–3b shows the corresponding logic circuit *implemented* (built) with TTL gates. As you can see, the circuit is more complicated than the original.

In fact, we can factor the foregoing equation as follows:

$$Y = (\overline{A} + A)B + B = B + B = B$$

Since $Y = B$, we don't need a logic circuit. All we need is a wire connecting the $B$ input to the $Y$ output.

## ☐ EXAMPLE 2–3

A *logic clip* is a device that you can attach to a 14- or 16-pin DIP. This troubleshooting tool contains 16 light-emitting diodes (LEDs) that monitor the state of the pins. When a pin voltage is high, the corresponding LED lights up. If the pin voltage is low, the LED is dark.

Suppose you have built the circuit of Fig. 2–3b, but it doesn't work correctly. When you connect a logic clip to the 7408, you get the readings of Fig. 2–4a (a black circle means an LED is off, and a white one means it's on). When you connect the clip to the 7432, you get the indications of Fig. 2–4b. Which of the gates is faulty?

## ☐ SOLUTION

When you use a logic clip, all you have to do is look at the inputs and output to isolate a faulty gate. For instance, Fig. 2–4a applies to a 7408 (quad 2-input AND gate). The first

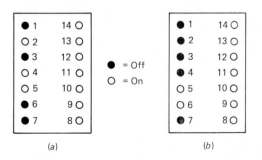

(a)                                   (b)

FIG. 2–4.  Logic clip

TABLE 2-1 *Fundamental Products for Two Inputs*

| A | B | Fundamental Product |
|---|---|---|
| 0 | 0 | $\overline{A}\,\overline{B}$ |
| 0 | 1 | $\overline{A}B$ |
| 1 | 0 | $A\overline{B}$ |
| 1 | 1 | $AB$ |

AND gate (pins 1 to 3) is all right because

Pin 1—low
Pin 2—high
Pin 3—low

A 2-input AND gate is supposed to have a low output if any input is low.

The second AND gate (pins 4 to 6) is defective. Why? Because

Pin 4—high
Pin 5—high
Pin 6—low

Something is wrong with this AND gate because it produces a low output even though both inputs are high.

If you check Fig. 2–4b (the 7432), all OR gates are normal. For instance, the first OR gate (pins 1 to 3) is all right because it produces a low output when the 2 inputs are low. The second OR gate (pins 4 to 6) is working correctly since it produces a high output when 1 input is high.

## 2-2 SUM-OF-PRODUCTS METHOD

Figure 2–5 shows the four possible ways to AND two input signals that are in complemented and uncomplemented form. These outputs are called *fundamental products*. Table 2–1 lists each fundamental product next to the input conditions producing a high output. For instance, $\overline{A}\,\overline{B}$ is high when $A$ and $B$ are low; $\overline{A}B$ is high when $A$ is low and $B$ is high; and so on.

The idea of fundamental products applies to three or more input variables. For example, assume three input variables: $A$, $B$, $C$, and their complements. There are eight ways to AND three input variables and their complements, resulting in fundamental products of

$$\overline{A}\,\overline{B}\,\overline{C},\ \overline{A}\,\overline{B}C,\ \overline{A}B\overline{C},\ \overline{A}BC,\ A\overline{B}\,\overline{C},\ A\overline{B}C,\ AB\overline{C},\ ABC$$

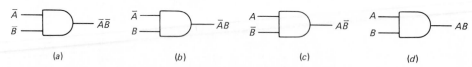

FIG. 2–5.  ANDing two variables and their complements

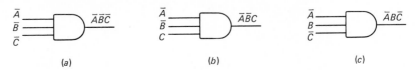

**FIG. 2–6.** Examples of ANDing three variables and their complements

Figure 2–6a shows the first fundamental product, Fig. 2–6b the second, and Fig. 2–6c the third. (For practice, draw the gates for the remaining fundamental products.)

Table 2–2 summarizes the fundamental products by listing each one next to the input condition that results in a high output. For instance, when $A = 1$, $B = 0$, and $C = 0$, the fundamental product results in an output of

$$Y = A\overline{B}\,\overline{C} = 1 \cdot \overline{0} \cdot \overline{0} = 1$$

## Sum-of-Products Equation

Here is how to get the sum-of-products solution, given a truth table like Table 2–3. What you have to do is locate each output 1 in the truth table and write down the fundamental product. For instance, the first output 1 appears for an input of $A = 0$, $B = 1$, and $C = 1$. The corresponding fundamental product is $\overline{A}BC$. The next output 1 appears for $A = 1$, $B = 0$, and $C = 1$. The corresponding fundamental product is $A\overline{B}C$. Continuing like this, you can identify all the fundamental products, as shown in Table 2–4. To get the sum-of-products equation, all you have to do is OR the fundamental products of Table 2–4:

$$Y = \overline{A}BC + A\overline{B}C + AB\overline{C} + ABC \qquad (2\text{--}17)$$

## Logic Circuit

After you have a sum-of-products equation, you can derive the corresponding logic circuit by drawing an AND–OR network, or what amounts to the same thing, a NAND–NAND network. In Eq. (2–17) each product is the output of a 3-input AND gate. Furthermore, the logical sum $Y$ is the output of a 4-input OR gate. Therefore, we can draw the logic circuit as shown in Fig. 2–7. This AND–OR circuit is one solution to the design problem that we started with. In other words, the AND–OR circuit of Fig. 2–7 has the truth table given by Table 2–3.

**TABLE 2–2** *Fundamental Products for Three Inputs*

| A | B | C | Fundamental Products |
|---|---|---|---|
| 0 | 0 | 0 | $\overline{A}\,\overline{B}\,\overline{C}$ |
| 0 | 0 | 1 | $\overline{A}\,\overline{B}C$ |
| 0 | 1 | 0 | $\overline{A}B\overline{C}$ |
| 0 | 1 | 1 | $\overline{A}BC$ |
| 1 | 0 | 0 | $A\overline{B}\,\overline{C}$ |
| 1 | 0 | 1 | $A\overline{B}C$ |
| 1 | 1 | 0 | $AB\overline{C}$ |
| 1 | 1 | 1 | $ABC$ |

| A | B | C | Y |
|---|---|---|---|
| 0 | 0 | 0 | 0 |
| 0 | 0 | 1 | 0 |
| 0 | 1 | 0 | 0 |
| 0 | 1 | 1 | 1 |
| 1 | 0 | 0 | 0 |
| 1 | 0 | 1 | 1 |
| 1 | 1 | 0 | 1 |
| 1 | 1 | 1 | 1 |

**TABLE 2–3** *Design Truth Table*

**TABLE 2–4**

| A | B | C | Y |
|---|---|---|---|
| 0 | 0 | 0 | 0 |
| 0 | 0 | 1 | 0 |
| 0 | 1 | 0 | 0 |
| 0 | 1 | 1 | $1 \rightarrow \overline{A}BC$ |
| 1 | 0 | 0 | 0 |
| 1 | 0 | 1 | $1 \rightarrow A\overline{B}C$ |
| 1 | 1 | 0 | $1 \rightarrow AB\overline{C}$ |
| 1 | 1 | 1 | $1 \rightarrow ABC$ |

We cannot build the circuit of Fig. 2–7 because a 4-input OR gate is not available as a TTL *chip* (a synonym for integrated circuit). But a 4-input NAND gate is. Figure 2–8 shows the logic circuit as a NAND–NAND circuit with TTL pin numbers. Also notice how the inputs come from a *bus*, a group of wires carrying logic signals. In Fig. 2–8, the bus has six wires with logic signals *A*, *B*, *C*, and their complements. Microcomputers are bus-organized, meaning that the input and output signals of the logic circuits are connected to buses.

☐ EXAMPLE 2—4

Suppose a three-variable truth table has a high output for these input conditions: 000, 010, 100, and 110. What is the sum-of-products circuit?

☐ SOLUTION

Here are the fundamental products:

$$000 : \overline{A}\,\overline{B}\,\overline{C}$$
$$010 : \overline{A}B\overline{C}$$
$$100 : A\overline{B}\,\overline{C}$$
$$110 : AB\overline{C}$$

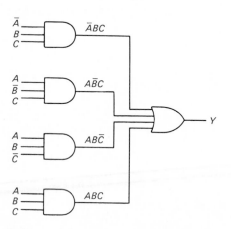

**FIG. 2–7.** AND–OR solution

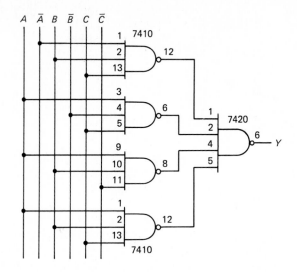

**FIG. 2–8.**

When you OR these products, you get

$$Y = \overline{A}\,\overline{B}\,\overline{C} + \overline{A}B\overline{C} + A\overline{B}\,\overline{C} + AB\overline{C}$$

The circuit of Fig. 2–8 will work if we reconnect the input lines to the bus as follows:

$\overline{A}$ : pins 1 and 3
$\overline{B}$ : pins 2 and 10
$\overline{C}$ : pins 13, 5, 11, and 13
$A$ : pins 9 and 1
$B$ : pins 4 and 2

☐ EXAMPLE 2–5
Simplify the Boolean equation in Example 2–4 and describe the logic circuit.

☐ SOLUTION
The Boolean equation is

$$Y = \overline{A}\,\overline{B}\,\overline{C} + \overline{A}B\overline{C} + A\overline{B}\,\overline{C} + AB\overline{C}$$

Since $\overline{C}$ is common to each term, factor as follows:

$$Y = (\overline{A}\,\overline{B} + \overline{A}B + A\overline{B} + AB)\overline{C}$$

Again, factor to get

$$Y = [\overline{A}(\overline{B} + B) + A(\overline{B} + B)]\overline{C}$$

Now, simplify as the foregoing as follows:

$$Y = [\overline{A}(1) + A(1)]\overline{C} = (\overline{A} + A)\overline{C}$$

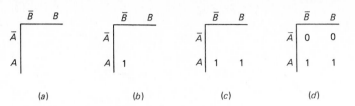

(a)    (b)    (c)    (d)

**FIG. 2–9.** Constructing a Karnaugh map

or

$$Y = \overline{C}$$

This final equation means that you don't even need a logic circuit. All you need is a wire connecting input $\overline{C}$ to output $Y$.

The lesson is clear. The AND–OR (NAND–NAND) circuit you get with the sum-of-products method is not necessarily as simple as possible. With algebra, you often can factor and reduce the sum-of-products equation to arrive at a simpler Boolean equation, which means a simpler logic circuit. A simpler logic circuit is preferred because it usually costs less to build and is more reliable.

## 2-3  TRUTH TABLE TO KARNAUGH MAP

A *Karnaugh map* is a visual display of the fundamental products needed for a sum-of-products solution. For instance, here is how to convert Table 2–5 into its Karnaugh map. Begin by drawing Fig. 2–9a. Note the variables and complements: the vertical column has $\overline{A}$ followed by $A$, and the horizontal row has $\overline{B}$ followed by $B$. Now, look for output 1s in Table 2–5. The first output 1 appears for $A = 1$ and $B = 0$. The fundamental product for this input condition is $A\overline{B}$. Enter this fundamental product on the Karnaugh map as shown in Fig. 2–9b. This 1 represents the product $A\overline{B}$ because the 1 is in row $A$ and column $\overline{B}$.

Similarly, Table 2–5 has an output 1 appearing for inputs of $A = 1$ and $B = 1$. The fundamental product is $AB$, which can be entered on the Karnaugh map as shown in Fig. 2–9c. The final step in drawing the Karnaugh map is to enter 0s in the remaining spaces (see Fig. 2–9d).

**TABLE 2–5**

| A | B | Y |
|---|---|---|
| 0 | 0 | 0 |
| 0 | 1 | 0 |
| 1 | 0 | 1 |
| 1 | 1 | 1 |

**TABLE 2–6**

| A | B | C | Y |
|---|---|---|---|
| 0 | 0 | 0 | 0 |
| 0 | 0 | 1 | 0 |
| 0 | 1 | 0 | 1 |
| 0 | 1 | 1 | 0 |
| 1 | 0 | 0 | 0 |
| 1 | 0 | 1 | 0 |
| 1 | 1 | 0 | 1 |
| 1 | 1 | 1 | 1 |

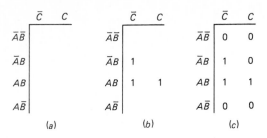

**FIG. 2–10.**  Three-variable Karnaugh map

## Three-Variable Maps

Here is how to draw a Karnaugh map for Table 2–6. First, draw the blank map of Fig. 2–10a. The vertical column is labeled $\bar{A}\,\bar{B}$, $\bar{A}B$, $AB$, and $A\bar{B}$. With this order, only one variable changes from complemented to uncomplemented form (or vice versa) as you move downward.

Next, look for output 1s in Table 2–6. Output 1s appear for $ABC$ inputs of 010, 110, and 111. The fundamental products for these input conditions are $\bar{A}B\bar{C}$, $AB\bar{C}$, and $ABC$. Enter 1s for these products on the Karnaugh map (Fig. 2–10b).

The final step is to enter 0s in the remaining spaces (Fig. 2–10c).

## Four-Variable Maps

Many digital computers and systems process 4-bit numbers. For instance, some digital chips will work with nibbles like 0000, 0001, 0010, and so on. For this reason, logic circuits are often designed to handle 4 input variables (or their complements). This is why you must know how to draw a four-variable Karnaugh map.

Here is an example. Suppose you have a truth table like Table 2–7. Start by drawing a blank map like Fig. 2–11a. Notice the order. The vertical column is $\bar{A}\,\bar{B}$, $\bar{A}B$, $AB$, and $A\bar{B}$. The horizontal row is $\bar{C}\,\bar{D}$, $\bar{C}D$, $CD$, and $C\bar{D}$. In Table 2–7, you have output 1s appearing for $ABCD$ inputs of 0001, 0110, 0111, and 1110. The fundamental products for these input conditions are $\bar{A}\,\bar{B}\,\bar{C}D$, $\bar{A}BC\bar{D}$, $\bar{A}BCD$, and $ABC\bar{D}$. After entering 1s on the Karnaugh map, you have Fig. 2–11b. The final step of filling in 0s results in the complete map of Fig. 2–11c.

(a)

|  | $\bar{C}\bar{D}$ | $\bar{C}D$ | $CD$ | $C\bar{D}$ |
|---|---|---|---|---|
| $\bar{A}\bar{B}$ |  |  |  |  |
| $\bar{A}B$ |  |  |  |  |
| $AB$ |  |  |  |  |
| $A\bar{B}$ |  |  |  |  |

(b)

|  | $\bar{C}\bar{D}$ | $\bar{C}D$ | $CD$ | $C\bar{D}$ |
|---|---|---|---|---|
| $\bar{A}\bar{B}$ |  | 1 |  |  |
| $\bar{A}B$ |  |  | 1 | 1 |
| $AB$ |  |  |  | 1 |
| $A\bar{B}$ |  |  |  |  |

(c)

|  | $\bar{C}\bar{D}$ | $\bar{C}D$ | $CD$ | $C\bar{D}$ |
|---|---|---|---|---|
| $\bar{A}\bar{B}$ | 0 | 1 | 0 | 0 |
| $\bar{A}B$ | 0 | 0 | 1 | 1 |
| $AB$ | 0 | 0 | 0 | 1 |
| $A\bar{B}$ | 0 | 0 | 0 | 0 |

**FIG. 2–11.**  Constructing a four-variable Karnaugh map

**TABLE 2–7**

| $A$ | $B$ | $C$ | $D$ | $Y$ |
|---|---|---|---|---|
| 0 | 0 | 0 | 0 | 0 |
| 0 | 0 | 0 | 1 | 1 |
| 0 | 0 | 1 | 0 | 0 |
| 0 | 0 | 1 | 1 | 0 |
| 0 | 1 | 0 | 0 | 0 |
| 0 | 1 | 0 | 1 | 0 |
| 0 | 1 | 1 | 0 | 1 |
| 0 | 1 | 1 | 1 | 1 |
| 1 | 0 | 0 | 0 | 0 |
| 1 | 0 | 0 | 1 | 0 |
| 1 | 0 | 1 | 0 | 0 |
| 1 | 0 | 1 | 1 | 0 |
| 1 | 1 | 0 | 0 | 0 |
| 1 | 1 | 0 | 1 | 0 |
| 1 | 1 | 1 | 0 | 1 |
| 1 | 1 | 1 | 1 | 0 |

# 2-4 PAIRS, QUADS, AND OCTETS

Look at Fig. 2–12*a*. The map contains a pair of 1s that are horizontally adjacent (next to each other). The first 1 represents the product $ABCD$; the second 1 stands for the product $ABC\overline{D}$. As we move from the first 1 to the second 1, only one variable goes from uncomplemented to complemented form ($D$ to $\overline{D}$); the other variables don't change form ($A$, $B$, and $C$ remain uncomplemented). Whenever this happens, you can *eliminate the variable that changes form.*

## Proof

The sum-of-products equation corresponding to Fig. 2–12*a* is

$$Y = ABCD + ABC\overline{D}$$

which factors into

$$Y = ABC(D + \overline{D})$$

| | $\overline{C}\overline{D}$ | $\overline{C}D$ | $CD$ | $C\overline{D}$ |
|---|---|---|---|---|
| $\overline{A}\overline{B}$ | 0 | 0 | 0 | 0 |
| $\overline{A}B$ | 0 | 0 | 0 | 0 |
| $AB$ | 0 | 0 | 1 | 1 |
| $A\overline{B}$ | 0 | 0 | 0 | 0 |

(*a*)

| | $\overline{C}\overline{D}$ | $\overline{C}D$ | $CD$ | $C\overline{D}$ |
|---|---|---|---|---|
| $\overline{A}\overline{B}$ | 0 | 0 | 0 | 0 |
| $\overline{A}B$ | 0 | 0 | 0 | 0 |
| $AB$ | 0 | 0 | 1 | 1 |
| $A\overline{B}$ | 0 | 0 | 0 | 0 |

(*b*)

**FIG. 2–12.** Horizontally adjacent ones

Since $D$ is ORed with its complement, the equation simplifies to

$$Y = ABC$$

In general, a pair of horizontally adjacent 1s like those of Fig. 2–12a means the sum-of-products equation will have a variable and a complement that drop out as shown above.

For easy identification, we will encircle a pair of adjacent 1s as shown in Fig. 2–12b. In this way, we can tell at a glance that one variable and its complement will drop out of the corresponding Boolean equation. In other words, an encircled pair of 1s like those of Fig. 2–12b no longer stand for the ORing of two separate products, $ABCD$ and $ABC\overline{D}$. Rather, the encircled pair is visualized as representing a single reduced product $ABC$.

Here is another example. Figure 2–13a shows a pair of 1s that are vertically adjacent. These 1s correspond to $ABC\overline{D}$ and $\overline{A}BC\overline{D}$. Notice that only one variable changes from uncomplemented to complemented form ($B$ to $\overline{B}$). Therefore, $B$ and $\overline{B}$ can be factored and eliminated algebraically, leaving a reduced product of $AC\overline{D}$.

## More Examples

Whenever you see a pair of horizontally or vertically adjacent 1s, you can eliminate the variable that appears in both complemented and uncomplemented form. The remaining variables (or their complements) will be the only ones appearing in the single-product term corresponding to the pair of 1s. For instance, a glance at Fig. 2–13b indicates that $B$ goes from complemented to uncomplemented form when we move from the upper to the lower 1; the other variables remain the same. Therefore, the encircled pair of 1s in Fig. 2–13b represents the product $\overline{A}CD$. Likewise, given the pair of 1s in Fig. 2–13c, the only change is from $\overline{D}$ to $D$. So the encircled pair of 1s stands for the product $A\overline{B}\,\overline{C}$.

If more than one pair exists on a Karnaugh map, you can OR the simplified products to get the Boolean equation. For instance, the lower pair of Fig. 2–13d represents the

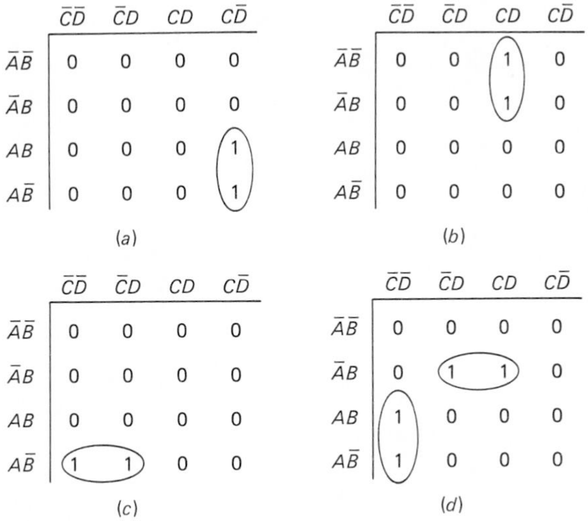

FIG. 2–13. Examples of pairs

simplified product $A\overline{C}\,\overline{D}$; the upper pair stands for $\overline{A}BD$. The corresponding Boolean equation for this map is

$$Y = A\overline{C}\,\overline{D} + \overline{A}BD$$

## Quad

A quad is a group of four 1s that are horizontally or vertically adjacent. The 1s may be end-to-end, as shown in Fig. 2–14$a$, or in the form of a square, as in Fig. 2–14$b$. When you see a quad, always encircle it because it leads to a simpler product. In fact, a quad eliminates *two variables and their complements*.

Here is why a quad eliminates two variables and their complements. Visualize the four 1s of Fig. 2–14$a$ as two pairs (see Fig. 2–14$c$). The first pair represents $AB\overline{C}$; the second pair stands for $ABC$. The Boolean equation for these two pairs is

$$Y = AB\overline{C} + ABC$$

This factors into

$$Y = AB(\overline{C} + C)$$

which reduces to

$$Y = AB$$

So, the quad of Fig. 2–14$a$ represents a product whose two variables and their complements have dropped out.

A similar proof applies to any quad. You can visualize it as two pairs whose Boolean equation leads to a single product involving only two variables or their complements. There's no need to go through the algebra each time. Merely step through the different 1s in the quad and determine which two variables go from complemented to uncomplemented form (or vice versa); these are the variables that drop out.

For instance, look at the quad of Fig. 2–14$b$. Pick any 1 as a starting point. When you move horizontally, $D$ is the variable that changes form. When you move vertically, $B$ changes form. Therefore, the remaining variables ($A$ and $C$) are the only ones appearing in the simplified product. In other words, the simplified equation for the quad of Fig. 2–14$b$ is

$$Y = AC$$

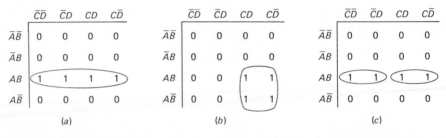

**FIG. 2–14.** Examples of quads

## The Octet

Besides pairs and quads, there is one more group of adjacent 1s to look for: the *octet*. This is a group of eight 1s like those of Fig. 2–15a. An octet like this eliminates *three variables and their complements*. Here's why. Visualize the octet as two quads (see Fig. 2–15b). The equation for these two quads is

$$Y = A\overline{C} + AC$$

After factoring,

$$Y = A(\overline{C} + C)$$

But this reduces to

$$Y = A$$

So the octet of Fig. 2–15a means three variables and their complements drop out of the corresponding product.

A similar proof applies to any octet. From now on don't bother with the algebra. Merely step through the 1s of the octet and determine which three variables change form. These are the variables that drop out.

# 2-5 KARNAUGH SIMPLIFICATIONS

As you know, a pair eliminates one variable and its complement, a quad eliminates two variables and their complements, and an octet eliminates three variables and their complements. Because of this, after you draw a Karnaugh map, encircle the octets first, the quads second, and the pairs last. In this way, the greatest simplification results.

## An Example

Suppose you have translated a truth table into the Karnaugh map shown in Fig. 2–16a. First, look for octets. There are none. Next, look for quads. When you find them, encircle them. Finally, look for and encircle pairs. If you do this correctly, you arrive at Fig. 2–16b.

The pair represents the simplified product $\overline{A}\,\overline{B}D$, the lower quad stands for $A\overline{C}$, and the quad on the right represents $C\overline{D}$. By ORing these simplified products, we get the

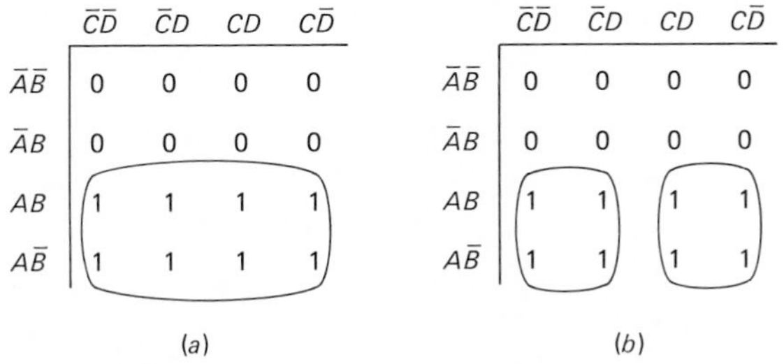

FIG. 2–15.   Example of octet

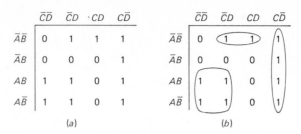

| | $\overline{C}\overline{D}$ | $\overline{C}D$ | $CD$ | $C\overline{D}$ | | | $\overline{C}\overline{D}$ | $\overline{C}D$ | $CD$ | $C\overline{D}$ |
|---|---|---|---|---|---|---|---|---|---|---|
| $\overline{A}\overline{B}$ | 0 | 1 | 1 | 1 | | $\overline{A}\overline{B}$ | 0 | 1 | 1 | 1 |
| $\overline{A}B$ | 0 | 0 | 0 | 1 | | $\overline{A}B$ | 0 | 0 | 0 | 1 |
| $AB$ | 1 | 1 | 0 | 1 | | $AB$ | 1 | 1 | 0 | 1 |
| $A\overline{B}$ | 1 | 1 | 0 | 1 | | $A\overline{B}$ | 1 | 1 | 0 | 1 |
| | | (a) | | | | | | (b) | | |

**FIG. 2–16.** Encircling octets, quads, and pairs

Boolean equation corresponding to the entire Karnaugh map:

$$Y = \overline{A}\,\overline{B}D + A\overline{C} + C\overline{D} \qquad (2\text{--}18)$$

## Overlapping Groups

You are allowed to use the same 1 more than once. Figure 2–17a illustrates this idea. The 1 representing the fundamental product $AB\overline{C}D$ is part of the pair and part of the octet. The simplified equation for the overlapping groups is

$$Y = A + B\overline{C}D \qquad (2\text{--}19)$$

It is valid to encircle the 1s as shown in Fig. 2–17b, but then the isolated 1 results in a more complicated equation:

$$Y = A + \overline{A}B\overline{C}D$$

So, always overlap groups if possible. That is, use the 1s more than once to get the largest groups you can.

## Rolling the Map

Another thing to know about is rolling. Look at Fig. 2–18a. The pairs result in this equation:

$$Y = B\overline{C}\,\overline{D} + BC\overline{D} \qquad (2\text{--}20)$$

Visualize picking up the Karnaugh map and rolling it so that the left side touches the right side. If you are visualizing correctly, you will realize the two pairs actually form a quad.

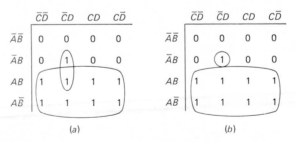

| | $\overline{C}\overline{D}$ | $\overline{C}D$ | $CD$ | $C\overline{D}$ | | | $\overline{C}\overline{D}$ | $\overline{C}D$ | $CD$ | $C\overline{D}$ |
|---|---|---|---|---|---|---|---|---|---|---|
| $\overline{A}\overline{B}$ | 0 | 0 | 0 | 0 | | $\overline{A}\overline{B}$ | 0 | 0 | 0 | 0 |
| $\overline{A}B$ | 0 | 1 | 0 | 0 | | $\overline{A}B$ | 0 | 1 | 0 | 0 |
| $AB$ | 1 | 1 | 1 | 1 | | $AB$ | 1 | 1 | 1 | 1 |
| $A\overline{B}$ | 1 | 1 | 1 | 1 | | $A\overline{B}$ | 1 | 1 | 1 | 1 |
| | | (a) | | | | | | (b) | | |

**FIG. 2–17.** Overlapping groups

| | $\bar{C}\bar{D}$ | $\bar{C}D$ | $CD$ | $C\bar{D}$ |
|---|---|---|---|---|
| $\bar{A}\bar{B}$ | 0 | 0 | 0 | 0 |
| $\bar{A}B$ | 1 | 0 | 0 | 1 |
| $AB$ | 1 | 0 | 0 | 1 |
| $A\bar{B}$ | 0 | 0 | 0 | 0 |

(a)

| | $\bar{C}\bar{D}$ | $\bar{C}D$ | $CD$ | $C\bar{D}$ |
|---|---|---|---|---|
| $\bar{A}\bar{B}$ | 0 | 0 | 0 | 0 |
| $\bar{A}B$ | 1 | 0 | 0 | 1 |
| $AB$ | 1 | 0 | 0 | 1 |
| $A\bar{B}$ | 0 | 0 | 0 | 0 |

(b)

**FIG. 2–18.** Rolling the Karnaugh map

To indicate this, draw half circles around each pair, as shown in Fig. 2–18b. From this viewpoint, the quad of Fig. 2–18b has the equation

$$Y = B\bar{D} \qquad (2–21)$$

Why is rolling valid? Because Eq. (2–20) can be algebraically simplified to Eq. (2–21). The proof is to start with Eq. (2–20):

$$Y = B\bar{C}\,\bar{D} + BC\bar{D}$$

This factors into

$$Y = B\bar{D}(\bar{C} + C)$$

which reduces to

$$Y = B\bar{D}$$

But this final equation is the one that represents a rolled quad like Fig. 2–18b. Therefore, 1s on the edges of a Karnaugh map can be grouped with 1s on opposite edges.

## More Examples

If possible, roll and overlap to get the largest groups you can find. For instance, Fig. 2–19a shows an inefficient way to encircle groups. The octet and pair have a Boolean equation of

$$Y = \bar{C} + BC\bar{D}$$

You can do better by rolling and overlapping as shown in Fig. 2–19b; the Boolean equation now is

$$Y = \bar{C} + B\bar{D}$$

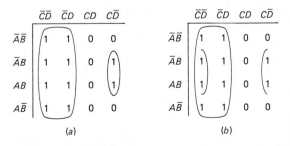

(a)                    (b)

**FIG. 2–19.** Rolling and overlapping

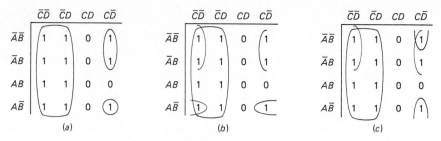

**FIG. 2–20.** Different ways of encircling groups

Here is another example. Figure 2–20*a* shows an inefficient grouping of 1s; the corresponding equation is

$$Y = \overline{C} + \overline{A}C\overline{D} + A\overline{B}C\overline{D}$$

If we roll and overlap as shown in Fig. 2–20*b*, the equation is simpler:

$$Y = \overline{C} + \overline{A}\,\overline{D} + A\overline{B}\,\overline{D}$$

It is possible to group the 1s as shown in Fig. 2–20*c*. The equation now becomes

$$Y = \overline{C} + \overline{A}\,\overline{D} + \overline{B}C\overline{D} \qquad (2\text{–}22)$$

Compare this with the preceding equation. As you can see, the equations are comparable in simplicity. Either grouping (Fig. 2–20*b* or *c*) is valid; therefore, you can use whichever you like.

## Eliminating Redundant Groups

After you have finished encircling groups, eliminate any *redundant* group. This is a group whose 1s are already used by other groups. Here is an example. Given Fig. 2–21*a*,

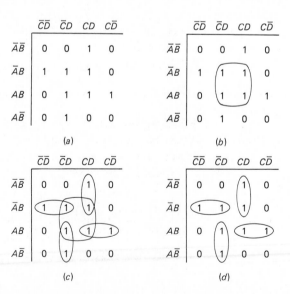

**FIG. 2–21.** Eliminating an unnecessary group

encircle the quad to get Fig. 2–21b. Next, group the remaining 1s into pairs by overlapping (Fig. 2–21c). In Fig. 2–21c, all the 1s of the quad are used by the pairs. Because of this, the quad is redundant and can be eliminated to get Fig. 2–21d. As you see, all the 1s are covered by the pairs. Figure 2–21d contains one less product than Fig. 2–21c; therefore, Fig. 2–21d is the most efficient way to group the 1s.

## Conclusion

Here is a summary of the Karnaugh-map method for simplifying Boolean equations:

1. Enter a 1 on the Karnaugh map for each fundamental product that produces a 1 output in the truth table. Enter 0s elsewhere.
2. Encircle the octets, quads, and pairs. Remember to roll and overlap to get the largest groups possible.
3. If any isolated 1s remain, encircle each.
4. Eliminate any redundant group.
5. Write the Boolean equation by ORing the products corresponding to the encircled groups.

☐ EXAMPLE 2–6

What is the simplified Boolean equation for the Karnaugh map of Fig. 2–22a?

☐ SOLUTION

There are no octets, but there is a quad as shown in Fig. 2–22b. By overlapping, we can find two more quads (see Fig. 2–22c). We can encircle the remaining 1 by making it part of an overlapped pair (Fig. 2–22d). Finally, there are no redundant groups.

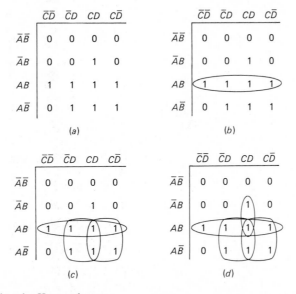

FIG. 2–22.   Using the Karnaugh map

The horizontal quad of Fig. 2–22d corresponds to a simplified product $AB$. The square quad on the right corresponds to $AC$, while the one on the left stands for $AD$. The pair represents $BCD$. By ORing these products, we get the simplified Boolean equation:

$$Y = AB + AC + AD + BCD \qquad (2\text{–}23)$$

☐ EXAMPLE 2–7

What is the NAND–NAND circuit for Eq. (2–23)?

☐ SOLUTION

We need three 2-input NAND gates (7400) and a 3-input NAND gate (7410) driving a 4-input NAND gate (7420), as shown in Fig. 2–23a. Alternatively, we can convert a 4-input NAND gate to a 3-input NAND gate by tying pins 4 and 5 of a 7420 together, as shown in Fig. 2–23b. In this way, we need only two chips, a 7400 and a 7420.

## 2·6 DON'T·CARE CONDITIONS

In some digital systems, certain input conditions never occur during normal operation; therefore, the corresponding output never appears. Since the output never appears, it is indicated by an $X$ in the truth table. For instance, Table 2–8 shows a truth table where the output is low for all input entries from 0000 to 1000, high for input entry 1001, and an $X$ for 1010 through 1111. The $X$ is called a *don't-care* condition. Whenever you see an $X$ in a truth table, you can let it equal either 0 or 1, whichever produces a simpler logic circuit.

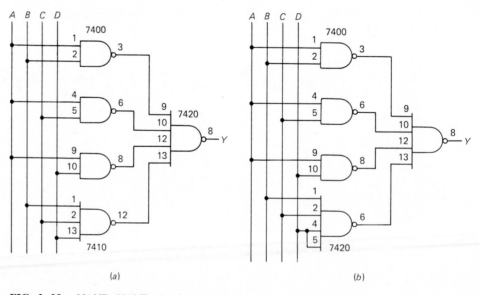

(a)                                        (b)

**FIG. 2–23.** NAND–NAND circuit

**TABLE 2–8** *Truth Table with Don't-Care Conditions*

| A | B | C | D | Y |
|---|---|---|---|---|
| 0 | 0 | 0 | 0 | 0 |
| 0 | 0 | 0 | 1 | 0 |
| 0 | 0 | 1 | 0 | 0 |
| 0 | 0 | 1 | 1 | 0 |
| 0 | 1 | 0 | 0 | 0 |
| 0 | 1 | 0 | 1 | 0 |
| 0 | 1 | 1 | 0 | 0 |
| 0 | 1 | 1 | 1 | 0 |
| 1 | 0 | 0 | 0 | 0 |
| 1 | 0 | 0 | 1 | 1 |
| 1 | 0 | 1 | 0 | X |
| 1 | 0 | 1 | 1 | X |
| 1 | 1 | 0 | 0 | X |
| 1 | 1 | 0 | 1 | X |
| 1 | 1 | 1 | 0 | X |
| 1 | 1 | 1 | 1 | X |

Figure 2–24*a* shows the Karnaugh map of Table 2–8 with don't cares for all inputs from 1010 to 1111. These don't cares are like wild cards in poker because you can let them stand for whatever you like. Figure 2–24*b* shows the most efficient way to encircle the 1. Notice two crucial ideas. First, the 1 is included in a quad, the largest group you can find if you visualize all *X*'s as 1s. Second, after the 1 has been encircled, all *X*'s outside the quad are visualized as 0s. In this way, the *X*'s are used to the best possible advantage. As already mentioned, you are free to do this because don't cares correspond to input conditions that never appear.

The quad of Fig. 2–24*b* results in a Boolean equation of

$$Y = AD$$

The logic circuit for this is an AND gate with inputs of *A* and *D*, as shown in Fig. 2–24*c*. You can check this logic circuit by examining Table 2–8. The possible inputs are from 0000 to 1001; in this range a high *A* and a high *D* produce a high *Y* only for input condition 1001.

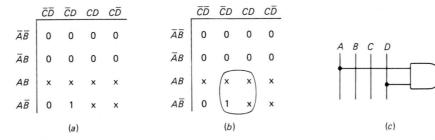

**FIG. 2–24.** Don't care conditions

Remember these ideas about don't-care conditions:

1. Given the truth table, draw a Karnaugh map with 0s, 1s, and don't cares.
2. Encircle the actual 1s on the Karnaugh map in the largest groups you can find by treating the don't cares as 1s.
3. After the actual 1s have been included in groups, disregard the remaining don't cares by visualizing them as 0s.

☐ EXAMPLE 2—8

Suppose Table 2–8 has a high output for an input of 0000, low outputs for 0001 to 1001, and don't cares for 1010 to 1111. What is the simplest logic circuit with this truth table?

☐ SOLUTION

The truth table has a 1 output only for the input condition 0000. The corresponding fundamental product is $\overline{A}\,\overline{B}\,\overline{C}\,\overline{D}$. Figure 2–25a shows the Karnaugh map with a 1 for the fundamental product, 0s for inputs 0001 to 1001, and $X$'s for inputs 1010 to 1111. In this case, the don't cares are of no help. The best we can do is to encircle the isolated 1, while treating the don't cares as 0s. So, the Boolean equation is

$$Y = \overline{A}\,\overline{B}\,\overline{C}\,\overline{D}$$

Figure 2–25b shows the logic circuit. The 4-input AND gate produces a high output only for the input condition $A = 0$, $B = 0$, $C = 0$, and $D = 0$.

☐ EXAMPLE 2—9

A truth table has low outputs for inputs of 0000 to 0110, a high output for 0111, low outputs for 1000 to 1001, and don't cares for 1010 to 1111. Show the simplest logic circuit for this truth table.

☐ SOLUTION

Figure 2–26a is the Karnaugh map. The most efficient encircling is to group the 1s into a pair using the don't care as shown. Since this is the largest group possible, all remaining don't cares are treated as 0s. The equation for the pair is

$$Y = BCD$$

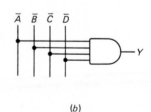

| | $\overline{C}\overline{D}$ | $\overline{C}D$ | $CD$ | $C\overline{D}$ |
|---|---|---|---|---|
| $\overline{A}\overline{B}$ | (1) | 0 | 0 | 0 |
| $\overline{A}B$ | 0 | 0 | 0 | 0 |
| $AB$ | x | x | x | x |
| $A\overline{B}$ | 0 | 0 | x | x |

(a)

(b)

**FIG. 2–25.** Decoding 0000

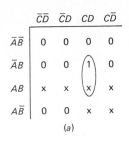

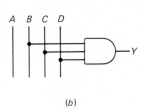

(a)                                                          (b)

**FIG. 2–26.** Decoding 0111

and Fig. 2–26b is the logic circuit. This 3-input AND gate produces a high output only for an input of $A = 0$, $B = 1$, $C = 1$, and $D = 1$ because the input possibilities range only from 0000 to 1001.

## 2·7 PRODUCT-OF-SUMS METHOD

With the sum-of-products method the design starts with a truth table that summarizes the desired input-output conditions. The next step is to convert the truth table into an equivalent sum-of-products equation. The final step is to draw the AND–OR network or its NAND–NAND equivalent.

The product-of-sums method is similar. Given a truth table, you identify the fundamental sums needed for a logic design. Then by ANDing these sums, you get the product-of-sums equation corresponding to the truth table. But there are some differences between the two approaches. With the sum-of-products method, the fundamental product produces an output 1 for the corresponding input condition. But with the product-of-sums method, the fundamental sum produces an output 0 for the corresponding input condition. The best way to understand this distinction is with an example.

### Converting a Truth Table to an Equation

Suppose you are given a truth table like Table 2–9 and you want to get the product-of-sums equation. What you have to do is locate each output 0 in the truth table and write down its fundamental sum. In Table 2–9, the first output 0 appears for $A = 0$, $B = 0$, and $C = 0$. The fundamental sum for these inputs is $A + B + C$. Why? Because this produces an output zero for the corresponding input condition:

$$Y = A + B + C = 0 + 0 + 0 = 0$$

**TABLE 2–9**

| $A$ | $B$ | $C$ | $Y$ |
|-----|-----|-----|-----|
| 0 | 0 | 0 | $0 \rightarrow A + B + C$ |
| 0 | 0 | 1 | 1 |
| 0 | 1 | 0 | 1 |
| 0 | 1 | 1 | $0 \rightarrow A + \bar{B} + \bar{C}$ |
| 1 | 0 | 0 | 1 |
| 1 | 0 | 1 | 1 |
| 1 | 1 | 0 | $0 \rightarrow \bar{A} + \bar{B} + C$ |
| 1 | 1 | 1 | 1 |

The second output 0 appears for the input condition of $A = 0$, $B = 1$, and $C = 1$. The fundamental sum for this is $A + \overline{B} + \overline{C}$. Notice that $B$ and $C$ are complemented because this is the only way to get a logical sum of 0 for the given input conditions:

$$Y = A + \overline{B} + \overline{C} = 0 + \overline{1} + \overline{1} = 0 + 0 + 0 = 0$$

Similarly, the third output 0 occurs for $A = 1$, $B = 1$, and $C = 0$; therefore, its fundamental sum is $\overline{A} + \overline{B} + C$:

$$Y = \overline{A} + \overline{B} + C = \overline{1} + \overline{1} + 0 = 0 + 0 + 0 = 0$$

Table 2–9 shows all the fundamental sums needed to implement the truth table. Notice that each variable is complemented when the corresponding input variable is a 1; the variable is uncomplemented when the corresponding input variable is 0. To get the product-of-sums equation, all you have to do is AND the fundamental sums:

$$Y = (A + B + C)(A + \overline{B} + \overline{C})(\overline{A} + \overline{B} + C) \qquad (2\text{–}24)$$

This is the product-of-sums equation for Table 2–9.

## Logic Circuit

After you have a product-of-sums equation, you can get the logic circuit by drawing an OR–AND network, or if you prefer, a NOR–NOR network. In Eq. (2–24) each sum represents the output of a 3-input OR gate. Furthermore, the logical product $Y$ is the output of a 3-input AND gate. Therefore, you can draw the logic circuit as shown in Fig. 2–27.

A 3-input OR gate is not available as a TTL chip. So, the circuit of Fig. 2–27 is not practical. With De Morgan's first theorem, however, you can replace the OR–AND circuit of Fig. 2–27 by the NOR–NOR circuit of Fig. 2–28.

☐ EXAMPLE 2–10

Suppose a truth table has a low output for the first three input conditions: 000, 001, and 010. If all other outputs are high, what is the product-of-sums circuit?

☐ SOLUTION

The product-of-sums equation is

$$Y = (A + B + C)(A + B + \overline{C})(A + \overline{B} + C)$$

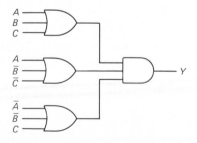

**FIG. 2–27.** Product-of-sums circuit

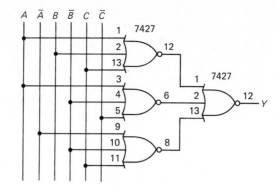

**FIG. 2–28.**

The circuit of Fig. 2–28 will work if we reconnect the input lines as follows:

$$A : \text{pins 1, 3, and 9}$$
$$B : \text{pins 2 and 4}$$
$$C : \text{pins 13 and 11}$$
$$\overline{B} : \text{pin 10}$$
$$\overline{C} : \text{pin 5}$$

# 2-8 PRODUCT-OF-SUMS SIMPLIFICATION

After you write a product-of-sums equation, you can simplify it with Boolean algebra. Alternatively, you may prefer simplification based on the Karnaugh map. There are several ways of using the Karnaugh map. In this section, we will discuss one of the easiest methods.

## Sum-of-Products Circuit

Suppose the design starts with a truth table like Table 2–10. The first thing to do is to draw the Karnaugh map in the usual way to get Fig. 2–29a. The encircled groups allow us to write a sum-of-products equation:

$$Y = \overline{A}\,\overline{B} + AB + AC$$

Figure 2–29b shows the corresponding NAND–NAND circuit.

## Complementary Circuit

To get a product-of-sums circuit, begin by complementing each 0 and 1 on the Karnaugh map of Fig. 2–29a. This results in the complemented map shown in Fig. 2–29c. The encircled 1s allow us to write the following sum-of-products equation:

$$\overline{Y} = \overline{A}B + A\overline{B}\,\overline{C}$$

Why is this $\overline{Y}$ instead of $Y$? Because complementing the Karnaugh map is the same as complementing the output of the truth table, which means the sum-of-products equation for Fig. 2–29c is for $\overline{Y}$ instead of $Y$.

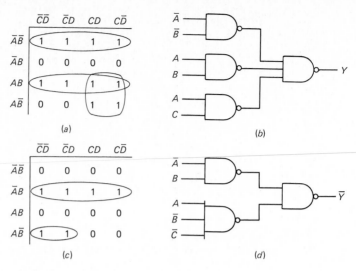

FIG. 2–29. Deriving the sum-of-products circuit

Figure 2–29d shows the corresponding NAND–NAND circuit for $\overline{Y}$. This circuit does not produce the desired output; it produces the complement of the desired output.

## Finding the NOR–NOR Circuit

What we want to do next is to get the product-of-sums solution, the NOR–NOR circuit that produces the original truth table of Table 2–10. De Morgan's first theorem tells us NAND gates can be replaced by bubbled OR gates; therefore, we can replace Fig. 2–29d

TABLE 2–10

| A | B | C | D | Y |
|---|---|---|---|---|
| 0 | 0 | 0 | 0 | 1 |
| 0 | 0 | 0 | 1 | 1 |
| 0 | 0 | 1 | 0 | 1 |
| 0 | 0 | 1 | 1 | 1 |
| 0 | 1 | 0 | 0 | 0 |
| 0 | 1 | 0 | 1 | 0 |
| 0 | 1 | 1 | 0 | 0 |
| 0 | 1 | 1 | 1 | 0 |
| 1 | 0 | 0 | 0 | 0 |
| 1 | 0 | 0 | 1 | 0 |
| 1 | 0 | 1 | 0 | 1 |
| 1 | 0 | 1 | 1 | 1 |
| 1 | 1 | 0 | 0 | 1 |
| 1 | 1 | 0 | 1 | 1 |
| 1 | 1 | 1 | 0 | 1 |
| 1 | 1 | 1 | 1 | 1 |

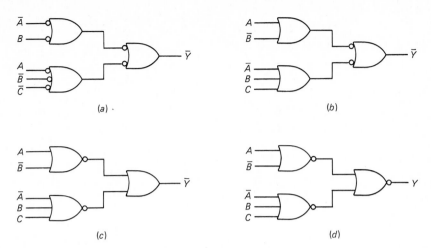

(a) ·

(b)

(c)

(d)

**FIG. 2–30.** Deriving the product-of-sums circuit

by Fig. 2–30a. A bus with each variable and its complement is usually available in a digital system. So, instead of connecting $\overline{A}$ and $B$ to a bubbled OR gate, as shown in Fig. 2–30a, we can connect $A$ and $\overline{B}$ to an OR gate, as shown in Fig. 2–30b. In a similar way, instead of connecting $A$, $\overline{B}$, and $\overline{C}$ to a bubbled OR gate, we have connected $\overline{A}$, $B$, and $C$ to an OR gate. In short, Fig. 2–30b is equivalent to Fig. 2–30a.

The next step toward a NOR–NOR circuit is to convert Fig. 2–30b into Fig. 2–30c, which is done by sliding the bubbles to the left from the output gate to the input gates. This changes the input OR gates to NOR gates. The final step is to use a NOR gate on the output to produce $Y$ instead of $\overline{Y}$, as shown in the NOR–NOR circuit of Fig. 2–30d.

From now on, you don't have to go through every step in changing a complementary NAND–NAND circuit to an equivalent NOR–NOR circuit. Instead, you can apply the duality theorem as described in the following.

## Duality

An earlier section introduced the duality theorem of Boolean algebra. Now we are ready to apply this theorem to logic circuits. Given a logic circuit, we can find its dual circuit as follows: Change each AND gate to an OR gate, change each OR gate to an AND gate, and complement all input-output signals. An equivalent statement of duality is this: Change each NAND gate to a NOR gate, change each NOR gate to a NAND gate, and complement all input-output signals.

Compare the NOR–NOR circuit of Fig. 2–30d with the NAND–NAND circuit of Fig. 2–29d. NOR gates have replaced NAND gates. Furthermore, all input and output signals have been complemented. This is an application of the duality theorem. From now on, you can change a complementary NAND–NAND circuit (Fig. 2–29d) into its dual NOR–NOR circuit (Fig. 2–30d) by changing all NAND gates to NOR gates and complementing all signals.

## Points to Remember

Here is a summary of the key ideas in the preceding discussion:

1. Convert the truth table into a Karnaugh map. After grouping the 1s, write the sum-of-products equation and draw the NAND–NAND circuit. This is the sum-of-products solution for $Y$.
2. Complement the Karnaugh map. Group the 1s, write the sum-of-products equation, and draw the NAND–NAND circuit for $\overline{Y}$. This is the complementary NAND–NAND circuit.
3. Convert the complementary NAND–NAND circuit to a dual NOR–NOR circuit by changing all NAND gates to NOR gates and complementing all signals. What remains is the product-of-sums solution for $Y$.
4. Compare the NAND–NAND circuit (step 1) with the NOR–NOR circuit (step 3). You can use whichever circuit you prefer, usually the one with fewer gates.

□ EXAMPLE 2–11

Show the sum-of-products and product-of-sums circuits for the Karnaugh map of Fig. 2–31a.

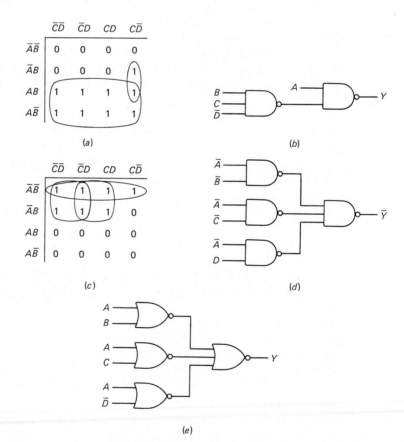

FIG. 2–31.

□   SOLUTION

The Boolean equation for Fig. 2–31a is

$$Y = A + BC\overline{D}$$

Figure 2–31b is the sum-of-products circuit.

After complementing and simplifying the Karnaugh map, we get Fig. 2–31c. The Boolean equation for this is

$$Y = \overline{A}\,\overline{B} + \overline{A}\,\overline{C} + \overline{A}D$$

Figure 2–31d is the sum-of-products circuit for the $\overline{Y}$. As shown earlier, we can convert the dual circuit into a NOR–NOR equivalent circuit to get Fig. 2–31e.

The two design choices are Fig. 2–31b and 2–31e. Fig. 2–31e is simpler.

## S U M M A R Y

Every Boolean equation has a dual form obtained by changing OR to AND, AND to OR, 0 to 1, and 1 to 0. With Boolean algebra you may be able to simplify a Boolean equation, which implies a simplified logic circuit.

Given a truth table, you can identify the fundamental products that produce output 1s. By ORing these products, you get a sum-of-products equation for the truth table. A sum-of-products equation always results in an AND–OR circuit or its equivalent NAND–NAND circuit.

The Karnaugh method of simplification starts by converting a truth table into a Karnaugh map. Next, you encircle all the octets, quads, and pairs. This allows you to write a simplified Boolean equation and to draw a simplified logic circuit. When a truth table contains don't cares, you can treat the don't cares as 0s or 1s, whichever produces the greatest simplification.

One way to get a product-of-sums circuit is to complement the Karnaugh map and write the simplified Boolean equation for $\overline{Y}$. Next, you draw the NAND–NAND circuit for $\overline{Y}$. Finally, you change the NAND–NAND circuit into a NOR–NOR circuit by changing all NAND gates to NOR gates and complementing all signals.

## G L O S S A R Y

*Chip* An integrated circuit. A piece of semiconductor material with a micro-miniature circuit on its surface.

*Don't Care Condition* An input-output condition that never occurs during normal operation. Since the condition never occurs, you can use an $X$ on the Karnaugh map. This $X$ can be a 0 or a 1, whichever you prefer.

*Dual Circuit* Given a logic circuit, you can find its dual as follows. Change each AND (NAND) gate to an OR (NOR) gate, change each OR (NOR) gate to an AND (NAND) gate, and complement all input-output signals.

*Logic Clip* A device attached to a 14- or 16-pin DIP. The LEDs in this troubleshooting tool indicate the logic states of the pins.

*Karnaugh Map* A drawing that shows all the fundamental products and the corresponding output values of a truth table.

*Octet* Eight adjacent 1s in a 2 × 4 shape on a Karnaugh map.

*Overlapping Groups* Using the same 1 more than once when looping the 1s of a Karnaugh map.

*Pair* Two horizontally or vertically adjacent 1s on a Karnaugh map.

*Product-of-Sums Equation* The logical product of those fundamental sums that produce output 1s in the truth table. The corresponding logic circuit is an OR–AND circuit, or the equivalent NOR–NOR circuit.

*Quad* Four horizontal, vertical, or rectangular 1s on a Karnaugh map.

*Redundant Group* A group of 1s on a Karnaugh map that are all part of other groups. You can eliminate any redundant group.

*Sum-of-Products Equation* The logical sum of those fundamental products that produce output 1s in the truth table. The corresponding logic circuit is an AND–OR circuit, or the equivalent NAND–NAND circuit.

# P R O B L E M S

*SECTION 2-1:*

**2–1.**   Draw the logic circuit for

$$Y = A\overline{B}C + ABC$$

Next, simplify the equation with Boolean algebra and draw the simplified logic circuit.

**2–2.**   Draw the logic circuit for

$$Y = (\overline{A} + B + C)(A + B + \overline{C})$$

Use Boolean algebra to simplify the equation. Then draw the corresponding logic circuit.

**2–3.**   In Fig. 2–32a, the output NAND gate acts like a 2-input gate because pins 10 and 11 are tied together. Suppose a logic clip is connected to the 7410. Which of the three gates is defective if the logic clip displays the data of Fig. 2–32b?

**2–4.**   If a logic clip displays the states of Fig. 2–32c for the circuit of Fig. 2–32a, which of the gates is faulty?

**2–5.**   The circuit of Fig. 2–32a has trouble. If Fig. 2–33 is the timing diagram, which of the following is the trouble:
   **a.** Upper NAND gate is defective.
   **b.** Pin 6 is shorted to +5 V.
   **c.** Pin 9 is grounded.
   **d.** Pin 8 is shorted to +5 V.

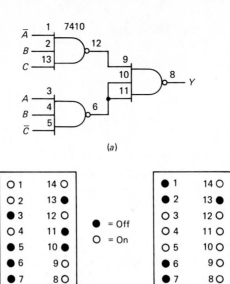

(a)

(b)                    (c)

● = Off
○ = On

FIG. 2–32.

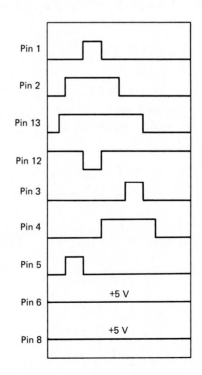

FIG. 2–33.

| A | B | C | D | Y |
|---|---|---|---|---|
| 0 | 0 | 0 | 0 | 0 |
| 0 | 0 | 0 | 1 | 1 |
| 0 | 0 | 1 | 0 | 0 |
| 0 | 0 | 1 | 1 | 0 |
| 0 | 1 | 0 | 0 | 0 |
| 0 | 1 | 0 | 1 | 1 |
| 0 | 1 | 1 | 0 | 0 |
| 0 | 1 | 1 | 1 | 0 |
| 1 | 0 | 0 | 0 | 0 |
| 1 | 0 | 0 | 1 | 0 |
| 1 | 0 | 1 | 0 | 1 |
| 1 | 0 | 1 | 1 | 1 |
| 1 | 1 | 0 | 0 | 1 |
| 1 | 1 | 0 | 1 | 1 |
| 1 | 1 | 1 | 0 | 0 |
| 1 | 1 | 1 | 1 | 0 |

**TABLE 2–11**

| A | B | C | D | Y |
|---|---|---|---|---|
| 0 | 0 | 0 | 0 | 0 |
| 0 | 0 | 0 | 1 | 1 |
| 0 | 0 | 1 | 0 | 1 |
| 0 | 0 | 1 | 1 | 1 |
| 0 | 1 | 0 | 0 | 0 |
| 0 | 1 | 0 | 1 | 0 |
| 0 | 1 | 1 | 0 | 0 |
| 0 | 1 | 1 | 1 | 1 |
| 1 | 0 | 0 | 0 | 1 |
| 1 | 0 | 0 | 1 | 1 |
| 1 | 0 | 1 | 0 | 1 |
| 1 | 0 | 1 | 1 | 0 |
| 1 | 1 | 0 | 0 | 0 |
| 1 | 1 | 0 | 1 | 1 |
| 1 | 1 | 1 | 0 | 0 |
| 1 | 1 | 1 | 1 | 0 |

**TABLE 2–12**

*SECTION 2-2:*

**2–6.** What is the sum-of-products circuit for the truth table of Table 2–11?

**2–7.** Simplify the sum-of-products equation in Prob. 2–6 as much as possible and draw the corresponding logic circuit.

**2–8.** A digital system has a 4-bit input from 0000 to 1111. Design a logic circuit that produces a high output whenever the equivalent decimal input is greater than 13.

**2–9.** We need a circuit with 2 inputs and 1 output. The output is to be high only when 1 input is high. If both inputs are high, the output is to be low. Draw a sum-of-products circuit for this.

*SECTION 2-3:*

**2–10.** Draw the Karnaugh map for Table 2–11.

**2–11.** Draw the Karnaugh map for Table 2–12.

*SECTION 2-4:*

**2–12.** Draw the Karnaugh map for Table 2–11. Then encircle all the octets, quads, and pairs you can find.

**2–13.** Repeat Prob. 2–12 for Table 2–12.

*SECTION 2-5:*

**2–14.** What is the simplified Boolean equation for the Karnaugh map of Table 2–11? The logic circuit?

**2–15.** Given Table 2–12, use Karnaugh simplification and draw the simplified logic circuit.

**2–16.** Table 2–13 shows a special code known as the *Gray code*. For each binary input *ABCD*, there is a corresponding Gray-code output. What is the simplified sum-of-products equation for $Y_3$? For $Y_2$? For $Y_1$? For $Y_0$? Draw a logic circuit that converts a 4-bit binary input to a Gray-code output.

**TABLE 2–13** *Gray Code*

| A | B | C | D | $Y_3$ | $Y_2$ | $Y_1$ | $Y_0$ |
|---|---|---|---|-------|-------|-------|-------|
| 0 | 0 | 0 | 0 | 0 | 0 | 0 | 0 |
| 0 | 0 | 0 | 1 | 0 | 0 | 0 | 1 |
| 0 | 0 | 1 | 0 | 0 | 0 | 1 | 1 |
| 0 | 0 | 1 | 1 | 0 | 0 | 1 | 0 |
| 0 | 1 | 0 | 0 | 0 | 1 | 1 | 0 |
| 0 | 1 | 0 | 1 | 0 | 1 | 1 | 1 |
| 0 | 1 | 1 | 0 | 0 | 1 | 0 | 1 |
| 0 | 1 | 1 | 1 | 0 | 1 | 0 | 0 |
| 1 | 0 | 0 | 0 | 1 | 1 | 0 | 0 |
| 1 | 0 | 0 | 1 | 1 | 1 | 0 | 1 |
| 1 | 0 | 1 | 0 | 1 | 1 | 1 | 1 |
| 1 | 0 | 1 | 1 | 1 | 1 | 1 | 0 |
| 1 | 1 | 0 | 0 | 1 | 0 | 1 | 0 |
| 1 | 1 | 0 | 1 | 1 | 0 | 1 | 1 |
| 1 | 1 | 1 | 0 | 1 | 0 | 0 | 1 |
| 1 | 1 | 1 | 1 | 1 | 0 | 0 | 0 |

*SECTION 2-6:*

**2–17.** Suppose the last six entries of Table 2–11 are changed to don't cares. Using the Karnaugh map, show the simplified logic circuit.

**2–18.** Assume the first six entries of Table 2–12 are changed to don't cares. What is the simplified logic circuit?

**2–19.** Suppose the inputs 1010 through 1111 only appear when there is trouble in a digital system. Design a logic circuit that detects the presence of any nibble input from 1010 to 1111.

*SECTION 2-7:*

**2–20.** Draw the unsimplified product-of-sums circuit for Table 2–11.

**2–21.** Repeat Prob. 2–20 for Table 2–12.

**2–22.** Draw a NOR–NOR circuit for this Boolean expression:

$$Y = (\overline{A} + \overline{B} + \overline{C})(\overline{A} + B + \overline{C})(A + B + \overline{C})$$

*SECTION 2-8:*

**2–23.** What is the simplified NOR–NOR circuit for Table 2–11?

**2–24.** Draw the simplified NOR–NOR circuit for Table 2–12.

**2–25.** Figure 2–34 shows all the input waveforms for the timing diagram of Fig. 2–31*e*. Draw the waveform for the output *Y*.

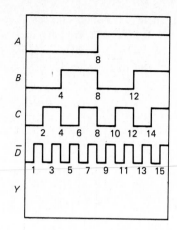

**FIG. 2–34.**

**2–26.** You are given the following Boolean equation:

$$Y = \overline{A}B\overline{C}D + \overline{A}BC\overline{D} + A\overline{B}C\overline{D}$$

Show the simplified NAND–NAND circuit for this. Also, show the simplified NOR–NOR circuit.

**2–27.** Table 2–14 is the truth table of a *full adder,* a logic circuit with two outputs called the *CARRY* and the *SUM.* What is the simplified NAND–NAND circuit for the *CARRY* output? For the *SUM* output?

**TABLE 2–14** *Full-Adder Truth Table*

| A | B | C | Carry | Sum |
|---|---|---|-------|-----|
| 0 | 0 | 0 | 0 | 0 |
| 0 | 0 | 1 | 0 | 1 |
| 0 | 1 | 0 | 0 | 1 |
| 0 | 1 | 1 | 1 | 0 |
| 1 | 0 | 0 | 0 | 1 |
| 1 | 0 | 1 | 1 | 0 |
| 1 | 1 | 0 | 1 | 0 |
| 1 | 1 | 1 | 1 | 1 |

**2–28.** Repeat Prob. 2–23 using NOR–NOR circuits.

# 3

# DATA·PROCESSING CIRCUITS

This chapter is about logic circuits that process binary data. The chapter begins with a discussion of multiplexers, which are circuits that can select one of many inputs. Then you will see how multiplexers are used as a design alternative to the sum-of-products solution. This will be followed by an examination of a variety of circuits, such as demultiplexers, decoders, encoders, exclusive–OR gates, parity checkers, and read-only memories. The chapter ends with a discussion of programmable array logic.

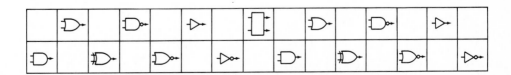

# 3-1 MULTIPLEXERS

Multiplex means *many into one*. A *multiplexer* is a circuit with many inputs but only one output. By applying control signals, we can steer any input to the output. Figure 3–1 illustrates the general idea. The circuit has $n$ input signals, $m$ control signals, and 1 output signal. One of the popular multiplexers is the 16-to-1 multiplexer which has 16 input bits, 4 control bits, and 1 output bit.

## 16-to-1 Multiplexer

Figure 3–2 shows a 16-to-1 multiplexer, which is also called a *data selector* because the output bit depends on the input data bit that is selected. The input bits are labeled $D_0$ to $D_{15}$. Only one of these is transmitted to the output. Which one depends on the value of *ABCD,* the control input. For instance, when

$$ABCD = 0000$$

the upper AND gate is enabled while all other AND gates are disabled. Therefore, data bit $D_0$ is transmitted to the output, giving

$$Y = D_0$$

If $D_0$ is low, $Y$ is low; if $D_0$ is high, $Y$ is high. The point is that $Y$ depends only on the value of $D_0$.

If the control nibble is changed to

$$ABCD = 1111$$

all gates are disabled except the bottom AND gate. In this case, $D_{15}$ is the only bit transmitted to the output, and

$$Y = D_{15}$$

As you can see, the control nibble determines which of the input data bits is transmitted to the output.

## The 74150

Try to visualize the 16-input OR gate of Fig. 3–2 changed to a NOR gate. What effect does this have on the operation of the circuit? Almost none. All that happens is we get the

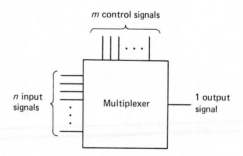

**FIG. 3–1.** Multiplexer

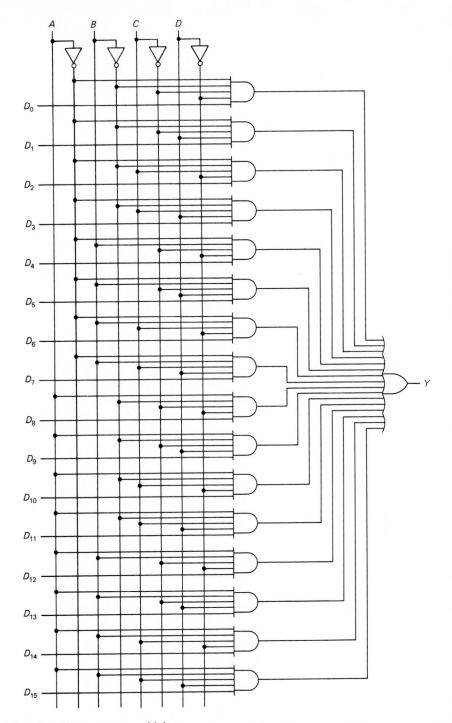

**FIG. 3–2.** Sixteen-to-one multiplexer

complement of the selected data bit rather than the data bit itself. For instance, when $ABCD = 0111$, the output is

$$Y = \overline{D}_7$$

This is the boolean equation for a typical transistor-transistor logic (TTL) multiplexer because it has an inverter on the output that produces the complement of the selected data bit.

The 74150 is a 16-to-1 TTL multiplexer with the pin diagram shown in Fig. 3–3. Pins 1 to 8 and 16 to 23 are for the input data bits $D_0$ to $D_{15}$. Pins 11, 13, 14, and 15 are for the control bits $ABCD$. Pin 10 is the output; and it equals the complement of the selected data bit. Pin 9 is for the *strobe,* an input signal that disables or enables the multiplexer. As shown in Table 3–1, a low strobe enables the multiplexer, so that output $Y$ equals the complement of the input data bit:

$$Y = \overline{D}_n$$

where $n$ is the decimal equivalent of $ABCD$. On the other hand, a high strobe disables the multiplexer and forces the output into the high state. With a high strobe, the value of $ABCD$ doesn't matter.

## Multiplexer Logic

Digital design usually begins with a truth table. The problem is to come up with a logic circuit that has the same truth table. In Chapter 2, you saw two standard methods for implementing a truth table: the sum-of-products and the product-of-sums solutions. Now you are ready for a third method: the *multiplexer solution.*

For example, suppose we want a logic circuit with a truth table like Table 3–2. To use a 74150 to implement this table, complement each $Y$ output to get the corresponding data input:

**TABLE 3–1** *74150 Truth Table*

| Strobe | A B C D | Y |
|--------|---------|---|
| L | L L L L | $\overline{D}_0$ |
| L | L L L H | $\overline{D}_1$ |
| L | L L H L | $\overline{D}_2$ |
| L | L L H H | $\overline{D}_3$ |
| L | L H L L | $\overline{D}_4$ |
| L | L H L H | $\overline{D}_5$ |
| L | L H H L | $\overline{D}_6$ |
| L | L H H H | $\overline{D}_7$ |
| L | H L L L | $\overline{D}_8$ |
| L | H L L H | $\overline{D}_9$ |
| L | H L H L | $\overline{D}_{10}$ |
| L | H L H H | $\overline{D}_{11}$ |
| L | H H L L | $\overline{D}_{12}$ |
| L | H H L H | $\overline{D}_{13}$ |
| L | H H H L | $\overline{D}_{14}$ |
| L | H H H H | $\overline{D}_{15}$ |
| H | X X X X | H |

**TABLE 3–2**

| A | B | C | D | Y |
|---|---|---|---|---|
| 0 | 0 | 0 | 0 | 1 |
| 0 | 0 | 0 | 1 | 0 |
| 0 | 0 | 1 | 0 | 1 |
| 0 | 0 | 1 | 1 | 1 |
| 0 | 1 | 0 | 0 | 1 |
| 0 | 1 | 0 | 1 | 1 |
| 0 | 1 | 1 | 0 | 0 |
| 0 | 1 | 1 | 1 | 0 |
| 1 | 0 | 0 | 0 | 1 |
| 1 | 0 | 0 | 1 | 1 |
| 1 | 0 | 1 | 0 | 1 |
| 1 | 0 | 1 | 1 | 1 |
| 1 | 1 | 0 | 0 | 1 |
| 1 | 1 | 0 | 1 | 1 |
| 1 | 1 | 1 | 0 | 0 |
| 1 | 1 | 1 | 1 | 1 |

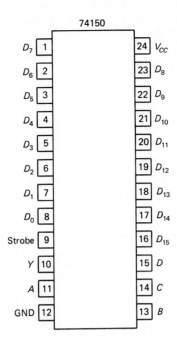

74150

| | | | | |
|---|---|---|---|---|
| $D_7$ | 1 | | 24 | $V_{CC}$ |
| $D_6$ | 2 | | 23 | $D_8$ |
| $D_5$ | 3 | | 22 | $D_9$ |
| $D_4$ | 4 | | 21 | $D_{10}$ |
| $D_3$ | 5 | | 20 | $D_{11}$ |
| $D_2$ | 6 | | 19 | $D_{12}$ |
| $D_1$ | 7 | | 18 | $D_{13}$ |
| $D_0$ | 8 | | 17 | $D_{14}$ |
| Strobe | 9 | | 16 | $D_{15}$ |
| $Y$ | 10 | | 15 | $D$ |
| $A$ | 11 | | 14 | $C$ |
| GND | 12 | | 13 | $B$ |

**FIG. 3–3.** Pinout diagram of 74150

$$D_0 = \overline{1} = 0$$
$$D_1 = \overline{0} = 1$$
$$D_2 = \overline{1} = 0$$

and so forth, up to

$$D_{15} = \overline{1} = 0$$

Next, wire the data inputs of 74150 as shown in Fig. 3–4, so that they equal the foregoing values. In other words, $D_0$ is grounded, $D_1$ is connected to $+5$ V, $D_2$ is grounded, and so on. In each case, the data input is the complement of the desired $Y$ output of Table 3–2.

Figure 3–4 is the multiplexer design solution. It has the same truth table given in Table 3–2. If in doubt, analyze it as follows for each input condition. When $ABCD = 0000$, $D_0$ is the selected input in Fig. 3–4. Since $D_0$ is low, $Y$ is high. When $ABCD = 0001$, $D_1$ is selected. Since $D_1$ is high, $Y$ is low. If you check the remaining input possibilities, you will see that the circuit has the truth table given in Table 3–2.

## Bubbles on Signal Lines

Data sheets often show inversion bubbles on some of the signal lines. For instance, notice the bubble on pin 10, the output of Fig. 3–4. This bubble is a reminder that the output is the complement of the selected data bit.

Also notice the bubble on the strobe input (pin 9). As discussed earlier, the multiplexer is active (enabled) when the strobe is low and inactive (disabled) when it is high. Because of this, the strobe is called an *active-low signal*; it causes something to happen

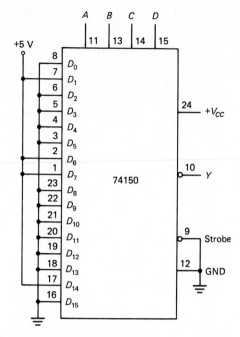

**FIG. 3–4.** Using a 74150 for multiplexer logic

when it is low rather than when it is high. Most schematic diagrams use bubbles to indicate active-low signals. From now on, whenever you see a bubble on an input pin, remember that it means the signal is active-low.

## Universal Logic Circuit

The circuit of Fig. 3–4 is an example of multiplexer logic. You ground an input data pin when the corresponding output is high in the truth table. You connect an input data pin to a high voltage if the corresponding output is low in the truth table. The 74150 is sometimes called a *universal logic circuit* because it can be used as a design solution for any four-variable truth table. In other words, by changing the input data bits in Fig. 3–4, we can use the same integrated circuit (IC) to generate thousands of different truth tables.

## Nibble Multiplexers

Sometimes we want to select one of two input nibbles. In this case, we can use a nibble multiplexer like the one shown in Fig. 3–5. The input nibble on the left is $A_3A_2A_1A_0$ and the one on the right is $B_3B_2B_1B_0$. The control signal labeled *SELECT* determines which input nibble is transmitted to the output. When SELECT is low, the four NAND gates on the left are activated; therefore,

$$Y_3Y_2Y_1Y_0 = A_3A_2A_1A_0$$

When SELECT is high, the four NAND gates on the right are active, and

$$Y_3Y_2Y_1Y_0 = B_3B_2B_1B_0$$

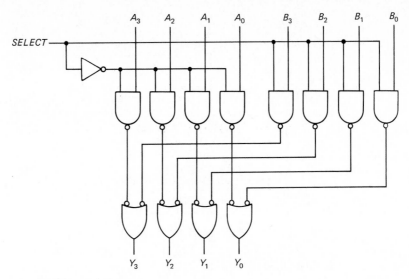

**FIG. 3–5.** Nibble multiplexer

Figure 3–6*a* shows the pinout diagram of a 74157, a nibble multiplexer with a SELECT input as previously described. When SELECT is low, the left nibble is steered to the output. When SELECT is high, the right nibble is steered to the output. The 74157 also includes a strobe input. As before, the strobe must be low for the multiplexer to work properly. When the strobe is high, the multiplexer is inoperative. Figure 3–6*b* shows how to draw a 74157 on a schematic diagram. The bubble on pin 15 tells us that STROBE is an active-low input.

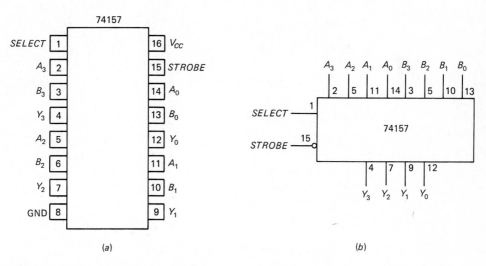

**FIG. 3–6.** Pinout diagrams of 74157

□ EXAMPLE 3—1

For what $X$ inputs is the strobe signal low in Fig. 3–7?

□ SOLUTION

The NAND–NAND circuit is equivalent to an AND–OR circuit. If you check through all 16 input possibilities ($X_3X_2X_1X_0$ = 0000 to 1111), you get a low STROBE for the following $X$ inputs: 0000, 0001, 0010, 0100, 0101, 0110, 1000, 1001, and 1010.

□ EXAMPLE 3—2

In Fig. 3–7, what does $Y$ equal for each of these input conditions:

   **a.** $ABCD$ = 0111 and $X_3X_2X_1X_0$ = 0011.
   **b.** $ABCD$ = 1001 and $X_3X_2X_1X_0$ = 0110.
   **c.** $ABCD$ = 1111 and $X_3X_2X_1X_0$ = 0001.

□ SOLUTION

   **a.** When $X_3X_2X_1X_0$ = 0011, the STROBE is high and the 74150 is inactive. In this case, $Y$ is high (see Table 3–1).
   **b.** When $X_3X_2X_1X_0$ = 0110, the STROBE is low. Now, the 74150 is active. Since $ABCD$ = 1001, data bit $D_9$ is selected. Because $D_9$ is grounded, $Y$ is high.
   **c.** When $X_3X_2X_1X_0$ = 0001, the STROBE is low. With $ABCD$ = 1111, $D_{15}$ is selected, so that $Y$ is high.

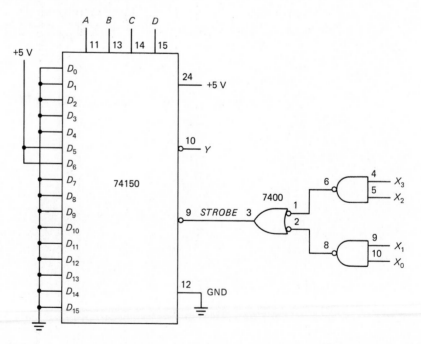

**FIG. 3–7.**

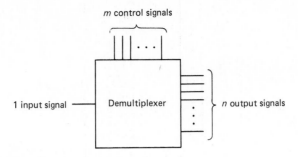

m control signals

1 input signal — Demultiplexer

n output signals

**FIG. 3–8.** Demultiplexer

## 3·2 DEMULTIPLEXERS

Demultiplex means *one into many*. A *demultiplexer* is a logic circuit with one input and many outputs. By applying control signals, we can steer the input signal to one of the output lines. Figure 3–8 illustrates the general idea. The circuit has 1 input signal, $m$ control signals, and $n$ output signals.

### 1-to-16 Demultiplexer

Figure 3–9 shows a 1-to-16 demultiplexer. The input bit is labeled $D$. This data bit ($D$) is transmitted to the data bit of the output lines. But which one? Again, this depends on the value of *ABCD,* the control input. When

$$ABCD = 0000$$

the upper AND gate is enabled while all other AND gates are disabled. Therefore, data bit $D$ is transmitted only to the $Y_0$ output, giving

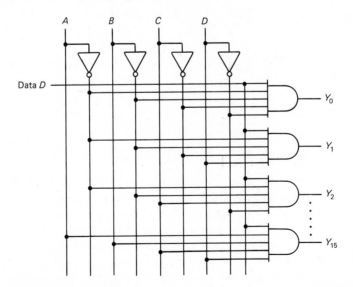

**FIG. 3–9.** 1-to-16 demultiplexer

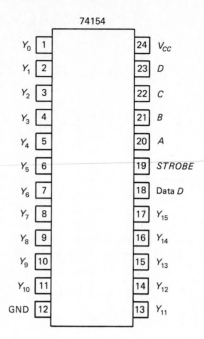

**FIG. 3–10.** Pinout diagram of 74154

$$Y_0 = D$$

If $D$ is low, $Y_0$ is low. If $D$ is high, $Y_0$ is high. As you can see, the value of $Y_0$ depends on the value of $D$. All other outputs are in the low state.

If the control nibble is changed to

$$ABCD = 1111$$

all gates are disabled except the bottom AND gate. Then, $D$ is transmitted only to the $Y_{15}$ output, and

$$Y_{15} = D$$

## The 74154

The 74154 is a 1-to-16 demultiplexer with the pin diagram of Fig. 3–10. Pin 18 is for the input data $D$, and pins 20 to 23 are for the control bits $ABCD$. Pins 1 to 11 and 13 to 17 are for the output bits $Y_0$ to $Y_{15}$. Pin 19 is for the strobe, again an active-low input. Finally, pin 24 is for $V_{CC}$ and pin 12 for ground.

Table 3–3 shows the truth table of a 74154. First, notice the STROBE input. It has to be low to activate the 74154. When the STROBE is low, the control input $ABCD$ determines which of the output lines is low when the data input is low. When the data input is

TABLE 3–3 *74154 Truth Table*

| Strobe | Data | A | B | C | D | $Y_0$ | $Y_1$ | $Y_2$ | $Y_3$ | $Y_4$ | $Y_5$ | $Y_6$ | $Y_7$ | $Y_8$ | $Y_9$ | $Y_{10}$ | $Y_{11}$ | $Y_{12}$ | $Y_{13}$ | $Y_{14}$ | $Y_{15}$ |
|---|---|---|---|---|---|---|---|---|---|---|---|---|---|---|---|---|---|---|---|---|---|
| L | L | L | L | L | L | L | H | H | H | H | H | H | H | H | H | H | H | H | H | H | H |
| L | L | L | L | L | H | H | L | H | H | H | H | H | H | H | H | H | H | H | H | H | H |
| L | L | L | L | H | L | H | H | L | H | H | H | H | H | H | H | H | H | H | H | H | H |
| L | L | L | L | H | H | H | H | H | L | H | H | H | H | H | H | H | H | H | H | H | H |
| L | L | L | H | L | L | H | H | H | H | L | H | H | H | H | H | H | H | H | H | H | H |
| L | L | L | H | L | H | H | H | H | H | H | L | H | H | H | H | H | H | H | H | H | H |
| L | L | L | H | H | L | H | H | H | H | H | H | L | H | H | H | H | H | H | H | H | H |
| L | L | L | H | H | H | H | H | H | H | H | H | H | L | H | H | H | H | H | H | H | H |
| L | L | H | L | L | L | H | H | H | H | H | H | H | H | L | H | H | H | H | H | H | H |
| L | L | H | L | L | H | H | H | H | H | H | H | H | H | H | L | H | H | H | H | H | H |
| L | L | H | L | H | L | H | H | H | H | H | H | H | H | H | H | L | H | H | H | H | H |
| L | L | H | L | H | H | H | H | H | H | H | H | H | H | H | H | H | L | H | H | H | H |
| L | L | H | H | L | L | H | H | H | H | H | H | H | H | H | H | H | H | L | H | H | H |
| L | L | H | H | L | H | H | H | H | H | H | H | H | H | H | H | H | H | H | L | H | H |
| L | L | H | H | H | L | H | H | H | H | H | H | H | H | H | H | H | H | H | H | L | H |
| L | L | H | H | H | H | H | H | H | H | H | H | H | H | H | H | H | H | H | H | H | L |
| L | H | X | X | X | X | H | H | H | H | H | H | H | H | H | H | H | H | H | H | H | H |
| H | L | X | X | X | X | H | H | H | H | H | H | H | H | H | H | H | H | H | H | H | H |
| H | H | X | X | X | X | H | H | H | H | H | H | H | H | H | H | H | H | H | H | H | H |

high, all output lines are high. Similarly, when the STROBE is high, all output lines are high.

Figure 3–11 shows how to draw a 74154 on a schematic diagram. There is one input data bit (pin 18) under the control of nibble *ABCD*. The data bit is automatically steered to the output line whose subscript is the decimal equivalent of *ABCD*. Again, the bubble on the STROBE pin indicates an active-low input.

☐ EXAMPLE 3–3

In Fig. 3–12, what does the $Y_{12}$ output equal for each of the following conditions:

   **a.** *R* is high, *T* is high, *ABCD* = 0110.
   **b.** *R* is low, *T* is high, *ABCD* = 1100.
   **c.** *R* is high, *T* is high, *ABCD* = 1100.

☐ SOLUTION

   **a.** Since *R* and *T* are both high, the STROBE is low and the 74154 is active. Because *ABCD* = 0110, the input data is steered to the $Y_6$ output line (pin 7). The $Y_{12}$ output remains in the high state (see Table 3–3).
   **b.** In this case, the STROBE is high and the 74154 is inactive. The $Y_{12}$ output is high.
   **c.** With *R* and *T* both high, the STROBE is low and the 74154 is active. Since *ABCD* = 1100, the two pulses on the data input are steered to the $Y_{12}$ output (pin 14).

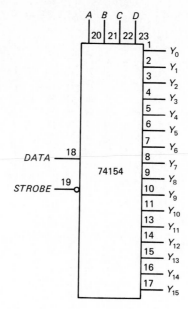

**FIG. 3–11.** Logic diagram of 74154

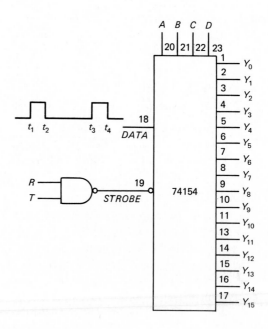

**FIG. 3–12.**

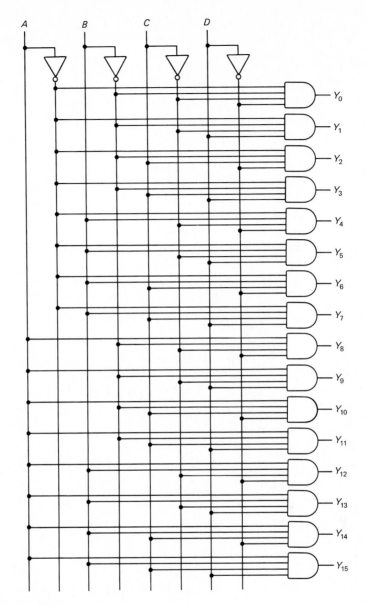

**FIG. 3–13.** 1-of-16 decoder

## 3-3  1-OF-16 DECODER

A *decoder* is similar to a demultiplexer, with one exception—there is no data input. The only inputs are the control bits *ABCD,* which are shown in Fig. 3–13. This logic circuit is called a *1-of-16 decoder* because only 1 of the 16 output lines is high. For instance, when

$ABCD$ is 0001, only the $Y_1$ AND gate has all inputs high; therefore, only the $Y_1$ output is high. If $ABCD$ changes to 0100, only the $Y_4$ AND gate has all inputs high; as a result, only the $Y_4$ output goes high.

If you check the other $ABCD$ possibilities (0000 to 1111), you will find that the subscript of the high output always equals the decimal equivalent of $ABCD$. For this reason, the circuit is sometimes called a *binary-to-decimal decoder*. Because it has 4 input lines and 16 output lines, the circuit is also known as a *4 line-to-16 line decoder*.

Normally, you would not build a decoder with separate inverters and AND gates as shown in Fig. 3–13. Instead, you would use an IC such as the 74154. The 74154 is called a *decoder-demultiplexer* because it can be used either as a decoder or as a demultiplexer. You saw how to use a 74154 as a demultiplexer in Sec. 3–2. To use this same IC as a decoder, all you have to do is ground the DATA and STROBE inputs as shown in Fig. 3–14. Then, the selected output line is in the low state (see Table 3–3). This is why bubbles are shown on the output lines. They remind us that the output line is low when it is active or selected. For instance, if the binary input is

$$ABCD = 0111$$

then the $Y_7$ output is low, while all other outputs are high.

## ☐ EXAMPLE 3—4

Figure 3–15 illustrates *chip expansion*. We have expanded two 74154s to get a 1-of-32 decoder. Here is the way the circuit works. Bit $X$ drives the first 74154, and the complement of $X$ drives the second 74154. When $X$ is low, the first 74154 is active and the second is inactive. The $ABCD$ input drives both decoders but only the first is active; therefore, only one output line on the first decoder is in the low state.

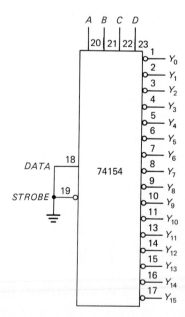

**FIG. 3–14.** Using 74154 as decoder

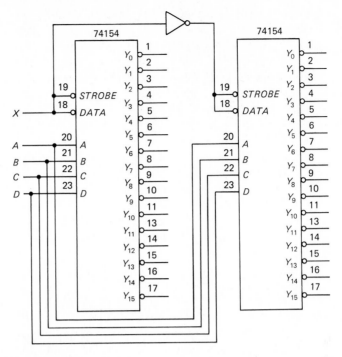

**FIG. 3–15.** Chip expansion

On the other hand, when $X$ is high, the first 74154 is disabled and the second one is enabled. This means the $ABCD$ input is decoded into a low output from the second decoder. In effect, the circuit of Fig. 3–15 acts like a 1-of-32 decoder.

Most people add bubbles to their logic circuits whenever the low state is the active state. In Fig. 3–15, all output lines are high, except the decoded output line; therefore, we have included a bubble on each output line. This tells anyone looking at the schematic diagram that the active output line is in the low state rather than the high state. Similarly, we have added bubbles to the STROBE and DATA inputs of each 74154 to indicate active-low inputs.

## 3-4 BCD-TO-DECIMAL DECODERS

BCD is an abbreviation for *binary-coded decimal*. The BCD code expresses each digit in a decimal number by its nibble equivalent. For instance, decimal number 429 is changed to its BCD form as follows:

$$\begin{array}{ccc} 4 & 2 & 9 \\ \downarrow & \downarrow & \downarrow \\ 0100 & 0010 & 1001 \end{array}$$

To anyone using the BCD code, 0100 0010 1001 is equivalent to 429.

As another example, here is how to convert the decimal number 8963 to its BCD form:

$$\begin{array}{cccc} 8 & 9 & 6 & 3 \\ \downarrow & \downarrow & \downarrow & \downarrow \\ 1000 & 1001 & 0110 & 0011 \end{array}$$

Again, we have changed each decimal digit to its binary equivalent.

Some early computers processed BCD numbers. This means that the decimal numbers were changed into BCD numbers, which the computer then added, subtracted, etc. The final answer was converted from BCD back to decimal numbers.

Here is an example of how to convert from the BCD form back to the decimal number:

$$\begin{array}{ccc} 0101 & 0111 & 1000 \\ \downarrow & \downarrow & \downarrow \\ 5 & 7 & 8 \end{array}$$

As you can see, 578 is the decimal equivalent of 0101 0111 1000.

One final point should be considered. Notice that BCD digits are from 0000 to 1001. All combinations above this (1010 to 1111) cannot exist in the BCD code because the highest decimal digit being coded is 9.

## BCD-to-Decimal Decoder

The circuit of Fig. 3–16 is called a *1-of-10 decoder* because only 1 of the 10 output lines is high. For instance, when $ABCD$ is 0011, only the $Y_3$ AND gate has all high inputs; therefore, only the $Y_3$ output is high. If $ABCD$ changes to 1000, only the $Y_8$ AND gate has all high inputs; as a result, only the $Y_8$ output goes high.

If you check the other $ABCD$ possibilities (0000 to 1001), you will find that the subscript of the high output always equals the decimal equivalent of the input BCD digit. For this reason, the circuit is also called a *BCD-to-decimal converter*.

## The 7445

Typically, you would not build a decoder with separate inverters and AND gates, as shown in Fig. 3–16. Instead, you would use a TTL IC like the 7445 of Fig. 3–17. Pin 16 connects to the supply voltage $V_{CC}$ and pin 8 is grounded. Pins 12 to 15 are for the BCD input, while pins 1 to 7 and 9 to 11 are for the outputs. This IC is functionally equivalent to the one in Fig. 3–16, except that the active output line is in the low state. All other output lines are in the high state, as shown in Table 3–4. Notice that an invalid BCD input (1010 to 1111) forces all output lines into the high state.

☐ EXAMPLE 3–5

The decoded outputs of a 7445 can be connected to light-emitting diodes (LEDs), as shown in Fig. 3–18. If each resistance is 1 kΩ and each LED has a forward voltage drop of 2 V, how much current is there through a LED when it is conducting?

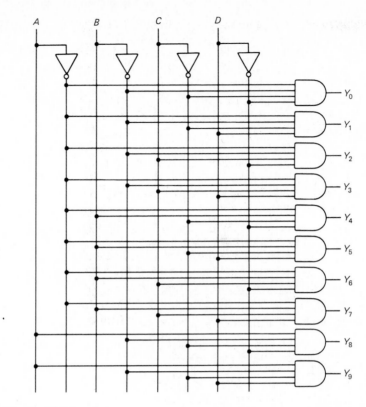

**FIG. 3–16.** 1-of-10 decoder

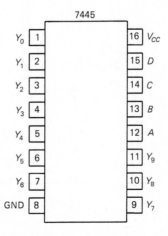

**FIG. 3–17.** Pinout diagram of 7445

**TABLE 3-4** *7445 Truth Table*

| No. | A | B | C | D | $Y_0$ | $Y_1$ | $Y_2$ | $Y_3$ | $Y_4$ | $Y_5$ | $Y_6$ | $Y_7$ | $Y_8$ | $Y_9$ |
|---|---|---|---|---|---|---|---|---|---|---|---|---|---|---|
| | | Inputs | | | | | | | Outputs | | | | | |
| 0 | L | L | L | L | L | H | H | H | H | H | H | H | H | H |
| 1 | L | L | L | H | H | L | H | H | H | H | H | H | H | H |
| 2 | L | L | H | L | H | H | L | H | H | H | H | H | H | H |
| 3 | L | L | H | H | H | H | H | L | H | H | H | H | H | H |
| 4 | L | H | L | L | H | H | H | H | L | H | H | H | H | H |
| 5 | L | H | L | H | H | H | H | H | H | L | H | H | H | H |
| 6 | L | H | H | L | H | H | H | H | H | H | L | H | H | H |
| 7 | L | H | H | H | H | H | H | H | H | H | H | L | H | H |
| 8 | H | L | L | L | H | H | H | H | H | H | H | H | L | H |
| 9 | H | L | L | H | H | H | H | H | H | H | H | H | H | L |
| Invalid | H | L | H | L | H | H | H | H | H | H | H | H | H | H |
| Invalid | H | L | H | H | H | H | H | H | H | H | H | H | H | H |
| Invalid | H | H | L | L | H | H | H | H | H | H | H | H | H | H |
| Invalid | H | H | L | H | H | H | H | H | H | H | H | H | H | H |
| Invalid | H | H | H | L | H | H | H | H | H | H | H | H | H | H |
| Invalid | H | H | H | H | H | H | H | H | H | H | H | H | H | H |

☐ SOLUTION

When an output is in the low state, you can approximate the output voltage as zero. Therefore, the current through a LED is

$$I = \frac{5\ V - 2\ V}{1\ k\Omega} = 3\ mA$$

☐ EXAMPLE 3–6

The LEDs of Fig. 3–18 are numbered 0 through 9. Which of the LEDs is lit for each of the following conditions:

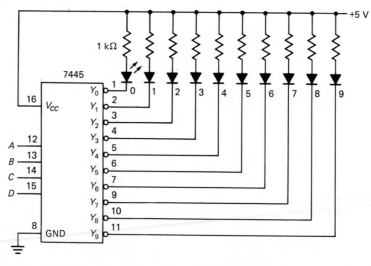

**FIG. 3–18.**

**a.** $ABCD = 0101$.
**b.** $ABCD = 1001$.
**c.** $ABCD = 1100$.

☐ SOLUTION

**a.** When $ABCD = 0101$, the decoded output line is $Y_5$. Since $Y_5$ is approximately grounded, LED 5 lights up. All other LEDs remain off because the other outputs are high.
**b.** When $ABCD = 1001$, LED 9 is on.
**c.** $ABCD = 1100$ is an invalid input. Therefore, none of the LEDs is on because all output lines are high (see Table 3–4).

## 3-5 SEVEN-SEGMENT DECODERS

A LED emits radiation when forward-biased. Why? Because free electrons recombine with holes near the junction. As the free electrons fall from a higher energy level to a lower one, they give up energy in the form of heat and light. By using elements like gallium, arsenic, and phosphorus, a manufacturer can produce LEDs that emit red, green, yellow, blue, orange, and infrared (invisible) light. LEDs that produce visible radiation are useful in test instruments, pocket calculators, etc.

### Seven-Segment Indicator

Figure 3–19*a* shows a *seven-segment indicator*; i.e., seven LEDs labeled *a* through *g*. By forward-biasing different LEDs, we can display the digits 0 through 9 (see Fig. 3–19*b*). For instance, to display a 0, we need to light up segments *a*, *b*, *c*, *d*, *e*, and *f*. To light up a 5, we need segments *a*, *c*, *d*, *f*, and *g*.

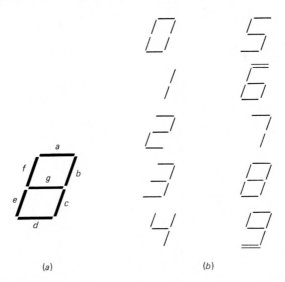

(a)                    (b)

**FIG. 3–19.** Seven-segment indicator

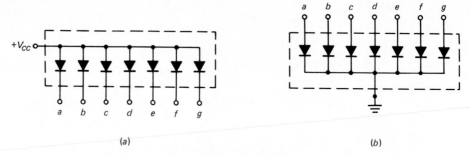

**FIG. 3–20.** (a) Common-anode type, (b) Common-cathode type.

Seven-segment indicators may be the common-anode type where all anodes are connected together (Fig. 3–20a) or the common-cathode type where all cathodes are connected together (Fig. 3–20b). With the common-anode type of Fig. 3–20a, you have to connect a current-limiting resistor between each LED and ground. The size of this resistor determines how much current flows through the LED. (Typical LED current is between 1 and 50 mA.) The common-cathode type of Fig. 3–20b uses a current-limiting resistor between each LED and $+V_{CC}$.

## The 7446

A seven-segment *decoder-driver* is an IC decoder that can be used to drive a seven-segment indicator. There are two types of decoder-drivers, corresponding to the common-anode and common-cathode indicators. Each decoder-driver has 4 input pins (the BCD input) and 7 output pins (the *a* through *g* segments).

Figure 3–21a shows a 7446 driving a common-anode indicator. Logic circuits inside the 7446 convert the BCD input to the required output. For instance, if the BCD input is 0111, the internal logic (not shown) of the 7446 will force LEDs *a*, *b*, and *c* to conduct because the corresponding transistors go into saturation. As a result, digit 7 will appear on the seven-segment indicator.

Notice the current-limiting resistors between the seven-segment indicator and the 7446 of Fig. 3–21a. You have to connect these external resistors to limit the current in each segment to a safe value between 1 and 50 mA, depending on how bright you want the display to be.

## The 7448

Figure 3–21b is the alternative decoding approach. Here, a 7448 drives a common-cathode indicator. Again, internal logic converts the BCD input to the required output. For example, when a BCD input of 0100 is used, the internal logic forces LEDs *b*, *c*, *f*, and *g* to conduct. The seven-segment indicator then displays a 4. Unlike the 7446 that requires external current-limiting resistors, the 7448 has its own current-limiting resistors on the chip.

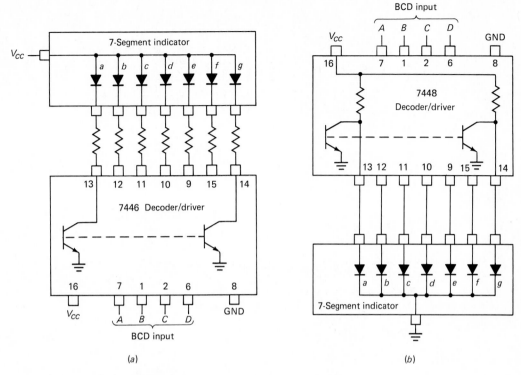

**FIG. 3–21.** (a) 7446 decoder-driver, (b) 7448 decoder-driver

## 3-6 ENCODERS

An encoder converts an active input signal into a coded output signal. Figure 3–22 illustrates the general idea. There are *n* input lines, only one of which is active. Internal logic within the encoder converts this active input to a coded binary output with *m* bits.

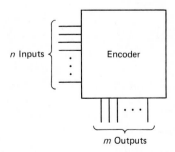

**FIG. 3–22.** Encoder

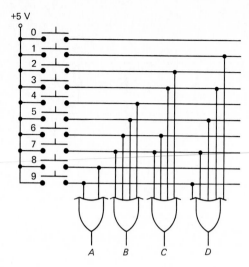

**FIG. 3–23.** Decimal-to-bcd encoder

## Decimal-to-BCD Encoder

Figure 3–23 shows a common type of encoder—the *decimal-to-BCD encoder*. The switches are push-button switches like those of a pocket calculator. When button 3 is pressed, the $C$ and $D$ OR gates have high inputs; therefore, the output is

$$ABCD = 0011$$

If button 5 is pressed, the output becomes

$$ABCD = 0101$$

When switch 9 is pressed,

$$ABCD = 1001$$

## The 74147

Figure 3–24$a$ is the pinout diagram for a 74147, a decimal-to-BCD encoder. The decimal input, $X_1$ to $X_9$, connect to pins 1 to 5, and 10 to 13. The BCD output comes from pins 14, 6, 7, and 9. Pin 16 is for the supply voltage, and pin 8 is grounded. The label NC on pin 15 means *no connection* (the pin is not used).

Figure 3–24$b$ shows how to draw a 74147 on a schematic diagram. As usual, the bubbles indicate active-low inputs and outputs. Table 3–5 is the truth table of a 74147. Notice the following. When all $X$ inputs are high, all outputs are high. When $X_9$ is low, the $ABCD$ output is LHHL (equivalent to 9 if you complement the bits). When $X_8$ is the only low input, $ABCD$ is LHHH (equivalent to 8 if the bits are complemented). When $X_7$ is the only low input, $ABCD$ becomes HLLL (equivalent to 7 if the bits are complemented). Continue like this through the rest of the truth table and you can see that an active-low decimal input is being converted to a complemented BCD output.

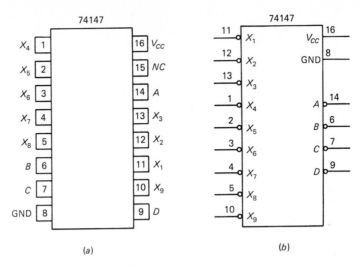

FIG. 3–24. (a) Pinout diagram of 74147, (b) Logic diagram

Incidentally, the 74147 is called a *priority encoder* because it gives priority to the highest-order input. You can see this by looking at Table 3–5. If all inputs $X_1$ through $X_9$ are low, the highest of these, $X_9$, is encoded to get an output of LHHL. In other words, $X_9$ has priority over all others. When $X_9$ is high, $X_8$ is next in line of priority and gets encoded if it is low. Working your way through Table 3–5, you can see that the highest active-low from $X_9$ to $X_0$ has priority and will control the encoding.

□ EXAMPLE 3–7

What is the *ABCD* output of Fig. 3–25 when the button 6 is pressed?

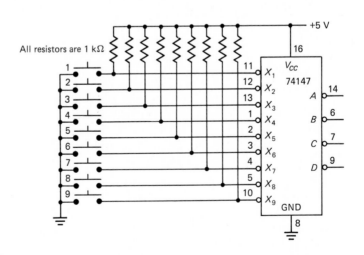

FIG. 3–25.

**TABLE 3–5** *74147 Truth Table*

| Inputs | | | | | | | | | Outputs | | | |
|---|---|---|---|---|---|---|---|---|---|---|---|---|
| $X_1$ | $X_2$ | $X_3$ | $X_4$ | $X_5$ | $X_6$ | $X_7$ | $X_8$ | $X_9$ | $A$ | $B$ | $C$ | $D$ |
| H | H | H | H | H | H | H | H | H | H | H | H | H |
| X | X | X | X | X | X | X | X | L | L | H | H | L |
| X | X | X | X | X | X | X | L | H | L | H | H | H |
| X | X | X | X | X | X | L | H | H | H | L | L | L |
| X | X | X | X | X | L | H | H | H | H | L | L | H |
| X | X | X | X | L | H | H | H | H | H | L | H | L |
| X | X | X | L | H | H | H | H | H | H | L | H | H |
| X | X | L | H | H | H | H | H | H | H | H | L | L |
| X | L | H | H | H | H | H | H | H | H | H | L | H |
| L | H | H | H | H | H | H | H | H | H | H | H | L |

☐ SOLUTION

When all switches are open, the $X_1$ to $X_9$ inputs are pulled up to the high state ($+5$ V). A glance at Table 3–5 indicates that the $ABCD$ output is HHHH at this time.

When switch 6 is pressed, the $X_6$ input is grounded. Therefore, all $X$ inputs are high, except for $X_6$. Table 3–5 indicates that the $ABCD$ output is HLLH, which is equivalent to 6 when the output bits are complemented.

## 3·7 EXCLUSIVE–OR GATES

The *exclusive–OR gate* has a high output only when an odd number of inputs is high. Figure 3–26 shows how to build an exclusive–OR gate. The upper AND gate forms the product $\overline{A}B$, while the lower one produces $A\overline{B}$. Therefore, the output of the OR gate is

$$Y = \overline{A}B + A\overline{B}$$

Here is what happens for different inputs. When $A$ and $B$ are low, both AND gates have low outputs; therefore, the final output $Y$ is low. If $A$ is low and $B$ is high, the upper AND gate has a high output, so the OR gate has high output. Likewise, a high $A$ and a low $B$ result in a final output that is high. If both inputs are high, both AND gates have low outputs and the final output is low.

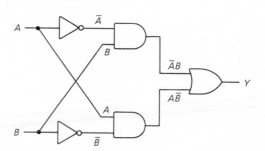

**FIG. 3–26.** Exclusive–OR gate

**TABLE 3–6** *Exclusive–OR Truth Table*

| A | B | C |
|---|---|---|
| 0 | 0 | 0 |
| 0 | 1 | 1 |
| 1 | 0 | 1 |
| 1 | 1 | 0 |

Table 3–6 shows the truth table for a 2-input exclusive–OR gate. The output is high when A or B is high, but not when both are high. This is why the circuit is known as an exclusive–OR gate. In other words, the output is a 1 *only* when the inputs are different.

Figure 3–27 shows the symbol for a 2-input exclusive–OR gate. Whenever you see this symbol, remember the action—the output is high if either input is high, but not when both are high. Stated another way, the inputs must be different to get a high output.

## Four Inputs

Figure 3–28a shows a pair of exclusive–OR gates driving an exclusive-OR gate. If all inputs (A to D) are low, the input gates have low outputs, so the final gate has a low output. If A to C are low and D is high, the upper gate has a low output, the lower gate has a high output, and output gate has a high output.

If we continue analyzing the circuit operation for the remaining input possibilities, we can work out Table 3–7. The following is an important property of this truth table. Each ABCD input with an odd number of 1s produces an output 1. For instance, the first ABCD entry to produce an output 1 is 0001; this has an odd number of 1s. The next ABCD entry to produce an output 1 is 0010; again, an odd number of 1s. An output 1 also occurs for these ABCD inputs: 0100, 0111, 1000, 1011, 1101, and 1110, each of which has an odd number of 1s.

**TABLE 3–7** *4-Input Exclusive–OR Gate*

| Comment | A | B | C | D | Y |
|---------|---|---|---|---|---|
| Even | 0 | 0 | 0 | 0 | 0 |
| Odd  | 0 | 0 | 0 | 1 | 1 |
| Odd  | 0 | 0 | 1 | 0 | 1 |
| Even | 0 | 0 | 1 | 1 | 0 |
| Odd  | 0 | 1 | 0 | 0 | 1 |
| Even | 0 | 1 | 0 | 1 | 0 |
| Even | 0 | 1 | 1 | 0 | 0 |
| Odd  | 0 | 1 | 1 | 1 | 1 |
| Odd  | 1 | 0 | 0 | 0 | 1 |
| Even | 1 | 0 | 0 | 1 | 0 |
| Even | 1 | 0 | 1 | 0 | 0 |
| Odd  | 1 | 0 | 1 | 1 | 1 |
| Even | 1 | 1 | 0 | 0 | 0 |
| Odd  | 1 | 1 | 0 | 1 | 1 |
| Odd  | 1 | 1 | 1 | 0 | 1 |
| Even | 1 | 1 | 1 | 1 | 0 |

**FIG. 3–27.** Logic symbol for exclusive–OR gate

Figure 3–28*a* illustrates the logic for a 4-input exclusive–OR gate. In this book, we will use the abbreviated symbol given in Fig. 3–28*b* to represent a 4-input exclusive–OR gate. When you see this symbol, remember the action—the gate produces an output 1 when the *ABCD* input has an odd number of 1s.

## Any Number of Inputs

Using 2-input exclusive–OR gates as building blocks, you can produce exclusive–OR gates with any number of inputs. For example, Fig. 3–29*a* shows a pair of exclusive–OR gates. There are 3 inputs and 1 output. If you analyze this circuit, you will find it produces an output 1 only when the 3-bit input has an odd number of 1s. Figure 3–29*b* shows an abbreviated symbol for a 3-input exclusive–OR gate.

As another example, Fig. 3–29*c* shows a circuit with 6 inputs and 1 output. If you analyze the circuit, you will find it produces an output 1 only when the 6-bit input has an odd number of 1s. Figure 3–29*d* shows an abbreviated symbol for a 6-input exclusive–OR gate.

In general, you can build an exclusive–OR gate with any number of inputs. Such a gate always produces an output 1 only when the *n*-bit input has an odd number of 1s.

## 3-8 PARITY GENERATORS-CHECKERS

*Even parity* means an *n*-bit input has an even number of 1s. For instance, 110011 has even parity because it contains four 1s. *Odd parity* means an *n*-bit input has an odd number of 1s. For example, 110001 has odd parity because it contains three 1s.

Here are two more examples:

$$1111\ 0000\ 1111\ 0011 \qquad \text{even parity}$$
$$1111\ 0000\ 1111\ 0111 \qquad \text{odd parity}$$

The first binary number has even parity because it contains ten 1s; the second binary number has odd parity because it contains eleven 1s. Incidentally, longer binary numbers are much easier to read if they are split into nibbles, or groups of four, as done here.

### Parity Checker

Exclusive–OR gates are ideal for checking the parity of a binary number because they produce an output 1 when the input has an odd number of 1s. Therefore, an even-parity input to an exclusive–OR gate produces a low output, while an odd-parity input produces a high output.

For instance, Fig. 3–30 shows a 16-input exclusive–OR gate. A 16-bit number drives the input. The exclusive–OR gate produces an output 1 because the input has odd parity (an odd number of 1s). If the 16-bit input changes to another value, the output becomes 0 for even-parity numbers and 1 for odd-parity numbers.

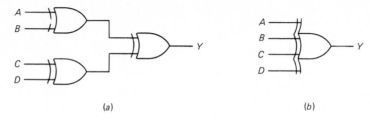

(a)                                                   (b)

**FIG. 3–28.** Four-input exclusive OR gate

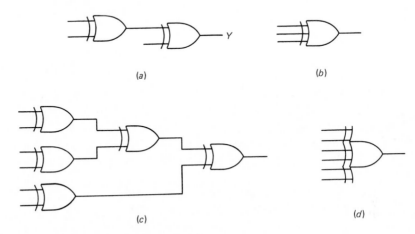

(a)                                                   (b)

(c)                                                   (d)

**FIG. 3–29.** Exclusive–OR gate with several inputs

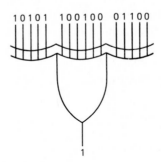

**FIG. 3–30.** Exclusive–OR gate with 16 inputs

## Parity Generation

In a computer, a binary number may represent an instruction that tells the computer to add, subtract, and so on; or the binary number may represent data to be processed like a number, letter, etc. In either case, you sometimes will see an extra bit added to the original binary number to produce a new binary number with even or odd parity.

For instance, Fig. 3–31 shows this 8-bit binary number:

$$X_7X_6X_5X_4 \ X_3X_2X_1X_0$$

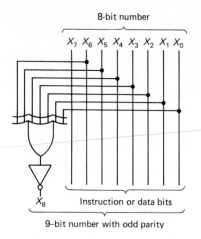

8-bit number

$X_7$ $X_6$ $X_5$ $X_4$ $X_3$ $X_2$ $X_1$ $X_0$

$X_8$        Instruction or data bits

9-bit number with odd parity

**FIG. 3–31.**   Odd-parity generation

Suppose this number equals 0100 0001. Then, the number has even parity, which means the exclusive–OR gate produces an output of 0. Because of the inverter,

$$X_8 = 1$$

and the final 9-bit output is 1 0100 0001. Notice that this has odd parity.

Suppose we change the 8-bit input to 0110 0001. Now, it has odd parity. In this case, the exclusive–OR gate produces an output 1. But the inverter produces a 0, so that the final 9-bit output is 0 0110 0001. Again, the final output has odd parity.

The circuit given in Fig. 3–31 is called an *odd-parity generator* because it always produces a 9-bit output number with odd parity. If the 8-bit input has even parity, a 1 comes out of the inverter to produce a final output with odd parity. On the other hand, if the 8-bit input has odd parity, a 0 comes out of the inverter, and the final 9-bit output again has odd parity. (To get an even-parity generator, delete the inverter.)

## Application

What is the practical application of parity generation and checking? Because of transients, noise, and other disturbances, 1-bit errors sometimes occur when binary data is transmitted over telephone lines or other communication paths. One way to check for errors is to use an odd-parity generator at the transmitting end and an odd-parity checker at the receiving end. If no 1-bit errors occur in transmission, the received data will have odd parity. But if one of the transmitted bits is changed by noise or any other disturbance, the received data will have even parity.

For instance, suppose we want to send 0100 0011. With an odd-parity generator like Fig. 3–31, the data to be transmitted will be 0 0100 0011. This data can be sent over telephone lines to some destination. If no errors occur in transmission, the odd-parity checker at the receiving end will produce a high output, meaning the received number has odd parity. On the other hand, if a 1-bit error does creep into the transmitted data, the odd-parity checker will have a low output, indicating the received data is invalid.

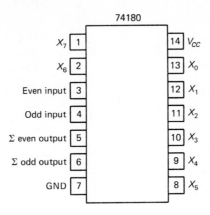

**FIG. 3–32.** Pinout diagram of 74180

One final point should be made. Errors are rare to begin with. When they do occur, they are usually 1-bit errors. This is why the method described here catches almost all of the errors that occur in transmitted data.

## The 74180

Figure 3–32 shows the pinout diagram for a 74180, which is a TTL parity generator-checker. The input data bits are $X_7$ to $X_0$; these bits may have even or odd parity. The EVEN INPUT (pin 3) and the ODD INPUT (pin 4) control the operation of the chip as shown in Table 3–8. The symbol $\Sigma$ stands for *summation*. In the left input column of Table 3–8, $\Sigma$ of H's (highs) refers to the parity of the input data $X_7$ to $X_0$. Depending on how you set up the values of the EVEN and ODD INPUTS, the $\Sigma$ EVEN and $\Sigma$ ODD OUTPUTS may be low or high.

For instance, suppose EVEN INPUT is high and ODD INPUT is low. When the input data has even parity (the first entry of Table 3–8), the $\Sigma$ EVEN OUTPUT is high and the $\Sigma$ ODD OUTPUT is low. When the input data has odd parity, the $\Sigma$ EVEN OUTPUT is low and the $\Sigma$ ODD OUTPUT is high.

If you change the control inputs, you change the operation. Assume that the EVEN INPUT is low and the ODD INPUT is high. When the input data has even parity, the $\Sigma$ EVEN OUTPUT is low and the $\Sigma$ ODD OUTPUT is high. When the input data has odd

**TABLE 3–8** *74180 Truth Table*

| Inputs | | | Outputs | |
|---|---|---|---|---|
| $\Sigma$ OF H's at $X_7$ to $X_0$ | Even | Odd | $\Sigma$ even | $\Sigma$ odd |
| Even | H | L | H | L |
| Odd | H | L | L | H |
| Even | L | H | L | H |
| Odd | L | H | H | L |
| X | H | H | L | L |
| X | L | L | H | H |

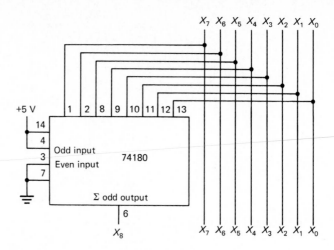

**FIG. 3–33.** Using a 74180 to generate odd-parity

parity, the $\Sigma$ EVEN OUTPUT is high and the $\Sigma$ ODD OUTPUT is low.

The 74180 can be used to detect even or odd parity. It can also be set up to generate even or odd parity.

☐ **EXAMPLE 3—8**

Show how to connect a 74180 to a generate a 9-bit output with odd parity.

☐ **SOLUTION**

Figure 3–33 shows one solution. The ODD INPUT (pin 4) is connected to +5 V, and the EVEN INPUT (pin 3) is grounded. Suppose the input data $X_7 \cdots X_0$ has even parity. Then, the third entry of Table 3–8 tells us the $\Sigma$ ODD OUTPUT (pin 6) is high. Therefore, the 9-bit number $X_8 \cdots X_0$ coming out of the circuit has odd parity.

On the other hand, suppose $X_7 \cdots X_0$ has odd parity. Then the fourth entry of Table 3–8 says that the $\Sigma$ ODD OUTPUT is low. Again, the 9-bit number $X_8 \cdots X_0$ coming out the bottom of Fig. 3–33 has odd parity.

The following conclusion may be drawn. Whether the input data has even or odd parity, the 9-bit number being generated in Fig. 3–33 always has odd parity.

## 3-9 READ-ONLY MEMORY

A *read-only memory* (which is abbreviated ROM and rhymes with Mom) is an IC that can store thousands of binary numbers representing computer instructions and data. Some of the smaller ROMs are also used to implement truth tables. In other words, we can use a ROM instead of sum-of-products circuit to generate any boolean function.

### Diode ROM

Suppose we want to build a circuit that stores the binary numbers shown in Table 3–9 (page 108). To keep track of where the numbers are stored, we will assign *addresses*. For instance, we want to store 0111 at address 0, 1000 at address 1, 1011 at address 2, and so

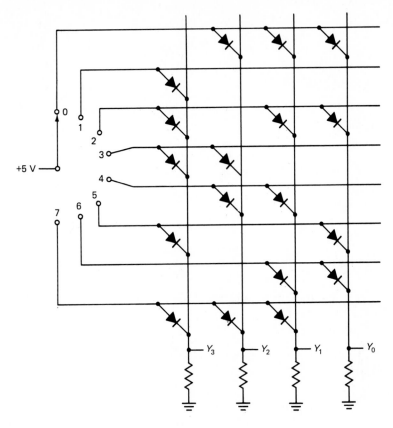

**FIG. 3–34.** Diode ROM

forth. Figure 3–34 shows one way to store the nibbles given in Table 3–9. When the switch is in position 0 (address 0), the upper row of diodes is turned on, so that the output of the ROM is

$$Y_3Y_2Y_1Y_0 = 0111$$

When the switch is moved to position 1, the second row is activated and

$$Y_3Y_2Y_1Y_0 = 1000$$

As you move the switch to the remaining positions or addresses, you get a $Y_3 \cdots Y_0$ output that matches the nibbles given in Table 3–9.

## On-Chip Decoding

Rather than switch-select the addresses as shown in Fig. 3–34, a manufacturer uses *on-chip decoding*. Figure 3–35 illustrates the idea. The 3-input pins ($A$, $B$, and $C$) supply the binary address of the stored number. Then, a 1-of-8 decoder produces a high output to one of the diode rows. For instance, if

$$ABC = 100$$

| TABLE 3–9 *Diode ROM* | | | | | | |
|---|---|---|---|---|---|---|
| Address | Nibble | | | | | |
| 0 | 0111 | | | | | |
| 1 | 1000 | | | | | |
| 2 | 1011 | | | | | |
| 3 | 1100 | | | | | |
| 4 | 0110 | | | | | |
| 5 | 1001 | | | | | |
| 6 | 0011 | | | | | |
| 7 | 1110 | | | | | |

**TABLE 3–10** *Truth Table*

| $A$ | $B$ | $C$ | $Y_3$ | $Y_2$ | $Y_1$ | $Y_0$ |
|---|---|---|---|---|---|---|
| 0 | 0 | 0 | 0 | 1 | 1 | 1 |
| 0 | 0 | 1 | 1 | 0 | 0 | 0 |
| 0 | 1 | 0 | 1 | 0 | 1 | 1 |
| 0 | 1 | 1 | 1 | 1 | 0 | 0 |
| 1 | 0 | 0 | 0 | 1 | 1 | 0 |
| 1 | 0 | 1 | 1 | 0 | 0 | 1 |
| 1 | 1 | 0 | 0 | 0 | 1 | 1 |
| 1 | 1 | 1 | 1 | 1 | 1 | 0 |

the 1-of-8 decoder applies a high voltage to the $A \bar{B} \bar{C}$ line, and the ROM output is

$$Y_3 Y_2 Y_1 Y_0 = 0110$$

If you change the binary address to

$$ABC = 110$$

the ROM output changes to

$$Y_3 Y_2 Y_1 Y_0 = 0011$$

With on-chip decoding, $n$ inputs can select $2^n$ memory locations (stored numbers). For instance, we need 3 address lines to access 8 memory locations, 4 address lines for 16 memory locations, 8 address lines for 256 memory locations, and so on.

## Commercially Available ROMs

A binary number is sometimes called a *word*. In a computer, binary numbers or words represent instructions, alphabet letters, decimal numbers, etc. The circuit given in Fig. 3–35 is a 32-bit ROM organized as 8 words with 4 bits at each address (an 8 × 4 ROM). The ROM given in Fig. 3–35 is for instructional purposes only because you would not build this circuit with discrete components. Instead, you would select a commercially available ROM. For instance, here are some TTL ROMS:

**7488:** 256 bits organized as 32 × 8
**74187:** 1024 bits organized as 256 × 4
**74S370:** 2048 bits organized as 512 × 4

As you can see, the 7488 can store 32 words of 8 bits each, the 74187 can store 256 words of 4 bits each, and the 74S370 can store 512 words of 4 bits each. If you want to store bytes (words with 8 bits), then you can parallel the 4-bit ROMs. For example, two parallel 74187s can store 256 words of 8 bits each.

One way to change the stored numbers of a ROM is by adding or removing diodes. With discrete circuits, you would have to solder or unsolder diodes to change the stored nibbles. With integrated circuits, however, you can send a list of the data to be stored to an IC manufacturer, who then produces a *mask* (a photographic template of the circuit). This mask is used in the mass production of your ROMs. As a rule, ROMs are used only for large production runs (thousands or more) because of the manufacturing costs involved.

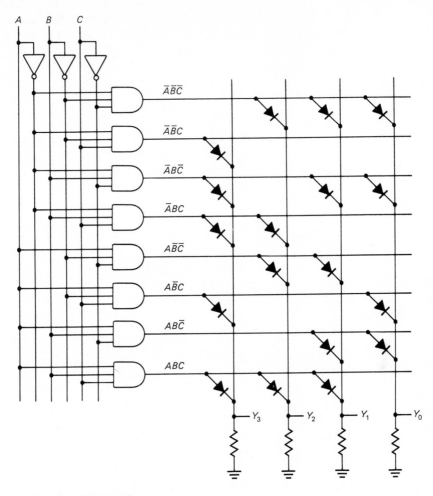

**FIG. 3–35.** On-chip decoding

## Generating Boolean Functions

Because the AND gates of Fig. 3–35 produce all the fundamental products and the diodes OR some of these products, the ROM is generating four boolean functions as follows:

$$Y_3 = \overline{A}\,\overline{B}\,C + \overline{A}\,B\,\overline{C} + \overline{A}\,B\,C + A\,\overline{B}\,C + ABC \tag{3-1}$$

$$Y_2 = \overline{A}\,\overline{B}\,\overline{C} + \overline{A}\,B\,C + A\,\overline{B}\,\overline{C} + ABC \tag{3-2}$$

$$Y_1 = \overline{A}\,\overline{B}\,\overline{C} + \overline{A}\,B\,\overline{C} + A\,\overline{B}\,\overline{C} + A\,B\,\overline{C} + ABC \tag{3-3}$$

$$Y_0 = \overline{A}\,\overline{B}\,\overline{C} + \overline{A}\,B\,\overline{C} + A\,\overline{B}\,C + A\,B\,\overline{C} \tag{3-4}$$

What this means is that you can use a ROM instead of a logic circuit to implement a truth table.

For instance, suppose you start with a truth table like the one in Table 3–10. There are four outputs: $Y_3$, $Y_2$, $Y_1$, and $Y_0$. A sum-of-products solution would lead to four

AND–OR circuits, one for Eq. (3–1), a second for Eq. (3–2), and so on. The ROM solution is different. With a ROM you have to store the binary numbers of Table 3–9 (same as Table 3–10) at the indicated addresses. When this is done, the ROM given in Fig. 3–35 is equivalent to a sum-of-products circuit. In other words, you can use the ROM instead of an AND–OR circuit to generate the desired truth table.

## Programmable ROMs

A *programmable ROM* (PROM) allows the user instead of the manufacturer to store the data. An instrument called a *PROM programmer* stores the words by "burning in." Here is an example of how a PROM programmer works. Originally, all diodes are connected at the cross points. For instance, in Fig. 3–35 there would be a total of 32 diodes (8 rows and 4 columns). Each of these diodes has a *fusible link* (a small fuse). The PROM programmer sends destructively high currents through all diodes to be removed. In this way, only the desired diodes remain connected after programming a PROM. Programming like this is permanent because the data cannot be erased after it has been burned in.

Here are some commercially available PROMs:

**74S188:** 256 bits organized as $32 \times 8$
**74S287:** 1024 bits organized as $256 \times 4$
**74S472:** 4096 bits organized as $512 \times 8$

PROMs such as these are useful for small production runs. For instance, if you are building only a few hundred units (or maybe even just one), you would choose a PROM rather than a ROM.

Since PROMs are useful in many applications, manufacturers produce these chips in high volume. Furthermore, the PROM is a universal logic solution. Why? Because the AND gates generate all the fundamental products; the user can then OR these products as needed to generate any boolean output. One disadvantage of PROMs is the limit on number of input variables; typically, PROMs have 8 inputs or less.

## Erasable PROMS

The *erasable PROM* (EPROM) uses metal-oxide-semiconductor field-effect transistors (MOSFETs). Data is stored with an EPROM programmer. Later, data can be erased with ultraviolet light. The light passes through a quartz window in the IC package. When it strikes the chip, the ultraviolet light releases all stored charges. The effect is to wipe out the stored contents. In other words, the EPROM is ultraviolet-light-erasable and electrically reprogrammable.

The EPROM is useful in project development. With an EPROM, the designer can modify the contents until the stored data is perfect. When the design is finalized, the data can be burned into PROMs (small production runs) or sent to an IC manufacturer who produces ROMs (large production runs).

## 3-10  PROGRAMMABLE ARRAY LOGIC

*Programmable array logic* (PAL) is a programmable array of logic gates on a single chip. PALs are another design solution, similar to a sum-of-products solution, product-of-sums solution, and multiplexer logic. Before discussing this alternative approach to logic design, let us discuss several key ideas about the PROM.

## Simplified Drawing of PROM

Look at Fig. 3–35 again. Assume this figure represents a PROM that has been burned in with the diodes at the places shown. The AND–gate array produces all the fundamental products, $\overline{A}\,\overline{B}\,\overline{C}$ to $ABC$. Then, depending on which diodes are connected, some of these fundamental products are ORed by the diodes to get the four outputs, $Y_3$ to $Y_0$.

It is very difficult to draw larger PROMs because too many diodes are involved. For this reason, drawing a PROM like the one given in Fig. 3–35 is streamlined, as shown in Fig. 3–36. In this new drawing, the $X$s indicate connections to the gate inputs. Furthermore, each AND gate has 3 inputs, indicated by the $X$s on its input line. Similarly, each OR gate has several inputs given by the $X$s on its input line.

Notice the following. The input side of Fig. 3–36 is a fixed AND array, meaning the inputs to the AND gates are not programmable in a PROM. On the other hand, the output side of the circuit is a programmable OR array because of the fusible links at each diode location. A fixed AND array and programmable OR array is characteristic of all PROMs. With this approach, the user can OR any desired combination of fundamental products.

## Programming a PAL

A PAL is different from a PROM because it has a programmable AND array and a fixed OR array. For instance, Fig. 3–37 shows a PAL with 4 inputs and 4 outputs. The $X$s on

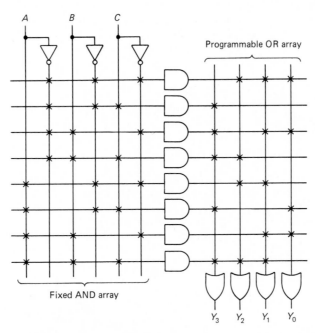

**FIG. 3–36.** Streamlined drawing of PROM

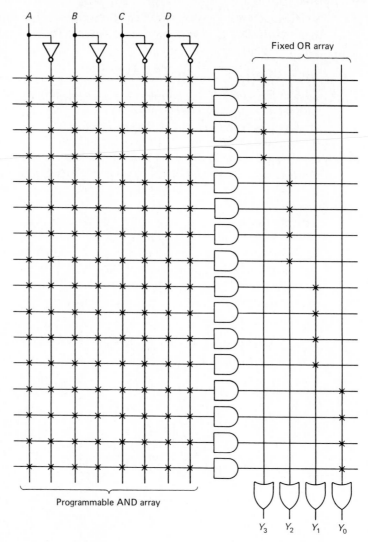

**FIG. 3-37.** Structure of PAL

the input side are fusible links, while the $X$s on the output side are fixed connections. With a PROM programmer, we can burn in the desired fundamental products, which are then ORed by the fixed output connections.

Here is an example of how to program a PAL. Suppose we want to generate the following boolean functions:

$$Y_3 = \overline{A}\,B\,\overline{C}\,D + \overline{A}\,B\,C\,\overline{D} + \overline{A}\,B\,C\,D + A\,B\,C\,\overline{D} \qquad (3\text{-}5)$$

$$Y_2 = \overline{A}\,B\,C\,\overline{D} + \overline{A}\,B\,C\,D + ABCD \qquad (3\text{-}6)$$

$$Y_1 = \overline{A}\,B\,\overline{C} + \overline{A}\,B\,C + A\,\overline{B}\,C + A\,B\,\overline{C} \qquad (3\text{-}7)$$

$$Y_0 = ABCD \qquad (3\text{-}8)$$

Start with Eq. (3–5). The first desired product is $\overline{A}\overline{B}\overline{C}D$. On the top input line of Fig. 3–37, we have to remove the first $X$, the fourth $X$, the fifth $X$, and the eighth $X$. Then, the top AND gate has an output of $\overline{A}\,B\,\overline{C}\,D$.

By removing $X$s on the next three input lines, we can make the top four AND gates produce the fundamental products of Eq. (3–5). The fixed OR connections on the output side imply that the first OR gate produces an output of

$$Y_3 = \overline{A}\,B\,\overline{C}\,D + \overline{A}\,B\,C\,\overline{D} + \overline{A}\,B\,C\,D + A\,B\,C\,\overline{D}$$

Similarly, we can remove $X$s as needed to generate $Y_2$, $Y_1$, and $Y_0$. Figure 3–38 shows how the PAL looks after the necessary $X$s have been removed. If you examine this circuit, you will see that it produces the $Y$ outputs given by Eqs. (3–5) to (3–8).

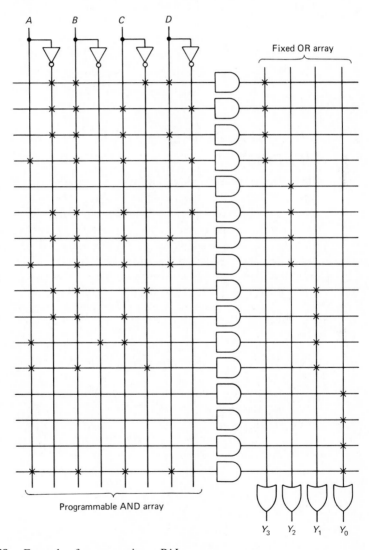

**FIG. 3–38.** Example of programming a PAL

## Commercially Available PALs

The PAL given in Fig. 3–38 is hypothetical. Commercially available PALs typically have more inputs. For instance, here is a sample of some TTL PALs available from National Semiconductor Corporation:

**10H8:** 10 input and 8 output AND–OR
**16H2:** 6 input and 2 output AND–OR
**14L4:** 14 input and 4 output AND–OR–INVERT

For these chip numbers, H stands for active-high output and L for active-low output. The 10H8 and the 16H2 produce active-high outputs because they are AND–OR PALs. The 14L4, on the other hand, produces an active-low output because it is an AND–OR–INVERT circuit (one that has inverters at the final outputs).

Unlike PROMS, PALs are not a universal logic solution. Why? Because only some of the fundamental products can be generated and ORed at the final outputs. Nevertheless, PALs have enough flexibility to produce all kinds of complicated logic functions. Furthermore, PALs have the advantage of 16 inputs compared to the typical limit of 8 inputs for PROMs.

## 3-11 TROUBLESHOOTING WITH A LOGIC PROBE

Chapter 2 introduced the logic clip, a device that connects to a 14- or 16-pin IC. The logic clip contains 16 LEDs that monitor the state of the pins. When a pin voltage is high, the corresponding LED lights up. When the pin voltage is low, the LED is dark.

Figure 3–39 shows a *logic probe*, which is another troubleshooting tool you will find helpful in diagnosing faulty circuits. When you touch the probe tip to the output node as shown, the device lights up for a high state and goes dark for a low state. For instance, if either $A$ or $B$, or both, are low, then $Y$ is high and the probe lights up. On the other hand, if $A$ and $B$ are both high, $Y$ is low and the probe is dark.

Among other things, the probe is useful for locating short circuits that occur in manufacturing. For example, during the stuffing and soldering of printed-circuit boards, an undesirable splash of solder may connect two adjacent traces (conducting lines). Known as a solder bridge, this kind of trouble can short-circuit a node to the ground or to

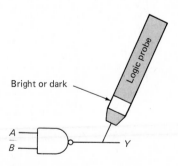

**FIG. 3–39.**  Using a logic probe

the supply voltage. The node is then stuck in a low or high state. The probe helps you to find short-circuited nodes because it stays in one state, no matter how the inputs are changing.

## S U M M A R Y

A multiplexer is a circuit with many inputs but only one output. The 16-to-1 multiplexer has 16 input bits, 4 control bits, and 1 output bit. The 4 control bits select and steer 1 of the 16 inputs to the output. The multiplexer is a universal logic circuit because it can generate any truth table.

A demultiplexer has one input and many outputs. By applying control signals, we can steer the input signal to one of the output lines. A decoder is similar to a demultiplexer, except that there is no data input. The control bits are the only input. They are decoded by activating one of the output lines.

BCD is an abbreviation for binary-coded decimal. The BCD code expresses each digit in a decimal number by its nibble equivalent. A BCD-to-decimal decoder converts a BCD input to its equivalent decimal value. A seven-segment decoder converts a BCD input to an output suitable for driving a seven-segment indicator.

An encoder converts an input signal into a coded output signal. An example is the decimal-to-BCD encoder. An exclusive–OR gate has a high output only when an odd number of inputs are high. Exclusive–OR gates are useful in parity generators-checkers.

A ROM is a read-only memory. Smaller ROMs are used to implement truth tables. ROMs are expensive because they require a mask for programming. PROMs are user-programmable and ideal for small production runs. EPROMs are not only user-programmable, but they are also erasable and reprogrammable during the design and development cycle. PALs are chips that are programmable arrays of logic. Unlike the PROM with its fixed AND array and programmable OR array, a PAL has a programmable AND array and a fixed OR array. The PAL has the advantage of having up to 16 inputs in commercially available devices.

## G L O S S A R Y

*Active Low* The low state is the one that causes something to happen rather than the high state.

*BCD* A binary-coded decimal.

*Data Selector* A synonym for multiplexer.

*Decoder* A circuit that is similar to a demultiplexer, except there is no data input. The control input bits produce one active output line.

*Demultiplexer* A circuit with one input and many outputs.

*EPROM* An erasable programmable read-only memory. With this device, the user can erase the stored contents with ultraviolet light and electrically store new data. EPROMs are useful during project development where programs and data are being perfected.

*Even Parity* A binary number with an even number of 1s.

*Exclusive–OR Gate* A gate that produces a high output only when an odd number of inputs is high.

*LED* A light-emitting diode.

*Logic Probe* A troubleshooting device that indicates the state of a signal line.

*Multiplexer* A circuit with many inputs but only one output.

*Odd Parity* A binary number with an odd number of 1s.

*PAL* A programmable array logic (sometimes written PLA, which stands for programmable logic array). In either case, it is a chip with a programmable AND array and a fixed OR array.

*Parity Generation* An extra bit that is generated and attached to a binary number, so that the new number has either even or odd parity.

*PROM* A programmable read-only memory. A type of chip that allows the user to program it with a PROM programmer that burns fusible links at the diode cross points. Once the data is stored, the programming is permanent. PROMs are useful for small production runs.

*ROM* A read-only memory. An IC that can store many binary numbers at locations called addresses. ROMs are expensive to manufacture and are used only for large production runs where the cost of the mask can be recovered by sales.

*Strobe* An input that disables or enables a circuit.

# P R O B L E M S

*SECTION 3-1:*

**3–1.** In Fig. 3–2, if $ABCD = 1001$, what does $Y$ equal?

**3–2.** In Fig. 3–4, if $ABCD = 1100$, what does $Y$ equal?

**3–3.** We want to implement Table 2–12 of the preceding chapter using multiplexer logic. Show a circuit, similar to the one in Fig. 3–4, that can do the job.

**3–4.** Show how to connect a 74150 to implement this boolean equation:

$$Y = \overline{A}\,B\,\overline{C}\,D + A\,\overline{B}\,C\,\overline{D} + ABC\overline{D}$$

**3–5.** Draw a circuit with four 74150s that has a truth table like the one in Table 3–11.

| A | B | C | D | $Y_3$ | $Y_2$ | $Y_1$ | $Y_0$ |
|---|---|---|---|---|---|---|---|
| 0 | 0 | 0 | 0 | 0 | 0 | 1 | 0 |
| 0 | 0 | 0 | 1 | 1 | 0 | 1 | 0 |
| 0 | 0 | 1 | 0 | 0 | 0 | 0 | 0 |
| 0 | 0 | 1 | 1 | 0 | 1 | 0 | 0 |
| 0 | 1 | 0 | 0 | 0 | 0 | 0 | 0 |
| 0 | 1 | 0 | 1 | 0 | 0 | 0 | 0 |
| 0 | 1 | 1 | 0 | 1 | 0 | 0 | 0 |
| 0 | 1 | 1 | 1 | 1 | 0 | 0 | 0 |
| 1 | 0 | 0 | 0 | 0 | 1 | 0 | 1 |
| 1 | 0 | 0 | 1 | 0 | 0 | 0 | 1 |
| 1 | 0 | 1 | 0 | 0 | 0 | 0 | 0 |
| 1 | 0 | 1 | 1 | 0 | 0 | 0 | 0 |
| 1 | 1 | 0 | 0 | 0 | 0 | 0 | 0 |
| 1 | 1 | 0 | 1 | 0 | 1 | 0 | 1 |
| 1 | 1 | 1 | 0 | 1 | 0 | 1 | 0 |
| 1 | 1 | 1 | 1 | 0 | 0 | 1 | 0 |

TABLE 3–11

**TABLE 3–12** *Gray Code*

| A | B | C | D | $Y_3$ | $Y_2$ | $Y_1$ | $Y_0$ |
|---|---|---|---|---|---|---|---|
| 0 | 0 | 0 | 0 | 0 | 0 | 0 | 0 |
| 0 | 0 | 0 | 1 | 0 | 0 | 0 | 1 |
| 0 | 0 | 1 | 0 | 0 | 0 | 1 | 1 |
| 0 | 0 | 1 | 1 | 0 | 0 | 1 | 0 |
| 0 | 1 | 0 | 0 | 0 | ·1 | 1 | 0 |
| 0 | 1 | 0 | 1 | 0 | 1 | 1 | 1 |
| 0 | 1 | 1 | 0 | 0 | 1 | 0 | 1 |
| 0 | 1 | 1 | 1 | 0 | 1 | 0 | 0 |
| 1 | 0 | 0 | 0 | 1 | 1 | 0 | 0 |
| 1 | 0 | 0 | 1 | 1 | 1 | 0 | 1 |
| 1 | 0 | 1 | 0 | 1 | 1 | 1 | 1 |
| 1 | 0 | 1 | 1 | 1 | 1 | 1 | 0 |
| 1 | 1 | 0 | 0 | 1 | 0 | 1 | 0 |
| 1 | 1 | 0 | 1 | 1 | 0 | 1 | 1 |
| 1 | 1 | 1 | 0 | 1 | 0 | 0 | 1 |
| 1 | 1 | 1 | 1 | 1 | 0 | 0 | 0 |

**3–6.** Table 3–12 shows the Gray code. Show how four 74150s may be connected to convert from binary to Gray code.

## SECTION 3-2:

**3–7.** In Fig. 3–11, if $ABCD = 0101$, which is the active output line when the strobe is high? When it is low?

**3–8.** Input signals $R$ and $T$ are low in Fig. 3–12. Which is the active output line when $ABCD = 0011$? To have the $Y_9$ output line active, what input signals do you need?

**3–9.** Suppose a logic probe shows that pin 19, given in Fig. 3–12, is always high. Which of the following may possibly cause trouble:
 **a.** Pin 20 is grounded.
 **b.** Pin 18 has a sine wave instead of pulses.
 **c.** The $R$ input is grounded.
 **d.** The $T$ input is connected to +5 V.

## SECTION 3-3:

**3–10.** Are the output signals of Fig. 3–14 active low or active high? For the IC to decode the $ABCD$ input, does the strobe have to be low or high?

**3–11.** In Fig. 3–15, suppose $X = 1$ and $ABCD = 0110$. Which is the active chip and which is the active output line?

## SECTION 3-4:

**3–12.** Convert the following decimal numbers into their BCD equivalents:
 **a.** 32
 **b.** 634
 **c.** 4898

**3–13.** Convert the following BCD numbers into their decimal equivalents:
  **a.** 0110 0111
  **b.** 1000 0001 0011
  **c.** 0111 0010 0101 1001

**3–14.** In Fig. 3–16, what is the high output line when $ABCD = 0101$?

**3–15** In Fig. 3–18, which is the low output when $ABCD = 0111$?

**3–16.** Figure 3–40 shows a group of chips numbered 0 through 9. Each chip has an active-low STROBE input. Which chip is active for each of these conditions:
  **a.** $ABCD = 0000$.
  **b.** $ABCD = 0010$.
  **c.** $ABCD = 1001$.

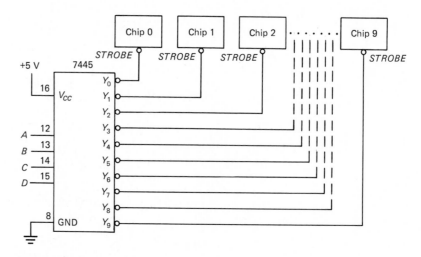

**FIG. 3–40.**

**3–17.** The $ABCD$ input of Fig. 3–40 initially equals 1111. For this condition, all output waveforms start high in the timing diagram of Fig. 3–41. Another circuit not shown is supposed to produce the following input values of $ABCD$: 0000, 0001, 0010, 0011, 0100, 0101, 0110, 0111, 1000, and 1001.

The timing diagram tells us that something is wrong with the logic circuit of Fig. 3–40. Which of the following is a possible trouble:
  **a.** Pin 16 is not connected to the supply voltage.
  **b.** Pin 8 is open.
  **c.** Pin 12 is short-circuited to the ground.
  **d.** Pin 15 is short-circuited to +5 V.

*SECTION 3-5:*

**3–18.** In Fig. 3–19, which of the segments have to be active to display the following digits:
  **a.** 2      **c.** 8
  **b.** 6

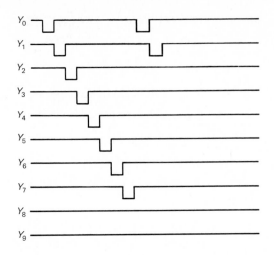

**FIG. 3–41.**

**3–19.** In Fig. 3–21a, $V_{CC}$ = +5 V, all resistors are 1 kΩ, and each LED has a voltage drop of 2 V. Approximately how much current is there through an active segment?

*SECTION 3-6:*

**3–20.** In Fig. 3–23, what is the output when button 7 is pressed? When button 3 is pressed?

**3–21.** In Fig. 3–25, if button 8 is pressed, which is the input pin that goes into the low state? What does the *ABCD* output equal?

*SECTION 3-7:*

**3–22.** In Fig. 3–29d, what does Y equal for each of the following inputs:
   **a.** 000110
   **b.** 011001
   **c.** 011111
   **d.** 111100

**3–23.** In Fig. 3–30, what does Y equal for the following inputs:
   **a.** 1111 0000 1111 0000
   **b.** 0101 1010 1100 0111
   **c.** 1110 1011 1101 0001
   **d.** 0001 0101 0011 0110

**3–24.** In Fig. 3–42, the 8-bit register is a logic circuit that stores byte $A_7 \cdots A_0$. What does byte $Y_7 \cdots Y_0$ equal for each of these conditions:
   **a.** $A_7 \cdots A_0$ = 1000 0111 and INVERT = 0.
   **b.** $A_7 \cdots A_0$ = 0011 1100 and INVERT = 1.
   **c.** $A_7 \cdots A_0$ = 1111 0000 and INVERT = 0.
   **d.** $A_7 \cdots A_0$ = 1110 0001 and INVERT = 1.

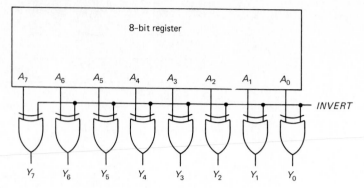

FIG. 3–42.

**3–25.** In Fig. 3–43, each register is a logic circuit that stores a 6-bit number. The left register stores $A_5 \cdots A_0$ and the right register stores $B_5 \cdots B_0$. What value does the output signal labeled *EQUAL* have for each of these:

   **a.** $A_7 \cdots A_0$ is less than $B_7 \cdots B_0$.
   **b.** $A_7 \cdots A_0$ equals $B_7 \cdots B_0$.
   **c.** $A_7 \cdots A_0$ is greater than $B_7 \cdots B_0$.

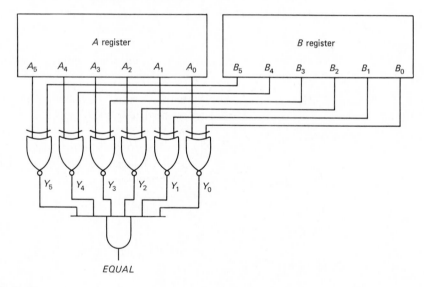

FIG. 3–43.

*SECTION 3-8:*

   **3–26.** In Fig. 3–44, what does $X_8$ equal for each of the following $X_7 \cdots X_0$ inputs:

     **a.** 0000 1111              **c.** 1111 0001
     **b.** 1010 1110              **d.** 1011 1100

   **3–27.** In Fig. 3–44, what changes can you make to get a 9-bit output with even parity?

**3–28.** In Fig. 3–44, assuming the circuit is working all right, what will the logic probe indicate for each of the following:

**a.** Input data has even parity.
**b.** Input data has odd parity.
**c.** Pins 3 and 4 are grounded.

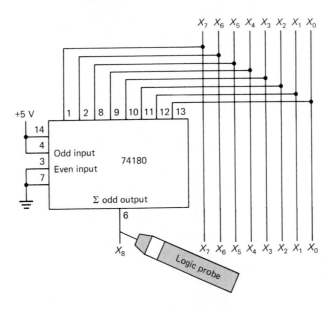

**FIG. 3–44.**

*SECTION 3-9:*

**3–29.** Suppose a ROM has 8 input address lines. How many memory locations does it have?

**3–30.** Two 74S370s are connected in parallel. To address all memory locations, how many bits must the binary address have?

**3–31.** In Fig. 3–35, if $ABC = 011$, what does $Y_3 Y_2 Y_1 Y_0$ equal?

**3–32.** Draw a ROM circuit similar to the one in Fig. 3–35 that produces these outputs:

$$Y_3 = \overline{A}\,B\,\overline{C}$$
$$Y_2 = A\,\overline{B}\,C + ABC$$
$$Y_1 = A\,\overline{B}\,C + \overline{A}\,B\,C + ABC$$
$$Y_0 = \overline{A}\,B\,\overline{C} + \overline{A}\,B\,C + A\,B\,\overline{C} + ABC$$

*SECTION 3-10:*

**3–33.** Draw PROM circuit similar to the one in Fig. 3–36 that generates the $Y_3$ to $Y_0$ output given in Table 3–11.

**3–34.** What is the boolean equation for $Y_3$ in Fig. 3–45? For $Y_2$? For $Y_1$? For $Y_0$?

**3–35.** Draw a 4-input and 4-output PAL circuit that has the truth table of Table 3–11.

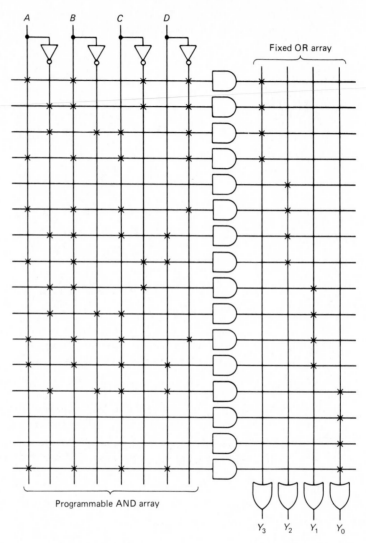

FIG. 3–45.

# 4

# NUMBERS
# SYSTEMS
# AND CODES

On hearing the word *number*, most of us immediately think of the familiar decimal number system with its 10 digits:

$$0, 1, 2, 3, 4, 5, 6, 7, 8, 9$$

Modern computers do not process decimal numbers. Instead, they work with binary numbers which use only the digits 0 and 1. This creates a problem. People do not like working with binary numbers because they are very long when representing larger decimal quantities. Therefore, *octal* and *hexadecimal* numbers are now widely used to compress long strings of binary numbers.

This chapter discusses binary, octal, and hexadecimal numbers. Besides learning to count in these new number systems, you will learn how to convert from decimal to binary, octal, and hexadecimal, and vice versa. The chapter also examines some of the binary codes used in digital electronics. You will also see how a logic pulser and a logic probe can be used to locate stuck states in a logic circuit.

## 4-1 WHY BINARY NUMBERS ARE USED

Almost all digital computers and systems are based on binary (two-state) operation. For instance, the switch shown in Fig. 4–1a can be open or closed; therefore, a switch is one example of a natural binary device.

The magnetic core shown in Fig. 4–1b is another example of two-state operation. With the right-hand rule, conventional current into the wire produces clockwise flux; reverse the current, and you get a counterclockwise flux. Because there are only two possible directions for the flux, the magnetic core is a binary or two-state device. Earlier computers used thousands of magnetic cores to store the binary instructions and data needed during a calculation.

A punched card (Fig. 4–1c) is another example of the two-state concept. A hole in the card is one possibility, and no hole is the other. Using a prearranged code, a card-punch machine with a keyboard can produce a stack of punched cards containing binary instructions and data. Some computers use punched cards to enter the instructions and data into a computer.

Tape recorders can magnetize some points on the tape while leaving other points unmagnetized. When a prearranged code is used, a row of points can represent either a coded instruction or data (Fig. 4–1d). In this way, a reel of tape can store thousands of binary instructions and data for later use in a computer or other digital system.

Even the transistor circuits used in digital computers and systems are designed to operate in either of two states, typically at cutoff or saturation. As a result, the output of a transistor circuit is either a low or a high voltage. When you look at digital signals with an oscilloscope, you find that each voltage is either low or high. These digital signals change during a computer run, so that they appear like pulses (Fig. 4–1e).

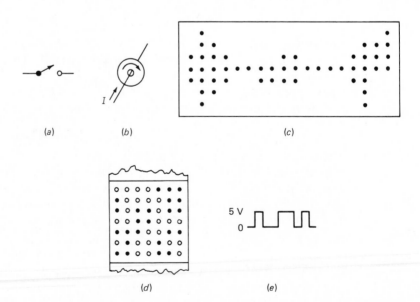

FIG. 4–1. Examples of using binary numbers

In conclusion, switches, cores, punched cards, tape, transistors, and almost all other digital components are based on binary operation. This is why it is convenient to use binary numbers when analyzing or designing digital circuits. Besides knowing how to count with binary numbers, it is necessary to learn how to convert from binary to decimal, and vice versa.

## 4-2 BINARY-TO-DECIMAL CONVERSION

Table 4–1 lists the binary numbers from 0000 to 1111. But how do you convert larger binary numbers into their decimal values? For instance, what does binary 101001 represent in decimal numbers? This section shows how to convert a binary number quickly and easily into its decimal equivalent.

### Positional Notation and Weights

We can express any decimal *integer* (a whole number) in units, tens, hundreds, thousands, and so on. For instance, decimal number 2945 may be written as

$$2945 = 2000 + 900 + 40 + 5$$

In powers of 10, this becomes

$$2945 = 2(10^3) + 9(10^2) + 4(10^1) + 5(10^0)$$

The decimal number system is an example of *positional notation*; each digit position has a *weight* or value. With decimal numbers, the weights are units, tens, hundreds, thousands, and so on. The sum of all the digits multiplied by their weights gives the total amount being represented. In the foregoing example, the 2 is multiplied by a weight of 1000, the 9 by a weight of 100, the 4 by a weight of 10, and the 5 by a weight of 1; the total is

$$2000 + 900 + 40 + 5 = 2945$$

**TABLE 4–1** *Binary Numbers*

| Binary | Decimal |
|--------|---------|
| 0000 | 0 |
| 0001 | 1 |
| 0010 | 2 |
| 0011 | 3 |
| 0100 | 4 |
| 0101 | 5 |
| 0110 | 6 |
| 0111 | 7 |
| 1000 | 8 |
| 1001 | 9 |
| 1010 | 10 |
| 1011 | 11 |
| 1100 | 12 |
| 1101 | 13 |
| 1110 | 14 |
| 1111 | 15 |

## Binary Weights

In a similar way, we can rewrite any binary number in terms of weights. For instance, binary number 111 becomes

$$111 = 100 + 10 + 1 \tag{4-1}$$

In decimal numbers, this may be rewritten as

$$7 = 4 + 2 + 1 \tag{4-2}$$

Writing a binary number as shown in Eq. (4–1) is the same as splitting its decimal equivalent into units, 2s, and 4s as indicated by Eq. (4–2). In other words, each digit position in a binary number has a weight. The least significant digit (the one on the right) has a weight of 1. The second position from the right has a weight of 2; the next, 4; and then 8, 16, 32, and so forth. These weights are in ascending powers of 2; therefore, we can write the foregoing equation as

$$7 = 1(2^2) + 1(2^1) + 1(2^0)$$

Whenever you look at a binary number, you can find its decimal equivalent as follows:

1. When there is a 1 in a digit position, add the weight of that position.
2. When there is a 0 in a digit position, disregard the weight of that position. For example, binary number 101 has a decimal equivalent of

$$4 + 0 + 1 = 5$$

As another example, binary number 1011 is equivalent to

$$8 + 0 + 2 + 1 = 11$$

Still another example is 11001, which is equivalent to

$$16 + 8 + 0 + 0 + 1 = 25$$

## Streamlined Method

We can streamline binary-to-decimal conversion by the following procedure:

1. Write the binary number.
2. Directly under the binary number write 1, 2, 4, 8, 16, . . . , working from right to left.
3. If a zero appears in a digit position, cross out the decimal weight for that position.
4. Add the remaining weights to obtain the decimal equivalent.

As an example of this approach, let us convert binary 101 to its decimal equivalent:

**Step 1** 1 0 1
**Step 2** 4 2 1
**Step 3** 4 ~~2~~ 1
**Step 4** 4 + 1 = 5

As another example, notice how quickly 10101 is converted to its decimal equivalent:

$$1 \quad 0 \quad 1 \quad 0 \quad 1$$
$$16 \quad \not{8} \quad 4 \quad \not{2} \quad 1 \rightarrow 21$$

## Fractions

So far, we have discussed binary *integers* (whole numbers). How are binary fractions converted into corresponding decimal equivalents? For instance, what is the decimal equivalent of 0.101? In this case, the weights of digit positions to the right of the binary point are given by $\frac{1}{2}$, $\frac{1}{4}$, $\frac{1}{8}$, $\frac{1}{16}$, and so on. In powers of 2, the weights are

$$2^{-1} \quad 2^{-2} \quad 2^{-3} \quad 2^{-4} \quad \text{etc.}$$

or in decimal form:

$$0.5 \quad 0.25 \quad 0.125 \quad 0.0625 \quad \text{etc.}$$

Here is an example. Binary fraction 0.101 has a decimal equivalent of

$$0.1 \quad 0 \quad 1$$
$$0.5 + 0 + 0.125 = 0.625$$

Another example, the decimal equivalent of 0.1101 is

$$0.1 \quad 1 \quad 0 \quad 1$$
$$0.5 + \quad 0.25 + 0 + 0.0625 = 0.8125$$

## Mixed Numbers

For *mixed* numbers (numbers with an integer and a fractional part), handle each part according to the rules just developed. The weights for a mixed number are

$$\text{etc.} \quad 2^3 \quad 2^2 \quad 2^1 \quad 2^0 \quad . \quad 2^{-1} \quad 2^{-2} \quad 2^{-3} \quad \text{etc.}$$
$$\uparrow$$
$$\text{Binary point}$$

## Computer Diagram

During our discussion of binary numbers, it will help to have a basic idea of what is inside a computer. Look at Fig. 4–2 for each of the following sections:

**Input** This transfers into the computer a sequence of instructions called a *program* and a set of numbers called *data*. In microcomputers the input section includes a keyboard that converts letters and numbers into strings of binary data.

**Memory** Here is where the program and data are stored before a calculation begins. Also, during calculations the memory stores intermediate answers, similar to the way we use a scratch pad during a long calculation. The memory of a microcomputer includes read-only memory (ROM) and random-access memory (RAM) (discussed in a later chapter).

**Control** This part of the computer eliminates the human operator from the calculation process. The control section tells the other parts of the computer how to process the data.

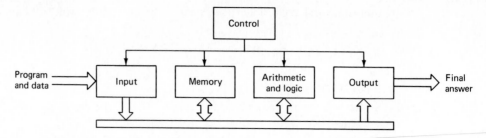

**FIG. 4–2.** Basic parts of a computer

**Arithmetic and logic** This section adds, subtracts, multiplies, and divides. It also performs logic like ORing, ANDing, and so on.

**Output** This transfers the final answer to the outside world. The output section of a microcomputer includes a video display to allow the user to see the processed data.

For future reference, Table 4–2 lists powers of 2 and their decimal equivalents and the number of K. The abbreviation K stands for 1024. Therefore, 1K means 1024, 2K stands for 2048, 4K represents 4096, and so on. Many personal computers have 64K memories; this means that they can store up to 65,536 bytes in the memory section shown in Fig. 4–2.

☐ EXAMPLE 4–1

Convert binary 110.001 to a decimal number.

☐ SOLUTION

$$
\begin{array}{ccccccc}
1 & 1 & 0 & . & 0 & 0 & 1 \\
4 & 2 & \cancel{X} & & \cancel{0.5} & \cancel{0.25} & 0.125 \rightarrow 6.125
\end{array}
$$

**TABLE 4–2** *Powers of 2*

| Powers of 2 | Decimal Equivalent | Abbreviation |
|---|---|---|
| $2^0$ | 1 | |
| $2^1$ | 2 | |
| $2^2$ | 4 | |
| $2^3$ | 8 | |
| $2^4$ | 16 | |
| $2^5$ | 32 | |
| $2^6$ | 64 | |
| $2^7$ | 128 | |
| $2^8$ | 256 | |
| $2^9$ | 512 | |
| $2^{10}$ | 1,024 | 1K |
| $2^{11}$ | 2,048 | 2K |
| $2^{12}$ | 4,096 | 4K |
| $2^{13}$ | 8,192 | 8K |
| $2^{14}$ | 16,384 | 16K |
| $2^{15}$ | 32,768 | 32K |
| $2^{16}$ | 65,536 | 64K |

□ EXAMPLE 4—2

What is the decimal value of binary 1011.11?

□ SOLUTION

$$
\begin{array}{cccc cc}
1 & 0 & 1 & 1 & . & 1 & 1 \\
8 & \not{4} & 2 & 1 & & 0.5 & 0.25 \rightarrow 11.75
\end{array}
$$

□ EXAMPLE 4—3

A computer has a 256K memory. What is the decimal equivalent of 256K?

□ SOLUTION

$$256 \times 1024 = 262,144$$

This means that the computer can store 262,144 bytes in its memory.

## 4-3 DECIMAL-TO-BINARY CONVERSION

One way to convert a decimal number into its binary equivalent is to reverse the process described in the preceding section. For instance, suppose that you want to convert decimal 9 into the corresponding binary number. All you need to do is express 9 as a sum of powers of 2, and then write 1s and 0s in the appropriate digit positions:

$$9 = 8 + 0 + 0 + 1$$
$$\rightarrow 1001$$

As another example:

$$25 = 16 + 8 + 0 + 0 + 1$$
$$\rightarrow 11001$$

### Double Dabble

A popular way to convert decimal numbers to binary numbers is the *double-dabble* method. In the double-dabble method you progressively divide the decimal number by 2, writing down the remainder after each division. The remainders, taken in reverse order, form the binary number. The best way to understand the method is to go through an example step by step.

Here is how to convert decimal 13 to its binary equivalent:

**Step 1** Divide 13 by 2, writing your work like this:

$$
\begin{array}{r}
6 \\
2\overline{)13}
\end{array}
\qquad 1 \rightarrow \text{(first remainder)}
$$

The quotient is 6 with a remainder of 1.

**Step 2** Divide 6 by 2 to get

$$
\begin{array}{r}
3 \\
2\overline{)6} \\
2\overline{)13}
\end{array}
\qquad
\begin{array}{l}
0 \rightarrow \text{(second remainder)} \\
1 \rightarrow \text{(first remainder)}
\end{array}
$$

This division gives 3 with a remainder of 0.

**Step 3** Again you divide by 2:

$$
\begin{array}{ll}
\quad 1 & 1 \rightarrow \text{(third remainder)} \\
2\overline{)3} & 0 \rightarrow \text{(second remainder)} \\
2\overline{)6} & 1 \rightarrow \text{(first remainder)} \\
2\overline{)13}
\end{array}
$$

Here you get a quotient of 1 with a remainder of 1.

**Step 4** One more division gives

$$
\begin{array}{ll}
\quad 0 & 1 \\
2\overline{)1} & 1 \\
2\overline{)3} & 0 \qquad \text{Read down} \\
2\overline{)6} & 1 \\
2\overline{)13}
\end{array}
$$

In this final division, 2 does not divide into 1; therefore, the quotient is 0 with a remainder of 1.

Whenever you arrive at a quotient of 0 with a remainder of 1, the conversion is finished. The remainders when read downward give the binary equivalent. In this example, binary 1101 is equivalent to decimal 13.

There is no need to keep writing down 2 before each division because you are always dividing by 2. Here is an efficient way to show the conversion of decimal 13 to its binary equivalent:

$$
\begin{array}{ll}
\quad 0 & 1 \\
\quad 1 & 1 \\
\quad 3 & 0 \qquad \text{Read down} \\
\quad 6 & 1 \\
2\overline{)13}
\end{array}
$$

## Fractions

As far as fractions are concerned, you *multiply* by 2 and record a carry in the integer position. The carries read downward are the binary fraction. As an example, 0.85 converts to binary as follows:

$$
\begin{array}{ll}
0.85 \times 2 = 1.7 = 0.7 & \text{with a carry of } 1 \\
0.7 \times 2 = 1.4 = 0.4 & \text{with a carry of } 1 \\
0.4 \times 2 = 0.8 = 0.8 & \text{with a carry of } 0 \\
0.8 \times 2 = 1.6 = 0.6 & \text{with a carry of } 1 \qquad \text{Read down} \\
0.6 \times 2 = 1.2 = 0.2 & \text{with a carry of } 1 \\
0.2 \times 2 = 0.4 = 0.4 & \text{with a carry of } 0
\end{array}
$$

Reading the carries downward gives binary fraction 0.110110. In this case, we stopped the conversion process after getting six binary digits. Because of this, the answer is an approximation. If more accuracy is needed, continue multiplying by 2 until you have as many digits as necessary for your application.

## Useful Equivalents

Table 4–3 shows some decimal-binary equivalences. This will be useful for future reference. The table has an important property that you should be aware of. Whenever a binary number has all 1s (consists of only 1s), you can find its decimal equivalent by using this formula:

$$\text{Decimal} = 2^n - 1 \qquad (4\text{--}1)$$

where $n$ is the number of bits. For instance, 1111 has 4 bits; therefore, its decimal equivalent is

$$\text{Decimal} = 2^4 - 1 = 16 - 1 = 15$$

As another example, 1111 1111 has 8 bits, so

$$\text{Decimal} = 2^8 - 1 = 256 - 1 = 255$$

☐ EXAMPLE 4–4

Convert decimal 23.6 to a binary number.

☐ SOLUTION

Split decimal 23.6 into an integer of 23 and a fraction of 0.6, and apply double dabble to each part.

```
   0        1   ⌉
   1        0   │
   2        1   │   Read down
   5        1   │
  11        1   ↓
2)23
```

**TABLE 4–3** *Decimal-Binary Equivalences*

| Decimal | Binary |
|---------|--------|
| 1 | 1 |
| 3 | 11 |
| 7 | 111 |
| 15 | 1111 |
| 31 | 1 1111 |
| 63 | 11 1111 |
| 127 | 111 1111 |
| 255 | 1111 1111 |
| 511 | 1 1111 1111 |
| 1,023 | 11 1111 1111 |
| 2,047 | 111 1111 1111 |
| 4,095 | 1111 1111 1111 |
| 8,191 | 1 1111 1111 1111 |
| 16,383 | 11 1111 1111 1111 |
| 32,767 | 111 1111 1111 1111 |
| 65,535 | 1111 1111 1111 1111 |

and

$$0.6 \times 2 = 1.2 = 0.2 \quad \text{with a carry of 1}$$
$$0.2 \times 2 = 0.4 = 0.4 \quad \text{with a carry of 0}$$
$$0.4 \times 2 = 0.8 = 0.8 \quad \text{with a carry of 0} \qquad \text{Read down}$$
$$0.8 \times 2 = 1.6 = 0.6 \quad \text{with a carry of 1}$$
$$0.6 \times 2 = 1.2 = 0.2 \quad \text{with a carry of 1}$$

The binary number is 10111.10011. This 10-bit number is an approximation of decimal 21.6 because we terminated the conversion of the fractional part after 5 bits.

## EXAMPLE 4—5

The Macintosh computer processes binary numbers that are 32 bits long. If a 32-bit number has all 1s, what is its decimal equivalent?

## SOLUTION

$$\text{Decimal} = 2^{32} - 1 = (2^8)(2^8)(2^8)(2^8) - 1$$
$$= (256)(256)(256)(256) - 1 = 4,294,967,295$$

## 4-4 OCTAL NUMBERS

The base of a number system equals the number of digits it uses. The decimal number system has a base of 10 because it uses the digits 0 to 9. The binary number system has a base of 2 because it uses only the digits 0 and 1. The *octal* number system has a base of 8. Although we can use any eight digits, it is customary to use the first eight decimal digits:

$$0, 1, 2, 3, 4, 5, 6, 7$$

(There is no 8 or 9 in the octal number code.) These digits, 0 through 7, have exactly the same physical meaning as decimal symbols; that is, 2 stands for ●●, 5 symbolizes ●●●●●, and so on.

### Octal Odometer

The easiest way to learn how to count in octal numbers is to use an *octal odometer*. This hypothetical device is similar to the odometer of a car, except that each display wheel contains only eight digits, numbered 0 to 7. When a wheel turns from 7 back to 0, it sends a carry to the next-higher wheel.

Initially, an octal odometer shows

0000      (zero)

The next 7 miles produces readings of

0001      (one)
0002      (two)
0003      (three)
0004      (four)
0005      (five)
0006      (six)
0007      (seven)

At this point, the least-significant wheel has run out of digits. Therefore, the next mile forces a reset and carry to obtain

0010    (eight)

The next 7 miles produces these readings: 0011, 0012, 0013, 0014, 0015, 0016, and 0017. Once again, the least-significant wheel has run out of digits. So the next mile results in a reset and carry:

0020    (sixteen)

Subsequent miles produce readings of 0021, 0022, 0023, 0024, 0025, 0026, 0027, 0030, 0031, and so on.

You should have the idea by now. Each mile advances the least-significant wheel by one. When this wheel runs out of octal digits, it resets and carries. And so on for the other wheels. For instance, if the odometer reading is 6377, the next octal number is 6400.

## Octal-to-Decimal Conversion

How do we convert octal numbers to decimal numbers? In the octal number system each digit position corresponds to a power of 8 as follows:

$$8^3 \quad 8^2 \quad 8^1 \quad 8^0 \quad . \quad 8^{-1} \quad 8^{-2} \quad 8^{-3}$$

↑
Octal point

Therefore, to convert from octal to decimal, multiply each octal digit by its weight and add the resulting products.

For instance, octal 23 converts to decimal like this:

$$2(8^1) + 3(8^0) = 16 + 3 = 19$$

Here is another example. Octal 257 converts to

$$2(8^2) + 5(8^1) + 7(8^0) = 128 + 40 + 7 = 175$$

## Decimal-to-Octal Conversion

How do you convert in the opposite direction, that is, from decimal to octal? *Octal dabble*, a method similar to double dabble, is used with octal numbers. Instead of dividing by 2 (the base of binary numbers), you divide by 8 (the base of octal numbers) writing down the remainders after each division. The remainders in reverse order form the octal number. As an example, convert decimal 175 as follows:

| | |
|---|---|
| 0 | 2 → (third remainder) |
| 8)2 | 5 → (second remainder) |
| 8)21 | 7 → (first remainder) |
| 8)175 | |

You can condense these steps by writing

| | | |
|---|---|---|
| 0 | 2 | |
| 2 | 5 | Read down |
| 21 | 7 | |
| 8)175 | | |

## Fractions

With decimal fractions, multiply instead of divide, writing the carry into the integer position. An example of this is to convert decimal 0.23 into an octal fraction.

$$
\begin{array}{ll}
0.23 \times 8 = 1.84 = 0.84 & \text{with a carry of 1} \\
0.84 \times 8 = 6.72 = 0.72 & \text{with a carry of 6} \\
0.72 \times 8 = 5.76 = 0.76 & \text{with a carry of 5} \\
\multicolumn{2}{c}{\text{etc.}}
\end{array}
\qquad \text{Read down}
$$

The carries read downward give the octal fraction 0.165. We terminated after three places; if more accuracy were required, we would continue multiplying to obtain more octal digits.

## Octal-to-Binary Conversion

Because 8 (the base of octal numbers) is the third power of 2 (the base of binary numbers), you can convert from octal to binary as follows: change each octal digit to its binary equivalent. For instance, change octal 23 to its binary equivalent as follows:

$$
\begin{array}{cc}
2 & 3 \\
\downarrow & \downarrow \\
010 & 011
\end{array}
$$

Here, each octal digit converts to its binary equivalent (2 becomes 010, and 3 becomes 011). The binary equivalent of octal 23 is 010 011, or 010011. Often, a space is left between groups of 3 bits; this makes it easier to read the binary number.

As another example, octal 3574 converts to binary as follows:

$$
\begin{array}{cccc}
3 & 5 & 7 & 4 \\
\downarrow & \downarrow & \downarrow & \downarrow \\
011 & 101 & 111 & 100
\end{array}
$$

Hence binary 011101111100 is equivalent to octal 3574. Notice how much easier the binary number is to read if we leave a space between groups of 3 bits: 011 101 111 100.

Mixed octal numbers are no problem. Convert each octal digit to its equivalent binary value. Octal 34.562 becomes

$$
\begin{array}{ccccc}
3 & 4 & . & 5 & 6 & 2 \\
\downarrow & \downarrow & & \downarrow & \downarrow & \downarrow \\
011 & 100 & . & 101 & 110 & 010
\end{array}
$$

## Binary-to-Octal Conversion

Conversion from binary to octal is a reversal of the foregoing procedures. Simply remember to group the bits in threes, starting at the binary point; then convert each group of three to its octal equivalent (0s are added at each end, if necessary). For instance, binary number 1011.01101 converts as follows:

$$
\begin{array}{ccccc}
1011.01101 \rightarrow 001 & 011.011 & 010 \\
\downarrow & \downarrow \quad \downarrow & \downarrow \\
1 & 3 \quad . \quad 3 & 2
\end{array}
$$

Start at the binary point and, working both ways, separate the bits into groups of three. When necessary, as in this case, add 0s to complete the outside groups. Then convert each group of three into its binary equivalent. Therefore:

$$1011.01101 = 13.32$$

The simplicity of converting octal to binary and vice versa has many advantages in digital work. For one thing, getting information into and out of a digital system requires less circuitry because it is easier to read and print out octal numbers than binary numbers. Another advantage is that large decimal numbers are more easily converted to binary if first converted to octal and then to binary, as shown in Example 4–6.

☐ EXAMPLE 4—6

What is the binary equivalent of decimal 363?

☐ SOLUTION

One approach is double dabble. Another approach is octal dabble, followed by octal-to-binary conversion. Here is how the second method works:

$$
\begin{array}{cc}
0 & 5 \; | \\
5 & 5 \; \downarrow \quad \text{Read down} \\
\underline{45} & 3 \; \downarrow \\
8\overline{)363} &
\end{array}
$$

Next, convert octal 553 to its binary equivalent:

$$
\begin{array}{ccc}
5 & 5 & 3 \\
\downarrow & \downarrow & \downarrow \\
101 & 101 & 011
\end{array}
$$

The double-dabble approach would produce the same answer, but it is tedious because you have to divide by 2 nine times before the conversion terminates.

## 4-5  HEXADECIMAL NUMBERS

*Hexadecimal* numbers are used extensively in microprocessor work. To begin with, they are much shorter than binary numbers. This makes them easy to write and remember. Furthermore, you can mentally convert them to binary whenever necessary.

The hexadecimal number system has a base of 16. Although any 16 digits may be used, everyone uses 0 to 9 and A to F as shown in Table 4–4. In other words, after reaching 9 in the hexadecimal system, you continue counting as follows:

A, B, C, D, E, F

### Hexadecimal Odometer

The easiest way to learn how to count in hexadecimal numbers is to use a hexadecimal odometer. This hypothetical device is similar to the odometer of a car, except that each display wheel has 16 digits, numbered 0 to F. When a wheel turns from F back to 0, it sends a carry to the next higher wheel.

**TABLE 4-4** *Hexadecimal Digits*

| Decimal | Binary | Hexadecimal |
|---------|--------|-------------|
| 0 | 0000 | 0 |
| 1 | 0001 | 1 |
| 2 | 0010 | 2 |
| 3 | 0011 | 3 |
| 4 | 0100 | 4 |
| 5 | 0101 | 5 |
| 6 | 0110 | 6 |
| 7 | 0111 | 7 |
| 8 | 1000 | 8 |
| 9 | 1001 | 9 |
| 10 | 1010 | A |
| 11 | 1011 | B |
| 12 | 1100 | C |
| 13 | 1101 | D |
| 14 | 1110 | E |
| 15 | 1111 | F |

Initially, an hexadecimal odometer shows

0000     (zero)

The next 9 miles produces readings of

0001     (one)
0002     (two)
0003     (three)
0004     (four)
0005     (five)
0006     (six)
0007     (seven)
0008     (eight)
0009     (nine)

The next 6 miles gives

000A     (ten)
000B     (eleven)
000C     (twelve)
000D     (thirteen)
000E     (fourteen)
000F     (fifteen)

At this point, the least-significant wheel has run out of digits. Therefore, the next mile forces a reset and carry to obtain

0010     (sixteen)

The next 15 miles produces these readings: 0011, 0012, 0013, 0014, 0015, 0016, 0017, 0018, 0019, 001A, 001B, 001C, 001D, 001E, and 001F. Once again, the least-significant wheel has run out of digits. So, the next mile results in a reset and carry:

0020    (thirty-two)

Subsequent miles produce readings of 0021, 0022, 0023, 0024, 0025, 0026, 0027, 0028, 0029, 002A, 002B, 002C, 002D, 002E, and 002F.

You should have the idea by now. Each mile advances the least-significant wheel by one. When this wheel runs out of hexadecimal digits, it resets and carries, and so on for the other wheels. For instance, here are three more examples:

| Number | Next number |
|--------|-------------|
| 835C | 835D |
| A47F | A480 |
| BFFF | C000 |

## Hexadecimal-to-Binary Conversion

To convert a hexadecimal number to a binary number, convert each hexadecimal digit to its 4-bit equivalent using the code given in Table 4–4. For instance, here's how 9AF converts to binary:

9        A        F

↓        ↓        ↓

1001   1010   1111

As another example, C5E2 converts like this:

C      5      E      2

↓      ↓      ↓      ↓

1100  0101  1110  0010

## Binary-to-Hexadecimal Conversion

To convert in the opposite direction, from binary to hexadecimal, again use the code from Table 4–4. Here are two examples. Binary 1000 1100 converts as follows:

1000   1100

↓       ↓

8       C

Binary 1110 1000 1101 0110 converts like this:

1110  1000  1101  0110

↓      ↓      ↓      ↓

E      8      D      6

In both these conversions, we start with a binary number and wind up with the equivalent hexadecimal number.

## Hexadecimal-to-Decimal Conversion

How do we convert hexadecimal numbers to decimal numbers? In the hexadecimal number system each digit position corresponds to a power of 16. The weights of the digit positions in a hexadecimal number are as follows:

$$16^3 \quad 16^2 \quad 16^1 \quad 16^0 \quad . \quad 16^{-1} \quad 16^{-2} \quad 16^{-3}$$

↑<br>Hexadecimal point

Therefore, to convert from hexadecimal to decimal, multiply each hexadecimal digit by its weight and add the resulting products.

Here's an example. Hexadecimal F8E6.39 converts to decimal as follows:

$$
\begin{aligned}
F8E6 &= F(16^3) + 8(16^2) + E(16^1) + 6(16^0) + 3(16^{-1}) + 9(16^{-2}) \\
&= 15(16^3) + 8(16^2) + 14(16^1) + 6(16^0) + 3(16^{-1}) + 9(16^{-2}) \\
&= 61{,}440 + 2048 + 224 + 6 + 0.1875 + 0.0352 \\
&= 63{,}718.2227
\end{aligned}
$$

## Decimal-to-Hexadecimal Conversion

One way to convert from decimal to hexadecimal is the *hex dabble*. The idea is to divide successively by 16, writing down the remainders. Here's a sample of how it's done. To convert decimal 2479 to hexadecimal, the first division is

$$\frac{154}{16 \overline{)2479}} \qquad 15 \rightarrow F$$

In this first division, we get a quotient of 154 with a remainder of 15 (equivalent to F). The next step is

$$
\begin{array}{c}
9 \\
154 \\
\hline
16\overline{)2479}
\end{array}
\qquad
\begin{array}{l}
10 \rightarrow A \\
15 \rightarrow F
\end{array}
$$

Here we obtain a quotient of 9 with a remainder of 10 (same as A). The final step is

$$
\begin{array}{c}
0 \\
9 \\
154 \\
\hline
16\overline{)2479}
\end{array}
\qquad
\begin{array}{l}
9 \rightarrow 9 \\
10 \rightarrow A \\
15 \rightarrow F
\end{array}
\qquad \text{Read down} \downarrow
$$

Therefore, hexadecimal 9AF is equivalent to decimal 2479.

Notice how similar hex dabble is to double dabble. Notice also that remainders greater than 9 have to be changed to hexadecimal digits (10 becomes A, 15 becomes F, etc.).

## Using Appendix 1

A typical microcomputer can store up to 65,535 bytes. The decimal addresses of these bytes are from 0 to 65,535. The equivalent binary addresses are from

0000 0000 0000 0000

to

1111 1111 1111 1111

The first 8 bits are called the *upper byte*, and the second 8 bits are the *lower byte*.

If you have to do many conversions between binary, hexadecimal, and decimal, learn to use Appendix 1. It has four headings: *binary*, *hexadecimal*, *upper byte*, and *lower byte*. For any decimal number between 0 and 255, you would use the binary, hexadecimal, and lower byte columns. Here is the recommended way to use Appendix 1. Suppose you want to convert binary 0001 1000 to its decimal equivalent. First, mentally convert to hexadecimal:

$$0001\ 1000 \rightarrow 18 \qquad \text{(mental conversion)}$$

Next, look up hexadecimal 18 in Appendix 1 and read the corresponding decimal value from the lower-byte column:

$$18 \rightarrow 24 \qquad \text{(look up in Appendix 1)}$$

For another example, binary 1111 0000 converts like this:

$$1111\ 0000 \rightarrow F0 \rightarrow 240$$

The reason for mentally converting from binary to hexadecimal is that you can more easily locate a hexadecimal number in Appendix 1 than a binary number. Once you have the hexadecimal equivalent, you can read the lower-byte column to find the decimal equivalent.

When the decimal number is greater than 255, you have to use both the upper byte and the lower byte in Appendix 1. For instance, suppose you want to convert this binary number to its decimal equivalent:

$$1110\ 1001\ 0111\ 0100$$

First, convert the upper byte to its decimal equivalent as follows:

$$1110\ 1001 \rightarrow E9 \rightarrow 59{,}648 \qquad \text{(upper byte)}$$

Second, convert the lower byte to its decimal equivalent:

$$0111\ 0100 \rightarrow 74 \rightarrow 116 \qquad \text{(lower byte)}$$

Finally, add the upper and lower bytes to obtain the total decimal value:

$$59{,}648 + 116 = 59{,}764$$

Therefore, binary 1110 1001 0111 0100 is equivalent to decimal 59,764.

Once you get used to working with Appendix 1, you will find it to be a quick and easy way to convert between the number systems. Because it covers the decimal numbers from 0 to 65,535, Appendix 1 is extremely useful for microcomputers where the typical memory addresses are over the same decimal range.

☐ EXAMPLE 4—7

A computer memory can store thousands of binary instructions and data. A typical microcomputer has 65,536 addresses, each storing a byte. Suppose that the first 16 addresses contain these bytes:

```
0011 1100
1100 1101
0101 0111
0010 1000
1111 0001
0010 1010
1101 0100
0100 0000
0111 0111
1100 0011
1000 0100
0010 1000
0010 0001
0011 1010
0011 1110
0001 1111
```

Convert these bytes to their hexadecimal equivalents.

☐ SOLUTION

Here are the stored bytes and their hexadecimal equivalents:

| Memory contents | Hexadecimal equivalents |
| --- | --- |
| 0011 1100 | 3C |
| 1100 1101 | CD |
| 0101 0111 | 57 |
| 0010 1000 | 28 |
| 1111 0001 | F1 |
| 0010 1010 | 2A |
| 1101 0100 | D4 |
| 0100 0000 | 40 |
| 0111 0111 | 77 |
| 1100 0011 | C3 |
| 1000 0100 | 84 |
| 0010 1000 | 28 |
| 0010 0001 | 21 |
| 0011 1010 | 3A |
| 0011 1110 | 3E |
| 0001 1111 | 1F |

What is the point of this example? When discussing the contents of a computer memory, we can use either binary numbers or hexadecimal numbers. For instance, we can say that the first address contains 0011 1100, or we can say that it contains 3C. Either way, we obtain the same information. But notice how much easier it is to say, write, and think 3C than it is to say, write, and think 0011 1100. In other words, hexadecimal numbers are much easier for people to work with.

Convert the hexadecimal numbers of the preceding example to their decimal equivalents.

□ SOLUTION

The first address contains 3C, which converts like this:

$$3(16^1) + C(16^0) = 48 + 12 = 60$$

Even easier, look up the decimal equivalent of 3C in Appendix 1, and you get 60. Either by powers of 16 or with reference to Appendix 1, we can convert the other memory contents to get the following:

| Memory contents | Hexadecimal equivalents | Decimal equivalents |
| --- | --- | --- |
| 0011 1100 | 3C | 60 |
| 1100 1101 | CD | 205 |
| 0101 0111 | 57 | 87 |
| 0010 1000 | 28 | 40 |
| 1111 0001 | F1 | 241 |
| 0010 1010 | 2A | 42 |
| 1101 0100 | D4 | 212 |
| 0100 0000 | 40 | 64 |
| 0111 0111 | 77 | 119 |
| 1100 0011 | C3 | 195 |
| 1000 0100 | 84 | 132 |
| 0010 1000 | 28 | 40 |
| 0010 0001 | 21 | 33 |
| 0011 1010 | 3A | 58 |
| 0011 1110 | 3E | 62 |
| 0001 1111 | 1F | 31 |

□ EXAMPLE 4—9

Convert decimal 65,535 to its hexadecimal and binary equivalents.

□ SOLUTION

Use hex dabble as follows:

$$
\begin{array}{ll}
\phantom{000}0 & 15 \rightarrow F \\
\phantom{00}15 & 15 \rightarrow F \\
\phantom{0}255 & 15 \rightarrow F \\
\underline{4095} & 15 \rightarrow F \\
16\overline{)65,535} &
\end{array}
$$

Read down

Therefore, decimal 65535 is equivalent to hexadecimal FFFF.

Next, convert from hexadecimal to binary as follows:

$$
\begin{array}{cccc}
F & F & F & F \\
\downarrow & \downarrow & \downarrow & \downarrow \\
1111 & 1111 & 1111 & 1111
\end{array}
$$

This means that hexadecimal FFFF is equivalent to binary 1111 1111 1111 1111.

☐ EXAMPLE 4—10

Show how to use Appendix 1 to convert decimal 56,000 to its hexadecimal and binary equivalents.

☐ SOLUTION

The first thing to do is to locate the largest decimal number equal to 56,000 or less in Appendix 1. The number is 55,808, which converts like this:

$$55{,}808 \rightarrow DA \qquad \text{(upper byte)}$$

Next, you need to subtract this upper byte from the original number:

$$56{,}000 - 55{,}808 = 192 \qquad \text{(difference)}$$

This difference is always less than 256 and represents the lower byte, which Appendix 1 converts as follows:

$$192 \rightarrow C0$$

Now, combine the upper and lower byte to obtain

$$DAC0$$

which you can mentally convert to binary:

$$DAC0 \rightarrow 1101\ 1010\ 1100\ 0000$$

☐ EXAMPLE 4—11

Convert Table 4–3 into a new table with three column headings: "Decimal," "Binary," and "Hexadecimal."

☐ SOLUTION

This is easy. Convert each group of bits to its hexadecimal equivalent as shown in Table 4–5.

## 4-6 THE ASCII CODE

To get information into and out of a computer, we need to use some kind of *alphanumeric* code (one for letters, numbers, and other symbols). At one time, manufacturers used their own alphanumeric codes, which led to all kinds of confusion. Eventually, industry settled on an input-output code known as the *American Standard Code for Information Interchange* (abbreviated ASCII). This code allows manufacturers to standardize computer hardware such as keyboards, printers, and video displays.

**TABLE 4–5** *Decimal-Binary-Hexadecimal Equivalences*

| Decimal | Binary | Hexadecimal |
|---|---|---|
| 1 | 1 | 1 |
| 3 | 11 | 3 |
| 7 | 111 | 7 |
| 15 | 1111 | F |
| 31 | 1 1111 | 1F |
| 63 | 11 1111 | 3F |
| 127 | 111 1111 | 7F |
| 255 | 1111 1111 | FF |
| 511 | 1 1111 1111 | 1FF |
| 1,023 | 11 1111 1111 | 3FF |
| 2,047 | 111 1111 1111 | 7FF |
| 4,095 | 1111 1111 1111 | FFF |
| 8,191 | 1 1111 1111 1111 | 1FFF |
| 16,383 | 11 1111 1111 1111 | 3FFF |
| 32,767 | 111 1111 1111 1111 | 7FFF |
| 65,535 | 1111 1111 1111 1111 | FFFF |

## Using the Code

The ASCII (pronounced ask'-ee) code is a 7-bit code whose format is

$$X_6X_5X_4X_3X_2X_1X_0$$

where each $X$ is a 0 or a 1. For instance, the letter A is coded as

$$1000001$$

For easier reading, we can leave a space as follows:

$$100\ 0001$$

Table 4–6 shows the ASCII code. Read the table similar to the way you read a graph. For instance, the letter A has an $X_6X_5X_4$ of 100 and an $X_3X_2X_1X_0$ of 0001. Therefore, the ASCII code is

$$100\ 0001 \quad (A)$$

Table 4–6 includes the ASCII code for lowercase letters. The letter a is coded as

$$110\ 0001 \quad (a)$$

More examples are

$$110\ 0010 \quad (b)$$
$$110\ 0011 \quad (c)$$
$$110\ 0100 \quad (d)$$

and so on.

Also study the punctuation and mathematical symbols. Some examples are

$$010\ 0100 \quad (\$)$$
$$010\ 1011 \quad (+)$$
$$011\ 1101 \quad (=)$$

**TABLE 4–6** *ASCII Code*

| $X_3X_2X_1X_0$ | $X_6X_5X_4$ | | | | | |
|---|---|---|---|---|---|---|
| | 010 | 011 | 100 | 101 | 110 | 111 |
| 0000 | SP | 0 | @ | P | | p |
| 0001 | ! | 1 | A | Q | a | q |
| 0010 | " | 2 | B | R | b | r |
| 0011 | # | 3 | C | S | c | s |
| 0100 | $ | 4 | D | T | d | t |
| 0101 | % | 5 | E | U | e | u |
| 0110 | & | 6 | F | V | f | v |
| 0111 | ' | 7 | G | W | g | w |
| 1000 | ( | 8 | H | X | h | x |
| 1001 | ) | 9 | I | Y | i | y |
| 1010 | * | : | J | Z | j | z |
| 1011 | + | ; | K | | k | |
| 1100 | , | < | L | | l | |
| 1101 | − | = | M | | m | |
| 1110 | • | > | N | | n | |
| 1111 | / | ? | O | | o | |

In Table 4–6, SP stands for space (blank). Hitting the space bar of an ASCII keyboard sends this into a microcomputer:

$$010\ 0000 \quad \text{(space)}$$

## Parity Bit

The ASCII code is used for sending digital data over telephone lines. As mentioned in the preceding chapter, 1-bit errors may occur in transmitted data. To catch these errors, a parity bit is usually transmitted along with the original bits. Then a parity checker at the receiving end can test for even or odd parity, whichever parity has been prearranged between the sender and the receiver. Since ASCII code uses 7 bits, the addition of a parity bit to the transmitted data produces an 8-bit number in this format:

$$X_7X_6X_5X_4\ X_3X_2X_1X_0$$
$$\uparrow$$
Parity bit

This is an ideal length because most digital equipment is set up to handle bytes of data.

☐ EXAMPLE 4–12

With an ASCII keyboard, each keystroke produces the ASCII equivalent of the designated character. Suppose that you type PRINT X. What is the output of an ASCII keyboard?

☐ SOLUTION

The sequence is as follows: P (101 0000), R (101 0010), I (100 1001), N (100 1110), T (101 0100), space (010 000), X (101 1000).

□ EXAMPLE 4-13

A computer sends a message to another computer using an odd-parity bit. Here is the message in ASCII code, plus the parity bit:

$$1100\ 1000$$
$$0100\ 0101$$
$$0100\ 1100$$
$$0100\ 1100$$
$$0100\ 1111$$

What do these numbers mean?

□ SOLUTION

First, notice that each 8-bit number has odd parity, an indication that no 1-bit errors occurred during transmission. Next, use Table 4–6 to translate the ASCII characters. If you do this correctly, you get a message of HELLO.

## 4-7 THE EXCESS-3 CODE

The excess-3 code is an important 4-bit code sometimes used with binary-coded decimal (BCD) numbers. To convert any decimal number into its excess-3 form, add 3 to each decimal digit, and then convert the sum to a BCD number. For example, here is how to convert 12 to an excess-3 number. First, add 3 to each decimal digit:

$$
\begin{array}{cc}
1 & 2 \\
+3 & +3 \\
\hline
4 & 5
\end{array}
$$

Second, convert the sum to BCD form:

$$
\begin{array}{cc}
4 & 5 \\
\downarrow & \downarrow \\
0100 & 0101
\end{array}
$$

So, 0100 0101 in the excess-3 code stands for decimal 12.

Take another example; convert 29 to an excess-3 number:

$$
\begin{array}{cc}
2 & 9 \\
+3 & +3 \\
\hline
5 & 12 \\
\downarrow & \downarrow \\
0101 & 1100
\end{array}
$$

After adding 9 and 3, do *not* carry the 1 into the next column; instead, leave the result intact as 12, and then convert as shown. Therefore, 0101 1100 in the excess-3 code stands for decimal 29.

Table 4–7 shows the excess-3 code. In each case, the excess-3 code number is 3 greater than the BCD equivalent. The reason for this is that it simplifies BCD addition. Incidentally, if you need an integrated circuit (IC) that converts from excess 3 to decimal,

**TABLE 4-7** *Excess-3 Code*

| Decimal | BCD | Excess-3 |
|---------|------|----------|
| 0 | 0000 | 0011 |
| 1 | 0001 | 0100 |
| 2 | 0010 | 0101 |
| 3 | 0011 | 0110 |
| 4 | 0100 | 0111 |
| 5 | 0101 | 1000 |
| 6 | 0110 | 1001 |
| 7 | 0111 | 1010 |
| 8 | 1000 | 1011 |
| 9 | 1001 | 1100 |

look at the data sheet of a 7443. This transistor-transistor logic (TTL) chip has four input lines for the excess-3 input and 10 output lines for the decoded decimal output.

## 4-8 THE GRAY CODE

Table 4–8 shows the *Gray code*, along with the corresponding binary numbers. Each Gray-code number differs from the preceding number by a single bit. For instance, in going from decimal 7 to 8, the Gray-code numbers change from 0100 to 1100; these numbers differ only in the most-significant bit. As another example, decimal numbers 13 and 14 are represented by Gray-code numbers 1011 and 1001; these numbers differ in only one digit position (the second position from the right). So it is with the entire Gray code; every number differs by only 1 bit from the preceding number.

Besides the excess-3 and Gray codes, there are other binary-type codes. Appendix 4 lists some of these codes for future reference. Incidentally, the BCD code is sometimes referred to as the *8421 code* because the weights of the digit positions from left to right are

**TABLE 4-8** *Gray Code*

| Decimal | Gray Code | Binary |
|---------|-----------|--------|
| 0 | 0000 | 0000 |
| 1 | 0001 | 0001 |
| 2 | 0011 | 0010 |
| 3 | 0010 | 0011 |
| 4 | 0110 | 0100 |
| 5 | 0111 | 0101 |
| 6 | 0101 | 0110 |
| 7 | 0100 | 0111 |
| 8 | 1100 | 1000 |
| 9 | 1101 | 1001 |
| 10 | 1111 | 1010 |
| 11 | 1110 | 1011 |
| 12 | 1010 | 1100 |
| 13 | 1011 | 1101 |
| 14 | 1001 | 1110 |
| 15 | 1000 | 1111 |
| . . . | . . . | . . . |

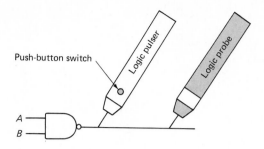

Push-button switch

A —

B —

**FIG. 4–3.** Using a logic pulser and a logic probe

8, 4, 2, and 1. As shown in Appendix 4, there are many other weighted codes such as the 7421, 6311, 5421, and so on.

# 4-9 TROUBLESHOOTING WITH A LOGIC PULSER

Figure 4–3 shows a typical *logic pulser*, a troubleshooting tool that generates a brief voltage pulse when its push-button switch is pressed. Because of its design, the logic pulser (on the left) senses the original state of the node and produces a voltage pulse of the opposite polarity. When this happens, the logic probe (on the right) blinks, indicating a temporary change of output state.

## Thévenin Circuit

Figure 4–4a shows the Thévenin equivalent circuit for a typical logic pulser. The Thévenin voltage is a pulse with an amplitude of 5 V; the polarity automatically adjusts to the original state of the test node. As shown, the Thévenin resistance or output impedance is only 2 Ω. This Thévenin resistance is representative; the exact value depends on the particular logic pulser being used.

Typically, a TTL gate has an output resistance between 12 (low state) and 70 Ω (high state). When a logic pulser drives the output of a NAND gate, the equivalent circuit appears as shown in Fig. 4–4b. Because of the low output impedance (2 Ω) of the logic pulser, most of the voltage pulse appears across the load (12 to 70 Ω). Therefore, the output is briefly driven into the opposite voltage state.

## Testing Any Node

You can use a logic pulser to drive any node in a circuit, whether input or output. Almost always, the load impedance of the node being driven is larger than the output impedance of the logic pulser. For this reason, the logic pulser can usually change the state of any node in a logic circuit. Also, the pulse width is kept very short (fractions of a microsecond) to avoid damaging the circuit being tested. (*Note:* Power dissipation is what damages ICs. A brief voltage pulse produces only a small power dissipation.)

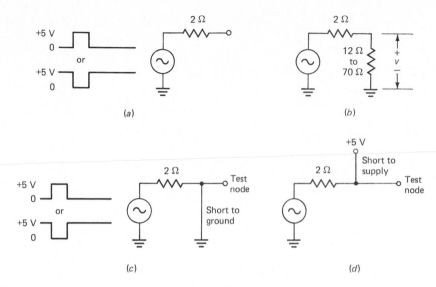

**FIG. 4–4.** (a) Thévenin equivalent of logic pulser (b) Logic pulser driving NAND–gate output (c) Node stuck in low state (d) Node struck in high state

## Stuck Nodes

When is a logic pulser unable to change the state of a node? When the test node is shorted to ground or to the supply voltage. For instance, Fig. 4–4c shows the test node shorted to ground. In this case, all the voltage pulse is dropped across the internal impedance of the logic pulser; therefore the test node is stuck at 0 V, the low state.

On the other hand, the test node may be shorted to the supply voltage as shown in Fig. 4–4d. Most power supplies are regulated and have impedances in fractions of 1 Ω. For this reason, most of the voltage pulse is again dropped across the output impedance of the logic pulser, which means that the test node is stuck at +5 V.

## Finding Stuck Nodes

If a circuit is faulty, you can use a logic pulser and logic probe to locate stuck nodes. Here's how. Touch both the logic pulser and the logic probe to a node as shown in Fig. 4–3. If the node is stuck in either state, the logic pulser will be unable to change the state. So, if the logic probe does not blink, you have a stuck node. Then, you can look for solder bridges on any trace connected to the stuck node, or possibly replace the IC having the stuck node.

## SUMMARY

To convert from binary to decimal numbers, add the weight of each bit position (1, 2, 4, 8, . . .) when there is a 1 in that position. With fractions, the binary weights are $\frac{1}{2}$, $\frac{1}{4}$, $\frac{1}{8}$, . . . , and so on. To convert from decimal to binary, use double dabble for integers and the multiply-by-2 method for fractions.

The base of a number system equals the number of digits it uses. The decimal number system has a base of 10, while the binary number system has a base of 2. The octal number system has a base of 8. A useful model for counting is the octal odometer. When a display wheel turns from 7 back to 0, it sends a carry to the next-higher wheel.

Hexadecimal numbers have a base of 16. The model for counting is the hexadecimal odometer, whose wheels reset and carry beyond F. Hexadecimal numbers are easy to convert mentally into their binary equivalents. For this reason, people prefer using hexadecimal numbers because they are much shorter than the corresponding binary numbers.

The ASCII code is an alphanumeric code widely used for transferring data into and out of a computer. This 7-bit code is used to represent alphabet letters, numbers, and other symbols. The excess-3 code and the Gray code are two other codes that are used.

A logic pulser can temporarily change the state of a node under test. If the original state is low, the logic pulser drives the node briefly into the high state. If the state is high, the logic pulser drives the node briefly into the low state. The output impedance of a logic pulser is so low that it can drive almost any normal node in a logic circuit. When a node is shorted to ground or to the supply voltage, the logic pulser is unable to change the voltage level; this is a confirmation of the shorted condition.

# GLOSSARY

*Base*  The number of digits or basic symbols in a number system. The decimal system has a base of 10 because it uses 10 digits. Binary has a base of 2, octal a base of 8, and hexadecimal a base of 16.

*Binary*  Refers to a number system with a base of 2, that is, containing two digits.

*Bit*  An abbreviated form of binary digit. Instead of saying that 10110 has five binary digits, we can say that it has 5 bits.

*Digit*  A basic symbol used in a number system. The decimal system has 10 digits, 0 through 9.

*Hexadecimal*  Refers to number system with a base of 16. The hexadecimal system has digits 0 through 9, followed by A through F.

*Logic Pulser*  A troubleshooting device that generates brief voltage pulses. The typical logic pulser has a push-button switch that produces a single pulse for each closure. More advanced logic pulsers can generate a pulse train with a specified number of pulses.

*Octal*  Refers to a number system with a base of 8, that is, one that uses 8 digits. Normally, these are 0, 1, 2, 3, 4, 5, 6, and 7.

*Weight*  Refers to the decimal value of each digit position of a number. For decimal numbers, the weights are 1, 10, 100, 1000, . . . , working from the decimal point to the left. For binary numbers the weights are 1, 2, 4, 8, . . . to the left of the binary point. With octal numbers, the weights become 1, 8, 64, . . . to the left of the octal point.

# PROBLEMS

*SECTION 4-1:*

**4-1.** The tape shown in Fig. 4–1*d* is magnetized at each dark dot and unmagnetized at each light dot. If a magnetized point represents a binary 1 and an unmagnetized point a binary 0, the first 7-bit number stored on the tape is 0000111. What are the remaining binary numbers stored on the tape?

*SECTION 4-2:*

**4-2.** Give the decimal equivalents for each of the following binary numbers:
   **a.** 110101
   **b.** 11001.011

**4-3.** Convert the following binary numbers to their decimal equivalents:
   **a.** 1011 1100
   **b.** 1111 1111

**4-4.** What is the decimal equivalent of 1000 1100 1011 0011?

**4-5.** A computer has 128K of memory. How many bytes does this represent?

*SECTION 4-3:*

**4-6.** Convert the following decimal numbers to binary numbers: 24, 65, and 106.

**4-7.** What binary number does decimal 268 stand for?

**4-8.** Convert decimal 108.364 to a binary number.

**4-9.** Calculate the binary equivalent for 5280.

*SECTION 4-4:*

**4-10.** Convert the following octal numbers to decimal equivalents:
   **a.** 65
   **b.** 216
   **c.** 4073

**4-11.** What is the decimal equivalent of octal 325.736?

**4-12.** Convert these decimal numbers to octal numbers:
   **a.** 4096
   **b.** 65535

**4-13.** What is the octal equivalent of decimal 324.987?

**4-14.** Convert the following octal numbers to binary numbers: 34, 567, 4673.

**4-15.** Convert the following binary numbers to octal numbers:
   **a.** 10101111
   **b.** 1101.0110111
   **c.** 1010011.101101

**4-16.** What are the hexadecimal numbers that follow each of these:

    **a.** ABCD

    **b.** 7F3F

    **c.** BEEF

**4-17.** Convert the following hexadecimal numbers to binary numbers:

    **a.** E5

    **b.** B4D

    **c.** 7AF4

**4-18.** Convert these binary numbers into hexadecimal numbers:

    **a.** 1000 1100

    **b.** 0011 0111

    **c.** 1111 0101 0110

**4-19.** Convert hexadecimal 2F59 to its decimal equivalent.

**4-20.** What is the hexadecimal equivalent of decimal 62359?

**4-21.** Give the value of $Y_3Y_2Y_1Y_0$ in Fig. 4-5 for each of these:

    **a.** All switches are open

    **b.** Switch 4 is closed

    **c.** Switch A is closed

    **d.** Switch F is closed

**4-22.** A computer has the following hexadecimal contents stored at the addresses shown:

| Address | Hexadecimal contents |
| --- | --- |
| 2000 | D5 |
| 2001 | AA |
| 2002 | 96 |
| 2003 | DE |
| 2004 | AA |
| 2005 | EB |

What are the binary contents at each address?

*SECTION 4-6:*

**4-23.** Give the ASCII code for each of these:

    **a.** 7    **c.** f

    **b.** W   **d.** y

**4-24.** Suppose that you type LIST with an ASCII keyboard. What is the binary output as you strike each letter?

**4-25.** In Example 4-13, a computer sends the word HELLO to another computer. The characters are coded in ASCII with an odd-parity bit. Here is how the word is stored in the memory of the receiving computer:

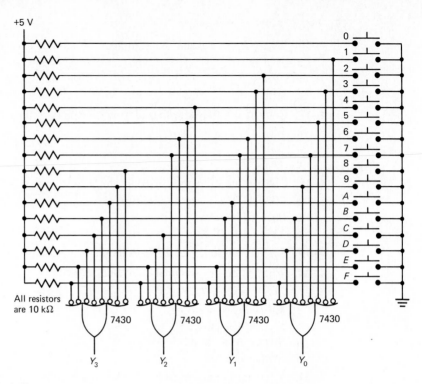

**FIG. 4–5.**

| Address | Alphanumeric | Hexadecimal contents |
|---------|--------------|----------------------|
| 2000 | H | C8 |
| 2001 | E | 45 |
| 2002 | L | 4C |
| 2003 | L | 4C |
| 2004 | O | 4F |

The transmitting computer then sends the word GOODBYE. Show how this word is stored in the receiving computer. Use a starting address of 2000 and include a parity bit.

*SECTION 4-7:*

**4–26.** Express decimal 5280 in excess-3 code.

**4–27.** Here is an excess-3 number:

0110 1001 1100 0111

What is the decimal equivalent?

*SECTION 4-8:*

**4–28.** What is the Gray code for decimal 8?

**4–29.** Convert Gray number 1110 to its BCD equivalent.

**4–30.** Figure 4–6 shows the decimal-to-BCD encoder discussed in Sec. 3–6 of the preceding chapter. Answer the following questions:

   **a.** If all switches are open and the logic pulser is inactive, what voltage level does the logic probe indicate?

   **b.** If switch 6 is closed and the logic pulser is inactive, what does the logic probe indicate?

   **c.** If all switches are open and the logic pulser is activated, what does the logic probe do?

**4–31.** The push-button switch of the logic pulser shown in Fig. 4–6 is pressed. Suppose that the logic probe is initially dark and remains dark. Indicate which of the following are possible sources of trouble:

   **a.** 74147 defective

   **b.** Pin 9 shorted to ground

   **c.** Pin 9 shorted to +5 V

   **d.** Pin 10 shorted to ground

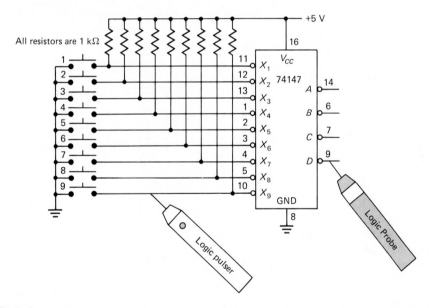

**FIG. 4–6.**

**4–32.** The instruction register shown in Fig. 4–7 is a logic circuit that stores a 16-bit number, $I_{15} \cdots I_0$. The first 4 bits, $I_{15} \cdots I_{12}$, are decoded by a 4- to 16-line decoder. Determine whether the logic probe indicates low, high, or blink for each of these conditions:

   **a.** $I_{15} \cdots I_{12} = 0000$ and logic pulser inactive

   **b.** $I_{15} \cdots I_{12} = 1000$ and logic pulser inactive

   **c.** $I_{15} \cdots I_{12} = 1000$ and logic pulser active

   **d.** $I_{15} \cdots I_{12} = 1111$ and logic pulser active

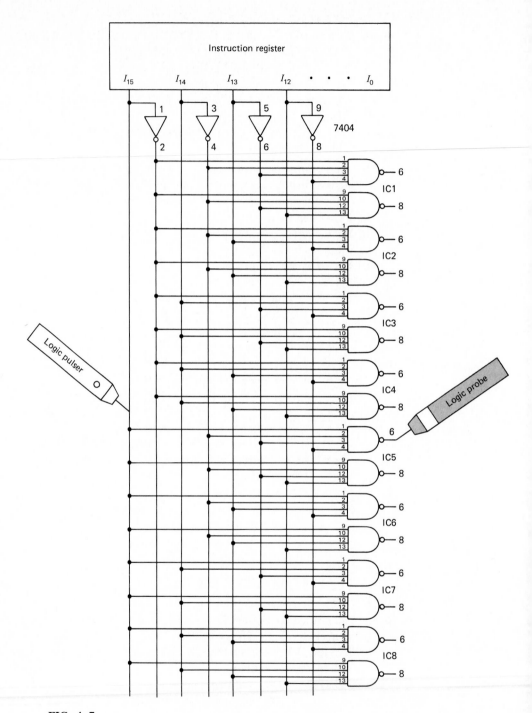

FIG. 4–7.

**4–33.** The logic pulser and logic probe shown in Fig. 4–7 are used to check the pins of the 7404 for stuck states. Suppose pin 8 is stuck in the high state. Indicate which of the following are possible sources of trouble:

**a.** No supply voltage anywhere in circuit
**b.** Pin 1 of IC2 shorted to ground
**c.** Pin 2 of IC4 shorted to the supply voltage
**d.** Pin 3 of IC5 shorted to ground
**e.** Pin 4 of IC8 shorted to the supply voltage

# 5

## ARITHMETIC
## CIRCUITS

By combining logic gates in the right way, we can build circuits that add and subtract. Since these circuits are electronic, they are fast. Typically, an addition is done in microseconds. This chapter begins with binary addition. Then, you will see two different methods for representing negative numbers. Finally, an 8-bit adder-subtracter is examined in detail. Besides giving you a preliminary idea of how a computer works, this chapter lays the foundation for later discussions of digital systems.

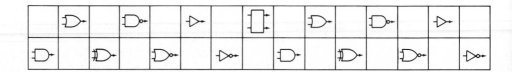

## 5-1 BINARY ADDITION

Numbers represent physical quantities. For instance, Table 5–1 shows the decimal digits and the corresponding amount of pebbles. Digit 2 stands for two pebbles (●●), 5 for five pebbles (●●●●●), and so on. Addition represents the combining of physical quantities. For instance:

$$2 + 3 = 5$$

symbolizes the combining of two pebbles with three pebbles to obtain a total of five pebbles. Symbolically, this is expressed

### Four Cases to Remember

Computer circuits don't process decimal numbers; they process binary numbers. Before you can understand how a computer performs arithmetic, you have to learn how to add binary numbers. Binary addition is the key to binary subtraction, multiplication, and division. So, let's begin with the four most basic cases of binary addition:

$$0 + 0 = 0 \qquad (5–1)$$

$$0 + 1 = 1 \qquad (5–2)$$

$$1 + 0 = 1 \qquad (5–3)$$

$$1 + 1 = 10 \qquad (5–4)$$

Equation (5–1) is obvious; so are Eqs. (5–2) and (5–3) because they are identical to decimal addition. The fourth case, however, may bother you. If so, you don't understand what Eq. (5–4) represents in the physical world. Equation (5–4) represents the combining of one pebble and one pebble to obtain a total of two pebbles:

$$● + ● = ●●$$

Since binary 10 stands for ●●, the binary equation

$$1 + 1 = 10$$

**TABLE 5–1** *The Decimal Digits*

| Pebbles | Symbol |
|:---:|:---:|
| None | 0 |
| ● | 1 |
| ●● | 2 |
| ●●● | 3 |
| ●●●● | 4 |
| ●●●●● | 5 |
| ●●●●●● | 6 |
| ●●●●●●● | 7 |
| ●●●●●●●● | 8 |
| ●●●●●●●●● | 9 |

makes perfect sense. From now on, remember that numbers, whether binary, decimal, octal, or hexadecimal are codes for physical amounts. If you're in doubt about the meaning of a numerical equation, convert the numbers to pebbles to see if the two sides of the equation are equal.

## Subscripts

The foregoing discussion brings up the idea of *subscripts*. Since we already have discussed four kinds of numbers (decimal, binary, octal, and hexadecimal), we have four different ways to code physical quantities. How do we know which code is being used? In other words, how can we tell when 10 is a decimal, binary, octal, or hexadecimal number?

Most of the time, it's clear from the discussion which kind of numbers are involved. For instance, if we have been discussing nothing but binary numbers for page after page, you can count on the next 10 being binary 10, which represents ●● in the physical world. On the other hand, if a discussion uses more than one type of number, it may be helpful to use subscripts for the base as follows:

$$2 \rightarrow \text{binary}$$
$$8 \rightarrow \text{octal}$$
$$10 \rightarrow \text{decimal}$$
$$16 \rightarrow \text{hexadecimal}$$

For instance, $11_2$ represents binary 11, $23_8$ stands for octal 23, $45_{10}$ for decimal 45, and $F4_{16}$ for hexadecimal F4. With the subscripts in mind, the following equations should make perfect sense:

$$1_2 + 1_2 = 10_2$$
$$7_8 + 1_8 = 10_8$$
$$9_{10} + 1_{10} = 10_{10}$$
$$F_{16} + 1_{16} = 10_{16}$$

## Larger Binary Numbers

Column-by-column addition applies to binary as well as decimal numbers. For example, suppose you have this problem in binary addition:

$$\begin{array}{r} 11100 \\ +11010 \\ \hline ? \end{array}$$

Start by adding the least-significant column to get

$$\begin{array}{r} 11100 \\ +11010 \\ \hline 0 \end{array}$$

Next, add the bits in the second column as follows:

$$11100$$
$$\underline{+11010}$$
$$10$$

The third column gives

$$11100$$
$$\underline{+11010}$$
$$110$$

The fourth column results in

Carry $\rightarrow$    1
$$11100$$
$$\underline{+\ 11010}$$
$$0110$$

Notice the carry into the final column; this carry occurs because $1 + 1 = 10$. As in decimal addition, you write down the 0 and carry the 1 to the next-higher column.

Finally, the last column gives

Carry $\rightarrow$    1
$$11100$$
$$\underline{+\ 11010}$$
$$110110$$

In the last column, $1 + 1 + 1 = 10 + 1 = 11$.

## 8-Bit Arithmetic

That's all there is to binary addition. If you can remember the four basic rules, you can add column by column to find the sum of two binary numbers, regardless of how long they may be. In first-generation microcomputers (Apple II, TRS-80, etc.), addition is done on two 8-bit numbers such as

$$A_7A_6A_5A_4\ A_3A_2A_1A_0$$
$$\underline{+B_7B_6B_5B_4\ B_3B_2B_1B_0}$$
$$?$$

The most-significant bit (MSB) of each number is on the left, and the least-significant bit (LSB) is on the right. For the first number, $A_7$ is the MSB and $A_0$ is the LSB. For the second number, $B_7$ is the MSB and $B_0$ is the LSB. Try to remember the abbreviations MSB and LSB because they are used frequently in computer discussions.

☐ EXAMPLE 5–1

Add these 8-bit numbers: 0101 0111 and 0011 0101. Then, show the same numbers in hexadecimal notation.

☐ SOLUTION

This is the problem:

$$
\begin{array}{r}
0101\ 0111 \\
+\,0011\ 0101 \\
\hline
?
\end{array}
$$

If you add the bits in each column as previously discussed, you will obtain

$$
\begin{array}{r}
0101\ 0111 \\
+\,0011\ 0101 \\
\hline
1000\ 1100
\end{array}
$$

Many people prefer hexadecimal notation because it's a faster code to work with. Expressed in hexadecimal numbers, the foregoing addition is

$$
\begin{array}{r}
57 \\
+\,35 \\
\hline
8C
\end{array}
$$

Often, the letter H is used to signify hexadecimal numbers, so the foregoing addition may be written as

$$
\begin{array}{r}
57H \\
+\,35H \\
\hline
8CH
\end{array}
$$

☐ EXAMPLE 5–2

Add these 16-bit numbers: 0000 1111 1010 1100 and 0011 1000 0111 1111. Show the corresponding hexadecimal and decimal numbers.

☐ SOLUTION

Start at the right and add the bits, column by column:

| Binary | Hexadecimal | Decimal |
|---|---|---|
| 0000 1111 1010 1100 | 0FACH | 4,012 |
| + 0011 1000 0111 1111 | + 387FH | + 14,463 |
| 0100 1000 0010 1011 | 482BH | 18,475 |

(*Note*: Remember Appendix 1; it takes most of the work out of conversions between number systems.)

☐ EXAMPLE 5–3

Repeat Example 5–2, showing how a first-generation microcomputer does the addition.

☐ SOLUTION

First-generation microcomputers like the Apple II have an 8-bit *microprocessor* (a digital IC that performs binary arithmetic on 8-bit numbers). To add 16-bit numbers, a first-generation microcomputer adds the lower 8 bits in one operation and then the upper 8 bits in another operation.

Here is how it works for numbers of the preceding example. The original problem is

$$
\begin{array}{cc}
\text{Upper bytes} & \text{Lower bytes} \\
\downarrow & \downarrow \\
0000\ 1111 & 1010\ 1100 \\
+0011\ 1000 & 0111\ 1111 \\
\hline
& ?
\end{array}
$$

The microcomputer starts by adding the lower bytes:

$$
\begin{array}{r}
1010\ 1100 \\
+0111\ 1111 \\
\hline
1\ 0010\ 1011
\end{array}
$$

Notice the carry into the final column. The microcomputer will store the lower byte (0010 1011). Then, it will do another addition of the upper bytes, plus the carry, as follows:

$$
\begin{array}{r}
1 \leftarrow \text{Carry} \\
0000\ 1111 \\
+\ 0011\ 1000 \\
\hline
0100\ 1000
\end{array}
$$

The microcomputer then stores the upper byte. To output the total answer, the microcomputer pulls the upper and lower sums out of its memory to get

$$0100\ 1000\ 0010\ 1011$$

which is equivalent to 482BH or 18,475, the same as we found in the preceding example.

## 5·2  BINARY SUBTRACTION

Let's begin with four basic cases of binary subtraction:

$$0 - 0 = 0 \qquad\qquad (5\text{--}5)$$

$$1 - 0 = 1 \qquad\qquad (5\text{--}6)$$

$$1 - 1 = 0 \qquad\qquad (5\text{--}7)$$

$$10 - 1 = 1 \qquad\qquad (5\text{--}8)$$

Equations (5–5) to (5–7) are easy to understand because they are identical to decimal subtraction. The fourth case will disturb you if you have lost sight of what it really means. Back in the physical world, Eq. (5–4) represents

$$\bullet\bullet - \bullet = \bullet$$

Two pebbles minus one pebble equals one pebble.

For larger binary numbers, subtract column by column, the same as you do with decimal numbers. This means that you sometimes have to borrow from the next-higher column. Here is an example:

$$
\begin{array}{r}
1101 \\
-\ 1010 \\
\hline
?
\end{array}
$$

Subtract the LSBs to get

$$
\begin{array}{r}
1101 \\
-\ 1010 \\
\hline
1
\end{array}
$$

To subtract the bits of the second column, borrow from the next-higher column to obtain

$$
\begin{array}{r}
\text{Borrow} \rightarrow\quad 1\phantom{000} \\
1001 \\
-\ 1010 \\
\hline
1
\end{array}
$$

In the second column from the right, subtract as follows: $10 - 1 = 1$, to get

$$
\begin{array}{r}
\text{Borrow} \rightarrow\quad 1\phantom{000} \\
1001 \\
-\ 1010 \\
\hline
11
\end{array}
$$

Then subtract the remaining columns:

$$
\begin{array}{r}
\text{Borrow} \rightarrow\quad 1\phantom{000} \\
1001 \\
-\ 1010 \\
\hline
0011
\end{array}
$$

After you get used to it, binary subtraction is no more difficult than decimal subtraction. In fact, it's easier because there are only four basic cases to remember.

☐ EXAMPLE 5—4

Show the binary subtraction of $125_{10}$ from $200_{10}$.

☐ SOLUTION

First, use Appendix 1 to convert the numbers as follows:

$$
\begin{aligned}
200 &\rightarrow \text{C8H} \rightarrow 1100\ 1000 \\
125 &\rightarrow \text{7DH} \rightarrow 0111\ 1101
\end{aligned}
$$

So, here is the problem:

$$
\begin{array}{r}
1100\ 1000 \\
-\ 0111\ 1101 \\
\hline
?
\end{array}
$$

Column-by-column subtraction gives:

$$
\begin{array}{r}
1100\ 1000 \\
-\ 0111\ 1101 \\
\hline
0100\ 1011
\end{array}
$$

In hexadecimal notation, the foregoing appears as

$$
\begin{array}{r}
\text{C8H} \\
-\ \text{7DH} \\
\hline
\text{4BH}
\end{array}
$$

## 5-3 UNSIGNED BINARY NUMBERS

In some applications, all data is either positive or negative. When this happens, you can forget about + and − signs, and concentrate on the *magnitude* (absolute value) of numbers. For instance, the smallest 8-bit number is 0000 0000, and the largest is 1111 1111. Therefore, the total range of 8-bit numbers is

$$0000\ 0000 \qquad \text{(00H)}$$

to

$$1111\ 1111 \qquad \text{(FFH)}$$

This is equivalent to a decimal 0 to 255. As you can see, we are not including + or − signs with these decimal numbers.

With 16-bit numbers, the total range is

$$0000\ 0000\ 0000\ 0000 \qquad \text{(0000H)}$$

to

$$1111\ 1111\ 1111\ 1111 \qquad \text{(FFFFH)}$$

which represents the magnitude of all decimal numbers from 0 to 65,535.

### Why Called Unsigned Binary

Data of the foregoing type is called *unsigned binary* because all of the bits in a binary number are used to represent the magnitude of the corresponding decimal number. You can add and subtract unsigned binary numbers, provided certain conditions are satisfied. The following examples will tell you more about unsigned binary numbers.

### Limits

First-generation microcomputers can process only 8 bits at a time. For this reason, there are certain restrictions you should be aware of. With 8-bit unsigned arithmetic, all magnitudes must be between 0 and 255. Therefore, each number being added or subtracted must be between 0 and 255. Also, the answer must fall in the range of 0 to 255. If any magnitudes are greater than 255, you should use 16-bit arithmetic, which means operating on the lower 8 bits first, then on the upper 8 bits (see Example 5–3).

## Overflow

In 8-bit arithmetic, addition of two unsigned numbers whose sum is greater than 255 causes an *overflow*, a carry into the ninth column. Most microprocessors have a logic circuit called a *carry flag*; this circuit detects a carry into the ninth column and warns you that the 8-bit answer is invalid (see Example 5–7).

□ EXAMPLE 5–5

Show how to add $150_{10}$ and $85_{10}$ with unsigned 8-bit numbers.

□ SOLUTION

With Appendix 1, we obtain

$$150 \rightarrow 96H \rightarrow 1001\ 0110$$
$$85 \rightarrow 55H \rightarrow 0101\ 0101$$

Next, we can add these unsigned numbers to get

$$
\begin{array}{r}
1001\ 0110 \\
+\ 0101\ 0101 \\
\hline
1110\ 1011
\end{array}
\qquad
\begin{array}{r}
96H \\
+\ 55H \\
\hline
EBH
\end{array}
$$

Again, Appendix 1 gives

$$1110\ 1011 \rightarrow EBH \rightarrow 235$$

□ EXAMPLE 5–6

Show how to subtract $85_{10}$ from $150_{10}$ with unsigned 8-bit numbers.

□ SOLUTION

Use the same binary numbers as in the preceding example, but subtract to get

$$
\begin{array}{r}
1001\ 0110 \\
-\ 0101\ 0101 \\
\hline
0100\ 0001
\end{array}
\qquad
\begin{array}{r}
96H \\
-\ 55H \\
\hline
41H
\end{array}
$$

Again, Appendix 1 gives

$$0100\ 0001 \rightarrow 41H \rightarrow 65$$

□ EXAMPLE 5–7

In the two preceding examples, everything was well behaved because both decimal answers were between 0 and 255. Now, you will see how an overflow can occur to produce an invalid 8-bit answer.

Show the addition of $175_{10}$ and $118_{10}$ using unsigned 8-bit numbers.

□ SOLUTION

$$
\begin{array}{r}
175 \\
+\ 118 \\
\hline
293
\end{array}
$$

The answer is greater than 255. Here is what happens when we try to add 8-bit numbers: Appendix 1 gives

$$175 \rightarrow \text{BFH} \rightarrow 1010\ 1111$$
$$118 \rightarrow 76\text{H} \rightarrow 0111\ 0110$$

An 8-bit microprocessor adds like this:

$$
\begin{array}{r}
1010\ 1111 \\
+\ 0111\ 0110 \\
\hline
\text{Overflow} \rightarrow\ 1\ 0010\ 0101
\end{array}
\qquad
\begin{array}{r}
\text{AFH} \\
+\ 76\text{H} \\
\hline
125\text{H}
\end{array}
$$

With 8-bit arithmetic, only the lower 8 bits are used. Appendix 1 gives

$$0010\ 0101 \rightarrow 25\text{H} \rightarrow 37$$

As you see, the 8-bit answer is wrong.

It is true that if you take the overflow into account, the answer is valid, but then you no longer are using 8-bit arithmetic. The point is that somebody (the programmer) has to worry about the possibility of an overflow and must take steps to correct the final answer when an overflow occurs. If you study assembly-language programming, you will learn more about overflows and what to do about them.

In summary, 8-bit arithmetic circuits can process decimal numbers between 0 and 255 only. If there is any chance of an overflow during an addition, the programmer has to write instructions that look at the carry flag and use 16-bit arithmetic to obtain the final answer. This means operating on the lower 8 bits, and then the upper 8 bits and the overflow (as done in Example 5–3).

## 5-4 SIGN-MAGNITUDE NUMBERS

What do we do when the data has positive and negative values? The answer is important because it determines how complicated the arithmetic circuits must be. The negative decimal numbers are −1, −2, −3, and so on. The magnitude of these numbers is 1, 2, 3, and so forth. One way to code these as binary numbers is to convert the magnitude to its binary equivalent and prefix the sign. With this approach, the sequence −1, −2, and −3 becomes −001, −010, and −011. Since everything has to be coded as strings of 0s and 1s, the + and − signs also have to be represented in binary form. For reasons given soon, 0 is used for the + sign and 1 for the − sign. Therefore, −001, −010, and −011 are coded as 1001, 1010, and 1011.

### Definition

The foregoing numbers contain a sign bit followed by magnitude bits. Numbers in this form are called *sign-magnitude* numbers. For larger decimal numbers, you need more than 4 bits. But the idea is still the same: the MSB always represents the sign, and the remaining bits always stand for the magnitude. Here are some examples of converting sign-magnitude numbers:

$$+7 \rightarrow 0000\ 0111$$
$$-16 \rightarrow 1001\ 0000$$
$$+25 \rightarrow 0000\ 0000\ 0001\ 1001$$
$$-128 \rightarrow 1000\ 0000\ 1000\ 0000$$

## Range of Sign-Magnitude Numbers

As you know, the unsigned 8-bit numbers cover the decimal range of 0 to 255. When you use sign-magnitude numbers, you reduce the largest magnitude from 255 to 127 because you need to represent both positive and negative quantities. For instance, the negative numbers are

$$1000\ 0001 \qquad (-1)$$

to

$$1111\ 1111 \qquad (-127)$$

The positive numbers are

$$0000\ 0001 \qquad (+1)$$

to

$$0111\ 1111 \qquad (+127)$$

The largest magnitude is 127, approximately half of what is for unsigned binary numbers. As long as your input data is in the range of $-127$ to $+127$, you can use 8-bit arithmetic. The programmer still must check sums for an overflow because all 8-bit answers are between $-127$ and $+127$.

If the data has magnitudes greater than 127, then 16-bit arithmetic may work. With 16-bit numbers, the negative numbers are from

$$1000\ 0000\ 0000\ 0001 \qquad (-1)$$

to

$$1111\ 1111\ 1111\ 1111 \qquad (-32, 767)$$

and the positive numbers are from

$$0000\ 0000\ 0000\ 0001 \qquad (+1)$$

to

$$0111\ 1111\ 1111\ 1111 \qquad (+32, 767)$$

Again, you can see that the largest magnitude is approximately half that of unsigned binary numbers. Unless you actually need $+$ and $-$ signs to represent your data, you are better off using unsigned binary.

The main advantage of sign-magnitude numbers is their simplicity. Negative numbers are identical to positive numbers, except for the sign bit. Because of this, you can easily find the magnitude by deleting the sign bit and converting the remaining bits to their decimal equivalents. Unfortunately, sign-magnitude numbers have limited use because they require complicated arithmetic circuits. If you don't have to add or subtract the data, sign-magnitude numbers are acceptable. For instance, sign-magnitude numbers are often used in *analog-to-digital* (A/D) conversions (explained in a later chapter).

## 5-5 2'S COMPLEMENT REPRESENTATION

There is a rather unusual number system that leads to the simplest logic circuits for performing arithmetic. Known as *2's complement representation*, this system dominates microcomputer architecture and programming.

### 1's Complement

The 1's complement of a binary number is the number that results when we complement each bit. Figure 5–1 shows how to produce the 1's complement with logic circuits. Since each bit drives an inverter, the 4-bit output is the 1's complement of the 4-bit input. For instance, if the input is

$$X_3X_2X_1X_0 = 1000$$

the 1's complement is

$$\overline{X}_3\overline{X}_2\overline{X}_1\overline{X}_0 = 0111$$

The same principle applies to binary numbers of any length: complement each bit to obtain the 1's complement. More examples of 1's complements are

$$1010 \rightarrow 0101$$
$$1110\ 1100 \rightarrow 0001\ 0011$$
$$0011\ 1111\ 0000\ 0110 \rightarrow 1100\ 0000\ 1111\ 1001$$

### 2's Complement

The 2's complement is the binary number that results when we add 1 to the 1's complement. As a formula:

$$2\text{'s complement} = 1\text{'s complement} + 1$$

For instance, to find the 2's complement of 1011, proceed like this:

$$1011 \rightarrow 0100 \qquad (1\text{'s complement})$$
$$0100 + 1 = 0101 \qquad (2\text{'s complement})$$

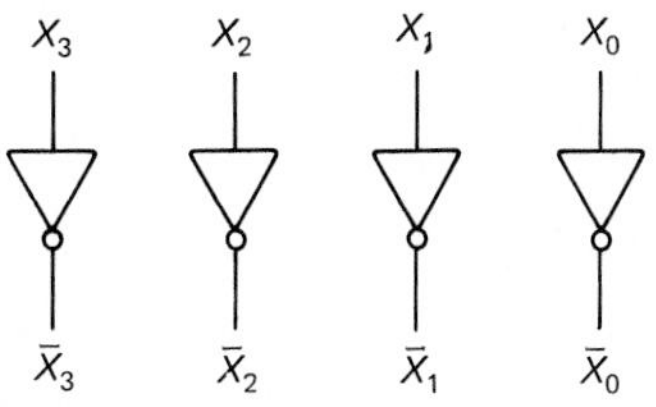

**FIG. 5–1.** Inverters produce the 1's complement

Instead of adding 1, you can visualize the next reading on a binary odometer. So, after obtaining the 1's complement 0100, ask yourself what comes next on a binary odometer. The answer is 0101.

Here are more examples of the 2's complements:

| Number | → | 1's complement | → | 2's complement |
|---|---|---|---|---|
| 1110 1100 | → | 0001 0011 | → | 0001 0100 |
| 1000 0001 | → | 0111 1110 | → | 0111 1111 |
| 0011 0110 | → | 1100 1001 | → | 1100 1010 |

## Back to the Odometer

The binary odometer is a marvelous way to understand 2's complement representation. By examining the numbers of a binary odometer, we can see how the typical microcomputer represents positive and negative numbers. With a binary odometer, all bits eventually reset to 0s. Some readings before and after a complete reset look like this:

| | |
|---|---|
| 1000 | (−8) |
| 1001 | (−7) |
| 1010 | (−6) |
| 1011 | (−5) |
| 1100 | (−4) |
| 1101 | (−3) |
| 1110 | (−2) |
| 1111 | (−1) |
| 0000 | ( 0) |
| 0001 | (+1) |
| 0010 | (+2) |
| 0011 | (+3) |
| 0100 | (+4) |
| 0101 | (+5) |
| 0110 | (+6) |
| 0111 | (+7) |

Binary 1101 is the reading 3 miles before reset, 1110 occurs 2 miles before reset, and 1111 indicates 1 mile before reset. Then, 0001 is the reading 1 mile after reset, 0010 occurs 2 miles after reset, and 0011 indicates 3 miles after reset.

## Positive and Negative Numbers

"Before" and "after" are synonymous with "negative" and "positive." Figure 5–2 illustrates this idea with the number line of basic algebra: 0 marks the origin, positive numbers are on the right, and negative numbers are on the left. The odometer readings are

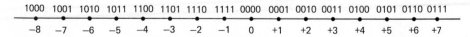

**FIG. 5–2.** Representing decimal numbers as 2's complements

the binary equivalents of decimal numbers: 1000 is the binary equivalent of $-8$, 1001 stands for $-7$, 1010 stands for $-6$, and so on.

The odometer readings in Fig. 5–2 demonstrate how positive and negative numbers are stored in a typical microcomputer. Here are two important ideas to notice about these odometer readings. First, the MSB is the sign bit: 0 represents a $+$ sign, and 1 stands for a $-$ sign. Second, the negative numbers in Fig. 5–2 are the 2's complements of the positive numbers, as you can see in the following:

| Magnitude | Positive | Negative |
|-----------|----------|----------|
| 1 | 0001 | 1111 |
| 2 | 0010 | 1110 |
| 3 | 0011 | 1101 |
| 4 | 0100 | 1100 |
| 5 | 0101 | 1011 |
| 6 | 0110 | 1010 |
| 7 | 0111 | 1001 |
| 8 | — | 1000 |

Except for the last entry, the positive and negative numbers are 2's complements of each other.

In other words, you can take the 2's complement of a positive binary number to find the corresponding negative binary number. For instance:

$$3 \rightarrow 0011$$
$$-3 \leftarrow 1101$$

After taking the 2's complement of 0011, we get 1101, which represents $-3$. The principle also works in reverse:

$$-7 \rightarrow 1001$$
$$+7 \leftarrow 0111$$

After taking the 2's complement of 1001, we obtain 0111, which represents $+7$.

What does the foregoing mean? It means that taking the 2's complement is equivalent to *negation*, changing the sign of the number. Why is this important? Because it's easy to build a logic circuit that produces the 2's complement. Whenever this circuit takes the 2's complement, the output is the negative of the input. This key idea leads to an incredibly simple arithmetic circuit that can add and subtract.

In summary, here are the things to remember about 2's complement representation:

1. Positive numbers always have a sign bit of 0, and negative numbers always have a sign bit of 1.
2. Positive numbers are stored in sign-magnitude form.
3. Negative numbers are stored as 2's complements.
4. Taking the 2's complement is equivalent to a sign change.

## Appendix 2

We need a fast way to express numbers in 2's complement representation. Appendix 2 lists all 8-bit numbers in positive and negative form. You will come to love this Appendix if you have to work a lot with negative numbers. By reading either the positive or negative column, you can quickly convert from decimal to the 2's complement representation, or vice versa.

Here are some examples of using Appendix 2 to convert from decimal to 2's complement representation:

$$+23 \rightarrow 17H \rightarrow 0001\ 0111$$
$$-48 \rightarrow DOH \rightarrow 1101\ 0000$$
$$-93 \rightarrow A3H \rightarrow 1010\ 0011$$

Of course, you can use Appendix 2 in reverse. Here are examples of converting from 2's complement representation to decimal:

$$0111\ 0111 \rightarrow 77H \rightarrow +119$$
$$1110\ 1000 \rightarrow E8H \rightarrow -24$$
$$1001\ 0100 \rightarrow 94H \rightarrow -108$$

A final point. Look at the last two entries in Appendix 2. As you see, $+127$ is the largest positive number in 2's complement representation, and $-128$ is the largest negative number. Similarly, in the 4-bit odometer discussed earlier, $+7$ was the largest positive number, and $-8$ was the largest negative number. The largest negative number has a magnitude that is one greater than the largest positive number. This *slight asymmetry* of 2's complement representation has no particular meaning, but it is something to keep in mind when we discuss overflows.

## ☐ EXAMPLE 5—8

A first-generation microcomputer stores 1 byte at each address or memory location. Show how the following decimal numbers are stored with the use of 2's complement representation: $+20$, $-35$, $+47$, $-67$, $-98$, $+112$, and $-125$. The first byte starts at address 2000.

## ☐ SOLUTION

With Appendix 2, we have

| Address | Binary contents | Hexadecimal contents | Decimal contents |
|---------|-----------------|----------------------|------------------|
| 2000 | 0001 0100 | 14H | $+20$ |
| 2001 | 1101 1101 | DDH | $-35$ |
| 2002 | 0010 1111 | 2DH | $+47$ |
| 2003 | 1011 1101 | BDH | $-67$ |
| 2004 | 1001 1110 | 9EH | $-98$ |
| 2005 | 0111 0000 | 70H | $+112$ |
| 2006 | 1000 0011 | 83H | $-125$ |

The computer actually stores binary 0001 0100 at address 2000. Instead of saying 0001 0100, however, we may prefer to say that it stores 14H. To anyone who knows the hexadecimal code, 14H means the same thing as 0001 0100, but 14H is much easier to say. To the person on the street who knows only the decimal code, we would say that $+20$ is stored at address 2000.

As you see, understanding computer operation requires knowledge of the different codes being used. Get this into your head, and you are on the way to understanding how computers work.

□ EXAMPLE 5—9

Express −19,750 in 2's complement representation. Then show how this number is stored starting at address 2000. Use hexadecimal notation to compress the data.

□ SOLUTION

The number −19,750 is outside the range of Appendix 2, so we have to fall back on Appendix 1. Start by converting the magnitude to binary form. With Appendix 1, we have

$$19,750 \rightarrow 4D26H \rightarrow 0100\ 1101\ 0010\ 0110$$

Now, take the 2's complement to obtain the negative value:

$$1011\ 0010\ 1101\ 1001 + 1 = 1011\ 0010\ 1101\ 1010$$

This means that

$$-19,750 \rightarrow 1011\ 0010\ 1101\ 1010$$

In hexadecimal notation, this is expressed

$$1011\ 0010\ 1101\ 1010 \rightarrow B2DAH$$

The memory of a first-generation microcomputer is organized in bytes. Each address or memory location contains 1 byte. Therefore, a first-generation microcomputer has to break a 16-bit number like B2DA into 2 bytes: an upper byte of B2 and a lower byte of DA. The lower byte is stored at the lower address and the upper byte, at the next-higher address like this:

| Address | Binary contents | Hexadecimal contents |
|---------|-----------------|----------------------|
| 2000    | 1101 1010       | DA                   |
| 2001    | 1011 0010       | B2                   |

The same approach, lower byte first and upper byte second, is used with first-generation microcomputers such as the Apple II and TRS-80.

## 5-6 2'S COMPLEMENT ARITHMETIC

Early computers used sign-magnitude numbers for positive and negative values. This led to complicated arithmetic circuits. Then an engineer discovered that 2's complement representation could simplify arithmetic *hardware*. (This refers to the electronic, magnetic, and mechanical devices of a computer.) Since then, 2's complement representation has become a universal code for processing positive and negative numbers.

### Help from the Binary Odometer

Addition and subtraction can be visualized in terms of a binary odometer. When you add a positive number, this is equivalent to advancing the odometer reading. When you add a negative number, this has the effect of turning the odometer backward. Likewise, subtraction of a positive number reverses the odometer, but subtraction of a negative number advances it. As you read the following discussion of addition and subtraction, keep the binary odometer in mind because it will help you to understand what's going on.

## Addition

Let us take a look at how binary numbers are added. There are four possible cases: both numbers positive, a positive number and a smaller negative number, a negative number and a smaller positive number, and both numbers negative. Let us go through all four cases for a complete coverage of what happens when a computer adds numbers.

**CASE 1** Both positive. Suppose that the numbers are +83 and +16. With Appendix 2, these numbers are converted as follows:

$$+83 \rightarrow 0101\ 0011$$
$$+16 \rightarrow 0001\ 0000$$

Then, here is how the addition appears:

$$\begin{array}{r} 83 \\ +\ 16 \\ \hline 99 \end{array} \qquad \begin{array}{r} 0101\ 0011 \\ +\ 0001\ 0000 \\ \hline 0110\ 0011 \end{array}$$

Nothing unusual happens here. Column-by-column addition produces a binary answer of 0110 0011. Mentally convert this to 63H. Now, look at Appendix 2 to get

$$63H \rightarrow 99$$

This agrees with the decimal sum.

**CASE 2** Positive and smaller negative. Suppose that the numbers are +125 and −68. With Appendix 2, we obtain

$$+125 \rightarrow 0111\ 1101$$
$$-68 \rightarrow 1011\ 1100$$

The computer will fetch these numbers from its memory and send them to an adding circuit. The numbers are then added column by column, including the sign bits to get

$$\begin{array}{r} 125 \\ +\ (-68) \\ \hline 57 \end{array} \qquad \begin{array}{r} 0111\ 1101 \\ +\ 1011\ 1100 \\ \hline 1\ 0011\ 1001 \rightarrow 0011\ 1001 \end{array}$$

With 8-bit arithmetic, you *disregard the final carry* into the ninth column. The reason is related to the binary odometer, which ignores final carries. In other words, when the eighth wheel resets, it does not generate a carry because there is no ninth wheel to receive the carry. You can convert the binary answer to decimal as follows:

$$0011\ 1001 \rightarrow 39H \qquad \text{(mental conversion)}$$
$$39H \rightarrow +57 \qquad \text{(look in Appendix 2)}$$

**CASE 3** Positive and larger negative. Let's use +37 and −115. Appendix 2 gives these 2's complement representations:

$$+37 \rightarrow 0010\ 0101$$
$$-115 \rightarrow 1000\ 1101$$

Then the addition looks like this:

$$
\begin{array}{r}
+37 \\
+\ (-115) \\
\hline
-78
\end{array}
\qquad
\begin{array}{r}
0010\ 0101 \\
+\ 1000\ 1101 \\
\hline
1011\ 0010
\end{array}
$$

Next, verify the binary answer as follows:

$$1011\ 0010 \rightarrow \text{B2H} \qquad \text{(mental conversion)}$$
$$\text{B2H} \rightarrow -78 \qquad \text{(look in Appendix 2)}$$

Incidentally, mentally converting to hexadecimal before reference to the appendix is an optional step. Most people find it easier to locate B2H in Appendix 2 than 1011 0010. It only saves a few seconds, but it adds up when you have to do a lot of binary-to-decimal conversions.

**CASE 4** Both negative. Assume that the numbers are $-43$ and $-78$. In 2's complement representation, the numbers are

$$-43 \rightarrow 1101\ 0101$$
$$-78 \rightarrow 1011\ 0010$$

The addition is

$$
\begin{array}{r}
-43 \\
+\ (-78) \\
\hline
-121
\end{array}
\qquad
\begin{array}{r}
1101\ 0101 \\
+\ 1011\ 0010 \\
\hline
1\ 1000\ 0111 \rightarrow 1000\ 0111
\end{array}
$$

Again, we ignore the final carry because it's meaningless in 8-bit arithmetic. The remaining 8 bits convert as follows:

$$1000\ 0111 \rightarrow 83\text{H}$$
$$83\text{H} \rightarrow -121$$

This agrees with the answer we obtained by direct decimal addition.

## Conclusion

We have exhausted the possibilities. In every case, 2's complement addition works. In other words, as long as positive and negative numbers are expressed in 2's complement representation, an adding circuit will automatically produce the correct answer. (This assumes the decimal sum is within the $-128$ to $+127$ range. If not, you get an overflow, which we will discuss later.)

## Subtraction

The format for subtraction is

$$
\begin{array}{r}
\text{Minuend} \\
-\ \text{Subtrahend} \\
\hline
\text{Difference}
\end{array}
$$

There are four cases: both numbers positive, a positive number and a smaller negative number, a negative number and a smaller positive number, and both numbers negative.

The question now is *how can we use an adding circuit* to do subtraction. By trickery, of course. From algebra, you already know that adding a negative number is equivalent to subtracting a positive number. If we take the 2's complement of the subtrahend, addition of the complemented subtrahend gives the correct answer. Remember that the 2's complement is equivalent to negation. One way to remove all doubt about this critical idea is to analyze the four cases that can arise during a subtraction.

**CASE 1** Both positive. Suppose that the numbers are $+83$ and $+16$. In 2's complement representation, these numbers appear as

$$+83 \rightarrow 0101\ 0011$$
$$+16 \rightarrow 0001\ 0000$$

To subtract $+16$ from $+83$, the computer will send the $+16$ to a 2's complementer circuit to produce

$$-16 \rightarrow 1111\ 0000$$

Then it will add $+83$ and $-16$ as follows:

$$
\begin{array}{r}
83 \\
+\ (-16) \\
\hline
67
\end{array}
\qquad
\begin{array}{r}
0101\ 0011 \\
+\ 1111\ 0000 \\
\hline
1\ 0100\ 0011 \rightarrow 0100\ 0011
\end{array}
$$

The binary answer converts like this:

$$0100\ 0011 \rightarrow 43\text{H}$$
$$43\text{H} \rightarrow +67$$

**CASE 2** Positive and smaller negative. Suppose that the minuend is $+68$ and the subtrahend is $-27$. In 2's complement representation, these numbers appear as

$$+68 \rightarrow 0100\ 0100$$
$$-27 \rightarrow 1110\ 0101$$

The computer sends $-27$ to a 2's complementer circuit to produce

$$+27 \rightarrow 0001\ 1011$$

Then it adds $+68$ and $+27$ as follows:

$$
\begin{array}{r}
68 \\
+\ 27 \\
\hline
95
\end{array}
\qquad
\begin{array}{r}
0100\ 0100 \\
+\ 0001\ 1011 \\
\hline
0101\ 1111
\end{array}
$$

The binary answer converts to decimal as follows:

$$0101\ 1111 \rightarrow 5\text{FH}$$
$$5\text{FH} \rightarrow +95$$

**CASE 3** Positive and larger negative. Let's use a minuend of $+14$ and a subtrahend of $-108$. Appendix 2 gives these 2's complement representations:

$$+14 \rightarrow 0000\ 1110$$
$$-108 \rightarrow 1001\ 0100$$

The computer produces the 2's complement of $-108$:

$$+108 \rightarrow 0110\ 1100$$

Then it adds the numbers like this:

$$
\begin{array}{r}
14 \\
+\ 108 \\
\hline
122
\end{array}
\qquad
\begin{array}{r}
0000\ 1110 \\
+\ 0110\ 1100 \\
\hline
0111\ 1010
\end{array}
$$

The binary answer converts to decimal like this:

$$0111\ 1010 \rightarrow 7\text{AH}$$
$$7\text{AH} \rightarrow +122$$

**CASE 4** Both negative. Assume that the numbers are $-43$ and $-78$. In 2's complement representation, the numbers are

$$-43 \rightarrow 1101\ 0101$$
$$-78 \rightarrow 1011\ 0010$$

First, take the 2's complement of $-78$ to get

$$+78 \rightarrow 0100\ 1110$$

Then add to obtain

$$
\begin{array}{r}
-43 \\
+\ 78 \\
\hline
35
\end{array}
\qquad
\begin{array}{r}
1101\ 0101 \\
+\ 0100\ 1110 \\
\hline
1\ 0010\ 0011 \rightarrow 0010\ 0011
\end{array}
$$

Then

$$0010\ 0011 \rightarrow 23\text{H}$$
$$23\text{H} \rightarrow +35$$

## Overflow

We have covered all cases of addition and subtraction. As shown, 2's complement arithmetic works and is the standard method used in microcomputers. In 8-bit arithmetic, the only thing that can go wrong is a sum outside the range of $-128$ to $+127$. When this happens, there is an overflow into the sign bit, causing a sign change. With the typical microcomputer, the programmer has to write instructions that check for this change in the sign bit.

Let's take a look at overflow problems. Assume that both input numbers are in the range of $-128$ to $+127$. If a positive and a negative number are being added, an overflow is impossible because the answer is always less than the larger of the two numbers being added. Trouble can arise only when the arithmetic circuit adds two positive numbers or two negative numbers. Then, it is possible for the sum to be outside the range of $-128$ to $+127$. (Subtraction is included in the foregoing discussion because the arithmetic circuit adds the complemented subtrahend.)

**CASE 1** Two positive numbers. Suppose that the numbers being added are $+100$ and $+50$. The decimal sum is $+150$, so an overflow occurs into the MSB position. This overflow forces the sign bit of the answer to change. Here is how it looks:

$$
\begin{array}{r}
100 \\
+\ \ 50 \\
\hline
150
\end{array}
\qquad
\begin{array}{r}
0110\ 0100 \\
+\ 0011\ 0010 \\
\hline
1001\ 0110
\end{array}
$$

The sign bit is negative, despite the fact that we added two positive numbers. Therefore, the overflow has produced an incorrect answer.

**CASE 2** Two negative numbers. Suppose that the numbers are $-85$ and $-97$. Then

$$
\begin{array}{r}
-85 \\
+\ (-97) \\
\hline
182
\end{array}
\qquad
\begin{array}{r}
1010\ 1011 \\
+\ 1001\ 1111 \\
\hline
1\ 0100\ 1010 \rightarrow 0100\ 1010
\end{array}
$$

The 8-bit answer is 0100 1010. The sign bit is positive, but we know that the right answer should contain a negative sign bit because we added two negative numbers.

## What to Do with an Overflow

Overflows are a *software* problem, not a hardware problem. (*Software* means a program or list of instructions telling the computer what to do.) The programmer must test for an overflow after each addition or subtraction. A change in the sign bit is easy to detect. All the programmer does is include instructions that compare the sign bits of the two numbers being added. When these are the same, the sign bit of the answer is compared to either of the preceding sign bits. If the sign bit is different, more instructions tell the computer to change the processing to 16-bit arithmetic. You will learn more about overflows, 16-bit arithmetic, and related topics if you study *assembly-language* programming.

☐ EXAMPLE 5—10

How would an 8-bit microcomputer process this:

$$
\begin{array}{r}
18,357 \\
-\ 12,618 \\
\hline
?
\end{array}
$$

☐ SOLUTION

It would use *double-precision* arithmetic, synonymous with 16-bit arithmetic. This arithmetic is used with 16-bit numbers in this form:

$$X_{15}X_{14}X_{13}X_{12} \quad X_{11}X_{10}X_9X_8 \quad X_7X_6X_5X_4 \quad X_3X_2X_1X_0$$

Numbers like these have an upper byte $X_{15} \cdots X_8$ and a lower byte $X_7 \cdots X_0$. To perform 16-bit arithmetic, an 8-bit microcomputer has to operate on each byte separately. The idea is similar to Example 5–3, where the lower bytes were added and then the upper bytes.

Here is how it is done. With Appendix 1, we have

$$18,357 \rightarrow 47\text{B5H} \rightarrow 0100\ 0111\ 1011\ 0101$$
$$12,618 \rightarrow 314\text{AH} \rightarrow 0011\ 0001\ 0100\ 1010$$

The 2's complement of 12,618 is

$$-12{,}618 \rightarrow \text{CEB6H} \rightarrow 1100\ 1110\ 1011\ 0110$$

The addition is carried out in two steps of 8-bit arithmetic. First, the lower bytes are added:

$$
\begin{array}{r}
1011\ 0101 \\
+\ 1011\ 0110 \\
\hline
1\ \ 0110\ 1011
\end{array}
\rightarrow X_8\ X_7 X_6 X_5 X_4\ X_3 X_2 X_1 X_0
$$

The computer will store $X_7 \cdots X_0$. The carry $X_8$ is used in the addition of the upper bytes. Now, the computer adds the upper bytes plus the carry as follows:

$$
\begin{array}{r}
1 \leftarrow X_8 \\
0100\ 0111 \\
+\ 1100\ 1110 \\
\hline
1\ 0001\ 0110
\end{array}
\rightarrow 0001\ 0110
$$

To obtain the final answer, the two 8-bit answers are combined:

$$0001\ 0110\ 0110\ 1011$$

Notice that the MSB is 0, which means that the answer is positive. With Appendix 1, we can convert this answer to decimal form:

$$0001\ 0110\ 0110\ 1011 \rightarrow \text{166BH} \rightarrow +5739$$

# 5·7 ARITHMETIC BUILDING BLOCKS

We are on the verge of seeing a logic circuit that performs 8-bit arithmetic on positive and negative numbers. But first we need to cover three basic circuits that will be used as building blocks. These building blocks are the half-adder, the full-adder, and the controller inverter. Once you understand how these work, it is only a short step to see how it all comes together, that is, how a computer is able to add and subtract binary numbers of any length.

## Half-Adder

When we add two binary numbers, we start with the least-significant column. This means that we have to add two bits with the possibility of a carry. The circuit used for this is called a *half-adder*. Figure 5–3 shows how to build a half-adder. The output of the exclusive–OR gate is called the *SUM*, while the output of the AND gate is the *CARRY*.

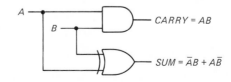

**FIG. 5–3.** Half-adder

**TABLE 5–2** *Half-Adder Truth Table*

| A | B | CARRY | SUM |
|---|---|-------|-----|
| 0 | 0 | 0 | 0 |
| 0 | 1 | 0 | 1 |
| 1 | 0 | 0 | 1 |
| 1 | 1 | 1 | 0 |

The AND gate produces a high output only when both inputs are high. The exclusive–OR gate produces a high output if either input, but not both, is high. Table 5–2 shows the truth table of a half-adder.

When you examine each entry in Table 5–2, you are struck by the fact that a half-adder performs binary addition. It does electronically what we do mentally when we add 2 bits. Here is the action, entry by entry:

**First Entry**
Input: $A = 0$ and $B = 0$
Human response: 0 plus 0 is 0 with a carry of 0
Half-adder response: SUM = 0 and CARRY = 0

**Second Entry**
Input: $A = 0$ and $B = 1$
Human response: 0 plus 1 is 1 with a carry of 0
Half-adder response: SUM = 1 and CARRY = 0

**Third Entry**
Input: $A = 1$ and $B = 0$
Human response: 1 plus 0 is 1 with a carry of 0
Half-adder response: SUM = 1 and CARRY = 0

**Fourth Entry**
Input: $A = 1$ and $B = 1$
Human response: 1 plus 1 is 0 with a carry of 1
Half-adder response: SUM = 0 and CARRY = 1

As you see, the half-adder mimics our brain processes in adding bits. The only difference is the half-adder is about a million times faster than we are.

## Full-Adder

For the higher-order columns, we have to use a full-adder, a logic circuit that can add 3 bits at a time. The third bit is the carry from a lower column. This implies that we need a logic circuit with three inputs and two outputs, similar to the full-adder shown in Fig. 5–4a. (Other designs are possible. This one is the simplest.)

Table 5–3 shows the truth table of a full-adder. You can easily check this truth table for its validity. For instance, CARRY is high in Fig. 5–4a when two or more of the *ABC* inputs are high; this agrees with the "CARRY" column in Table 5–3. Also, when an odd number of high *ABC* inputs drives the exclusive–OR gate, it produces a high output; this verifies the SUM column of the truth table.

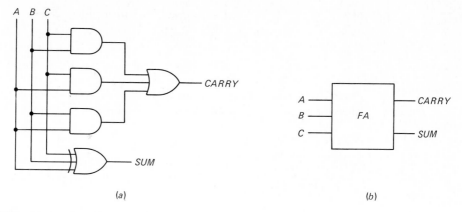

**FIG. 5–4.** Full-adder

When you examine each entry in Table 5–3, you can see that a full-adder performs binary addition on 3 bits. Here is the action for some selected entries to illustrate the similarity between human data processing and electronic data processing:

**First Entry**

Input: $A = 0$, $B = 0$, and $C = 0$

Human response: 0 plus 0 plus 0 is 0 with a carry of 0

Full-adder response: SUM = 0 and CARRY = 0

**Second Entry**

Input: $A = 0$, $B = 0$, and $C = 1$

Human response: 0 plus 0 plus 1 is 1 with a carry of 0

Full-adder response: SUM = 1 and CARRY = 0

**Fourth Entry**

Input: $A = 0$, $B = 1$, and $C = 1$

Human response: 0 plus 1 plus 1 is 0 with a carry of 1

Full-adder response: SUM = 0 and CARRY = 1

**Last Entry**

Input: $A = 1$, $B = 1$, and $C = 1$

Human response: 1 plus 1 plus 1 is 1 with a carry of 1

Full-adder response: SUM = 1 and CARRY = 1

**TABLE 5–3** *Full-Adder Truth Table*

| A | B | C | CARRY | SUM |
|---|---|---|-------|-----|
| 0 | 0 | 0 | 0 | 0 |
| 0 | 0 | 1 | 0 | 1 |
| 0 | 1 | 0 | 0 | 1 |
| 0 | 1 | 1 | 1 | 0 |
| 1 | 0 | 0 | 0 | 1 |
| 1 | 0 | 1 | 1 | 0 |
| 1 | 1 | 0 | 1 | 0 |
| 1 | 1 | 1 | 1 | 1 |

So you can see, the full-adder captures in electronic circuitry those brain processes that we use when adding 3 bits at a time. The full-adder can do more than a million additions per second. Besides that, it never gets tired or bored, or asks for a raise.

Incidentally, Fig. 5–4$b$ shows the block-diagram symbol of a full-adder. It has two inputs $A$ and $B$, plus a third input $C$, usually called the *CARRY IN* because it comes from a lower-order column. There are two outputs, *SUM* and *CARRY*. The output CARRY is called the *CARRY OUT* because it goes to the next higher column.

## Controlled Inverter

Figure 5–5 shows a *controlled inverter*. When INVERT is low, it transmits the 8-bit input to the output; when INVERT is high, it transmits the 1's complement. For instance, if the input number is

$$A_7 \cdots A_0 = 0110\ 1110$$

a low INVERT produces

$$Y_7 \cdots Y_0 = 0110\ 1110$$

But a high INVERT results in

$$Y_7 \cdots Y_0 = 1001\ 0001$$

The controlled inverter is important because it is a step in the right direction. During a subtraction, we first need to take the 2's complement of the subtrahend. Then we can add the complemented subtrahend to obtain the answer. With a controlled inverter, we can produce the 1's complement. There is an easy way to get the 2's complement, discussed in the next section. So, we now have all the building blocks: half-adder, full-adder, and controlled inverter.

## 5-8 THE ADDER-SUBTRACTER

We can connect full-adders as shown in Fig. 5–6 to add or subtract binary numbers. The circuit is laid out from right to left, similar to the way we add binary numbers. Therefore, the least-significant column is on the right, and the most-significant column is on the left. The boxes labeled FA are full-adders. (Some adding circuits use a half-adder instead of a full-adder in the least-significant column.)

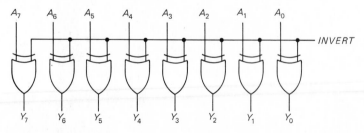

**FIG. 5–5.** Controlled inverter

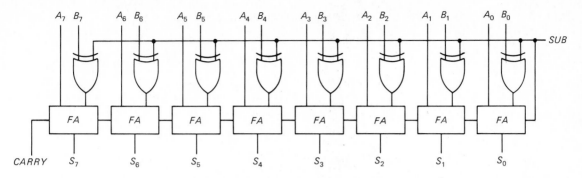

**FIG. 5–6.** Binary adder-subtractor

The CARRY OUT from each full-adder is the CARRY IN to the next-higher full-adder. The numbers being processed are $A_7 \cdots A_0$ and $B_7 \cdots B_0$, while the answer is $S_7 \cdots S_0$. With 8-bit arithmetic, the final carry is ignored for reasons given earlier. With 16-bit arithmetic, the final carry is the carry into the addition of the upper bytes.

## Addition

Here is how an addition appears:

$$
\begin{array}{r}
A_7A_6A_5A_4 \; A_3A_2A_1A_0 \\
+ \; B_7B_6B_5B_4 \; B_3B_2B_1B_0 \\
\hline
S_7S_6S_5S_4 \; S_3S_2S_1S_0
\end{array}
$$

During an addition, the *SUB* signal is deliberately kept in the low state. Therefore, the binary number $B_7 \cdots B_0$ passes through the controlled inverter with no change. The full-adders then produce the correct output sum. They do this by adding the bits in each column, passing carries to the next higher column, and so on. For instance, starting at the LSB, the full-adder adds $A_0$, $B_0$, and SUB. This produces a SUM of $S_0$ and a CARRY OUT to the next-higher full-adder. The next-higher full-adder then adds $A_1$, $B_1$, and the CARRY IN to produce $S_1$ and a CARRY OUT. A similar addition occurs for each of the remaining full-adders, and the correct sum appears at the output lines.

For instance, suppose that the numbers being added are $+125$ and $-67$. Then, $A_7 \cdots A_0 = 0111\ 1101$ and $B_7 \cdots B_0 = 1011\ 1101$. This is the problem:

$$
\begin{array}{r}
0111\ 1101 \\
+ \; 1011\ 1101 \\
\hline
?
\end{array}
$$

Since SUB $= 0$ during an addition, the CARRY IN to the least-significant column is 0:

$$
\begin{array}{r}
0 \leftarrow \text{SUB} \\
0111\ 1101 \\
+ \; 1011\ 1101 \\
\hline
?
\end{array}
$$

The first full-adder performs this addition:

$$0 + 1 + 1 = 0 \quad \text{with a carry of 1}$$

The CARRY OUT of the first full-adder is the CARRY IN to the second full-adder:

$$
\begin{array}{r}
1 \quad \leftarrow \text{Carry} \\
0111\ 1101 \\
+\ 1011\ 1101 \\
\hline
0
\end{array}
$$

In the second column

$$1 + 0 + 0 = 1 \quad \text{with a carry of 0}$$

The carry goes to the third full-adder:

$$
\begin{array}{r}
0 \quad \leftarrow \text{Carry} \\
0111\ 1101 \\
+\ 1011\ 1101 \\
\hline
10
\end{array}
$$

In a similar way, the remaining full-adders add their 3 input bits until we arrive at the last full-adder:

$$
\begin{array}{r}
1 \quad\quad\quad\quad \leftarrow \text{Carry} \\
0111\ 1101 \\
+\ 1011\ 1101 \\
\hline
0011\ 1010
\end{array}
$$

When the CARRY IN to the MSB appears, the full-adder produces

$$1 + 0 + 1 = 0 \quad \text{with a carry of 1}$$

The addition process ends with a final carry:

$$
\begin{array}{r}
0111\ 1101 \\
+\ 1011\ 1101 \\
\hline
1\ \ 0011\ 1010
\end{array}
$$

During 8-bit arithmetic, this last carry is ignored as previously discussed; therefore, the answer is

$$S_7 \cdots S_0 = 0011\ 1010$$

This answer is equivalent to decimal $+58$, which is the algebraic sum of the numbers we started with: $+125$ and $-67$.

## Subtraction

Here is how a subtraction appears:

$$
\begin{array}{r}
A_7A_6A_5A_4\ A_3A_2A_1A_0 \\
-\ B_7B_6B_5B_4\ B_3B_2B_1B_0 \\
\hline
S_7S_6S_5S_4\ S_3S_2S_1S_0
\end{array}
$$

*Full-Adder* A logic circuit with three inputs and two outputs. The circuit adds 3 bits at a time, giving a sum and a carry output.

*Half-Adder* A logic circuit with two inputs and two outputs. It adds 2 bits at a time, producing a sum and a carry output.

*Hardware* The electronic, magnetic, and mechanical devices used in a computer or digital system.

*LSB* Least-significant bit.

*Magnitude* The absolute or unsigned value of a number.

*Microprocessor* A digital IC that combines the arithmetic and control sections of a computer.

*MSB* Most-significant bit.

*Overflow* An unwanted carry that produces an answer outside the valid range of the numbers being represented.

*Software* A program or programs. The instructions that tell a computer how to process the data.

*2's Complement* The binary number that results when 1 is added to the 1's complement.

## PROBLEMS

*SECTION 5-1:*

**5-1.** Give the sum in each of the following:
  **a.** $3_8 + 7_8 = ?$
  **b.** $5_8 + 6_8 = ?$
  **c.** $4_{16} + C_{16} = ?$
  **d.** $8_{16} + F_{16} = ?$

**5-2.** Work out each of these binary sums:
  **a.** 0000 1111 + 0011 0111
  **b.** 0001 0100 + 0010 1001
  **c.** 0001 1000 1111 0110 + 0000 1111 0000 1000

**5-3.** Show the binary addition of $750_{10}$ and $538_{10}$ using 16-bit numbers.

*SECTION 5-2:*

**5-4.** Subtract the following: 0100 1111 − 0000 0101.

**5-5.** Show this subtraction in binary form: $47_{10} - 23_{10}$.

*SECTION 5-3:*

**5–6.** Indicate which of the following produces an overflow with 8-bit unsigned arithmetic:

  **a.** $45_{10} + 78_{10}$
  **b.** $34_8 + 56_8$
  **c.** $CF_{16} + 67_{16}$

*SECTION 5-4:*

**5–7.** Express each of the following in 8-bit sign-magnitude form:

  **a.** $+23$
  **b.** $+123$
  **c.** $-56$
  **d.** $-107$

**5–8.** Convert each of the following sign-magnitude numbers into decimal equivalents:

  **a.** 0011 0110
  **b.** 1010 1110
  **c.** 1111 1000
  **d.** 1000 1100 0111 0101

*SECTION 5-5:*

**5–9.** Express the 1's complement of each of the following in hexadecimal notation:

  **a.** 23H
  **b.** 45H
  **c.** C9H
  **d.** FDH

**5–10.** What is the 2's complement of each of these:

  **a.** 0000 1111
  **b.** 0101 1010
  **c.** 1011 1110
  **d.** 1111 0000 1111 0000

**5–11.** Use Appendix 2 to convert each of the following to 2's complement representation:

  **a.** $+78$
  **b.** $-23$
  **c.** $-90$
  **d.** $-121$

**5–12.** Decode the following numbers into decimal values, using Appendix 2:

  **a.** FCH
  **b.** 34H
  **c.** 9AH
  **d.** B4H

**5–13.** Show the 8-bit addition of these decimal numbers in 2's complement representation:

    **a.** +45, +56
    **b.** +89, −34
    **c.** +67, −98

**5–14.** Show the 8-bit subtraction of these decimal numbers in 2's complement representation:

    **a.** +54, +65
    **b.** +68, −43
    **c.** +16, −38
    **d.** −28, −65

*SECTION 5-7:*

**5–15.** Suppose that FD34H is the input to a 16-bit controlled inverter. What is the inverted output in hexadecimal notation? In binary?

*SECTION 5-8:*

**5–16.** Expressed in hexadecimal notation, the two input numbers in Fig. 5–6 are 7FH and 4DH. What is the output when SUB is low?

**5–17.** The input numbers in Fig. 5–6 are 0001 0010 and 1011 1111. What is the output when SUB is high?

*SECTION 5-9:*

**5–18.** Describe a program that multiples 9 × 7 using repeated addition.

# 6

# T T L
# C I R C U I T S

In 1964 Texas Instruments introduced *transistor-transistor logic* (TTL), a widely used family of digital devices. TTL is fast, inexpensive, and easy to use. In this chapter we discuss several types of TTL: standard, high-speed, low-power, Schottky, and low-power Schottky. You will learn about open-collector and three-state devices because these are used to build *buses*, the backbone of modern computers and digital systems. Since TTL uses active-low as well as active-high signals, *negative logic* may be used as well as positive logic. In turn, this leads to *assertion-level logic*, a way of drawing gates and labeling signals that enhances circuit analysis.

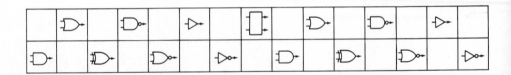

# 6-1 DIGITAL INTEGRATED CIRCUITS

Using advanced photographic techniques, a manufacturer can produce miniature circuits on the surface of a small piece of semiconductor material called a *chip*. The finished network is so small that you need a microscope to see the connections. Such a circuit is called an *integrated circuit* (IC) because the components—transistors, diodes, resistors—are an integral part of the chip. This is different from a discrete circuit, in which the components are individually connected during assembly.

## Technologies and Families

*Small-scale integration* (SSI) refers to ICs with fewer than 12 gates on the same chip. *Medium-scale integration* (MSI) means from 12 to 100 gates per chip. And *large-scale integration* (LSI) refers to more than 100 gates per chip.

The two basic techniques for manufacturing ICs are bipolar and metal-oxide semiconductor (MOS). The first fabricates bipolar transistors on a chip; the second, MOS field-effect transistors (MOSFETs). Bipolar technology is preferred for SSI and MSI because it is faster. MOS technology dominates the LSI field because more MOSFETs can be packed into the same chip area.

A digital family is a group of compatible devices with the same logic levels and supply voltages. The word *compatible* means that you can connect the output of one device to the input of another. With the devices of a digital family, you can create a wide variety of logic circuits.

## Bipolar Families

In the bipolar category are these basic families:

**Diode-transistor logic** (DTL)
**Transistor-transistor logic** (TTL)
**Emitter-coupled logic** (ECL)

DTL uses diodes and transistors; this design, once popular, is now obsolete. TTL uses transistors almost exclusively; it has become the most popular family of SSI and MSI chips. ECL, the fastest logic family, is used in high-speed applications.

## MOS Families

These families are in the MOS category:

**PMOS** p-Channel MOSFETs
**NMOS** n-Channel MOSFETs
**CMOS** Complimentary MOSFETs

PMOS, the oldest and slowest type, is becoming obsolete. NMOS dominates the LSI field in its use for microprocessors and memories. CMOS, a push-pull arrangement of n- and p-channel MOSFETs, is extensively used where low power consumption is needed, as in pocket calculators.

## 6-2 7400 DEVICES

The 7400 series, the line of TTL circuits introduced by Texas Instruments in 1964, has become the most widely used of all bipolar ICs. This TTL family contains a variety of SSI and MSI chips that allow you to build all kinds of digital circuits and systems.

### Standard TTL

Figure 6–1 shows a TTL NAND gate. The *multiple-emitter* input transistor is typical of the gates and other devices in the 7400 series. Each emitter acts like a diode; therefore, $Q_1$ and the 4-k $\Omega$ resistor act like a 2-input AND gate. The rest of the circuit inverts the signal so that the overall circuit acts like a 2-input NAND gate. The output transistors ($Q_3$ and $Q_4$) form a *totem-pole* connection (one *npn* in series with another); this kind of output stage is typical of most TTL devices. With a totem-pole output stage, either the upper or lower transistor is on. When $Q_3$ is on, the output is high; when $Q_4$ is on, the output is low.

The input voltages $A$ and $B$ are either low (ideally grounded) or high [ideally +5 volts (V)]. If $A$ or $B$ is low, the base of $Q_1$ is pulled down to approximately 0.7 V. This reduces the base voltage of $Q_2$ to almost zero. Therefore, $Q_2$ cuts off. With $Q_2$ open, $Q_4$ goes into cutoff, and the $Q_3$ base is pulled high. Since $Q_3$ acts as an emitter follower, the $Y$ output is pulled up to a high voltage.

On the other hand, when $A$ and $B$ are both high voltages, the emitter diodes of $Q_1$ stop conducting, and the collector diode goes into forward conduction. This forces $Q_2$ base to go high. In turn, $Q_4$ goes into saturation, producing a low output. Table 6–1 summarizes all input and output conditions.

Without diode $D_1$ in the circuit, $Q_3$ will conduct slightly when the output is low. To prevent this, the diode is inserted; its voltage drop keeps the base-emitter diode of $Q_3$ reverse-biased. In this way, only $Q_4$ conducts when the output is low.

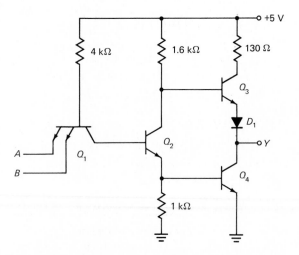

**FIG. 6–1.** Two-input TTL NAND gate

**TABLE 6–1** *Two-Input NAND Gate*

| A | B | y |
|---|---|---|
| 0 | 0 | 1 |
| 0 | 1 | 1 |
| 1 | 0 | 1 |
| 1 | 1 | 0 |

## Totem-Pole Output

Totem-pole transistors are used because they produce a *low output impedance*. Either $Q_3$ acts as an emitter follower (high output), or $Q_4$ is saturated (low output). When $Q_3$ is conducting, the output impedance is approximately 70 ohms ($\Omega$); when $Q_4$ is saturated, the output impedance is only 12 $\Omega$ (this can be calculated from information on the data sheet). Either way, the output impedance is low. This means the output voltage can change quickly from one state to the other because any stray output capacitance is rapidly charged or discharged through the low output impedance.

## Propagation Delay Time and Power Dissipation

Two quantities needed for later discussion are power dissipation and *propagation delay time*. A standard TTL gate has a power dissipation of about 10 milliwatts (mW). It may vary from this value because of signal levels, tolerances, etc., but on the average it is 10 mW per gate. The propagation delay time is the time it takes for the output of a gate to change after the inputs have changed. The propagation delay time of a TTL gate is approximately 10 nanoseconds (ns).

## Device Numbers

By varying the design of Fig. 6–1 manufacturers can alter the number of inputs and the logic function. With only few exceptions, the multiple-emitter inputs and the totem-pole outputs are used for different TTL devices. Table 6–2 lists some of the 7400 series TTL gates. For instance, the 7400 is a chip with four 2-input NAND gates in one package. Similarly, the 7402 has four 2-input NOR gates, the 7404 has six inverters, and so on.

**TABLE 6–2** *Standard TTL*

| Device Number | Description |
|---|---|
| 7400 | Quad 2-input NAND gates |
| 7402 | Quad 2-input NOR gates |
| 7404 | Hex inverter |
| 7408 | Quad 2-input AND gates |
| 7410 | Triple 3-input NAND gates |
| 7411 | Triple 3-input AND gates |
| 7420 | Dual 4-input NAND gates |
| 7421 | Dual 4-input AND gates |
| 7425 | Dual 4-input NOR gates |
| 7427 | Triple 3-input NOR gates |
| 7430 | 8-input NAND gate |
| 7486 | Quad 2-input XOR gates |

## 5400 Series

Any device in the 7400 series works over a temperature range of 0 to 70°C and over a supply range of 4.75 to 5.25 V. This is adequate for commercial applications. The 5400 series, developed for the military applications, has the same logic functions as the 7400 series, except that it works over a temperature range of −55 to 125°C and over a supply range of 4.5 to 5.5 V. Although 5400 series devices can replace 7400 series devices, they are rarely used commercially because of their much higher cost.

## High-Speed TTL

The circuit of Fig. 6–1 is called *standard* TTL. By decreasing the resistances a manufacturer can lower the internal time constants; this decreases the propagation delay time. The smaller resistances, however, increase the power dissipation. This design variation is known as *high-speed* TTL. Devices of this type are numbered 74H00, 74H01, 74H02, and so on. A high-speed TTL gate has a power dissipation around 22 mW and a propagation delay time of approximately 6 ns.

## Low-Power TTL

By increasing the internal resistances a manufacturer can reduce the power dissipation of TTL gates. Devices of this type are called *low-power* TTL and are numbered 74L00, 74L01, 74L02, etc. These devices are slower than standard TTL because of the larger internal time constants. A low-power TTL gate has a power dissipation of 1 mW and a propagation delay time of about 35 ns.

## Schottky TTL

With standard TTL, high-speed TTL, and low-power TTL, the transistors go into hard saturation, causing a surplus of carriers to be stored in the base. When you switch a transistor from saturation to cutoff, you have to wait for the extra carriers to flow out of the base. The delay is known as *saturation delay time*.

One way to reduce saturation delay time is with *Schottky* TTL. The idea is to fabricate a Schottky diode along with each bipolar transistor of a TTL circuit, as shown in Fig. 6–2. Because the Schottky diode has a forward voltage of only 0.25 to 0.4 V, it prevents the transistor from saturating fully. This virtually eliminates saturation delay time, which means better switching speed. These devices are numbered 74S00, 74S01, 74S02, and so forth.

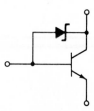

**FIG. 6–2.** Schottky diode prevents transistor saturation

Schottky TTL devices are very fast, capable of operating reliably at 100 megahertz (MHz). The 74S00 has a power dissipation around 20 mW per gate and a propagation delay time of approximately 3 ns.

## Low-Power Schottky TTL

By increasing internal resistances as well as using Schottky diodes, manufacturers have come up with a compromise between low power and high speed: *low-power Schottky* TTL. Devices of this type are numbered 74LS00, 74LS01, 74LS02, etc. A low-power Schottky gate has a power dissipation of around 2 mW and a propagation delay time of approximately 10 ns, as shown in Table 6–3.

## The Winner

Low-power Schottky TTL is the best compromise between power dissipation and saturation delay time. In other words, of the five TTL types listed in Table 6–3, low-power Schottky TTL has emerged as the favorite of digital designers. It is used for almost everything. When you must have more output current, you can fall back on standard TTL. Or, if your application requires faster switching speed, then Schottky TTL is useful. Low-power and high-speed TTL are rarely used, if at all.

## 6-3 TTL PARAMETERS

7400 series devices are guaranteed to work reliably over a temperature range of 0 to 70°C and over a supply range of 4.75 to 5.25 V. In the discussion that follows, *worst case* means the *parameters* (input current, output voltage, and so on) are measured under the worst conditions of temperature and voltage. This means maximum temperature and minimum voltage for some parameters, minimum temperature and maximum voltage for others, or whatever combination produces the worst values.

## Floating Inputs

When a TTL input is high (ideally +5 V), the emitter current is approximately zero (Fig. 6–3a). When a TTL input is *floating* (unconnected, as shown in Fig. 6–3b), no emitter current is possible because of the open circuit. Therefore, a floating TTL input is equivalent to a high output. Because of this, you sometimes see unused TTL inputs left unconnected; an open input allows the rest of the gate to function properly.

**TABLE 6–3** *TTL Power-Delay Values*

| Type | Power, mW | Delay, Time, ns |
|---|---|---|
| Low-power | 1 | 35 |
| Low-power Schottky | 2 | 10 |
| Standard | 10 | 10 |
| High-speed | 22 | 6 |
| Schottky | 20 | 3 |

There is a disadvantage to floating inputs. When you leave an input open, it acts as a small antenna. Therefore, it will pick up stray electromagnetic noise voltages. In some environments, the noise pickup is large enough to cause erratic operation of logic circuits. For this reason, most designers prefer to connect unused TTL inputs to the supply voltage.

For instance, Fig. 6–3c shows a 3-input NAND gate. The top input is unused, so it is connected to +5 V. A direct connection like this is all right with most Schottky devices (74S and 74LS) because their inputs can withstand supply overvoltages caused by switching transients. Since the top input is always high, it has no effect on the output. (*Note*: You don't ground the unused TTL input of Fig. 6–3c because then the output would remain stuck high, no matter what the values of *A* and *B*.)

Figure 6–3d shows an indirect connection to the supply through a resistor. This type of connection is used with standard, low-power, and high-speed TTL devices (74, 74L, and 74H). These older TTL devices have an absolute maximum input rating of +5.5 V. Beyond this level, the ICs may be damaged. The resistor is called a *pull-up* resistor because it serves to pull the input voltage up to a high. Most transients on the supply voltage are too short to charge the input capacitance through the pull-up resistor. Therefore, the input is protected against temporary overvoltages.

## Worst-Case Input Voltages

Figure 6–4a shows a TTL inverter with an input voltage of $V_I$ and an output voltage of $V_O$. When $V_I$ is 0 V (grounded), it is in the low state and is designated $V_{IL}$. With TTL devices, we can increase $V_{IL}$ to 0.8 V and still have a low-state input because the output remains in the high state. In other words, the low-state input voltage $V_{IL}$ can have any value from 0 to 0.8 V. TTL data sheets list the worst-case low input as

$$V_{IL,\mathrm{max}} = 0.8 \text{ V}$$

If the input voltage is greater than this, the output state is unpredictable.

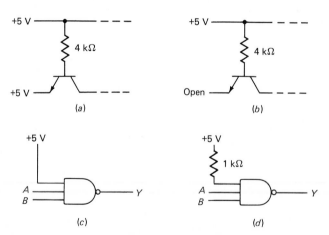

**FIG. 6–3.** (a) High input (b) Open is equivalent to high input (c) Direct connection to supply voltage (d) High input through a pull-up resistor

However, suppose $V_I$ is 5 V in Fig. 6–4a. This is a high input and can be designated $V_{IH}$. This voltage can decrease all the way down to 2 V without changing the output state. In other words, the high-stage input $V_{IH}$ is from 2 to 5 V; any input voltage in this range produces a low output voltage. Data sheets list the worst-case high input as

$$V_{IH,\min} = 2 \text{ V}$$

When the input voltage is less than this, the output state is again unpredictable.

Figure 6–4b summarizes these ideas. As you see, any input voltage less than 0.8 V is a valid low-state input. Any input greater than 2 V is a valid high-state input. Any input between 0.8 and 2 V is indeterminate because there is no guarantee that it will produce the correct output voltage.

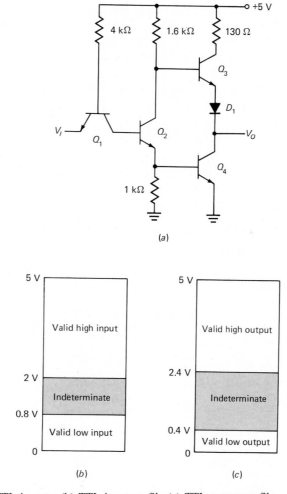

**FIG. 6–4.** (a) TTL inverter (b) TTL input profile (c) TTL output profile

## Worst-Case Output Voltages

Ideally, the low output state is 0 V, and the high output state is 5 V. We cannot attain these ideal values because of internal voltage drops inside TTL devices. For instance, when the output voltage is low in Fig. 6–4a, $Q_4$ is saturated and has a small voltage drop across it. With TTL devices, any output voltage from 0 to 0.4 V is considered a low output and is designated $V_{OL}$. This means the low-state output $V_{OL}$ of a TTL device may have any value between 0 and 0.4 V. Data sheets list the worst-case low output as

$$V_{OL,\text{max}} = 0.4 \text{ V}$$

When the output is high, $Q_3$ acts as an emitter follower. Because of the voltage drop across $Q_3$, $D_1$, and the 130-$\Omega$ resistor, the output voltage will be less than supply voltage. With TTL devices, the high-state output voltage is designated $V_{OH}$; it has a value between 2.4 and 3.9 V, depending on the supply voltage, temperature, and load. TTL data sheets list the worst-case high output as

$$V_{OH,\text{min}} = 2.4 \text{ V}$$

Figure 6–4c summarizes the output states. As shown, any output voltage less than 0.4 V is a valid low-state output, any output voltage greater than 2.4 V is a valid high-state output, and any output between 0.4 and 2.4 V is indeterminate under worst-case conditions.

## Profiles and Windows

The input characteristics of Fig. 6–4b are called the TTL input *profile*. Furthermore, each rectangular area in Fig. 6–4b can be thought of as a *window*. There is low window (0 to 0.8 V), an indeterminate window (0.8 to 2.0 V), and a high window (2.0 to 5 V).

Similarly, Fig. 6–4c is the TTL output profile. Here you see a low window from 0 to 0.4 V, an indeterminate window from 0.4 to 2.4 V, and a high window from 2.4 to 5 V.

## Values to Remember

We have discussed the low and high states for the input and output voltages. Here they are again as a reference for future discussions:

$$V_{IL,\text{max}} = 0.8 \text{ V}$$
$$V_{IH,\text{min}} = 2 \text{ V}$$
$$V_{OL,\text{max}} = 0.4 \text{ V}$$
$$V_{OH,\text{min}} = 2.4 \text{ V}$$

These are the worst-case values shown in Fig. 6–4b and c. On the input side, a voltage has to be less than 0.8 V to qualify as a low-state input, and it must be more than 2 V to be considered a high-state input. On the output side, the voltage has to be less than 0.4 V to be a low-state output and more than 2.4 V to be a high-state output.

## Compatibility

TTL devices are *compatible* because the low and high output windows fit inside the low and high input windows. Therefore the output of any TTL device is suitable for driving the input of another TTL device. For instance, Fig. 6–5a shows one TTL device driving another. The first device is called a *driver* and the second a *load*.

Figure 6–5b shows the output stage of the TTL driver connected to the input stage of the TTL load. The driver output is shown in the low state. Since any input less than 0.8 V is a low-state input, the driver output (0 to 0.4 V) is compatible with the load input requirements.

Similarly, Fig. 6–5c shows a high TTL output. The driver output (2.4 to 3.9 V) is compatible with the load input requirements (greater than 2 V).

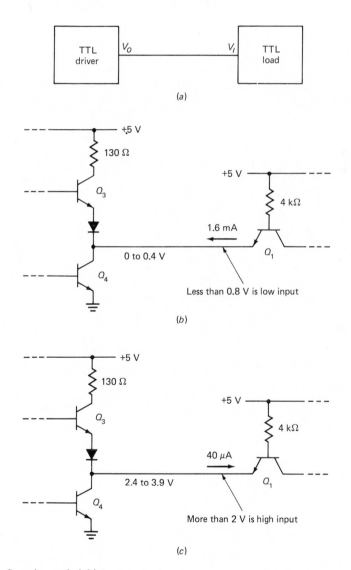

**FIG. 6–5.** Sourcing and sinking current

## Sourcing and Sinking

When a standard TTL output is low (Fig. 6–5b), an emitter current of approximately 1.6 milliamperes (mA) (worst case) exists in the direction shown. The conventional flow is from the emitter of $Q_1$ to the collector of $Q_4$. Because it is saturated, $Q_4$ acts as a *current sink*; conventional current flows through $Q_4$ to ground like water flowing down a sink.

However, when the standard TTL output is high (Fig. 6–5c), a reverse emitter current of 40 microamperes ($\mu A$) (worst-case) exists in the direction shown. Conventional current flows out of $Q_3$ to the emitter of $Q_1$. In this case, $Q_3$ is acting as a *source*.

Data sheets list the worst-case input currents:

$$I_{IL,\text{max}} = -1.6 \text{ mA} \qquad I_{IH,\text{max}} = 40 \text{ } \mu A$$

The minus sign indicates that the conventional current is out of the device; a plus sign means the conventional current is into the device. All data sheets use this notation, so do not be surprised when you see minus currents. The previous data tells us the maximum input current is 1.6 mA (outward) when an input is low and 40 $\mu A$ (inward) when an input is high.

## Noise Immunity

In the worst case, there is a difference of 0.4 V between the driver output voltages and required load input voltages. For instance, the worst-case low values are

$$V_{OL,\text{max}} = 0.4 \text{ V} \qquad \text{driver output}$$
$$V_{IL,\text{max}} = 0.8 \text{ V} \qquad \text{load input}$$

Similarly, the worst-case high values are

$$V_{OH,\text{min}} = 2.4 \text{ V} \qquad \text{driver output}$$
$$V_{IH,\text{min}} = 2 \text{ V} \qquad \text{load input}$$

In either case, the difference is 0.4 V. This difference is called *noise immunity*. It represents built-in protection against noise.

Why do we need protection against noise? The connecting wire between a TTL driver and load is equivalent to a small antenna that picks up stray noise signals. In the worst case, the low input to the TTL load is

$$V_{IL} = V_{OL} + V_{\text{noise}} = 0.4 \text{ V} + V_{\text{noise}}$$

and the high-stage input is

$$V_{IH} = V_{OH} - V_{\text{noise}} = 2.4 \text{ V} - V_{\text{noise}}$$

In most environments, the induced noise voltage is less than 0.4 V, and we get no false triggering of the TTL load.

For instance, Fig. 6–6a shows a low output from the TTL driver. If no noise voltage is induced on the connecting wire, the input voltage to the TTL load is 0.4 V, as shown. In a noisy environment, however, it is possible to have 0.4 V of induced noise on the connecting wire for either the low state (Fig. 6–6b) or the high state (Fig. 6–6c). Either way, the TTL load has an input that is on the verge of being unpredictable. The slightest additional noise voltage may produce a false change in the output state of the TTL load.

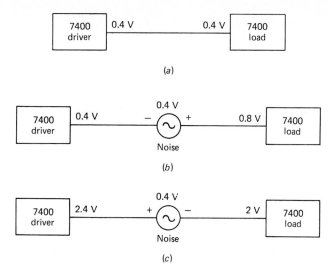

**FIG. 6–6.** (a) TTL driver and load (b) False triggering into high state (c) False triggering into low state

## Standard Loading

A TTL device can source current (high output) or sink current (low output). Data sheets of standard TTL devices indicate that any 7400 series device can sink up to 16 mA, designated

$$I_{OL,\max} = 16 \text{ mA}$$

and can source up to 400 μA, designated

$$I_{OH,\max} = -400 \ \mu\text{A}$$

(Again, a minus sign means that the conventional current is out of the device, and a plus sign means that it is into the device.) As discussed earlier, the worst-case TTL input currents are

$$I_{IL,\max} = -1.6 \text{ mA} \qquad I_{IH,\max} = 40 \ \mu\text{A}$$

Since the maximum output currents are 10 times larger than the input currents, we can connect up to 10 TTL emitters to any TTL output.

As an example, Fig. 6–7a shows a low output voltage (worst case). Notice that a single TTL driver is connected to 10 TTL loads (only the input emitters are shown). Here you see the TTL driver sinking 16 mA, the sum of the 10 TTL load currents. In the low state, the output voltage is guaranteed to be 0.4 V or less. If you try connecting more than 10 emitters, the output voltage may rise above 0.4 V under worst-case conditions. If this happens, the low-state operation is no longer reliable. Therefore, 10 TTL loads are the maximum that a manufacturer allows for guaranteed low-state operation.

Figure 6–7b shows a high output voltage (worst case) with the driver sourcing 400 μA for 10 TTL loads of 40 μA each. For this source current, the output voltage is guaranteed to be 2.4 V or greater under worst-case conditions. If you try to connect more than 10 TTL loads, you will exceed $I_{OH,\max}$ and high-state operation becomes unreliable.

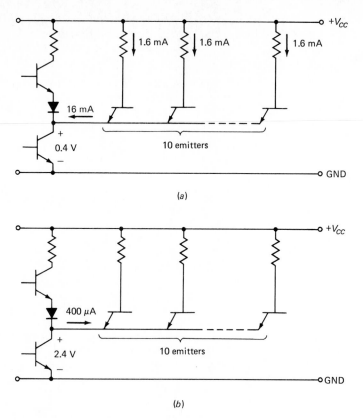

**FIG. 6–7.** (a) Low-state fanout (b) High-state fanout

## Loading Rules

Figure 6–8 shows the output-input profiles for different types of TTL. The output profiles are on the left, and the input profiles are on the right. These profiles are a concise summary of the voltages and currents for each TTL type. Start with the profiles of Fig. 6–8a; these are for standard TTL. On the left, you see the profile of output characteristics. The high output window is from 2.4 to 5 V with up to 400 $\mu$A of source current; the low output window is from 0 to 0.4 V with up to 16 mA of sink current. On the right, you see the input profile of a standard TTL device. The high window is from 2 to 5 V with an input current of 40 $\mu$A, while the low window is from 0 to 0.8 V with an input current of 1.6 mA.

Standard TTL devices are compatible because the low and high output windows fit inside the corresponding input window. In other words, 2.4 V is always large enough to be a high input to a TTL load, and 0.4 V is always small enough to be a low input. Furthermore, you can see at a glance that the available source current is 10 times the required high-state input current, and the available sink current is 10 times the required low-state input current. The maximum number of TTL loads that can be reliably driven under worst-case conditions is called the *fanout*. With standard TTL, the fanout is 10 because one TTL driver can drive 10 TTL loads.

The remaining figures all have identical voltage windows. The output states are always 0 to 0.4 and 2.4 to 5 V, while the input states are 0 to 0.8 and 2 to 5 V. For this reason, all the TTL types are compatible; this means you can use one type of TTL as a driver and another type as a load.

The only differences in the TTL types are the currents. You can see in Fig. 6–8a to d that the input and output currents differ for each TTL type. For instance, a low-power Schottky TTL driver (see Fig. 6–8d) can source 400 $\mu$A and sink 8 mA; a low-power Schottky load requires input currents of 20 $\mu$A (high state) and 0.36 mA (low state). These numbers are different from standard TTL (Fig. 6–8a) with its output currents of 400 $\mu$A and 16 mA and its input currents of 40 $\mu$A and 1.6 mA.

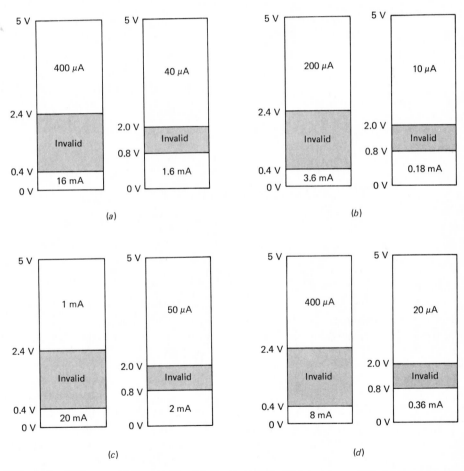

**FIG. 6–8.** TTL output/input profiles (a) Standard TTL (b) Low-power TTL (c) Schottky TTL (d) Low-power Schottky TTL

TABLE 6–4 *Fanouts*

| TTL Driver | TTL Load | | | | |
|---|---|---|---|---|---|
| | 74 | 74H | 74L | 74S | 74LS |
| 74 | 10 | 8 | 40 | 8 | 20 |
| 74H | 12 | 10 | 50 | 10 | 25 |
| 74L | 2 | 1 | 20 | 1 | 10 |
| 74S | 12 | 10 | 100 | 10 | 50 |
| 74LS | 5 | 4 | 40 | 4 | 20 |

Incidentally, the profiles of high-speed TTL are omitted in Fig. 6–8 because Schottky TTL has replaced high-speed TTL in virtually all applications. If you need high-speed TTL data, consult manufacturers' catalogs.

By analyzing Fig. 6–8a to d (plus the data sheets for high-speed TTL), we can calculate the fanout for all possible combinations. Table 6–4 summarizes these fanouts, which are useful if you ever have to mix TTL types.

Read Table 6–4 as follows. The TTL types have been abbreviated; 74 stands for 7400 series (standard), 74H for 74H00 series (high speed), and so forth. Drivers are on the left, and loads are on the right. Pick the driver, pick the load, and read the fanout at the intersection of the two. For instance, the fanout of a standard device (74) driving low-power Schottky devices (74LS) is 20. As another example, the fanout of a low-power device (74L) driving high-speed devices (74H) is only 1.

## 6-4 TTL OVERVIEW

Let's take a look at the logic functions available in the 7400 series. This overview will give you an idea of the variety of gates and circuits found in the TTL family. As a guide, Appendix 3 lists some of the 7400 series devices.

### NAND Gates

The NAND gate is the backbone of the 7400 series. All devices in this series are derived from the 2-input NAND gate shown in Fig. 6–1. To produce 3-, 4-, and 8-input NAND gates, the manufacturer uses 3-, 4-, and 8-emitter transistors. Because they are so basic, NAND gates are the least expensive devices in the 7400 series.

### NOR Gates

To get other logic functions the manufacturer modifies the basic NAND–gate design. For instance, Fig. 6–9 shows a 2-input NOR gate. Here $Q_5$ and $Q_6$ have been added to basic NAND–gate design. Since $Q_2$ and $Q_6$ are in parallel, we get the OR function, which is followed by inversion to get the NOR function.

When A and B are both low, the bases of $Q_1$ and $Q_5$ are pulled low; this cuts off $Q_2$ and $Q_6$. Then $Q_3$ acts as an emitter follower, and we get a high output.

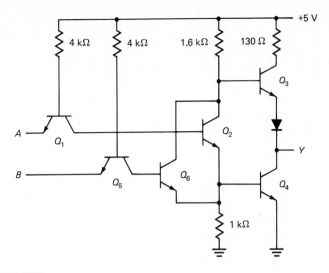

**FIG. 6–9.** TTL NOR gate

If $A$ or $B$ is high, $Q_1$ and $Q_5$ are cut off, forcing $Q_2$ or $Q_6$ to turn on. When this happens, $Q_4$ saturates and pulls the output down to a low voltage.

With more transistors, a manufacturer can produce 3- and 4-input NOR gates. (*Note*: A TTL 8-input NOR gate is not available.)

## AND and OR Gates

To produce the AND function, another common-emitter (CE) stage is inserted in the basic NAND–gate design. The extra inversion converts the NAND gate to an AND gate. The available TTL AND gates are the 7408 (quad 2-input), 7411 (triple 3-input), and 7421 (dual 4-input).

Similarly, another CE stage can be inserted in the NOR gate of Fig. 6–9; this converts the NOR gate to an OR gate. The only available TTL OR gate is the 7432 (quad 2-input).

## Buffer Drivers

An IC buffer can source and sink more current than a standard TTL gate. As an example, the 7437 is a quad 2-input NAND buffer, meaning four 2-input NAND gates optimized to get high output currents. Each gate has the following worst-case currents:

$$I_{IL} = -1.6 \text{ mA} \qquad I_{IH} = 40 \text{ } \mu\text{A}$$
$$I_{OL} = 48 \text{ mA} \qquad I_{OH} = -1.2 \text{ mA}$$

The input currents are the same as those of a 7400 (standard TTL NAND gate), but the output currents are 3 times as high. This means that a 7437 can drive heavier loads. In other words, the fanout of a 7437 is 3 times that of a 7400. Appendix 3 includes several other buffer drivers.

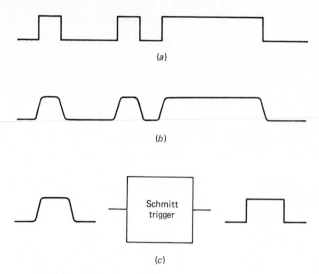

(a)

(b)

Schmitt
trigger

(c)

**FIG. 6–10.**  Schmitt trigger produces rectangular output

## Schmitt Triggers

When a computer is running, the gate outputs are rapidly switching from one state to another. If you look at these signals with an oscilloscope, you see signals that ideally represent rectangular waves like Fig. 6–10a. When digital signals are transmitted and later received, they are often corrupted by noise, attenuation, or other factors, and they may wind up looking like the ragged waveform shown in Fig. 6–10b. If you try to use these nonrectangular signals to drive a gate or other digital device, you may get unreliable operation.

This is where the *Schmitt trigger* comes in. It is designed to clean up ragged-looking pulses, producing almost vertical transitions between the high and low state, and vice versa (Fig. 6–10c). In other words, the Schmitt trigger produces a rectangular output, regardless of the input waveform.

The 7414 is a hex Schmitt-trigger inverter, meaning six Schmitt-trigger inverters in one package as in Fig. 6–11a. Notice the hysteresis symbol inside each inverter; it designates the Schmitt-trigger function. Two other TTL Schmitt triggers are available. The 7413 is a dual 4-input NAND Schmitt trigger, with two Schmitt-trigger gates as in Fig. 6–11b. The 74132 is a quad 2-input NAND Schmitt trigger, and it has four Schmitt triggers, as in Fig. 6–11c.

## 6·5  AND—OR—INVERT GATES

Figure 6–12a shows an AND–OR circuit. Figure 6–12b shows the De Morgan equivalent circuit, a NAND–NAND network. In either case, the boolean equation is

$$Y = AB + CD \qquad\qquad (6\text{--}1)$$

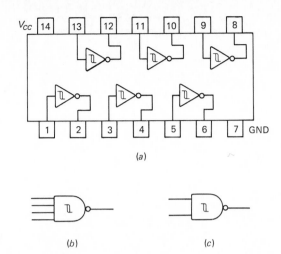

**FIG. 6–11.** (a) Hex Schmitt-trigger inverters (b) 4-input NAND Schmitt trigger (c) 2-input NAND Schmitt trigger

Since NAND gates are the preferred TTL gates, we would build the circuit of Fig. 6–12*b*. As you know, NAND–NAND circuits like this are important because with them you can build any desired logic circuit (discussed in Chap. 2).

## TTL Devices

AND–OR circuits are not easily derived from the basic NAND–gate design. But it is easy to get an AND–OR–INVERT circuit as in Fig. 6–12*c*. A variety of circuits like this are available as TTL chips. Because of the inversion, the output has an equation of

$$Y = \overline{AB + CD} \qquad (6\text{-}2)$$

Figure 6–13 shows the schematic diagram of an AND–OR–INVERT circuit. Here, $Q_1$, $Q_2$, $Q_3$, and $Q_4$ form the basic 2-input NAND gate of the 7400 series. By adding $Q_5$ and $Q_6$, we convert the basic NAND gate to an AND–OR–INVERT gate. Both $Q_1$ and $Q_5$ act as 2-input AND gates; $Q_2$ and $Q_6$ produce ORing and inversion. Because of this, the circuit is logically equivalent to Fig. 6–12*c*.

Table 6–5 lists the AND–OR–INVERT gates available in the 7400 series. In this table, *2-wide* means two AND gates across, *4-wide* means four AND gates across, and so on. For instance, the 7454 is a 2-input 4-wide AND–OR–INVERT gate as in Fig. 6–14*a*; each AND gate has two inputs (2-input), and there are four AND gates (4-wide). Figure 6–14*b* shows the 7464; it is a 2-2-3-4-input 4-wide AND–OR–INVERT gate.

**TABLE 6–5** *AND–OR–INVERT Gates*

| Device | Description |
|--------|-------------|
| 7451 | Dual 2-input 2-wide |
| 7454 | 2-input 4-wide |
| 7459 | Dual 2-3-input 2-wide |
| 7464 | 2-2-3-4-input 4-wide |

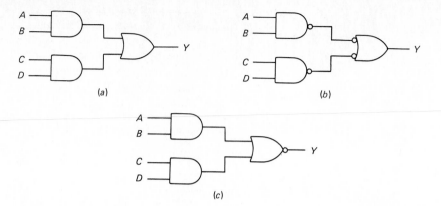

FIG. 6–12.  (a) AND–OR circuit (b) NAND–NAND circuit (c) AND–OR–INVERT circuit

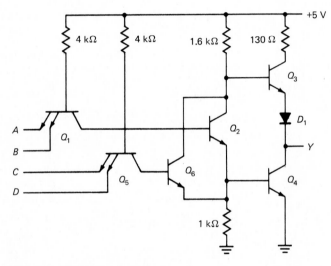

FIG. 6–13.  AND–OR–INVERT schematic diagram

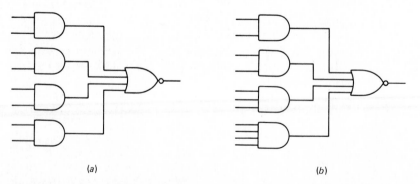

FIG. 6–14.  Examples of AND–OR circuits

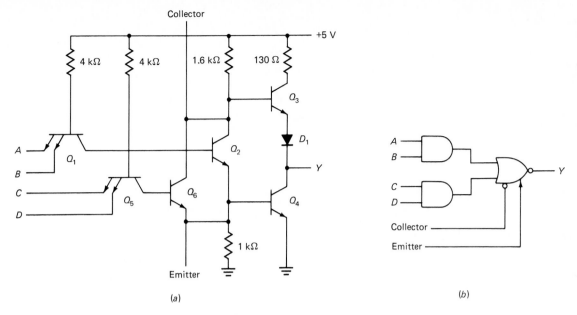

**FIG. 6–15.** (a) Expandable AND–OR–INVERT gate (b) Logic symbol

When we want the output given by Eq. (6–1), we can connect the output of a 2-input 2-wide AND–OR–INVERT gate to an inverter. This gives us the same output as an AND–OR circuit.

## Expandable AND–OR–INVERT Gates

The widest AND–OR–INVERT gate available in the 7400 series is 4-wide. What do we do when we need a 6- or 8-wide circuit? One solution is to use an *expandable* AND–OR–INVERT gate. Figure 6–15a shows the schematic diagram of an expandable AND–OR–INVERT gate. The only difference between this and the preceding AND–OR–INVERT gate (Fig. 6–13) is the collector and emitter pins brought outside the package. Since $Q_2$ and $Q_6$ are the key to the ORing operation, we are being given access to the internal ORing function. By connecting other gates to these new inputs, we can expand the width of the AND–OR–INVERT gate.

Figure 6–15b shows the logic symbol for an expandable AND–OR–INVERT gate. The arrow input represents the emitter, and the bubble stands for the collector. Table 6–6 lists the expandable AND–OR–INVERT gates in the 7400 series.

**TABLE 6–6** *Expandable AND–OR–INVERT Gates*

| Device | Description |
| --- | --- |
| 7450 | Dual 2-input 2-wide |
| 7453 | 2-input 4-wide |
| 7455 | 4-input 2-wide |

## Expanders

What do we connect to the collector and emitter inputs of an expandable gate? The output of an *expander* as in Fig. 6–16a. The input transistor acts as a 4-input AND gate. Figure 6–16b shows the symbol of a 4-input expander.

Visualize the outputs of Fig. 6–16a connected to the collector and emitter inputs of Figure 6–15a. Then $Q_8$ is in parallel with $Q_2$ and $Q_6$. Figure 6–16c shows the logic circuit. This means that the expander outputs are being ORed with the signals of the AND–OR–INVERT gate. In other words, Fig. 6–16c is equivalent to the AND–OR–INVERT circuit of Fig. 6–16d.

We can connect more expanders. Figure 6–16e shows two expanders driving the expandable gate. Now we have a 2-2-4-4-input 4-wide AND–OR–INVERT circuit.

The 7460 is a dual 4-input expander. The 7450, a dual expandable AND–OR–INVERT gate, is designed for use with up to four 7460 expanders. This means that we can add two more expanders in Fig. 6–16e to get 2-2-4-4-4-4-input 6-wide AND–OR–INVERT circuit.

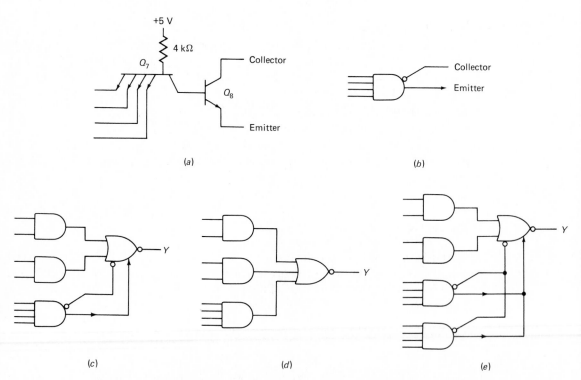

**FIG. 6–16.** (a) Expander (b) Symbol for expander (c) Expander driving expandable AND–OR–INVERT gate (d) AND-OR-INVERT circuit (e) Expandable AND-OR-INVERT with two expanders

# 6-6 OPEN-COLLECTOR GATES

Instead of a totem-pole output, some TTL devices have an *open-collector* output. This means they use only the lower transistor of a totem-pole pair. Figure 6–17*a* shows a 2-input NAND gate with an open-collector output. Because the collector of $Q_4$ is open, a gate like this will not work properly until you connect an external pull-up resistor, shown in Fig. 6–17*b*.

The outputs of open-collector gates can be wired together and connected to a common pull-up resistor. For instance, Fig. 6–17*c* shows three TTL devices connected to the pull-up resistor. This is known as *wire–OR* (some called it wire–AND). A connection like this has the advantage of combining the output of three devices without using a final OR gate (or AND gate). The combining is done by a direct connection of the three outputs to the lower end of the common pull-up resistor. This is very useful when many devices are wire–ORed together. For instance, in some systems the outputs of 16 open-collector devices are connected to a pull-up resistor.

The big disadvantage of open-collector gates is their slow switching speed. Why is it slow? Because the pull-up resistance is a few kilohms, which results in a relatively long time constant when it is multiplied by the stray output capacitance. The slow switching speed of open-collector TTL devices is worst when the output goes from low to high. Imagine all three transistors going into cutoff in Fig. 6–17*c*. Then any capacitance across

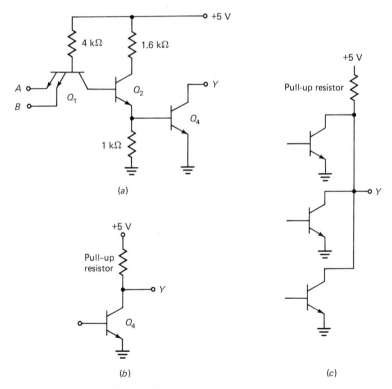

**FIG. 6–17.** Open-collector TTL (a) Circuit (b) Pull-up resistor (c) Open-collector outputs connected to a pull-up resistor

the output has to charge through the pull-up resistor. This charging produces a relatively slow exponential rise between the low and high state.

## 6·7 THREE-STATE TTL DEVICES

Using a common pull-up resistor with open-collector devices is called *passive pull-up* because the supply voltage pulls the output voltage up to a high level when all the transistors of Fig. 6–17c are cut off. There is another approach known as *active pull-up*. It uses a modified totem-pole output to speed up the charging of stray output capacitance. The effect is to dramatically lower the charging time constant, which means the output voltage can rapidly change from its low to its high stage.

### Why Standard TTL Will Not Work

If you try to wire–OR standard TTL gates, you will destroy one or more of the devices. Why? Look at Fig. 6–18 for an example of bad design. Notice that the output pins of two standard TTL devices are connected. If the output of the second device is low, $Q_4$ is saturated and appears approximately like a short circuit. If, at the same time, the output of the first device is in the high state, then $Q_1$ acts as an emitter follower that tries to pull the output voltage to a high level. Since $Q_1$ and $Q_4$ are both conducting heavily, only 130 $\Omega$ remains between the supply voltage and ground. The final result is an excessive current that destroys one of the TTL devices.

### Low DISABLE Input

As you have seen, wire–ORing standard TTL devices will not work because of destructive currents in the output stages. This inability to wire–OR ordinary totem-pole devices is what led to *three-state* TTL, a new breed of totem-pole devices introduced in the early 1970s. With three-state gates, we can connect totem-pole outputs directly without destroying any devices. The reason for wanting to use totem-pole outputs is to avoid the loss of speed that occurs with open-collector devices.

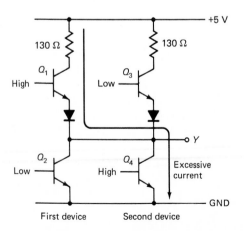

**FIG. 6–18.** Direct connection of TTL outputs produces excessive current

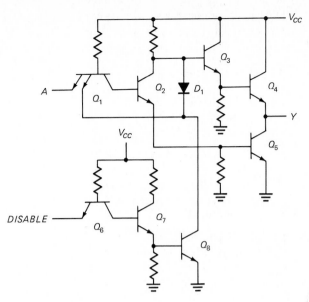

**FIG. 6–19.**  Three-state inverter

Figure 6–19 shows a simplified drawing for a three-state inverter. When DISABLE is low, the base and collector of $Q_6$ are pulled low. This cuts off $Q_7$ and $Q_8$. Therefore, the second emitter of $Q_1$ and the cathode of $D_1$ are floating. For this condition, the rest of the circuit acts as an inverter: a low $A$ input forces $Q_2$ and $Q_5$ to cut off, while $Q_3$ and $Q_4$ turn on, producing a high output. On the other hand, a high $A$ input forces $Q_2$ to turn on, which drives $Q_5$ into saturation and produces a low output. Table 6–7 summarizes the operation for low DISABLE.

## High DISABLE Input

When DISABLE is high, the base and collector of $Q_6$ go high, which turns on $Q_7$ and saturates $Q_8$. Ideally, the collector of $Q_8$ is pulled down to ground. This causes the base and collector of $Q_1$ to go low, cutting off $Q_2$ and $Q_5$. Also $Q_3$ is cut off because of the clamping action of $D_1$. In other words, the base of $Q_3$ is only 0.7 V above ground, which is insufficient to turn on $Q_3$ and $Q_4$.

With both $Q_4$ and $Q_5$ cut off, the $Y$ output is floating. Ideally, this means that the Thévenin impedance looking back into the $Y$ output approaches infinity. Table 6–7 summarizes the action for this high-impedance state. As shown, when DISABLE is high, input $A$ is a don't care because it has no effect on the $Y$ output. Furthermore, because of

**TABLE 6–7** *Three-State Inverter*

| Disable | A | Y |
|---------|---|------|
| 0 | 0 | 1 |
| 0 | 1 | 0 |
| 1 | X | Hi-Z |

the high output impedance, the output line appears to be disconnected from the rest of the gate. In effect, the output line is floating.

In conclusion, the output of Fig. 6–19 can be in one of three states: low, high, or floating.

## Logic Symbol

Figure 6–20*a* is an equivalent circuit for the three-state inverter. When DISABLE is low, the switch is closed and the circuit acts as an ordinary inverter. When DISABLE is high, the switch is open and the *Y* output is floating or disconnected.

Figure 6–20*b* shows the logic symbol for a three-state inverter. When you see this symbol, remember that a low DISABLE results in normal inverter action, but a high DISABLE floats the *Y* output.

## Three-State Buffer

By modifying the design, we can produce a three-state buffer, whose logic symbol is shown in Fig. 6–20*c*. When DISABLE is low, the circuit acts as a noninverting buffer, so that *Y* = *A*. But when DISABLE is high, the output floats. The three-state buffer is equivalent to an ordinary switch. When DISABLE is low, the switch is closed. When DISABLE is high, the switch is open.

The 74365 is an example of a commercially available three-state hex noninverting buffer. This IC contains six buffers with three-state outputs. It is ideal for organizing digital components around a *bus*, a group of wires that transmits binary numbers between registers.

## Bus Organization

Figure 6–21 shows some registers connected to a common bus. The three-state buffers control the flow of binary data between the registers. For instance, if we want the contents of register *A* to appear on the bus, all we have to do is make DISABLE low for register *A* but high for registers *B* and *C*. Then all the three-state switches on register *A* are closed, while all other three-state switches are open. As a result, only the contents of register *A* appear on the bus.

The idea in any bus-organized system is to make DISABLE high for all registers except the register whose contents are to appear on the bus. In this way, dozens of registers can time-share the same transmission path. Not only does this reduce the amount of wiring, but also it has simplified the architecture and design of computers and other digital systems.

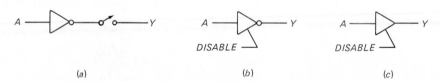

(a)        (b)        (c)

**FIG. 6–20.** Three-state logic diagrams (a) Equivalent circuit of inverter (b) Logic symbol of inverter (c) Logic symbol of buffer

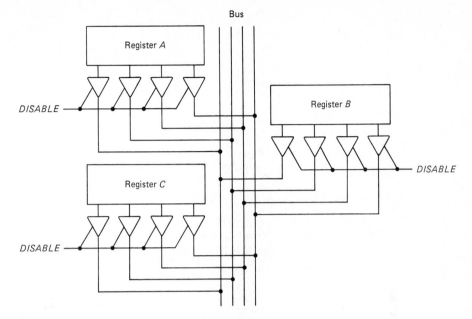

**FIG. 6–21.** Three-state bus control

# 6-8 EXTERNAL DRIVE FOR TTL LOADS

To drive a TTL load with an external source, you need to satisfy the TTL input requirements for voltage and current. For standard TTL in the low state, this means an input voltage between 0 and 0.8 V with a current of approximately 1.6 mA. In the high state, the voltage has to be from 2 to 5 V with a current of approximately 40 $\mu$A. Let us take a brief look at some of the ways to drive a TTL load.

## Switch Drive

Figure 6–22 shows the preferred method for driving a TTL input from a switch. With the switch open, the input is pulled up to +5 V. In the worst case, only 40 $\mu$A of input current exists. Therefore, the voltage appearing at the input pin is slightly less than the supply voltage because of the small voltage drop across the pull-up resistor:

$$V_{in} = 5 \text{ V} - (40 \text{ } \mu\text{A})(1 \text{ k}\Omega) = 4.96 \text{ V}$$

This is well above the minimum requirement of 2 V, which is fine because it means the noise immunity is excellent.

When the switch is closed, the input is pulled down to ground. In the worst case, the input current is 1.6 mA. This sink current creates no problem because it flows through the closed switch to ground. The noise immunity is fine because the input voltage is 0 V, well below the maximum allowable value of 0.8 V.

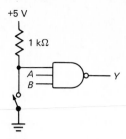

**FIG. 6–22.** Switch drive for TTL input

## Size of Pull-up Resistance

A pull-up resistance of 1 k$\Omega$ is nominal. You can use other values. Here are some of the factors to consider when you are selecting a pull-up resistor. In Fig. 6–22, the current drain with a closed switch is

$$I = \frac{5 \text{ V}}{1 \text{ k}\Omega} = 5 \text{ mA}$$

The smaller the pull-up resistance, the larger the current drain. At some point, too much current drain becomes a problem for the power supply, so you have to use a resistance that is large enough to keep the current drain to tolerable levels.

On the other hand, too large a pull-up resistance causes speed problems. The worst case occurs when the switch is opened. For instance, if the input capacitance is 10 picofarads (pF) in Fig. 6–22, the time constant is

$$RC = (1 \text{ k}\Omega)(10 \text{ pF}) = 10 \text{ ns}$$

The larger the pull-up resistance, the larger the time constant. A larger time constant means a slower switching speed because the input capacitance has to charge through the pull-up resistance.

Pull-up resistances between 1 and 10 k$\Omega$ are typical. They result in current drains and time constants that are acceptable in most applications.

## Transistor Drive

Figure 6–23a shows another way to drive a TTL load. This time, we are using a transistor switch to control the state of the TTL input. When $V_{\text{in}}$ is low, the transistor goes into cutoff and is equivalent to an open switch. Then the TTL input is pulled up to +5 V through a resistance of 1 k$\Omega$. When $V_{\text{in}}$ is high, the transistor goes into saturation and is equivalent to a closed switch. In this case, it easily sinks the 1.6 mA of input current.

The transistor inverts the control signal $V_{\text{in}}$. If this is objectionable, you can insert an inverter as shown in Fig. 6–23b. Now, the double inversion produces an in-phase control signal at the TTL input.

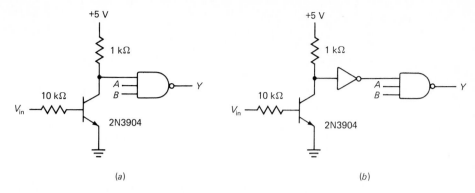

FIG. 6–23.   (a) Transistor drive for TTL input (b) Inverter eliminates transistor inversion.

## Op-Amp Drive

Sometimes, you want to use the output of an operational amplifier (op amp) to control a TTL input. Because op amps typically use split-supply voltages of $+15$ and $-15$ V, you have to be careful how you connect to the TTL load. Figure 6–24 shows one way to use the output of a 741 to control a TTL input. The output of the op amp ideally swings from $+15$ to $-15$ V. The positive swing closes the transistor switch, producing a TTL input of approximately 0 V. The negative swing drives the transistor into cutoff, producing a TTL input of $+5$ V.

Notice the diode in the base circuit. It protects the base against excessive reverse voltage. The data sheet of a 2N3904 indicates an absolute maximum base-emitter voltage rating of

$$V_{BE,\max} = -6 \text{ V}$$

Since the negative output of the op amp approaches $-15$ V, we need to use a protective diode as shown between the base and ground. This diode clamps the base voltage at approximately $-0.7$ V on the negative swing.

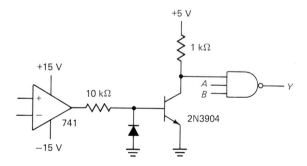

FIG. 6–24.   Op amp and transistor drive for TTL input

## Comparator Drive

Figure 6–25a shows the schematic diagram for a typical comparator, an IC that detects when the input voltage is positive or negative. Notice two things. First, a supply voltage of +15 V is typically used with this kind of device. Second, the comparator has an open-collector output transistor. This sink transistor can be connected to any supply voltage.

Figure 6–25b shows how to connect an LM339 (typical comparator) to a TTL load. Because of the open-collector output, we can connect the output pin of the comparator to a supply voltage of +5 V through a pull-up resistance of 1 kΩ. When $V_{in}$ is positive, the sink transistor goes into cutoff and the TTL input is pulled high. When $V_{in}$ is negative, the sink transistor goes into saturation and the TTL output is pulled low.

## 6-9 TTL DRIVING EXTERNAL LOADS

Because standard TTL can sink up to 16 mA, you can use a TTL driver to control an external load such as a relay, a LED, etc. Figure 6–26a illustrates the idea. When the TTL output is high, there is no load current. But when the TTL output is low, the lower end of $R_L$ is ideally grounded. This sets up a load current of approximately

$$I_L = \frac{5 \text{ V}}{R_L}$$

Since standard TTL can sink a maximum of 16 mA, the load resistance is limited to a minimum value of about

$$R_L = \frac{5 \text{ V}}{16 \text{ mA}} = 312 \ \Omega$$

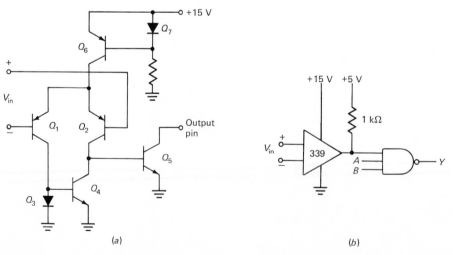

(a)                                             (b)

**FIG. 6–25.** (a) Schematic diagram of comparator (b) Interfacing an LM339 to a TTL input

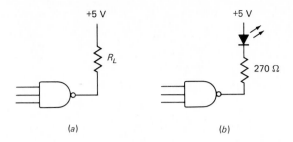

(a)                                 (b)

**FIG. 6–26.**   (a) TTL output drives load resistor (b) TTL output drives LED

## Driving a LED

Figure 6–26b is another example. Here a TTL circuit drives a LED. When the TTL output is high, the LED is dark. When the TTL output is low, the LED lights up. If the LED voltage drop is 2 V, the LED current for a low TTL output is approximately

$$I_L = \frac{5\text{ V} - 2\text{ V}}{270\ \Omega} = 11.1\text{ mA}$$

## Supply Voltage Different from +5 V

If you need to use a supply voltage different from +5 V, you can use an open-collector TTL device. For instance, Fig. 6–27c shows an open-collector gate driving a load resistor that is returned to +15 V. Since an open-collector device can sink a maximum of 16 mA, the minimum load resistance in Fig. 6–27a is slightly less than 1 kΩ.

If you want more than 16 mA of load current, you can use an external transistor, as shown in Fig. 6–27b. When the open-collector device has a low output, the external transistor goes into cutoff and the load current is zero. When the device has a high output, the external transistor goes into saturation and the load current is maximum.

## Hard Saturation

Here is an example of hard saturation (discussed in Chap. 1). If $R_B$ is 1 kΩ in Fig. 6–27b, the base current is approximately 15 mA for a high TTL output. A load resistance of

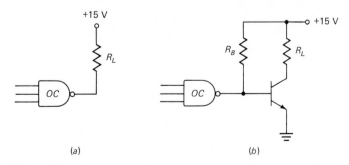

(a)                                 (b)

**FIG. 6–27.**   (a) Open-collector device allows a higher supply voltage (b) Transistor increases current drive

Positive OR

Negative AND

**FIG. 6–28.** Meaning of symbol depends on whether you use positive or negative logic

---

$100\ \Omega$ would result in a collector current of approximately 150 mA. So, design values for hard saturation are $R_B = 1\ \text{k}\Omega$ and $R_L = 100\ \Omega$. When the open-collector output is low, the TTL device sinks approximately 15 mA. When the open-collector output is high, 15 mA flows through the base.

## 6-10 POSITIVE AND NEGATIVE LOGIC

Up to now, we have used a binary 0 for low voltage and a binary 1 for high voltage. This is called *positive logic*. People are comfortable with positive logic because it feels right. But there is another code known as *negative logic* where binary 0 stands for high voltage and binary 1 for low voltage. Even though it seems unnatural, negative logic has many uses. The following discussion introduces some of the terminology and concepts for both types of logic.

### Positive and Negative Gates

An OR gate in a positive logic system becomes an AND gate in a negative logic system. Why? Look at the gate of Fig. 6–28. We have been calling it an OR gate. This is correct, provided we are using positive logic. Table 6–8 shows the truth for the gate of Fig. 6–28, no matter what you call it. In other words, if either input is high in Fig. 6–28, the output is high.

In a positive logic system, binary 0 stands for low and binary 1 for high. Therefore, we can convert Table 6–8 to Table 6–9. Notice that $Y$ is a 1 if either $A$ or $B$ is 1. This sounds like an OR gate. And it is, because we are using positive logic. To avoid ambiguity, we can refer to Fig. 6–28 as a positive OR gate because it performs the OR function with positive logic. (Some data sheets describe gates as positive OR gate, positive AND gate, etc.)

In a negative logic system, binary 1 stands for low and binary 0 for high. With this code, we can convert Table 6–8 to Table 6–10. Now, watch what happens. The output $Y$ is a 1 only when both $A$ and $B$ are 1. This sounds like an AND gate! And it is, because we

TABLE 6–8

| $A$ | $B$ | $Y$ |
|------|------|------|
| Low | Low | Low |
| Low | High | High |
| High | Low | High |
| High | High | High |

TABLE 6–9

| $A$ | $B$ | $Y$ |
|------|------|------|
| 0 | 0 | 0 |
| 0 | 1 | 1 |
| 1 | 0 | 1 |
| 1 | 1 | 1 |

**TABLE 6–10**

| A | B | Y |
|---|---|---|
| 1 | 1 | 1 |
| 1 | 0 | 0 |
| 0 | 1 | 0 |
| 0 | 0 | 0 |

are now using negative logic. In other words, gates are defined by the way they process the binary 0s and 1s. If you use binary 1 for low voltage and binary 0 for high voltage, then you have to refer to Fig. 6–8 as a negative AND gate.

As you see, the gate of Fig. 6–28 always produces a high output if either input is high. But what you call it depends on whether you see positive or negative logic. Use whichever name applies. With positive logic, call it a positive OR gate. With negative logic, call it a negative AND gate.

In a similar way, we can show the truth table of other gates with positive or negative logic. By analyzing the inputs and outputs in terms of 0s and 1s, you find these equivalences between the positive and negative logic:

$$\text{Positive OR} \leftrightarrow \text{negative AND}$$
$$\text{Positive AND} \leftrightarrow \text{negative OR}$$
$$\text{Positive NOR} \leftrightarrow \text{negative NAND}$$
$$\text{Positive NAND} \leftrightarrow \text{negative NOR}$$

Table 6–11 summarizes these gates and their definitions in terms of voltage levels. These definitions are always valid. If you get confused from time to time, refer to Table 6–11 to get back to the ultimate meaning of the basic gates.

## Positive True and Positive False

Boolean algebra originated as a mathematical approach to solving logic problems. Because of the historical connection, we refer to digital circuits as logic circuits. Furthermore, some of the historical terminology is still used with digital signals. For instance, with positive logic many people refer to high voltage as *positive true* and low voltage as *positive false*. As a code, here is how it looks:

$$\text{High voltage} \leftrightarrow \text{positive true}$$
$$\text{Low voltage} \leftrightarrow \text{positive false}$$

**TABLE 6–11** *Voltage Definitions of Basic Gates*

| Gate | Definition |
|---|---|
| Positive OR/negative AND | Output is high if any input is high. |
| Positive AND/negative OR | Output is high when all inputs are high. |
| Positive NOR/negative NAND | Output is low if any input is high. |
| Positive NAND/negative NOR | Output is low when all inputs are high. |

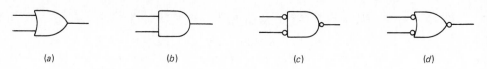

**FIG. 6–29.** (a) Positive OR (b) Positive AND (c) Negative AND (d) Negative OR

Instead of describing voltages as high or low, we can use positive true and positive false. For example, the output of Fig. 6–29a is positive true if either input is positive true. The output of Fig. 6–29b is positive true only when both inputs are positive true.

## Negative True and Negative False

With negative logic, things get turned around. Low voltage is described as negative true, and high voltage as negative false. For emphasis, here are the equivalences in code form:

$$\text{Low voltage} \leftrightarrow \text{negative true}$$
$$\text{High voltage} \leftrightarrow \text{negative false}$$

With this negative code, the output of Fig. 6–29a is negative true only when both inputs are negative true. Similarly, the output of Fig. 6–29b is negative true if either input is negative true.

The positive OR gate of Fig. 6–29a performs the AND function in a negative logic system because it produces a negative true output (low) only when both inputs are negative true (low). To emphasize the AND function of Fig. 6–29a in a negative logic system, many people use the symbol of Fig. 6–29c for a negative AND gate. This gate has the same low-high truth table as Fig. 6–29a, but it changes your viewpoint. Instead of saying the output is high if either input is high, you say the output is low only when both inputs are low.

Likewise, the positive AND gate of Fig. 6–29b performs the OR function in a negative logic system because it produces a negative true output if either input is negative true. Because of this, you often see a negative OR gate drawn as shown in Fig. 6–29d. Again, the emphasis is changed. Instead of saying the output is high only when both inputs are high, we say the output is low if either input is low.

## Equivalent Gates

Figure 6–30 shows different ways to draw the basic gates. You can redraw any logic circuit, using the equivalences shown here. Sometimes, this helps to simplify a logic circuit because double inversions on the same line cancel (discussed in Chap. 1). Besides simplifying circuits, some designers prefer to use bubbled inputs. As indicated, a positive OR gate is equivalent to a negative AND gate; a positive AND gate is equivalent to a negative OR gate; a positive NOR gate is equivalent to a bubbled AND gate; and a positive NAND gate is equivalent to a bubbled OR gate.

You can mix positive and negative true when describing logic circuits. For instance, look at Fig. 6–30c. For the gate on the left, you would say the output is negative true if

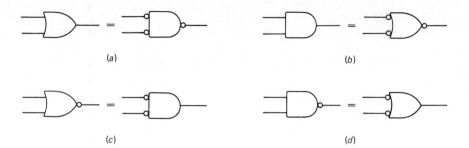

FIG. 6–30. Useful logic equivalences

either input is positive true. For the gate on the right, you would say the output is positive true only when both inputs are negative true.

Likewise, for the left gate of Fig. 6–30d, the output is negative true only when both inputs are positive true. For the right gate, the output is positive true if either input is negative true.

Here is how to remember the gate equivalences of Fig. 6–30. To get from any gate to its equivalent form, use this tried-and-true method:

1. Change the underlying gate from OR to AND or from AND to OR.
2. Invert all inputs and outputs.

For instance, the left gate of Fig. 6–30c is a positive NOR gate. The underlying gate is an OR gate. Change this to an AND gate; then invert all inputs and outputs. Inverting the input signals puts bubbles on both inputs. Inverting the output signal creates a double bubble, which means no bubble. Therefore, the final result is a bubbled AND gate (the right gate of Fig. 6–30c).

## Assertion-Level Logic

Why do we even bother with negative logic? The reason is related to the concept of *active-low* signals, introduced in Chap. 3. For instance, the 74150 multiplexer has an active-low strobe; this input turns on the chip only when it is low (negative true). As another example, the 74154 decoder has 16 output lines; the decoded output signal is low (negative true). In other words, all output lines have a high voltage, except the decoded output line. Besides TTL devices, microprocessor chips like the 8085, 6502, and so on have a lot of active-low input and output signals.

Many designers draw their logic circuits with bubbles on all pins with active-low signals and omit bubbles on all pins with active-high signals. This use of bubbles with active-low signals is called *assertion-level logic*. It means that you draw chips with the kind of input that causes something to happen, or with the kind of output that indicates something has happened. If a low input signal turns on a chip, you show a bubble on that input. If a low output is a sign of chip action, you draw a bubble on that output. Once you get used to assertion-level logic, you may prefer drawing logic circuits this way.

One final point. Sometimes you hear expressions such as "The inputs are asserted" or "What happens when the inputs are asserted?" An input is *asserted* when it is active. This means it may be low or high, depending on whether it is an active-low or active-high input. For instance, given a positive AND gate, all inputs must be asserted (high) to get a high output. As another example, the STROBE input of a TTL multiplexer must be asserted (low) to turn on the multiplexer. In short, you can equate the word *assert* with *activate*. You assert, or activate, the inputs of a gate or device to get something to happen.

## Labeling Signals

The way you label signal voltages can simplify the analysis of a logic circuit. Many designers use a label that represents a statement. When the statement is true, the signal voltage is high for positive logic and low for negative logic. Also, it is common practice to use an overbar on all signals that are active-low.

For instance, the sum out of an 8-bit adder-subtractor has the format of $S_7 \cdots S_0$. This binary number can vary from

$$S_7 \cdots S_0 = 0000\ 0000$$

to an unsigned maximum of

$$S_7 \cdots S_0 = 1111\ 1111$$

Suppose we want to detect when the sum equals the maximum value. Figure 6–31a shows one way to do it. The sum bits are the inputs to an 8-bit AND gate. The output signal of this AND gate is labeled THE SUM IS MAXIMUM. When this statement is true, the signal THE SUM IS MAXIMUM is a high voltage. Because of the inverter, the final output signal of Fig. 6–31a is

$$Y = \overline{\text{THE SUM IS MAXIMUM}}$$

Nobody uses the entire statement for a label because it is too long and would take up too much room on a logic diagram. Instead, each statement is abbreviated as a key word that jogs the memory. For instance, we can simplify the drawing of Fig. 6–31a by using MAXIMUM and $\overline{\text{MAXIMUM}}$ for the signal voltages, as shown in Fig. 6–31b. Now, when all the sum bits are high, the original statement is true and MAXIMUM is a high voltage (positive true). Likewise, when all the sum bits are high, the original statement is true and $\overline{\text{MAXIMUM}}$ is a low voltage (negative true). With this labeling system, active-high labels (no overbar) are positive true and active-low labels (overbar) are negative true whenever the statement is true.

There is no 8-input AND gate in the TTL series. But there is an 8-bit NAND gate. So, if we wanted to detect the maximum sum, we could build the logic circuit of Fig. 6–31c. Look at how clean the design and label are. When all sum bits are high, the output of the NAND gate is low, or negative true. This turns on the LED, indicating a maximum sum.

Figure 6–31c is an example of assertion-level logic. The inputs are active-high, and the output is active-low. Therefore, we draw a positive NAND gate rather than a negative NOR gate because we want unbubbled inputs and a bubbled output. Also, the label $\overline{\text{MAXIMUM}}$ has an overbar because it is an active-low signal.

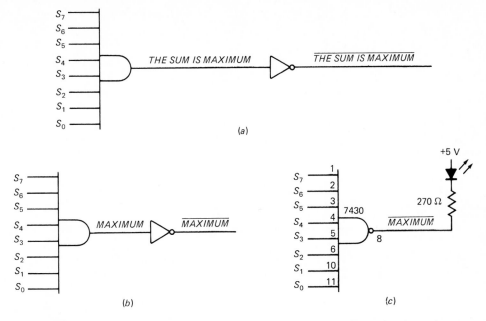

FIG. 6–31. (a) Complete statements (b) Abbreviated statements (c) Example of assertion-level logic

## Points to Remember

Here are some ideas that you should try to remember:

1. Positive true always represents a high voltage, and negative true always represents a low voltage.
2. If possible, draw basic gates with bubbles on active-low signal lines.
3. When a signal is active-low, use an overbar as a reminder that the signal voltage is negative true when the underlying statement is true.

☐ EXAMPLE 6—1

The sum out of an adder-subtractor may be zero (all bits low) or negative (sign bit high). Show how to detect either condition.

☐ SOLUTION

Figure 6–32 shows a design using assertion-level logic. The sum bits go to a bubbled AND gate (the same as positive NOR gate). When all the bits are low, output ZERO is high. Because of the inverter, the final output $\overline{\text{ZERO}}$ is active-low. Therefore, when the sum is zero, $\overline{\text{ZERO}}$ is negative true and the left LED lights up.

If the sum is negative, the sign bit is high. To turn on the right LED, we need an active-low signal. This is why an inverter is used to produce $\overline{\text{MINUS}}$. Whenever the sum is negative, $\overline{\text{MINUS}}$ is negative true and the right LED lights up.

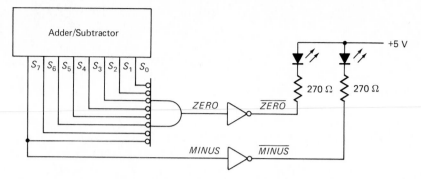

**FIG. 6–32.** Assertion-level logic diagram showing the detection of zero and minus accumulator contents

## S U M M A R Y

A chip is a small piece of semiconductor material with microminiature circuits on its surface. Small-scale integration (SSI) refers to chips with less than 12 gates. Medium-scale integration (MSI) means 12 to 100 gates per chip. Large-scale integration (LSI) refers to more than 100 gates on a chip.

The 7400 series is a line of standard TTL chips. This bipolar family contains a variety of compatible SSI and MSI devices. One way to recognize TTL design is the multiple-emitter input transistors and the totem-pole output transistors. The standard TTL chip has a power dissipation of about 10 mW per gate and a propagation delay time of around 10 ns.

By including a Schottky diode in parallel with the collector-base terminals, manufacturers produce Schottky TTL. This eliminates saturation delay time because it prevents the transistors from saturating. Numbered from 74S00, these devices have a power dissipation of 20 mW per gate and a propagation delay time of approximately 3 ns.

By increasing internal resistances and including Schottky diodes, manufacturers can produce low-power Schottky TTL devices (numbered from 74LS00). A low-power Schottky TTL gate has a power dissipation of around 2 mW per gate and a propagation delay time of approximately 10 ns. Low-power Schottky TTL is the most widely used of the TTL types.

A floating TTL input is equivalent to a high input. Do not use floating TTL inputs when you are operating in an electrically noisy environment. Floating inputs may pick up enough noise voltage to produce unwanted changes in the output states.

A standard TTL gate can sink 16 mA and source 400 $\mu$A. Since the maximum input currents are 1.6 mA (low state) and 40 $\mu$A (high state), standard TTL has a fanout of 10, meaning that one standard TTL gate can drive 10 others. Fanout has different values when you mix TTL types.

Open-collector devices have only the pull-down transistor; the pull-up transistor is omitted. Because of this, open-collector devices can be wire–ORed through a common pull-up resistor. This connection is inherently slow because the time constant is relatively long.

Three-state devices have replaced open-collector devices in most applications because they are much faster. These newer devices have a control input that can turn off the device. When this happens, the output floats and presents a high impedance to whatever it is connected to. Three-state devices are widely used for connecting to buses.

With positive logic, binary 1 represents high voltage and binary 0 represents low voltage. Also, positive true stands for high voltage and positive false for low voltage. With negative logic, binary 1 stands for low voltage and binary 0 for high voltage. In this system, negative true is equivalent to low voltage and negative false to high voltage.

With assertion-level logic, we draw gates and other devices with bubbled pins for active-low signals. Also, signal voltages are labeled with abbreviations of statements that describe circuit behavior. An overbar is used on a label whenever the signal is active-low.

## G L O S S A R Y

*Active-Low* Normally, a signal must be high to do something. Active-low refers to the opposite concept: a signal must be low to cause something to happen or to indicate that something has happened.

*Assert* To activate. If an input line has a bubble on it, you assert the input by making it low. If there is no bubble, you assert the input by making it high.

*Bipolar* Having two types of charge carriers; free electrons and holes.

*Bus* A group of wires that transmits binary data. The data bus of a first-generation microcomputer has eight wires, each carrying 1 bit. This means that the data bus can transmit 1 byte at a time. Typically, the byte represents an instruction or data word that is moved from one register to another.

*Chip* A small piece of semiconductor material. Sometimes chip refers to an IC device including its pins.

*Fanout* The maximum number of TTL loads that a TTL device can reliably drive.

*Low-Power Schottky TTL* A modification of standard TTL in which larger resistances and Schottky diodes are used. The larger resistances decrease the power dissipation, and the Schottky diodes increase the speed.

*Negative True* A signal is negative true when the voltage is low.

*Noise Immunity* The amount of noise voltage that causes unreliable operation. With TTL it is 0.4 V. As long as the noise voltages induced on connecting lines are less than 0.4 V, the TTL devices will work reliably.

*Positive True* A signal is positive true when the voltage is high.

*Saturation Delay Time* The time delay encountered when a transistor tries to come out of the saturation region. When the base drive switches from high to low, a transistor cannot instantaneously come out of hard saturation; extra carriers must first flow out of the base region.

*Schmitt Trigger* A digital circuit that produces a rectangular output. The input waveform may be sinusoidal, triangular, distorted, and so on. The output is always rectangular.

*Sink* A place where something is absorbed. When saturated, the lower transistor in a totem-pole output acts as a current sink because conventional charges flow through the transistor to ground.

*Source* The upper transistor of a totem-pole output acts as a source because conventional flow is out of the emitter into the load.

*Standard TTL* The basic TTL design. It has a power of dissipation of 10 mW per gate and a propagation delay time of 10 ns.

*Three-State TTL* A modified TTL design that allows us to connect outputs directly. Earlier computers used open-collector devices with their buses, but the passive pull-up severely limited the operating speed. By replacing open-collector devices with three-state devices, we can significantly reduce the switching time needed to change from the low state and the output state. The result is faster data changes on the bus, which is equivalent to speeding up the operation of a computer.

# PROBLEMS

**SECTION 6-1:**

**6–1.** For each of the following chips, identify whether the device is in the SSI, MSI, or LSI category:
  **a.** 234 gates      **c.** 8 gates
  **b.** 57 gates       **d.** 108 gates

**SECTION 6-2:**

**6–2.** Figure 6–1 shows typical resistance values at room temperature. Here $A$ is high, and $B$ is grounded. Allowing 0.7 V for the $V_{BE}$ drop, how much current is there through the 4 k$\Omega$?

**6–3.** Suppose you need a TTL device with a power dissipation of less than 5 mW per gate and a delay time of less than 20 ns. What TTL type would you choose?

**6–4.** Use the values of Table 6–3 to calculate the total propagation delay time of three cascaded gates for each of the following TTL types:
  **a.** Low-power          **d.** High-speed
  **b.** Low-power Schottky  **e.** Schottky
  **c.** Standard

**SECTION 6-3:**

**6–5.** What is the fanout of a 74S00 device when it drives low-power TTL loads?

**6–6.** What is the fanout of a low-power Schottky device driving standard TTL devices?

**6–7.** What is the fanout of a standard TTL device driving a 74LS device?

*SECTION 6-4:*

**6-8.** What is the fanout of a 7437 buffer when it drives standard TTL loads?

*SECTION 6-5:*

**6-9.** What is the output in Fig. 6–13 for these inputs?
   **a.** $ABCD = 0000$
   **b.** $ABCD = 0101$
   **c.** $ABCD = 1100$
   **d.** $ABCD = 1111$

**6-10.** Is the output $Y$ of Fig. 6–33 low or high for these conditions?
   **a.** Both switches open, $A$ is low
   **b.** Both switches closed, $A$ is high
   **c.** Left switch open, right switch closed, $A$ is low
   **d.** Left switch closed, right switch open, $A$ is high

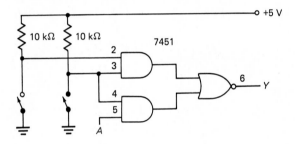

**FIG. 6–33.**

**6-11.** If all inputs are low in Fig. 6–16c, what is the output? If all inputs are high, what is the output?

**6-12.** What is the value of $Y$ in Fig. 6–34 for each of these?
   **a.** $ABCD = 0000$        **c.** $ABCD = 1000$
   **b.** $ABCD = 0101$        **d.** $ABCD = 1111$

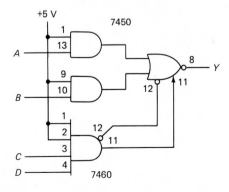

**FIG. 6–34.**

*SECTION 6-6:*

**6–13.** In Fig. 6–17a, $I_{OL,max} = 16$ mA. If three open-collector gates like these are wire–ORed together as shown in Fig. 6–17c, what is the minimum value of pull-up resistance needed to avoid destroying any device?

**6–14.** Suppose the total output capacitance is 20 pF in Fig. 6–17c. If the pull-up resistance equals 3.6 kΩ, what does the charging time constant equal?

*SECTION 6-7:*

**6–15.** You want the contents of register $B$ to appear on the bus of Fig. 6–21. What are the necessary DISABLE values?

*SECTION 6-8:*

**6–16.** In Fig. 6–35, what does output $Y$ equal when each switch is open? When either switch is closed?

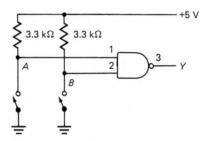

**FIG. 6–35.**

**6–17.** What is the current drain through the pull-up resistors when both switches are closed in Fig. 6–35? What is the time constant for each input when the switches are open?

**6–18.** In Fig. 6–36, what does the output $Y$ equal when either switch is open? When both are closed? This is not a preferred method of driving TTL loads. Try to figure out two reasons why this circuit is not as good as the circuit shown in Fig. 6–35.

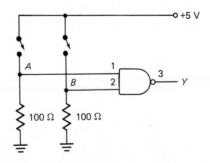

**FIG. 6–36.**

**6–19.** In Fig. 6–37*a*, the TTL output voltage is 0.4 V, and the LED voltage is 2 V. What is the sink current when the LED is lighted?

**6–20.** What is the LED current in Fig. 6–37*b* if the LED voltage drop is 2 V and the TTL output is high? If the TTL output is 0.4 V, what is the LED current?

## SECTION 6-10:

**6–21.** In Fig. 6–38, is each of the following an active-low or an active-high?

    **a.** Pin 1      **d.** Pin 5
    **b.** Pin 2      **e.** Pin 6
    **c.** Pin 3

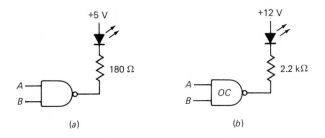

**FIG. 6–37.**

**6–22.** An 8085 microprocessor uses the following labels with assertion-level logic. Is each signal active-low or active-high?

    **a.** $\overline{\text{HOLD}}$      **e.** ALE
    **b.** $\overline{\text{RESET IN}}$      **f.** INTR
    **c.** $\overline{\text{RD}}$      **g.** $\overline{\text{INTA}}$
    **d.** $\overline{\text{WR}}$

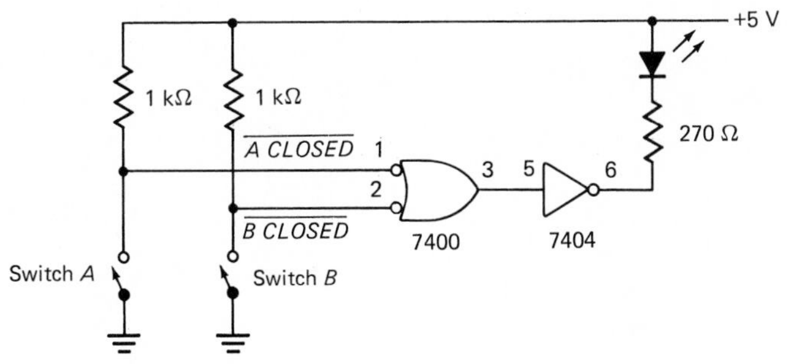

**FIG. 6–38.**

**6–23.** When switch *B* of Fig. 6–38 is closed, is the LED on or off? For this condition, is $\overline{\text{B CLOSED}}$ negative true or negative false?

**6–24.** Assume you have a 16-bit adder-subtractor that produces a sum $S_{15} \cdots S_0$. Design a logic circuit with TTL chips that turns on a LED when all the sum bits are high.

**6–25.** Do as in Prob. 6–24, except now turn on a LED when all sum bits are low.

**6–26.** In Fig. 6–38, a logic probe at pin 6 indicates a high state. If you trigger a logic pulser at pin 1 and the probe does not blink, which of these is a possible trouble?

    **a.** Pin 1 is shorted to ground.
    **b.** Pin 2 is shorted to +5 V.
    **c.** Pin 3 is shorted to ground.
    **d.** Pin 5 is shorted to +5 V.
    **e.** Pin 6 is shorted to ground.

# 7

## C M O S
## C I R C U I T S

Complementary metal-oxide semiconductor (CMOS) devices are chips that combine *p*-channel and *n*-channel MOSFETs in a push-pull arrangement. Because the input current of a MOSFET is much smaller than that of a bipolar transistor, cascaded CMOS devices have very low power dissipation compared with transistor-transistor logic (TTL) devices. This low dissipation explains why CMOS circuits are used in battery-powered equipment such as pocket calculators, digital wristwatches, and portable computers.

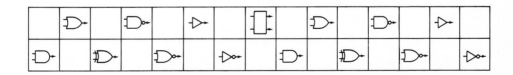

## 7-1 ENHANCEMENT-TYPE MOSFETS

The metal-oxide semiconductor FET (field-effect transistor), or MOSFET, has a source, gate, and drain. There are two types of MOSFETs: the depletion type and the enhancement type. Before we discuss CMOS circuits, it may be helpful to review the enhancement-type MOSFET because it is used in CMOS technology.

### Structure

Figure 7–1a shows an n-channel enhancement-type MOSFET. The p region, called the *substrate*, lies between two regions of n material. Free electrons can flow from the *source* (lower n region) to the *drain* (upper n region). A thin layer of silicon dioxide ($SiO_2$) is deposited on the left side of the substrate. Silicon dioxide is the same as glass, which is an insulator. The *gate* is a metallic electrode that controls the flow of free electrons between the source and the drain.

### Inversion Layer

Figure 7–1b shows the normal biasing polarities. When $V_{GS} = 0$, the $V_{DD}$ supply tries to force free electrons from the source to the drain, but the p substrate has only a few thermally produced electrons. Aside from these minority carriers and some surface leakage current, the current between the source and the drain is zero. For this reason, an enhancement-type MOSFET is also called a *normally off* MOSFET.

The gate and p substrate are like the two plates of a capacitor separated by a dielectric ($SiO_2$). When the gate is positive, it can attract enough free electrons from the source to form a thin layer of free electrons between the source and the drain. The effect is equivalent to creating a thin layer of n-type material next to the silicon dioxide. This layer of free electrons is called an *n-type inversion layer*. When this layer of free electrons appears, current flows easily between the source and the drain.

### Threshold Voltage

The minimum gate voltage that creates the n-type inversion layer is called the *threshold voltage* $V_{GS,\text{th}}$. When $V_{GS}$ is less than $V_{GS,\text{th}}$, zero current flows from the source to the drain. But when $V_{GS}$ is greater than $V_{GS,\text{th}}$, an n-type inversion layer connects the source

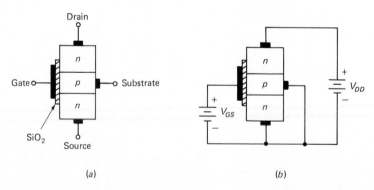

**FIG. 7–1** (a) n-channel enhancement-type MOSFET (b) Normal biasing polarities

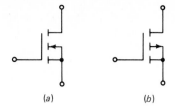

**FIG. 7–2** Schematic symbols for enhancement-type MOSFETs (a) n-channel (b) p-channel

to the drain, and we get current. The threshold voltage $V_{GS,\text{th}}$ can vary from less than 1 to more than 5 volts (V), depending on the particular device being used.

## Schematic Symbol

When $V_{GS} = 0$, the enhancement-type MOSFET is off because there is no conducting channel between the source and the drain. The schematic symbol of Fig. 7–2a has a broken line to indicate the normally off condition. A gate voltage greater than the threshold voltage creates an n-type inversion layer that connects the source and the drain. The arrow points to this inversion layer, which acts as an n channel when the device is conducting.

There is also a p-channel enhancement-type MOSFET. The schematic symbol of this complementary MOS transistor is similar, except that the arrow points outward, as shown in Fig. 7–2b. With a p-channel enhancement-type MOSFET, all voltages and currents are complementary to those of the n-channel enhancement-type MOSFET.

## 7-2 MOS INVERTERS

In this section we discuss three different ways to build an inverter. We use first a MOS driver and a passive load resistor, then two NMOS transistors, and finally p- and n-channel MOSFETs. This will show how a simple switching circuit evolves through NMOS technology to CMOS technology.

### Passive Load

Figure 7–3a shows a MOSFET driver and a passive load (resistor $R_D$). In this switching circuit, $v_{\text{in}}$ is either low or high, and the MOSFET acts as a switch that is either off or on. Figure 7–3b shows the load line, which emphasizes the switching action between the saturation and cutoff points. When $v_{\text{in}}$ is low, the MOSFET is cut off, and the operating point is at the lower end of the load line. When $v_{\text{in}}$ is high, the MOSFET is saturated, and the operating point is ideally at the upper end of the load line.

### Active Load

With MOSFET technology it is easier to fabricate a MOSFET than a resistor. For this reason, another MOS transistor can be used as the load resistance (see Fig. 7–4). Because the gate of the upper MOSFET is connected to the drain, this MOSFET is always conduct-

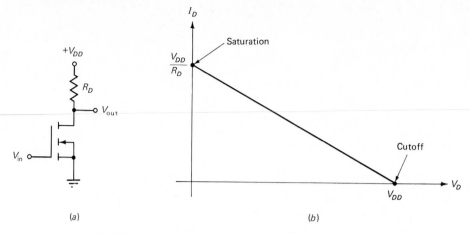

**FIG. 7–3** (*a*) MOSFET driver and passive load (*b*) Load line

ing, and it behaves as a resistor. By deliberate design, the upper MOSFET has a resistance that is greater than 10 times the on resistance of the lower MOSFET. For this reason, the upper MOSFET acts approximately as a resistor, and the lower MOSFET acts as a switch.

Using an MOS driver (lower MOSFET) and an MOS load (upper MOSFET) leads to much smaller integrated circuits because MOSFETs take up less room than resistors. This is why MOS technology dominates in computer applications; it allows many more gates on the surface of a chip than bipolar technology. In Fig. 7–4, both transistors are *n*-channel devices. Designs like these are an example of NMOS technology, which is used a lot for computer memory chips.

## CMOS Inverter

When $Q_2$ is saturated in Fig. 7–4, it has a resistance symbolized by $r_{DS,\text{on}}$, typically a few hundred ohms. Here $Q_1$ is always on and has an $r_{DS,\text{on}}$ at least 10 times greater. When $Q_1$

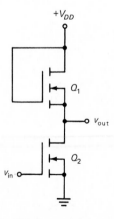

**FIG. 7–4** MOSFET driver and active load

and $Q_2$ are conducting, the same current flows through each. The product of this current and the supply voltage equals the power dissipation of the device with a low output.

One way to reduce the power dissipation is with a complementary MOSFET. For instance, Fig. 7–5a shows a CMOS inverter. Notice that $Q_1$ is a p-channel device and $Q_2$ is an n-channel device. This circuit is analogous to the push-pull bipolar amplifier of Fig. 7–5b. Because of the push-pull action, one device is normally on, and the other is off. Therefore, if $Q_1$ is on, $Q_2$ is off; if $Q_2$ is on, $Q_1$ is off.

When $v_{in}$ is low in Fig. 7–5a, $Q_2$ is off but $Q_1$ is on. This means the output voltage is high. On the other hand, when $v_{in}$ is high, $Q_2$ is on and $Q_1$ is off. In this case, the output voltage is low. Since the output voltage is always opposite in phase to the input voltage, the circuit is an inverter.

The CMOS inverter can be modified to build other CMOS logic circuits. The key advantage in using CMOS devices is the extremely low power consumption. Because both MOS transistors are in series, the current is the same through both. But the off transistor is the limiting factor because it ideally stops the current. Actually, there is small current through the off transistor called the *leakage current*. Since this current is typically in nanoamperes, the power dissipation is in nanowatts for either output state. Low power consumption is the reason for the popularity of CMOS devices in pocket calculators, digital wristwatches, and portable microcomputers.

The push-pull connection of Fig. 7–5a has another advantage over other MOS circuits. In either state, the output impedance is only a few hundred ohms. Because of this, the CMOS inverter can drive heavier loads than the circuits shown in Figs. 7–3 and 7–4.

The main disadvantage of CMOS devices is their slow speed. Typical propagation delay times are from 25 nanoseconds (ns) to more than 100 ns, depending on the particular device. By contrast, standard TTL devices have a typical delay time of only 10 ns. Remember that the total propagation delay time equals the sum of the individual delay times. Therefore, several cascaded CMOS gates can really slow things down. For instance, if four CMOS devices each have a propagation delay time $t_p = 50$ ns, the overall delay time is 200 ns.

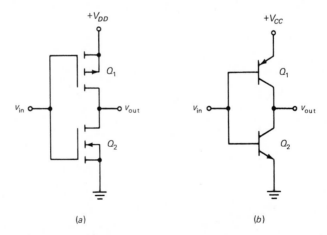

**FIG. 7–5**   (a) CMOS inverter (b) Analogous bipolar circuit

## 7-3 74C00 DEVICES

National Semiconductor Corporation pioneered the 74C00 series, a line of CMOS circuits that are pin-for-pin and function-for-function compatible with TTL devices of similar numbers. For instance, the 74C00 is a quad 2-input NAND gate, the 74C02 is a quad 2-input NOR gate, and so on. This CMOS family contains a variety of small-scale integration (SSI) and medium-scale integration (MSI) chips that allow you to replace many TTL designs by the comparable CMOS designs. This is useful if you are trying to build battery-powered equipment.

### NAND Gate

Figure 7–6 shows a CMOS NAND gate. The complementary design of input and output stages is typical of CMOS devices. Notice that $Q_1$ and $Q_2$ form one complementary connection; $Q_3$ and $Q_4$ form another. Visualize these transistors as switches. Then a low $A$ input will close $Q_1$ and open $Q_2$; a high $A$ input will open $Q_1$ and close $Q_2$. Similarly, a low $B$ input will open $Q_3$ and close $Q_4$; a high $B$ input will close $Q_3$ and open $Q_4$.

In Fig. 7–6, the $Y$ output is pulled up to the supply voltage when either $Q_1$ or $Q_4$ is conducting. The output is pulled down to ground only when $Q_2$ and $Q_3$ are conducting. If you keep this in mind, it simplifies the following discussion.

**CASE 1** Here $A$ is low and $B$ is low. Because $A$ is low, $Q_1$ is closed. Therefore, $Y$ is pulled high through the small resistance of $Q_1$.

**CASE 2** Now $A$ is low and $B$ is high. Since $A$ is still low, $Q_1$ remains closed and $Y$ stays in the high state.

**CASE 3** The $A$ input is high and the $B$ is low. Because $B$ is low, $Q_4$ is closed. This pulls $Y$ up to the supply voltage through the small resistance of $Q_4$.

**CASE 4** The $A$ is high, and the $B$ is high. When both inputs are high, $Q_2$ and $Q_3$ are closed, pulling the output down to ground.

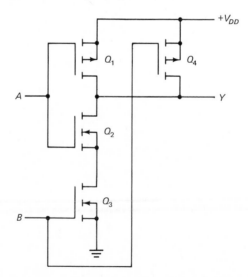

**FIG. 7–6** CMOS NAND gate

| **TABLE 7–1** *CMOS NAND Gate* | | | **TABLE 7–2** | | |
|---|---|---|---|---|---|
| A | B | Y | A | B | Y |
| Low | Low | High | Low | Low | High |
| Low | High | High | Low | High | Low |
| High | Low | High | High | Low | Low |
| High | High | Low | High | High | Low |

Table 7–1 summarizes all input-output possibilities. As you can see, this is the truth table of a positive NAND gate. The output is low only when all inputs are high.

To produce the positive AND function, we can connect the output of Fig. 7–6 to a CMOS inverter.

## NOR Gate

Figure 7–7 shows a CMOS NOR gate. The output goes high only when $Q_1$ and $Q_2$ are closed. The output goes low if either $Q_3$ or $Q_4$ is closed. There are the four possible cases:

**CASE 1** The $A$ is low, and the $B$ is low. For both inputs low, $Q_1$ and $Q_2$ are closed. Therefore, $Y$ is pulled high through the small series resistance of $Q_1$ and $Q_2$.

**CASE 2** The $A$ is low, and the $B$ is high. Because $B$ is high, $Q_3$ is closed, pulling the output down to ground.

**CASE 3** The $A$ is high, and the $B$ is low. With $A$ high, $Q_4$ is closed. The closed $Q_4$ pulls the output low.

**CASE 4** The $A$ is high, and the $B$ is high. Since $A$ is still high, $Q_4$ is still closed and the output remains low.

Table 7–2 summarizes these possibilities. As you can see, this is the truth table of a positive NOR gate. The output is low when any input is high.

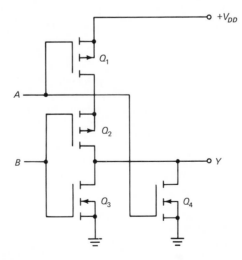

**FIG. 7–7** CMOS NOR gate

## Propagation Delay Time

A standard CMOS gate has a propagation delay time $t_p$ of approximately 25 to 100 ns, with the exact value depending on the power supply voltage and other factors. As you recall, $t_p$ is the time it takes for the output of a gate to change after its inputs have changed. When two or more CMOS gates are cascaded, you have to add the propagation delay times to get the total. For instance, if you cascade three CMOS gates each with a $t_p$ of 50 ns, then the total propagation delay time is 150 ns.

## Power Dissipation

The *static power dissipation* of a device is its average power dissipation when the output is constant. The static power dissipation of a CMOS gate is in nanowatts. For instance, a 74C00 has a power dissipation of approximately 10 nanowatts (nW) per gate. This dissipation equals the product of supply voltage and leakage current, both of which are dc quantities.

When a CMOS output changes from the low state to the high state (or vice versa), the average power dissipation increases. Why? The reason is that during a transition between states, there is a brief period when both MOSFETs are conducting. This produces a spike (quick rise and fall) in the supply current. Furthermore, during a transition, any stray capacitance across the output has to be charged before the output voltage can change. This capacitive charging draws additional current from the power supply. Since power equals the product of supply voltage and device current, the instantaneous power dissipation increases, which means the average power dissipation is higher.

The average power dissipation of a CMOS device whose output is continuously changing is called the *active power dissipation*. How large is the active power dissipation? This depends on the frequency at which the output is switching states. When the operating frequency increases, the current spikes occur more often and active power dissipation increases. Figure 7–8 shows the active power dissipation of a 74C00 versus frequency for a load capacitance of 50 picofarads (pF). As you see, the power dissipation per gate increases with frequency and supply voltage. For frequencies in the megahertz region, the gate dissipation approaches or exceeds 10 mW (TTL gate dissipation). For CMOS to have an advantage over TTL, you operate CMOS devices at lower frequencies.

Another way to reduce power dissipation is to decrease the supply voltage. But this has adverse effects because it increases propagation time and decreases noise immunity. Although CMOS devices can work over a range of 3 to 15 V, the best compromise for speed, noise immunity, and overall performance is a supply voltage from 9 to 12 V. From now on, we assume a supply voltage of 10 V, unless otherwise specified.

Incidentally, notice the use of $V_{CC}$ rather than $V_{DD}$ for the supply voltage. This is a carryover from TTL circuits. You will find $V_{CC}$ on the data sheets for 74C00 devices. Other CMOS devices such as the 4000 series (discussed soon) use $V_{DD}$ for the supply voltage.

## 54C00 Series

Any device in the 74C00 series works over a temperature range of $-40$ to $+85°C$. This is adequate for most commercial applications. The 54C00 series (for military applications)

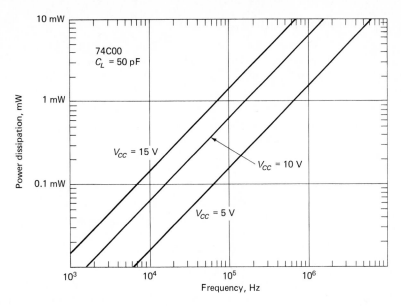

**FIG. 7–8** Active power dissipation of 74C00

works over a temperature range of $-55$ to $+125°C$. Although 54C00 devices can be substituted for 74C00 devices, they are rarely used commercially because of their much higher cost.

## 74HC00 Devices

The main disadvantage of CMOS devices is their relatively long propagation delay times. This places a limit on the maximum operating frequency of a system. The 74HC00 series is a CMOS series of devices that are pin-for-pin and function-for-function compatible with TTL devices. These devices have the advantage of higher speed (less propagation delay time).

## CD4000 Series

RCA was the first to introduce CMOS devices. The original devices were numbered from CD4000 upward. This 4000 series was soon replaced by the 4000A series (called conventional) and the 4000B series (called the buffered type). The 4000A and B series are widely used; they have many functions not available in the 74C00 series. The main disadvantage of 4000 devices is their lack of pin-for-pin and function-for-function compatibility with TTL.

## 7-4 CMOS CHARACTERISTICS

74C00 series devices are guaranteed to work reliably over a temperature range of $-40$ to $+85°C$ and over a supply range of 3 to 15 V. In the discussion that follows, *worst case* means the parameters are measured under the worst conditions of temperature and voltage.

## Floating Inputs

When a TTL input is floating, it is equivalent to a high input. You can use a floating TTL input to simulate a high input; but as already pointed out, it is better to connect unused TTL inputs to the supply voltage. This prevents the floating leads from picking up stray noise in the environment.

If you try to float a CMOS input, however, not only do you set up a possible noise problem, but, much worse, you produce excessive power dissipation. Because of the insulated gates, a floating input allows the gate voltage to drift into the linear region. When this happens, excessive current can flow through push-pull stages.

The absolute rule with CMOS devices, therefore, is to *connect all input pins*. Most of or all the inputs are normally connected to signal lines. If you happen to have an input that is unused, connect it to ground or the supply voltage, whichever prevents a stuck output state. For instance, with a positive NOR gate you should ground an unused input. Why? Because returning the unused NOR input to the supply voltage forces the output into a stuck low state. On the other hand, grounding an unused NOR input allows the other inputs to control the output.

With a positive NAND gate, you should connect an unused input to the supply voltage. If you try grounding an unused NAND input, you disable the gate because its output will stick in the high state. Therefore, the best thing to do with an unused NAND input is to tie it to the supply voltage. A direct connection is all right; CMOS inputs can withstand the full supply voltage.

## Easily Damaged

Because of the thin layer of silicon dioxide between the gate and the substrate, CMOS devices have a very high input resistance, approximately infinite. The insulating layer is kept as thin as possible to give the gate more control over the drain current. Because this layer is so thin, it is easily destroyed by excessive gate voltage.

Aside from directly applying an excessive gate voltage, you can destroy the thin insulating layer in more subtle ways. If you remove or insert a CMOS device into circuit while the power is on, transient voltages caused by inductive kickback and other effects may exceed the gate voltage rating. Even picking up a CMOS IC may deposit enough charge to exceed its gate voltage rating.

One way to protect against overvoltages is to include zener diodes across the input. By setting the zener voltage below the breakdown voltage of the insulating layer, manufacturers can prevent the gate voltage from becoming destructively high. Most CMOS ICs include this form of zener protection.

Figure 7–9 shows a typical *transfer characteristic* (input-output graph) of a CMOS inverter. When the input voltage is in the low state, the output voltage is in the high state. As the input voltage increases, the output remains in the high state until a threshold is reached. Somewhere near an input voltage of $V_{CC}/2$, the output will switch to the low state. Then any input voltage greater than $V_{CC}/2$ holds the output in the low state.

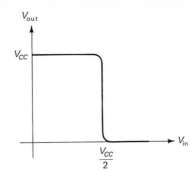

**FIG. 7-9**  Typical transfer characteristic of CMOS gate

This transfer characteristic is an improvement over TTL. Why? Because the indeterminate region is much smaller. As you can see, the input voltage has to be nearly equal to $V_{CC}/2$ before the CMOS output switches states. This implies that the noise immunity of CMOS devices ideally approaches $V_{CC}/2$. Typically, noise immunity is 45 percent of $V_{CC}$.

Also notice how much better defined the low and high output states are. When CMOS loads are used, the CMOS source and sink transistors have almost no voltage drop because there is almost no input current to a CMOS load. Therefore, the static currents are extremely small. For this reason, the high output voltage is approximately equal to $V_{CC}$, and the low output voltage is approximately at ground. Stated another way, the logic swing between the low and high output states approximately equals the supply voltage, a considerable advantage for CMOS over TTL.

## Compatibility

CMOS devices are compatible with one another because the output of any CMOS device can be used as the input to another CMOS device, as shown in Fig. 7-10a. For instance, Fig. 7-10b shows the output stage of a CMOS driver connected to the input stage of a CMOS load. The supply voltage is +10 V. Ideally, the ideal input switching level is +5 V. Since the CMOS driver has a low output, the CMOS load has a high input.

Similarly, Fig. 7-10c shows a high CMOS driver output. This is more than enough voltage to drive the CMOS load with a high-state input. In fact, the noise immunity typically approaches 4.5 V (from 45 percent of $V_{CC}$). Any noise picked up on the connecting line between devices would need a peak value of more than 4.5 V to cause unwanted switching action.

## Sourcing and Sinking

When a standard CMOS driver output is low (Fig. 7-10b), the input current to the CMOS load is only 1 microampere ($\mu$A) (worst case shown on data sheet). The input current is so low because of the insulated gates. This means that the CMOS driver has to sink only 1 $\mu$A. Similarly, when the driver output is high (Fig. 7-10c), the CMOS driver is sourcing 1 $\mu$A.

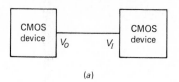

(a)

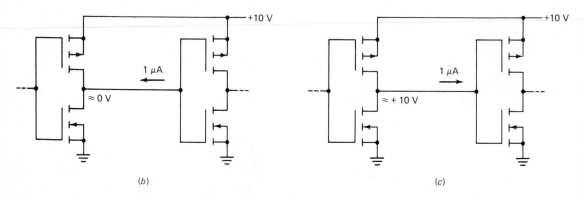

(b)                                          (c)

**FIG. 7–10** (*a*) Output of CMOS device can drive input of another CMOS device (*b*) Sink current (*c*) Source current

In symbols, here are the worst-case input currents for CMOS devices:

$$I_{IL,\text{max}} = -1 \ \mu A \qquad I_{IH,\text{max}} = 1 \ \mu A$$

We use these values to calculate the fanout.

### Fanout

The fanout of CMOS devices depends on the kind of load being driven. In Sec. 7–5, we discuss CMOS devices driving TTL devices. Now we want to concentrate on CMOS driving CMOS. Data sheets for 74C00 series devices give the following output currents for CMOS driving CMOS:

$$I_{OL,\text{max}} = 10 \ \mu A \qquad I_{OH,\text{max}} = -10 \ \mu A$$

Since the worst-case input current of a CMOS device is only 1 $\mu A$, a CMOS device can drive up to 10 CMOS loads. Therefore, you can use a fanout of 10 for CMOS-to-CMOS connections. This value is reliable under all operating conditions.

## 7-5 TTL-TO-CMOS INTERFACE

The word *interface* refers to the way a driving device is connected to a loading device. In this section, we discuss methods for interfacing CMOS devices to TTL devices. Recall that TTL devices need a supply voltage of 5 V, while CMOS devices can use any supply voltage from 3 to 15 V. Because the supply requirements differ, several interfacing schemes may be used. Here are a few of the more popular methods.

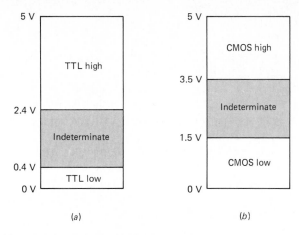

FIG. 7–11   (*a*) TTL output profile (*b*) CMOS input profile

## Supply Voltage at 5 V

One approach to TTL/CMOS interfacing is to use +5 V as the supply voltage for both the TTL driver and the CMOS load. In this case, the worst-case TTL output voltages (Fig. 7–11*a*) are almost compatible with the worst-case CMOS input voltages (Fig. 7–11*b*). *Almost*, but not quite. There is no problem with the TTL low-state window (0 to 0.4 V) because it fits inside the CMOS low-state window (0 to 1.5 V). This means the CMOS load always interprets the TTL low-state drive as a low.

The problem is in the TTL high state, which can be as low as 2.4 V (see Fig. 7–11*a*). If you try using a TTL high-state output as the input to a CMOS device, you get indeterminate action. The CMOS device needs at least 3.5 V for a high-state input (Fig. 7–11*b*). Because of this, you cannot get reliable operation by connecting a TTL output directly to a CMOS input. You have to do something extra to make the two different devices compatible.

What do you do? The standard solution is to use a pull-up resistor between the TTL driver and the CMOS load, as shown in Fig. 7–12. What effect does the pull-up resistor have? It has almost no effect on the low state, but it does raise the high state to approximately +5 V. For instance, when the TTL output is low, the lower end of the 3.3 kilohms (kΩ) is grounded (approximately). Therefore, the TTL driver sinks a current of roughly

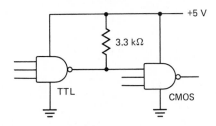

FIG. 7–12   TTL driver and CMOS load

$$I = \frac{5 \text{ V}}{3.3 \text{ k}\Omega} = 1.52 \text{ milliamperes (mA)}$$

When the TTL output is in the high state, however, the output voltage is pulled up to +5 V. Here is how it happens. As before, the upper totem-pole transistor actively pulls the output up to +2.4 V (worst case). Because of the pull-up resistor, however, the output rises above +2.4 V, which forces the upper totem-pole transistor into cutoff. The pull-up action is now passive because the supply voltage is pulling the output voltage up to +5 V through the pull-up resistor.

The gate capacitance of the CMOS load has to be charged through the pull-up resistor. This slows down the switching action. If speed is important, you can decrease the pull-up resistance. The minimum resistance is determined by the maximum sink current of the TTL device: $I_{OL,\text{max}} = 16$ mA. In the worst case, the supply voltage may be as high as 5.25 V, so the minimum resistance is

$$R_{\text{min}} = \frac{5.25 \text{ V}}{16 \text{ mA}} = 328 \text{ ohms } (\Omega)$$

The nearest standard value is 330 $\Omega$, which you should consider the absolute minimum value for the pull-up resistor. And you would use this only if switching speed were critical. In many applications, a pull-up resistance of 3.3 k$\Omega$ is fine.

Incidentally, the other inputs of the TTL driver and CMOS load (Fig. 7–12) are connected to signal lines not shown. Also, the use of 3-input gates is arbitrary. You can interface gates with any number of inputs. If more than one TTL chip is being interfaced to the CMOS load, connect each TTL driver to a separate pull-up resistor and CMOS input.

## Different Supply Voltages

CMOS performance deteriorates at lower voltages because the propagation delay time increases and the noise immunity decreases. Therefore, it is better to run CMOS devices with a supply voltage between 9 and 12 V. One way to use a higher supply voltage is with an open-collector TTL driver, as shown in Fig. 7–13. Recall that the output stage of an open-collector TTL device consists only of a sink transistor with a floating collector. In Fig. 7–13, this open collector is connected to a supply voltage of +12 V through a pull-up resistance of 6.8 k$\Omega$. Likewise, the CMOS device now has a supply voltage of +12 V.

When the TTL output is low, we can visualize a ground on the lower end of the pull-up resistor. Therefore, the TTL device has to sink approximately

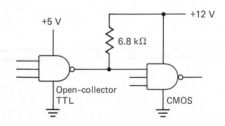

**FIG. 7–13** Open-collector TTL driver allows higher CMOS supply voltage

$$I_{\text{sink}} = \frac{12 \text{ V}}{6.8 \text{ k}\Omega} = 1.76 \text{ mA}$$

When the TTL output is high, the open-collector output rises passively to $+12$ V. In either case, the TTL outputs are compatible with the CMOS input states.

The passive pull-up in Fig. 7–13 produces slower switching action than before. For instance, with a gate input capacitance of 10 pF, the pull-up time constant is

$$RC = (6.8 \text{ k}\Omega)(10 \text{ pF}) = 68 \text{ ns}$$

If this is a problem, you can reduce the pull-up resistance to its minimum allowable value of

$$R_{\text{min}} = \frac{12 \text{ V}}{16 \text{ mA}} = 750 \text{ }\Omega$$

Then the pull-up time constant decreases to

$$RC = (750 \text{ }\Omega)(10 \text{ pF}) = 7.5 \text{ ns}$$

### External Transistor

Figure 7–14 shows a variation of the preceding scheme. Here, we are using a standard TTL device running off of $+5$ V. The TTL output goes to an external transistor with a pull-up resistance of 470 $\Omega$. The advantage of this approach is that we are not limited to a sink current of 16 mA. For instance, when the 2N3904 is saturated, the lower end of the 470 $\Omega$ is approximately grounded. Therefore, the sink current is

$$I_{\text{sink}} = \frac{12 \text{ V}}{470 \text{ }\Omega} = 25.5 \text{ mA}$$

The voltage divider between the TTL output and the base of the transistor increases the noise immunity to approximately 1.4 V because it takes 0.7 V to turn on the transistor. Also, the 68 pF is a *speed-up* capacitor; its charge is used to speed up the charging of the gate capacitance.

Another reason why you might need more sink current is to drive more than one CMOS load. For instance, suppose the collector of the 2N3904 drives eight CMOS de-

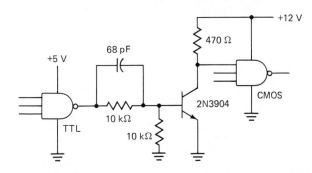

**FIG. 7–14** One way to increase sink current

vices. If each has a capacitance of 10 pF, then the total input capacitance is 80 pF. If the pull-up resistance is 470 $\Omega$, the pull-up time constant is

$$RC = (470\ \Omega)(80\ \text{pF}) = 37.6\ \text{ns}$$

This relatively fast (small) time constant is a direct result of the smaller pull-up resistance.

## CMOS Level Shifter

Figure 7–15 shows a 40109, called a *level shifter*. The input stage of the chip uses a supply voltage of +5 V, while the output stage uses +12 V. In other words, the input stage interfaces with TTL, and the output stage interfaces with CMOS.

In Fig. 7–15, a standard TTL device drives the level shifter. This produces active TTL pull-up to at least +2.4 V. Beyond this level, the pull-up resistor takes over and raises the voltage to +5 V, which ensures a valid high-state input to the level shifter. The output side of the level shifter connects to +12 V (this can be changed to any voltage from 3 to 15 V). Since the CMOS load runs off of +12 V, it has better propagation delay time and noise immunity.

In summary, TTL has to run off of +5 V, but CMOS does better with a supply voltage of +12 V. This is the reason for using a level shifter between the TTL driver and the CMOS load.

## 7-6 CMOS-TO-TTL INTERFACE

In this section, we discuss methods for interfacing CMOS devices to TTL devices. Again, the problem is to shift voltage levels until the CMOS output states fall inside the TTL input windows. Specifically, we have to make sure that the CMOS low-state output is always less than 0.8 V, the maximum allowable TTL low-state input voltage. Also, the CMOS high-state output must always be greater than 2 V, the minimum allowable TTL high-state input voltage.

### Supply Voltage at 5 V

One approach is to use +5 V as the supply voltage for the driver and the load, as shown in Fig. 7–16. A direct interface like this forces you to use a low-power Schottky TTL load (or two low-power TTL loads). Why? Because a low-power Schottky device has these worst-case input currents:

$$I_{IL,\text{max}} = -360\ \mu\text{A} \qquad I_{IH,\text{max}} = 20\ \mu\text{A}$$

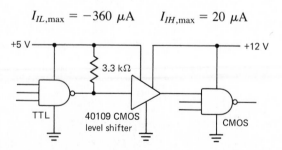

**FIG. 7–15**  CMOS level shifter allows the use of 5-V and 12-V supplies

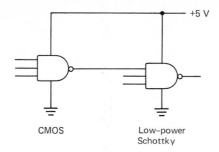

CMOS             Low–power<br>Schottky

**FIG. 7–16**   CMOS driver and low-power Schottky TTL load

Data sheets for 74C00 devices list these worst-case output currents for CMOS driving TTL:

$$I_{OL,\text{max}} = 360 \ \mu A \qquad I_{OH,\text{max}} = -360 \ \mu A$$

This tells us that a CMOS driver can sink 360 $\mu$A in the low state, exactly the input current for a low-power Schottky TTL device. On the other hand, the CMOS driver can source 360 $\mu$A, which is more than enough to handle the high-state input current (only 20 $\mu$A). So the sink current limits the CMOS/74LS fanout to 1.

    CMOS can also drive low-power TTL devices. The limiting factor again is the sink current. Low-power TTL has a worst-case low-state input current of 180 $\mu$A. Since a CMOS driver can sink 360 $\mu$A, it can drive two low-power TTL devices. Briefly stated, the CMOS/74L fanout is 2.

    CMOS cannot drive standard TTL directly because the latter requires a low-state input current of $-1.6$ mA, far too much current for a CMOS device to sink without entering the TTL indeterminate region. The problem is that the sink transistor of a CMOS device is equivalent to a resistance of approximately 1.11 k$\Omega$ (worst case). The CMOS output voltage equals the product of 1.6 mA and 1.11 k$\Omega$, which is 1.78 V. This is too large to be low-state TTL input.

## Using a CMOS Buffer

Figure 7–17 shows how to get around the fanout limitation just discussed. The CMOS driver now connects directly to a CMOS buffer, a chip with larger output currents. For instance, a 74C902 is a hex buffer, or six CMOS buffers in a single package. Each buffer has these worst-case output currents:

$$I_{OL,\text{max}} = 3.6 \ \text{mA} \qquad I_{OH,\text{max}} = 800 \ \mu A$$

Since a standard TTL load has low-state input current of 1.6 mA and a high-state input current of 40 $\mu$A, a 74C902 can drive two standard TTL loads. If you use one-sixth of a 74C902 in Fig. 7–17, the CMOS/TTL fanout is 2. Other available buffers are the CD4049A (inverting), CD4050A (noninverting), 74C901 (inverting), etc.

## Different Supply Voltages

CMOS buffers like the 74C902 can use a supply voltage of 3 to 15 V and an input voltage of $-0.3$ to 15V. The input voltage can be greater than the supply voltage without damag-

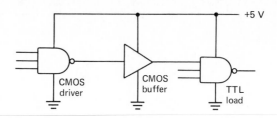

**FIG. 7-17**   CMOS buffer can drive standard TTL load

ing the device. For instance, you can use a high-state input of +12 V, even though the supply voltage is only +5 V.

Figure 7-18 shows how to use the previous idea to our advantage. Here, the supply pin of the CMOS driver is connected to +12 V. On the other hand, the supply pin of the CMOS buffer is connected to +5 V to produce the TTL interface. Therefore, the input to the CMOS buffer will be as much as +12 V, even if its supply voltage is only +5 V. The fanout of this interface is still two standard TTL loads.

## Open-Drain Interface

Recall open-collector TTL devices. The output stage consists of a sink transistor with a floating collector. Similar devices exist in the CMOS family. Known as *open-drain* devices, these have an output stage consisting only of a sink MOSFET. An example is the 74C906, a hex open-drain buffer.

Figure 7-19 shows how an open-drain CMOS buffer can be used as an interface between a CMOS driver and a TTL load. The supply voltage for most of the buffer is +12 V. The open drain, however, is connected to a supply voltage of +5 V through a pull-up resistance of 3.3 kΩ. This has the advantage that both the CMOS driver and the CMOS buffer run off of +12 V, except for the open-drain output which provides the TTL interface.

## Increasing Fanout

An open-drain buffer like that of Fig. 7-19 is limited to a fanout of two TTL loads because the maximum sink current is approximately 3.6 mA. One way to increase the fanout is with an external sink transistor, as shown in Fig. 7-20. When the CMOS output is low, the 2N3904 cuts off and its collector is pulled up to +5 V. However, when the CMOS output is high, the 2N3904 goes into saturation and acts as a sink.

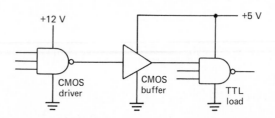

**FIG. 7-18**   CMOS driver runs better with 12-V supply

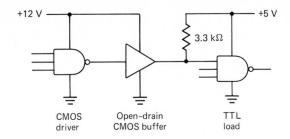

**FIG. 7–19**  Open-drain CMOS buffer increases sink current

The circuit of Fig. 7–20 is designed for a fanout of approximately 10 TTL loads. Here are the calculations. Visualize 10 TTL loads connected to the collector. Then the necessary output currents are

$$I_{OL,\max} = 16 \text{ mA} \qquad I_{OH,\max} = 400 \text{ }\mu\text{A}$$

When the 2N3904 is cut off, the collector voltage is pulled up to +5 V, less the voltage drop across the pull-up resistor. Since 400 $\mu$A of source current flows through the 3.3 k$\Omega$,

$$V_I = 5 \text{ V} - (400 \text{ }\mu\text{A})(3.3 \text{ k}\Omega) = 3.68 \text{ V}$$

If you really want to be precise, use the worst-case supply voltage of +4.75 V. Then

$$V_I = 4.75 \text{ V} - (400 \text{ }\mu\text{A})(3.3 \text{ k}\Omega) = 3.43 \text{ V}$$

This still leaves us with

$$\text{Noise immunity} = 3.43 \text{ V} - 2 \text{ V} = 1.43 \text{ V}$$

Next, consider the required low-state drive. When the 2N3904 is saturated, it has to sink 16 mA for the 10 TTL loads and approximately another 1.5 mA for the pull-up resistor. This means a total of 17.5 mA of collector current through the 2N3904. For hard saturation, the base current has to be one-tenth of the collector current; this implies a base current of approximately 1.75 mA. Allowing 0.7 V for the $V_{BE}$ of the 2N3904 and 1.11 k$\Omega$ for the conducting resistance of the CMOS source transistor, we can calculate the base current:

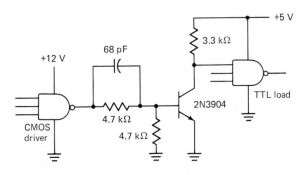

**FIG. 7–20**  Another way to increase sink current

$$I_B = \frac{12 \text{ V} - 0.7 \text{ V}}{4.7 \text{ k}\Omega + 1.1 \text{ k}\Omega} = 1.94 \text{ mA}$$

This is slightly more than the 1.75 mA required for hard saturation. Therefore, we have enough low-state drive for 10 TTL loads.

By modifying the design, you can increase the fanout further. The point of the circuit is that an external sink transistor allows you to drive more TTL loads than a CMOS buffer, which typically has a fanout of only two TTL loads.

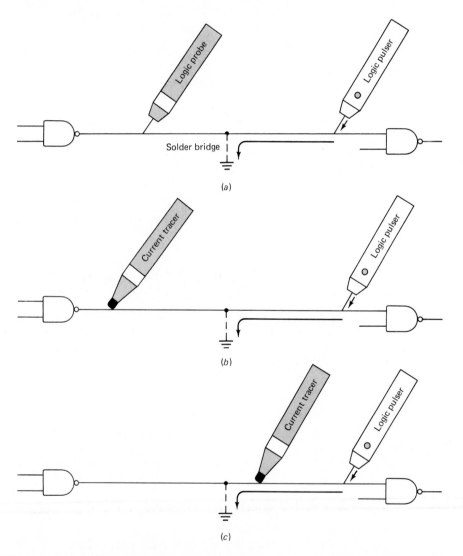

**FIG. 7–21** (*a*) Solder bridge shorts node to ground (*b*) Current tracer will not detect any current (*c*) Current tracer will detect current

# 7-7 TROUBLESHOOTING WITH A CURRENT TRACER

Figure 7–21a shows a solder bridge shorting a node to ground. When you trigger the logic pulser, the logic probe remains dark because the node is stuck in the low state. A logic pulser and logic probe will help you locate stuck nodes, but they cannot tell you the exact location of the short.

Figure 7–21b shows a *current tracer,* a troubleshooting tool than can detect current in a wire or circuit-board trace. Although it touches the wire, the current tracer does not make electric contract. Inside its blunt insulated tip is a small pick-up coil than can detect the magnetic field around a wire carrying current. Therefore, if there is any current through the wire, the current tracer lights up.

In Fig. 7–21b, each time you trigger the logic pulser, a conventional current flows through its tip to ground along the path shown. The current tracer will not detect this current because it is touching another part of the wire.

If you move the current tracer between the ground and the logic pulser as shown in Fig. 7–21c, the current tracer will light up. The critical position for the current tracer is directly over the short. Move left, and the current tracer goes out. Move right, and it turns on. When this happens, you know the current tracer is directly over the trouble. During troubleshooting, a visual check at this critical location usually reveals the nature of the trouble (a solder bridge, in this discussion).

That's the basic idea behind a current tracer. You use the logic pulser and logic probe to find stuck nodes. Then you use the current tracer to locate the exact position along a trace where the short is.

## SUMMARY

A CMOS inverter uses complementary MOSFETs in a push-pull arrangement. The key advantage of CMOS devices is the low power dissipation. The main disadvantage is the slow switching speed.

The 74C00 series is a line of CMOS circuits that are pin-for-pin and function-for-function compatible with TTL devices. The static power dissipation of 74C00 devices is approximately 10 nanowatts (nW) per gate. Active power dissipation is higher because of the current spikes during transitions. Lower supply voltages increase the propagation delay time and noise immunity. Higher supply voltages increase the power dissipation. The best compromise is a supply voltage from 9 to 12 V. The 74HC00 series is a line of high-speed CMOS devices. The CD4000 series is another line of CMOS devices with many functions not available in the 74C00 series.

CMOS devices are guaranteed to work reliably over a temperature range of −40 to +85°C and a supply range of 3 to 15 V. Unused inputs should be returned to the supply voltage or to ground, depending on which connection prevents a stuck output. A floating CMOS input is poor design because it produces large power dissipation. CMOS devices have a fanout of 10 when driving other CMOS devices. By using level shifting, CMOS devices can be interfaced with TTL devices.

*Active Load* A transistor that acts as a load for another transistor.

*Active Power Dissipation* The power dissipation of a device under switching conditions. It differs from static power dissipation because of the large current spikes during output transitions.

*CMOS Inverter* A push-pull connection of *p*- and *n*-channel MOSFETs.

*Compatibility* Ability of the output of one device to drive the input of another device.

*Interface* The way a driving device is connected to a loading device. All the circuitry between the output of a device and the input of another device.

*Noise Immunity* The amount of noise voltage that an input can tolerate without causing a false change in output state.

*Speed-up Capacitor* Whenever you use a voltage divider, the stray capacitance across the output resistor will slow down the switching speed unless you add a capacitor across the input resistor. This capacitor speeds up the charging of the stray capacitance across the output of the voltage divider.

*Static Power Dissipation* The product of dc voltage and current.

## P R O B L E M S

*SECTION 7-1:*

**7-1.** The MOSFET of Fig. 7–1*b* has a threshold voltage of 2 V. Is the device conducting or nonconducting for these conditions?
  **a.** $V_{GS} = 0$
  **b.** $V_{GS} = 1$ V
  **c.** $V_{GS} = +1.5$ V
  **d.** $V_{GS} = +2.5$ V

**7-2.** Is the threshold voltage positive or negative for each of these?
  **a.** *n*-channel MOSFET
  **b.** *p*-channel MOSFET
  **c.** Fig. 7–2*a*
  **d.** Fig. 7–2*b*

*SECTION 7-2:*

**7-3.** Visualize the dc load line of Fig. 7–22*a*. What is the cutoff voltage? What is the saturation current? If the threshold voltage is $+2$ V, is the device conducting or nonconducting when $V_{GS} = +3$ V?

**7-4.** The upper MOSFET of Fig. 7–22*b* has a conducting resistance of 20 kΩ. If the lower MOSFET has an on resistance of 1 kΩ, what is the drain current when $V_{IN}$ is less than the threshold voltage? When $V_{IN}$ is greater than the threshold voltage?

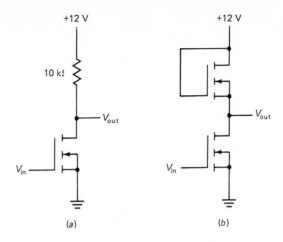

+12 V          +12 V

10 kΩ

$V_{out}$        $V_{out}$

$V_{in}$         $V_{in}$

(a)            (b)

**FIG. 7–22**

*SECTION 7-3:*

**7–5.** Three CMOS devices are cascaded. If each has a propagation delay time of 100 ns, what is the total propagation delay time?

**7–6.** A 74C00 has a load capacitance of 50 pF. If the supply voltage is 10 V, what is the active power dissipation per gate at each of the following frequencies?
   **a.** 1 kilohertz (kHz)
   **b.** 10 kHz
   **c.** 100 kHz

*SECTION 7-4:*

**7–7.** Figure 7–23a shows how to drive a CMOS device from a switch. As you see, the input does not float in either state. Is the output low or high when the switch is open? Is it low or high when the switch is closed?

**7–8.** Pin 13 is an unused input in Fig. 7–23b. As you see, it is grounded. Is the output low or high for each of these conditions?
   **a.** *A* and *B* both low

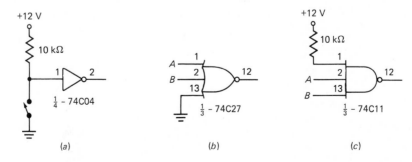

+12 V                                        +12 V

10 kΩ                                        10 kΩ

1 ▷ 2          A  1              A  1
               B  2    12           2    12
                  13               13
$\frac{1}{4}$ - 74C04    B         B
               $\frac{1}{3}$ - 74C27    $\frac{1}{3}$ - 74C11

(a)            (b)              (c)

**FIG. 7–23**

**b.** *A* low and *B* high

**c.** *A* high and *B* low

**d.** *A* and *B* both high

**7–9.** If pin 13 is returned to the supply voltage instead of grounded in Fig. 7–23*b*, the circuit is useless as a NOR gate. Why?

**7–10.** Pin 1 is an unused input in Fig. 7–23*c*. As you see, it is returned to the supply voltage through a pull-up resistor. Is the output low or high for each of these conditions?

**a.** *A* and *B* both low

**b.** *A* low and *B* high

**c.** *A* high and *B* low

**d.** *A* and *B* both high

**7–11.** If pin 1 is grounded instead of returned to the supply voltage in Fig. 7–23*c*, the circuit is useless as a NAND gate. Why?

*SECTION 7-5:*

**7–12.** If the CMOS input is low in Fig. 7–24, what is the sink current in the TTL driver? If the CMOS input is high, how much voltage drop is there across the 1.5 kΩ if the gate current is 1 $\mu$A (worst case)?

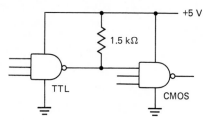

**FIG. 7–24**

**7–13.** Ideally, how much current does the open-collector driver of Fig. 7–25 have to sink when its output is low?

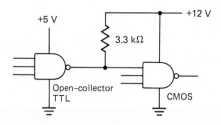

**FIG. 7–25**

**7–14.** If the gate input capacitance of Fig. 7–26 is 10 pF, what does the pull-up time constant equal? How much current will the 2N3904 sink?

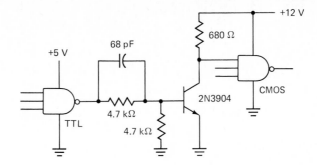

**FIG. 7–26**

*SECTION 7-6:*

**7–15.** Ideally, what is the sink current in Fig. 7–27? If the TTL load has a high-state input current of 40 $\mu$A, what is the voltage drop across the 2.2 k$\Omega$?

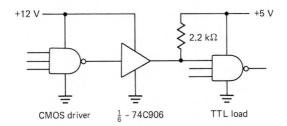

**FIG. 7–27**

**7–16.** If the input capacitance of the TTL load is 10 pF, what does the pull-up time constant equal in Fig. 7–27?

*SECTION 7-7:*

**7–17.** In Fig. 7–24, a logic probe indicates that the lower end of the pull-up resistor is stuck in the low state. Using a logic pulser and current tracer, you detect a current in the wire between this resistor and the TTL output. Which of the following is a possible trouble?

**a.** 1.5 k$\Omega$ shorted
**b.** 1.5 k$\Omega$ open
**c.** Open trace between the TTL output and the resistor
**d.** TTL sink transistor shorted

**7–18.** The lower end of the pull-up resistor (Fig. 7–24) is stuck in the high state. With a logic pulser and current tracer, you detect a current in the wire between this resistor and the CMOS input. Which is a possible trouble?

**a.** CMOS input trace shorted to supply voltage
**b.** CMOS input grounded
**c.** TTL output trace open
**d.** TTL output shorted to ground

# 8

The outputs of the digital circuits considered in previous chapters are dependent entirely on the inputs to these circuits; that is, if the input changes, the output also changes. However, there are requirements for a digital device or circuit whose output will remain unchanged, once set, even if there is a change in input. Such a device could be used, for example, to store a binary number. A flip-flop is one such circuit, and the characteristics of the most common types of flip-flops used in digital systems are considered in this chapter.

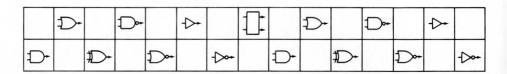

## 8·1  *RS* FLIP-FLOP

Any device or circuit that has two stable states is said to be *bistable*. For instance, a toggle switch has two stable states. It is either open or closed, depending on the position of the handle as shown in Fig. 8–1*a*. The switch is also said to have *memory* since it will remain as set until someone changes the position of the handle.

A flip-flop is a bistable electronic circuit that has two stable states—that is, its output is either 0 or +5 V dc as shown in Fig. 8–1*b*. The flip-flop also has memory since its output will remain as set until something is done to change it. As such, the flip-flop (or the switch) can be regarded as a memory device. In fact, any bistable device can be used to store one binary digit (bit). For instance, when the flip-flop has its output set at 0 V dc, it can be regarded as storing a logic 0 and when its output is set at +5 V dc, as storing a logic 1.

One of the easiest ways to construct a flip-flop is to connect two inverters in series as shown in Fig. 8–2*a*. The line connecting the output of inverter *B* (INV B) back to the input of inverter *A* (INV A) is referred to as the *feedback line*.

For the moment, remove the feedback line and consider $V_1$ as the input and $V_3$ as the output as shown in Fig. 8–2*b*. There are only two possible signals in a digital system, and in this case we will define 0 = 0 V dc and 1 = +5 V dc. If $V_1$ is set to 0 V dc, then $V_3$ will also be 0 V dc. Now, if the feedback line shown in Fig. 8–2*b* is reconnected, the ground can be removed from $V_1$, and $V_3$ will remain at 0 V dc. This is true since once the

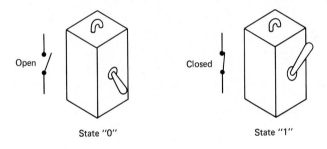

(*a*) Toggle switch

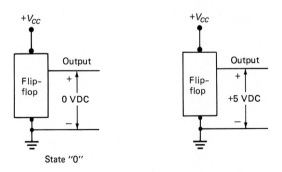

(*b*) Flip-flop

**FIG. 8–1**   Bistable devices

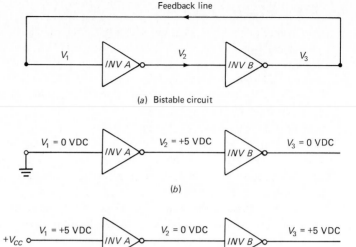

(a) Bistable circuit

(b)

(c)

**FIG. 8–2** Bistable circuit

input of INV A is grounded, the output of INV B will go low and can then be used to hold the input of INV A low by using the feedback line. This is one stable state—$V_3 = 0$ V dc.

Conversely, if $V_1$ is $+5$ V dc, $V_3$ will also be $+5$ V dc as seen in Fig. 8–2c. The feedback line can again be used to hold $V_1$ at $+5$ V dc since $V_3$ is also at $+5$ V dc. This is then the second stable state—$V_3 = +5$ V dc.

The basic flip-flop shown in Fig. 8–2a can be improved by replacing the inverters with either NAND or NOR gates. The additional inputs on these gates provide a convenient means for application of input signals to switch the flip-flop from one stable state to the other. Two 2-input NOR gates are connected in Fig. 8–3a to form a flip-flop circuit. Notice that if the two inputs labeled $R$ and $S$ are ignored, this circuit will function exactly as the one shown in Fig. 8–2a.

This circuit is redrawn in a more conventional form in Fig. 8–3b. The flip-flop actually has two outputs, defined in more general terms as $Q$ and $\overline{Q}$. It should be clear that regardless of the value of $Q$, its complement is $\overline{Q}$. There are two inputs to the flip-flop

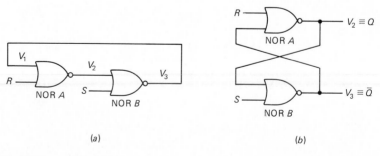

(a)

(b)

**FIG. 8–3** NOR–gate flip-flop

| R | S | Q | Action |
|---|---|---|---|
| 0 | 0 | Last value | No change |
| 0 | 1 | 1 | Set |
| 1 | 0 | 0 | Reset |
| 1 | 1 | ? | Forbidden |

**FIG. 8–4**  Truth table for a NOR–gate *RS* flip-flop

defined as $R$ and $S$. The input/output possibilities for this $RS$ flip-flop are summarized in the truth table in Fig. 8–4. To aid your understanding of the operation of this circuit, recall that a logic 1 at any input of a NOR gate forces its output to a logic 0.

The first input conduction in the truth table is $R = 0$ and $S = 0$. Since a 0 at the input of a NOR gate has no effect on its output, the flip-flop simply remains in its present state; that is, $Q$ remains unchanged.

The second input condition $R = 0$ and $S = 1$ forces the output of NOR gate *B* low. Both inputs to NOR gate *A* are now low, and the NOR-gate output must be high. Thus a 1 at the $S$ input is said to *SET* the flip-flop, and it switches to the stable state where $Q = 1$.

The third input condition is $R = 1$ and $S = 0$. This condition forces the output of NOR gate *A* low, and since both inputs to NOR gate *B* are now low, the output must be high. Thus a 1 at the $R$ input is said to *RESET* the flip-flop, and it switches to the stable state where $Q = 0$ (or $\overline{Q} = 1$).

The last input condition in the table, $R = 1$ and $S = 1$, is forbidden, as it forces the outputs of both NOR gates to the low state. In other words, both $Q = 0$ and $\overline{Q} = 0$ at the same time! But this violates the basic definition of a flip-flop that requires $Q$ to be the complement of $\overline{Q}$, and so it is generally agreed never to impose this input condition. Incidentally, if this condition is for some reason imposed, the resulting state of $Q$ is not predictable. That's why the truth table entry is a ?.

It is also important to remember that TTL gate inputs are quite noise-sensitive and therefore should never be left unconnected (floating). Each input must be connected either to the output of a prior circuit, or if unused, to GND or $+V_{CC}$.

☐ EXAMPLE 8–1

Use the pinout diagram for a 54/7427 triple 3-input NOR gate and show how to connect a simple *RS* flip-flop.

☐ SOLUTION

One possible arrangement is shown in Fig. 8–5. Notice that pins 3 and 4 are tied together, as are pins 10 and 11; thus no input leads are left unconnected and the two gates simply function as 2-input gates. The third NOR gate is not used. (It can be a spare or can be used elsewhere.)

The standard logic symbol for an *RS* flip-flop is shown in Fig. 8–6 along with its truth table. The truth table is necessary since it describes exactly how the flip-flop functions. An *RS* flip-flop is also called a ''latch,'' or a ''bistable multivibrator.''

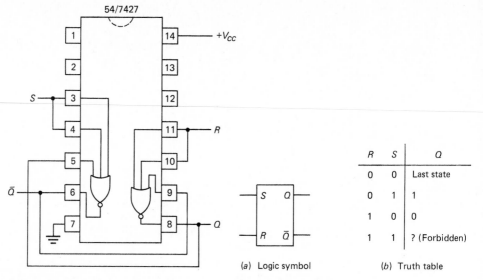

FIG. 8–5   54/7427

(a) Logic symbol

(b) Truth table

| R | S | Q |
|---|---|---|
| 0 | 0 | Last state |
| 0 | 1 | 1 |
| 1 | 0 | 0 |
| 1 | 1 | ? (Forbidden) |

FIG. 8–6   *RS* flip-flop

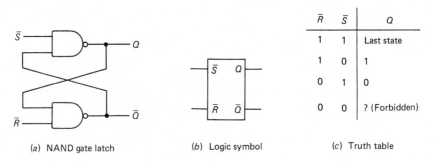

(a) NAND gate latch

(b) Logic symbol

(c) Truth table

| $\bar{R}$ | $\bar{S}$ | Q |
|---|---|---|
| 1 | 1 | Last state |
| 1 | 0 | 1 |
| 0 | 1 | 0 |
| 0 | 0 | ? (Forbidden) |

FIG. 8–7   $\overline{RS}$ flip-flop

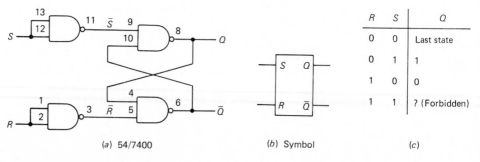

(a) 54/7400

(b) Symbol

(c)

| R | S | Q |
|---|---|---|
| 0 | 0 | Last state |
| 0 | 1 | 1 |
| 1 | 0 | 0 |
| 1 | 1 | ? (Forbidden) |

FIG. 8–8   An *RS* flip-flop (latch)

A slightly different latch can be constructed by using NAND gates as shown in Fig. 8–7. The truth table for this NAND–gate latch is different from that for the NOR–gate latch. We will call this latch an $\overline{RS}$ flip-flop. To understand how this circuit functions, recall that a low on any input to a NAND gate will force its output high. Thus a low on the $\overline{S}$ input will set the latch ($Q = 1$ and $\overline{Q} = 0$). A low on the $\overline{R}$ input will reset it ($Q = 0$). If both $\overline{R}$ and $\overline{S}$ are high, the flip-flop will remain in its previous state. Setting both $\overline{R}$ and $\overline{S}$ low simultaneously is forbidden since this forces both $Q$ and $\overline{Q}$ high.

◻ EXAMPLE 8–2

Show how to convert the $\overline{RS}$ flip-flop in Fig. 8–7 into an $RS$ flip-flop.

◻ SOLUTION

By placing an inverter at each input as shown in Fig. 8–8, the 2 inputs are now $R$ and $S$, and the resulting circuit behaves exactly as the $RS$ flip-flop in Fig. 8–6. A single 54/7400 (quad 2-input NAND gate) is used.

Simple latches as discussed in this section can be constructed from NAND or NOR gates or obtained as medium-scale integrated circuits (MSI). For instance, the 74LS279 is a *quad set-reset latch*. This pinout and truth table for this circuit are given in Fig. 8–9. Study the truth table carefully, and you will see that the latches behave exactly like the $\overline{RS}$ flip-flop discussed above.

# 8-2 CLOCKED *RS* FLIP-FLOP

Two different methods for constructing an $RS$ flip-flop were discussed in Sec. 8–1. The NOR–gate realization in Fig. 8–3b is an exact equivalent of the NAND–gate realization in Fig. 8–8a, and they both have the exact same symbol and truth table as given in Fig. 8–6. Both of these $RS$ flip-flops, or latches, are said to be "transparent"; that is, any

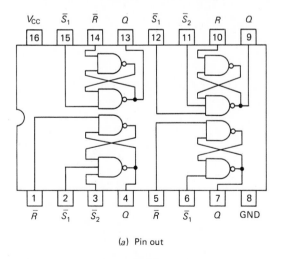

| $\overline{S}_1$ | $\overline{S}_2$ | $\overline{R}$ | $Q$ |
|---|---|---|---|
| 0 | 0 | 0 | ? Forbidden |
| 0 | x | 1 | 1 |
| x | 0 | 1 | 1 |
| 1 | 1 | 0 | 0 |
| 1 | 1 | 1 | Last state |

$x$ = Don't care

(a) Pin out      (b) Truth table

**FIG. 8–9** Quad SET–RESET latch

change in input information at $R$ or $S$ is transmitted immediately to the output at $Q$ and $\overline{Q}$ according to the truth table.

The addition of two AND gates at the $R$ and $S$ inputs as shown in Fig. 8–10 will result in a flip-flop that can be enabled or disabled. When the ENABLE input is low, the AND gate outputs must both be low and changes in neither $R$ nor $S$ will have any effect on the flip-flop output $Q$. The latch is said to be *disabled*.

When the ENABLE input is high, information at the $R$ and $S$ inputs will be transmitted directly to the outputs. The latch is said to be *enabled*. The output will change in response to input changes as long as the ENABLE is high. When the ENABLE input goes low, the output will retain the information that was present on the input when the high-to-low transition takes place.

In this fashion, it is possible to *strobe* or *clock* the flip-flop in order to store information (set it or reset it) at any time, and then hold the stored information for any desired period of time. This flip-flop is called a *clocked RS flip-flop*. The proper symbol and truth table are given in Fig. 8–10b. Notice that there are now three inputs—$R$, $S$, and the ENABLE or CLOCK input, labeled CLK. Notice also that the truth-table output is not simply $Q$, but $Q_{n+1}$. This is because we must consider two different instants in time: the time before the ENABLE goes low $Q_n$ and the time just after ENABLE goes low $Q_{n+1}$.

☐ EXAMPLE 8–3

Explain the meaning of $Q_n$ in the truth table in Fig. 8–10b.

☐ SOLUTION

For the flip-flop to operate properly, there must be a transition from low to high on the CLK input. While CLK is high, the information on $R$ and $S$ causes the latch to set or reset. Then when CLK transitions back to low, this information is retained in the latch. When this high-to-low transition occurred, both $R$ and $S$ inputs were low (0), and thus there was no change of state. In other words, the value of $Q$ at time $n + 1$ is the same as it was at time $n$. Remember that time $n$ occurs just before the high-to-low transition on CLK and time $n + 1$, just after the transition.

The logic diagrams shown in Fig. 8–11a and b illustrate two different methods for realizing a clocked $RS$ flip-flop. Both realizations are widely used in medium- and large-scale integrated circuits, and you will find them easy to recognize. You might like to examine the logic diagrams for a 54LS109 or a 54LS74, for instance.

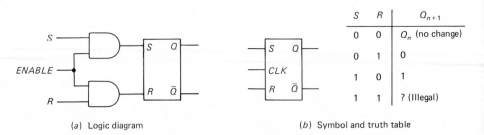

| $S$ | $R$ | $Q_{n+1}$ |
|-----|-----|-----------|
| 0 | 0 | $Q_n$ (no change) |
| 0 | 1 | 0 |
| 1 | 0 | 1 |
| 1 | 1 | ? (Illegal) |

(a) Logic diagram          (b) Symbol and truth table

**FIG. 8–10**   Clocked $RS$ flip-flop

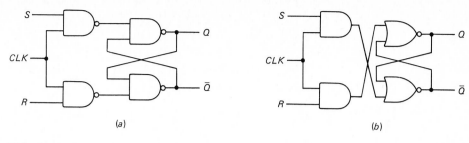

**FIG. 8–11**  Two different realizations for a clocked *RS* flip-flop

☐ EXAMPLE 8–4

Figure 8–12 shows the input waveforms *R*, *S*, and CLK applied to a clocked *RS* flip-flop. Explain the output waveform *Q*.

☐ SOLUTION

Initially, the flip-flop is reset ($Q = 0$). At time $t_1$ CLK goes high; the flip-flop is now enabled and it is immediately set ($Q = 1$) since $R = 0$ and $S = 1$. At time $t_2$ CLK goes low, and the flip-flop is disabled and latches in the stable state $Q = 1$.

Between $t_2$ and $t_3$ both *R* and *S* change states, but since CLK is low, the flip-flop is still disabled and *Q* remains at 1.

Between $t_3$ and $t_6$, the flip-flop will respond to any change in *R* and *S* since CLK is high. Thus at $t_3$ *Q* goes low, and at $t_4$ it goes back high. At $t_6$ the value $Q = 1$ is latched and no changes in *Q* occur between $t_6$ and $t_7$ even though both *R* and *S* change.

Between $t_7$ and $t_8$ no change in *Q* occurs since both *R* and *S* are low.

## 8-3 *D* FLIP-FLOP

The *RS* flip-flop has two data inputs, *R* and *S*. To store a high bit, you need a high *S*; to store a low bit, you need a high *R*. Generation of two signals to drive a flip-flop is a disadvantage in many applications. Furthermore, the forbidden condition of both *R* and *S*

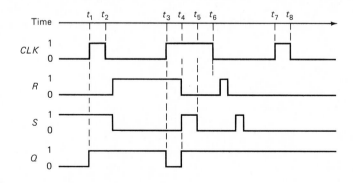

**FIG. 8–12**

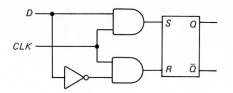

**FIG. 8–13**  A *D* flip-flop

high may occur inadvertently. This has led to the *D* flip-flop, a circuit that needs only a single data input.

Figure 8–13 shows a simple way to build a delay *D* flip-flop. This kind of flip-flop prevents the value of *D* from reaching the *Q* output until a clock pulse occurs. The action of the circuit is straightforward, as follows. When the clock is low, both AND gates are disabled; therefore, *D* can change value without affecting the value of *Q*. On the other hand, when the clock is high, both AND gates are enabled. In this case, *Q* is forced to equal the value of *D*. When the clock again goes low, *Q* retains or stores the last value of *D*.

There are many ways to design *D* flip-flops. In general, a *D* flip-flop is a bistable circuit whose *D* input is transferred to the output after a clock pulse is received. Figure 8–14 shows the logic symbol used in this book for any type of *D* flip-flop.

In this section we're talking about the kind of *D* flip-flop in which *Q* can follow the value of *D* while the clock is high. In other words, if the data bit changes while the clock in high, the last value of *D* before the clock returns low is the value of *D* that is stored. This kind of *D* flip-flop is often called a *D latch*.

Figure 8–14*b* shows the truth table for a *D* latch. While the clock (CLK) is low, *D* is a don't care; *Q* will remain latched in its last state. When the clock is high, *Q* takes on the last value of *D*. If *D* is changing while the clock is high, it is the last value of *D* that is stored.

The idea of data storage is illustrated in Fig. 8–15. Four *D* latches are driven by the same clock signal. When the clock goes high, input data is loaded into the flip-flops and appears at the output. Then when the clock goes low, the output retains the data. For instance, suppose that the data input is

$$D_3D_2D_1D_0 = 0111$$

When the clock goes high, this word is loaded into the *D* latches, resulting in an output of

$$Q_3Q_2Q_1Q_0 = 0111$$

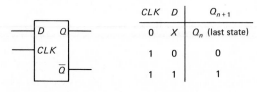

| CLK | D | $Q_{n+1}$ |
|-----|---|-----------|
| 0 | X | $Q_n$ (last state) |
| 1 | 0 | 0 |
| 1 | 1 | 1 |

(a) *D* flip–flop logic symbol.          (b) Truth table

**FIG. 8–14**  *D* flip-flop logic symbol

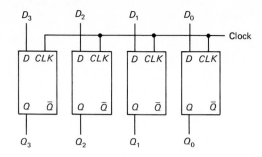

**FIG. 8–15** Storing a 4-bit word

After the clock goes low, the output data is retained, or stored. As long as the clock is low, the $D$ values can change without affecting the $Q$ values.

The 7475 in Fig. 8–16 is a TTL MSI circuit that contains four $D$ latches; it's called a *quad bistable latch*. The 7475 is ideal for handling 4-bit nibbles of data. If more than one 7475 is used, words of any length can be stored.

## 8-4 EDGE-TRIGGERED *D* FLIP-FLOP

Although the $D$ latch is used for temporary storage in electronic instruments, an even more popular kind of $D$ flip-flop is used in digital computers and systems. This kind of flip-flop samples the data bit at a unique point in time.

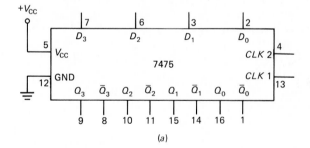

(a)

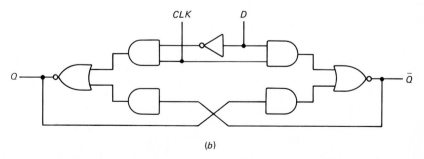

(b)

**FIG. 8–16** 4-bit bistable latch (a) Pinout (b) Logic diagram (each latch)

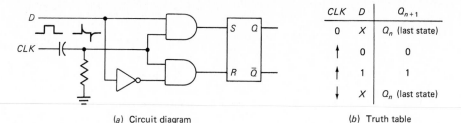

| CLK | D | $Q_{n+1}$ |
|:---:|:---:|:---:|
| 0 | X | $Q_n$ (last state) |
| ↑ | 0 | 0 |
| ↑ | 1 | 1 |
| ↓ | X | $Q_n$ (last state) |

(a) Circuit diagram    (b) Truth table

**FIG. 8–17**   *RC* differentiated clock input

Figure 8–17 shows an *RC* circuit at the input of a *D* latch. By deliberate design, the *RC* time constant is much smaller than the clock's pulse width. Because of this, the capacitor can charge fully when the clock goes high; this exponential charging produces a narrow positive voltage spike across the resistor. Later, the trailing edge of the pulse results in a narrow negative spike. (This is called *RC differentiation*.)

The narrow positive spike enables the AND gates for an instant; the narrow negative spike does nothing. The effect is to activate the AND gates during the positive spike, equivalent to sampling the value of *D* for an instant. At this unique point in time, *D* and its complement hit the flip-flop inputs, forcing *Q* to set or reset (unless *Q* already equals *D*).

This kind of operation is called *edge triggering* because the flip-flop responds only when the clock is in transition between its two voltage states. The triggering in Fig. 8–17 occurs on the positive-going edge of the clock; this is why it's referred to as *positive-edge triggering*.

The truth table in Fig. 8–17*b* summarizes the action of a positive-edge-triggered *D* flip-flop. When the clock is low, *D* is a don't care and *Q* is latched in its last state. On the leading edge of the clock, designated by the up arrow, the data bit is loaded into the flip-flop and *Q* takes on the values of *D*. On the trailing edge of the clock (the down arrow), *D* is a don't care and *Q* remains in its last state.

When power is first applied, flip-flops come up in random states. To get some computers started, an operator has to push a RESET button. This sends a CLEAR or RESET signal to all flip-flops. Also, it's necessary in some digital systems to preset (synonymous with set) certain flip-flops.

Figure 8–18 shows how to include both functions in a *D* flip-flop. The edge triggering is the same as previously described. Furthermore, the OR gates allow us to slip in a

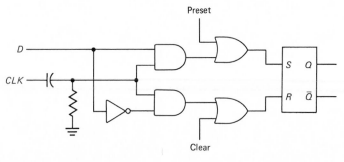

**FIG. 8–18**   Preset and Clear functions

CHAPTER 8

high PRESET or a high CLEAR when desired. A high PRESET forces $Q$ to equal 1; a high CLEAR resets $Q$ to 0.

The PRESET and CLEAR are called *asynchronous inputs* because they activate the flip-flop independently of the clock. On the other hand, the $D$ input is a synchronous input because it has an effect only when a clock edge occurs.

Integrated $D$ flip-flops do not use $RC$ circuits to obtain narrow spikes because capacitors are difficult to fabricate on a chip. Instead, a variety of direct-coupled designs is used. But the idea is the same: an edge-triggered $D$ flip-flop responds only during the brief instant the clock switches from one voltage level to another.

Figure 8–19*a* is the symbol for a positive-edge-triggered $D$ flip-flop. The clock input has a small triangle to serve as a reminder of edge triggering. When you see this symbol, remember what it means; the $D$ input is sampled and stored on the leading edge of the clock.

Sometimes, triggering on the negative edge is better suited to the application. In this case, an internal inverter can complement the clock pulse before it reaches the AND gates. This means that the trailing edge of the clock activates the gates, allowing data to be stored. Figure 8–19*b* is the symbol for a negative-edge-triggered $D$ flip-flop. The bubble and triangle symbolize the negative-edge triggering.

Figures 8–19*a* and *b* also include preset (PR) and clear (CLR) inputs. Figure 8–19*c* is another commercially available $D$ flip-flop (the 54/74175 or 54/74LS175). Besides having positive-edge triggering, it has an inverted CLEAR input. This means that a low CLR resets it. The 54/74175 has four of these $D$ flip-flops in a single 16-pin dual in-line package (DIP), and it's referred to as a *quad D-type flip-flop with clear*.

## 8-5 FLIP-FLOP SWITCHING TIME

Diodes and transistors cannot switch states immediately. It always takes a small amount of time to turn a diode on or off. Likewise, it takes time for a transistor to switch from saturation to cutoff, and vice versa. For bipolar diodes and transistors, the switching time is in the nanosecond region.

Switching time is the main cause of propagation delay, designated $t_p$. This represents the amount of time it takes for the output of a gate or flip-flop to change states after the input changes. For instance, if the data sheet of a $D$ flip-flop lists $t_p = 10$ ns, it takes about 10 ns for $Q$ to change states after $D$ has been sampled by the clock edge. Propagation delay is so small that it's negligible in many applications.

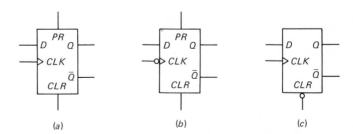

**FIG. 8–19** $D$ flip-flop symbols (*a*) Positive edge triggered (*b*) Negative edge triggered (*c*) Positive edge triggered with active low preset and clear

Stray capacitance at the $D$ input (plus other factors) makes it necessary for data bit $D$ to be at the input before the clock edge arrives. The setup time $t_{setup}$ is the minimum amount of time that the data bit must be present before the clock edge hits. For instance, if a $D$ flip-flop has a setup time of 15 ns, the data bit to be stored must be at the $D$ input at least 15 ns before the clock edge arrives; otherwise, the manufacturer does not guarantee correct sampling and storing.

Furthermore, data bit $D$ has to be held long enough for the internal transistors to switch states. Only after the transition is assured can we allow data bit $D$ to change. Hold time $t_{hold}$ is the minimum amount of time that data bit $D$ must be present after the clock edge arrives. For example, if $t_{setup} = 15$ ns and $t_{hold} = 5$ ns, the data bit has to be at the $D$ input at least 15 ns before the clock edge arrives and held at least 5 ns after the clock edge hits.

☐ EXAMPLE 8—5

Typical waveforms for setting a 1 in a positive-edge-triggered flip-flop are shown in Fig. 8–20. Discuss the timing.

☐ SOLUTION

The lower line in Fig. 8–20 is the time line with critical times marked on it. Prior to $t_1$, the data can be a 1 or a 0, or can be changing. This is shown by drawing lines for both high and low levels on $D$. From time $t_1$ to $t_2$, the data line $D$ must be held steady (in this case a 1). This is the setup time $t_{setup}$. Data is shifted into the flip-flop at time $t_2$ but does not appear at $Q$ until time $t_3$. The time from $t_2$ to $t_3$ is the propagation time $t_p$. In order to guarantee proper operation, the data line must be held steady from time $t_2$ until $t_4$; this is the hold time $t_{hold}$. After $t_4$, $D$ is free to change states—shown by the double lines.

## 8-6 JK FLIP-FLOP

Among other things, flip-flops can be used to build a counter, a circuit that counts the number of positive or negative clock edges driving its clock input. For purposes of counting, the $JK$ flip-flop is the ideal element to use.

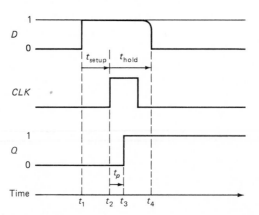

FIG. 8–20

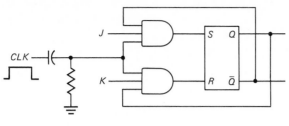

| CLK | J | K | $Q_{n+1}$ |
|-----|---|---|-----------|
| $X$ | 0 | 0 | $Q_n$ (last state) |
| ↑ | 0 | 1 | 0 |
| ↑ | 1 | 0 | 1 |
| ↑ | 1 | 1 | $\bar{Q}_n$ (toggle) |

(a) One way to implement a *JK* flip–flop        (b) Truth table

**FIG. 8–21** A *JK* flip-flop

Figure 8–21 shows one way to build a *JK* flip-flop. The variables *J* and *K* are called *control inputs* because they determine what the flip-flop does when a positive clock edge arrives. As before, the *RC* circuit has a short time constant, thus converting the rectangular clock pulse into narrow spikes. Because of the AND gates, the circuit is positive-edge-triggered.

When *J* and *K* are both low, both AND gates are disabled. Therefore, clock pulses have no effect. This first possibility is the initial entry in the truth table. As shown, when *J* and *K* are both 0s, *Q* retains its last value.

When *J* is low and *K* is high, the upper gate is disabled, so there's no way to set the flip-flop. The only possibility is reset. When *Q* is high, the lower gate passes a RESET trigger as soon as the next positive clock edge arrives. This forces *Q* to become low (the second entry in the truth table). Therefore, $J = 0$ and $K = 1$ means that the next positive clock edge resets the flip-flop (unless *Q* is already reset).

When *J* is high and *K* is low, the lower gate is disabled, so it's impossible to reset the flip-flop. But you can set the flip-flop as follows. When *Q* is low, $\bar{Q}$ is high; therefore, the upper gate passes a SET trigger on the next positive clock edge. This drives *Q* into the high state (the third entry in the truth table). As you can see, $J = 1$ and $K = 0$ means that the next positive clock edge sets the flip-flop (unless *Q* is already high).

When *J* and *K* are both high (notice that this is the forbidden state with an *RS* flip-flop), it's possible to set or reset the flip-flop. If *Q* is high, the lower gate passes a RESET trigger on the next positive clock edge. On the other hand, when *Q* is low, the upper gate passes a *SET* trigger on the next positive clock edge. Either way, *Q* changes to the complement of the last state (see the truth table). Therefore, $J = 1$ and $K = 1$ means the flip-flop will toggle on the next positive clock edge. "Toggle" means to switch to the opposite state.

Propagation delay prevents the *JK* flip-flop from racing (toggling more than once during a positive clock edge). Here's why. In Fig. 8–21, the outputs change after the positive clock edge has struck. By then, the new *Q* and $\bar{Q}$ values are too late to coincide with the positive spikes driving the AND gates. For instance, if $t_p = 20$ ns, the outputs change approximately 20 ns after the leading edge of the clock. If the clock spikes are narrower than 20 ns, the returning *Q* and $\bar{Q}$ arrive too late to cause false triggering.

Figure 8–22*a* shows a symbol for a *JK* flip-flop of any design. When you see this on a schematic diagram, remember that on the next positive clock edge, (1) low *J* and *K* have

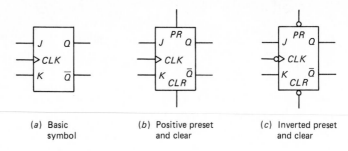

(a) Basic
symbol

(b) Positive preset
and clear

(c) Inverted preset
and clear

**FIG. 8–22** *JK* flip-flop symbols

no effect, (2) low *J* and high *K* produce a reset, high *J* and low *K* produce a set, and high *J* and *K* result in a toggle.

You can include OR gates in the design to accommodate PRESET and CLEAR as was done earlier. Figure 8–22*b* gives the symbol for a *JK* flip-flop with PR and CLR. Notice that it is positive-edge-triggered and requires a high PR to set it or a high CLR to reset it.

Figure 8–22*c* is another commercially available *JK* flip-flop. It is negative-edge-triggered and requires a low PR to set it or a low CLR to reset it. The 54LS/74LS76 described in the next section is one example of this type of flip-flop.

## 8·7 JK MASTER-SLAVE FLIP-FLOP

Figure 8–23 shows one way to build a *JK* master-slave flip-flop. It provides another way to avoid racing. Here's how it works. To begin with, the master is positive-edge-triggered and the slave, negative-edge-triggered. Therefore, the master responds to its *J* and *K* inputs before the slave. If $J = 1$ and $K = 0$, the master sets on the positive clock edge. The high *Q* output of the master drives the *J* input of the slave, so when the negative clock edge hits, the slave sets, copying the action of the master.

If $J = 0$ and $K = 1$, the master resets on the leading edge of the clock. The high $\overline{Q}$ output of the master goes to the *K* input of the slave. Therefore, the arrival of the clock's trailing edge forces the slave to reset. Again, the slave has copied the master.

If the master's *J* and *K* inputs are both high, it toggles on the positive clock edge and the slave then toggles on the negative clock edge. Regardless of what the master does, therefore, the slave copies it: if the master sets, the slave sets; if the master resets, the slave resets.

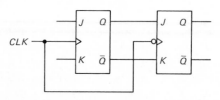

**FIG. 8–23** Master-slave flip-flop

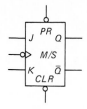

**FIG. 8–24** *JK* master-slave flip-flop symbol

Other designs are possible for a *JK* master-slave flip-flop. The basic idea remains the same: the slave copies the master on the negative clock edge. Figure 8–24 shows the symbol we will use for a *JK* master-slave flip-flop; it includes PRESET and CLEAR functions. The bubble on the clock input is a reminder that the output changes state on the negative clock edge.

Because of its great popularity in industry, the *JK* master-slave flip-flop is used as the main counting device in the remainder of this book. The 54LS/74LS76 is a dual *JK* master-slave flip-flop widely used in industry. Its pinout and truth table are given in Fig. 8–25.[1]

☐ EXAMPLE 8–6

The *JK* master-slave flip-flop in Fig. 8–26 has its *J* and *K* inputs tied to $+V_{CC}$, and a series of pulses (actually a square wave) are applied to its CLK input. Describe the waveform at *Q*.

☐ SOLUTION

Since $J = K = 1$, the flip-flop simply toggles each time the clock goes low. The waveform at *Q* has a period twice that of the CLK waveform. In other words, the frequency of *Q* is only one-half that of CLK. This circuit acts as a frequency divider—the output frequency is equal to the input frequency divided by 2.

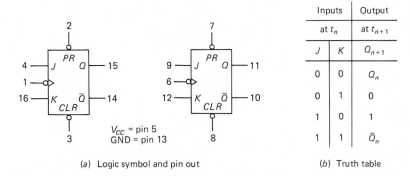

| Inputs at $t_n$ | | Output at $t_{n+1}$ |
|---|---|---|
| $J$ | $K$ | $Q_{n+1}$ |
| 0 | 0 | $Q_n$ |
| 0 | 1 | 0 |
| 1 | 0 | 1 |
| 1 | 1 | $\bar{Q}_n$ |

(a) Logic symbol and pin out          (b) Truth table

**FIG. 8–25** 54LS/74LS76 dual, *JK*, negative edge triggered flip-flop

More detailed information can be found on the data sheet included in the following laboratory manual: D. P. Leach, *Experiments in Digital Principles,* 3d ed., McGraw-Hill, New York, 1986

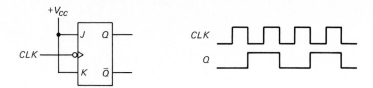

**FIG. 8–26**

## 8·8 SCHMITT TRIGGER

A Schmitt trigger is an electronic circuit that is used to detect whether a voltage has crossed over a given reference level. It has two stable states and is especially useful as a signal-conditioning device. Given a sinusoidal waveform, a triangular wave, or any other periodic waveform, the Schmitt trigger will produce a rectangular output that has sharp leading and trailing edges. Fast rise and fall times like this are desirable for all digital circuits.

Figure 8–27 shows the transfer function ($v_{out}$ versus $v_{in}$) for any Schmitt trigger. The value of $v_{in}$ that causes the output to jump from low to high is called the *positive-going threshold voltage* $V_{T+}$, The value of $v_{in}$ causing the output to switch from high to low is called the *negative-going threshold voltage* $V_{T-}$.

The output voltage is either high or low. When the output is low, it is necessary to raise the input to slightly more than $V_{T+}$ to produce switching action. The output will then switch to the high state and remain there until the input is reduced to slightly below $V_{T-}$. The output will then switch back to the low state. The arrows and the dashed lines show the switching action.

The difference between the two threshold voltages is known as *hysteresis*. It is possible to eliminate hysteresis by circuit design, but a small amount of hysteresis is desirable because it ensures rapid switching action over a wide temperature range. Hysteresis can

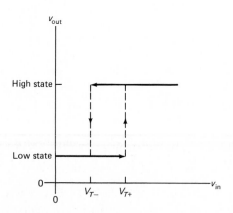

**FIG. 8–27**   Schmitt trigger transfer characteristic

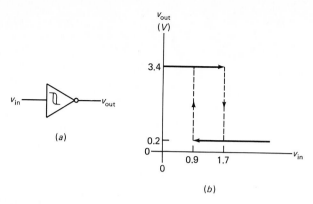

FIG. 8–28   7414 Schmitt-trigger inverter

also be a very beneficial feature. For instance, it can be used to provide noise immunity in certain applications (digital modems for example).

The TTL 7414 is a hex Schmitt-trigger inverter. The *hex* means there are six Schmitt-trigger circuits in one DIP. In Fig. 8–28, the standard logic symbol for one of the Schmitt-trigger inverters in a 7414 is shown along with a typical transfer characteristic. Because of the inversion, the characteristic curve is reversed from that shown in Fig. 8–27. Looking at the curve in Fig. 8–28*b*, when the input exceeds 1.7 V, the output will switch to the low state. When the input falls below 0.9 V, the output will switch back to the high state. The switching action is shown by the arrows and the dashed lines.

□ EXAMPLE 8—7

A sine wave with a peak of 6 V drives one of the inverters in a 7414. Sketch the output voltage.

□ SOLUTION

When the sinusoid exceeds 1.7 V, the output goes from high to low. The output stays in the low state until the input sinusoid drops below 0.9 V. Then the output jumps back to the high state. Figure 8–29 shows the input and output waveforms. This illustrates the signal-conditioning action of the Schmitt-trigger inverter. It has changed the sine wave into a rectangular pulse with fast rise and fall times. The same action would occur for any other periodic waveform.

## S U M M A R Y

A flip-flop is an electronic circuit that has two stable states. It is said to be bistable. A basic *RS* flip-flop, or latch can be constructed by connecting two NAND gates or two NOR gates in series with a feedback connection. A signal at the set input of an *RS* flip-flop

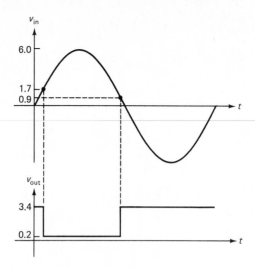

**FIG. 8–29** Schmitt-trigger input-output waveforms

will force the $Q$ output to become a 1, while a signal at the reset input will force $Q$ to become a 0.

A simple $RS$ flip-flop or latch is said to be transparent—that is, its output changes state whenever a signal appears at the $R$ or $S$ inputs. An $RS$ flip-flop can be modified to form a clocked $RS$ flip-flop whose output can change states only in synchronism with the applied clock.

An $RS$ flip-flop can also be modified to form a $D$ flip-flop. In a $D$ latch, the stored data may be changed while the clock is high. The last value of $D$ before the clock returns low is the data that is stored. With edge-triggered $D$ flip-flops, the data is sampled and stored on either the positive or negative clock edge.

The values of $J$ and $K$ determine what a $JK$ flip-flop does on the next clock edge. When both are low, the flip-flop retains its last state. When $J$ is low and $K$ is high, the flip-flop resets. When $J$ is high and $K$ is low, the flip-flop sets. When both are high, the flip-flcp toggles. In this last mode, the $JK$ flip-flop can be used as a frequency divider.

A Schmitt trigger is a bistable circuit that is widely used to change a periodic waveform into a rectangular waveform having very fast rise and fall times.

## G L O S S A R Y

*Asynchronous* Independent of clocking. The output can change without having to wait for a clock pulse.

*Bistable* Having two stable states.

*Bistable Multivibrator* Another term for an $RS$ flip-flop.

*Buffer Register* A group of memory elements, often flip-flops, that can store a binary word.

*Edge Triggering* A circuit responds only when the clock is in transition between its two voltage states.

*Flip-Flop* An electronic circuit that has two stable states.

*Hold Time* The minimum amount of time that data must be present after the clock trigger arrives.

*Latch* Another term for an *RS* flip-flop.

*Propagation Delay* The amount of time it takes for the output to change states after an input trigger.

*Schmitt Trigger* A bistable circuit used to produce a rectangular output waveform.

*Setup Time* The minimum amount of time required for data inputs to be present before the clock arrives.

*Synchronous* When outputs change states in time with a clock. A clock signal must be present in order for the outputs to change states.

# PROBLEMS

*SECTION 8-1:*

**8–1.** List as many bistable devices as you can think of—either electrical or mechanical. (*Hint:* Magnets, lamps, relays, etc.)

**8–2.** Redraw the NOR–gate flip-flop in Fig. 8–3*b* and label the logic level on each pin for $R = S = 0$. Repeat for $R = S = 1$, for $R = 0$ and $S = 1$, and for $R = 1$ and $S = 0$.

**8–3.** Redraw the NAND–gate flip-flop in Fig. 8–7*a* and label the logic level on each pin for $\overline{R} = \overline{S} = 0$. Repeat for $\overline{R} = \overline{S} = 1$, for $\overline{R} = 1$, and $\overline{S} = 0$, and for $\overline{R} = 0$ and $\overline{S} = 1$.

**8–4.** Redraw the NAND–gate flip-flop in Fig. 8–8*a* and label the logic level on each pin for $R = S = 0$. Repeat for $R = S = 1$, for $R = 0$ and $S = 1$, and for $R = 1$ and $S = 0$.

*SECTION 8-2:*

**8–5.** The waveforms in Fig. 8–30 drive the clocked *RS* flip-flop in Fig. 8–10. The clock signal goes from low to high at points $A$, $C$, $E$, and $G$. If $Q$ is low before point $A$ in time:
  **a.** At what point does $Q$ become a 1?
  **b.** When does $Q$ reset to 0?

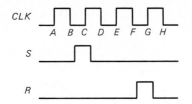

**FIG. 8–30**

**8–6.** Use the information in the preceding problem and draw the waveform at $Q$.

**8–7.** Prove that the flip-flop realizations in Fig. 8–11 are equivalent by writing the logic level present on every pin when $R = S = 0$ and the clock is high. Repeat for $R = S = 1$, for $R = 1$ and $S = 0$, and for $R = 0$ and $S = 1$. Describe what happens when the clock is low.

*SECTIONS 8-3, 8-4, AND 8-5:*

**8–8.** The waveforms in Fig. 8–31 drive a $D$ latch as shown in Fig. 8–14. What is the value of $D$ stored in the flip-flop after the clock pulse is over?

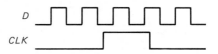

**FIG. 8–31**

**8–9.** A positive edge-triggered $D$ flip-flop has the input waveforms shown in Fig. 8–32. What is the value of $Q$ after the clock pulse?

**8–10.** A negative edge-triggered $D$ flip-flop is driven by the waveforms shown in Fig. 8–32. What is the value of $D$ stored in the flip-flop?

**8–11.** A $D$ flip-flop has the following data sheet information: setup time = 5 ns; hold time = 10 ns; propagation time = 15 ns.
   **a.** How far ahead of the triggering clock edge must the data be applied?
   **b.** How long after the clock edge must the data be present to ensure correct storage?
   **c.** How long after the clock edge before the output changes?

*SECTIONS 8-6 AND 8-7*

**8–12.** Redraw the $JK$ flip-flop in Fig. 8–21a. Connect $J = K = 1$. (This can be done by connecting the $J$ and $K$ inputs to $+V_{CC}$.) Now, begin with $Q = 1$, and show what logic level results on each pin after one positive clock pulse. Allow one more positive clock pulse and show the resulting logic level on every pin.

**8–13.** In the $JK$ flip-flop in Fig. 8–22a, $J = K = 1$. A 1-MHz square wave is applied to its CLK input. It has a propagation delay of 50 ns. Draw the input square wave and the output waveform expected at $Q$. Be sure to show the propagation delay time.

**8–14.** Repeat Prob. 8–13, but use the flip-flop in Fig. 8–22c.

**8–15.** In Prob. 8–13, what is the period of the clock? What are the period and frequency of the output waveform at $Q$?

**8–16.** Repeat Prob. 8–13, assuming that the CLK input has a frequency of 10 MHz.

**8–17.** Redraw the two flip-flops in Fig. 8–25 and show how to connect them such that a 500-kHz square wave applied to pin 1 will result in a 125-kHz square wave at pin 11. Give a complete wiring diagram (show each pin connection).

*SECTION 8-8:*

**8–18.** Draw the input and output waveforms for the Schmitt trigger in Fig. 8–28, assuming that the input voltage is $v = 10 \cos 1000t$.

**8–19.** Draw carefully the waveforms at points $A$, $B$, and $C$ in Fig. 8–32.

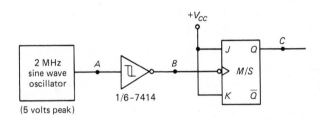

**FIG. 8–32**

# 9

# C L O C K S
# A N D   T I M E R S

The heart of every digital system is the system clock—it does, indeed, provide the heartbeat without which the system would cease to function. In this chapter we consider the characteristics of a digital clock signal as well as some typical clock circuits. A *monostable* is a basic digital timing circuit that is used in a wide variety of timing applications. We consider a number of different commercially available monostable circuits and examine some common applications.

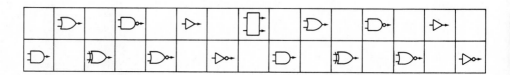

## 9-1 CLOCK WAVEFORMS

Since all logic operations in a synchronous machine occur in synchronism with a clock, the system clock becomes the basic timing unit. The system clock must provide a periodic waveform that can be used as a synchronizing signal. The square wave shown in Fig. 9–1a is a typical clock waveform used in a digital system. It should be noted that the clock need not be a perfectly symmetrical square wave as shown. It could simply be a series of positive (or negative) pulses as shown in Fig. 9–1b. This waveform could, of course, be considered as an asymmetrical square wave. The main requirement is simply that the clock be perfectly periodic.

Notice that the clock defines a basic timing interval during which logic operations must be performed. This basic timing interval is defined as a clock cycle time and is equal to one period of the clock waveform. Thus all logic elements, flip-flops, gates, and so on must complete their transitions in less than one clock cycle time.

☐ EXAMPLE 9—1

What is the clock cycle time for a system that uses a 500-kHz clock? A 2-MHz clock?

☐ SOLUTION

The clock cycle time is equal to one period of the clock. Therefore, the clock cycle time for a 500-kHz clock is $1/(500 \times 10^3) = 2~\mu s$. The clock cycle time for a 2-MHz clock is $1/(2 \times 10^6) = 0.5~\mu s$.

☐ EXAMPLE 9—2

The total propagation delay through a master-slave flip-flop is given as 100 ns. What is the maximum clock frequency that can be used with this flip-flop?

☐ SOLUTION

An alternate way of expressing the question is, how fast can the flip-flop operate? The flip-flop must complete its transition in less than one clock cycle time; therefore, the minimum clock cycle time must be 100 ns. So, the maximum clock frequency must be $1/(100 \times 10^{-9}) = 10$ MHz.

The clock waveform drawn above the time line in Fig. 9–2a is a perfect, ideal clock. What exactly are the characteristics that make up an ideal clock? First, the clock levels must be absolutely stable. When the clock is high, the level must hold a steady value of +5 V, as shown between points $a$ and $b$ on the time line. When the clock is low, the level must be an unchanging 0 V, as it is between points $b$ and $c$. In actual practice, the stability of the clock is much more important than the absolute value of the voltage level. For

(a)   (b)

**FIG. 9–1**  Ideal clock waveforms

instance, it might be perfectly acceptable to have a high level of +4.8 V instead of +5.0 V, provided it is a *steady*, *unchanging*, +4.8 V.

The second characteristic deals with the time required for the clock levels to change from high to low or vice versa. The transition of the clock from low to high at point *a* in Fig. 9–2*a* is shown by a vertical line segment. This implies a time of zero; that is, the transition occurs instantaneously—it requires zero time. The same is true of the transition time from high to low at point *b* in Fig. 9–2*a*. Thus an ideal clock has zero transition time.

A nearly perfect clock waveform might appear on an oscilloscope trace as shown in Fig. 9–2*b*. At first glance this would seem to be two horizontal traces composed of line segments. On closer examination, however, it can be seen that the waveform is exactly like the ideal waveform in Fig. 9–2*a* if the vertical segments are removed. The vertical segments might not appear on the oscilloscope trace because the transition times are so small (nearly zero) and the oscilloscope is not capable of responding quickly enough. The vertical segments can usually be made visible by either increasing the oscilloscope "intensity," or by reducing the "sweep time."

Figure 9–2*c* shows a portion of the waveform in Fig. 9–2*b* expanded by reducing the "sweep time" such that the transition times are visible. Clearly it requires some time for the waveform to transition from low to high—this is defined as the rise time $t_r$. The time required for transition from high to low is defined as the fall time $t_f$. It is customary to measure the rise and fall times from points on the waveform referred to as the *10 and 90 percent* points. In this case, a 100 percent level change is 5.0 V, so 10 percent of this is 0.5 V and 90 percent is 4.5 V. Thus the rise time is that time required for the waveform to travel from 0.5 up to 4.5 V. Similarly, the fall time is that time required for the waveform to transition from 4.5 down to 0.5 V.

Finally, the third requirement that defines an ideal clock is its frequency stability. The frequency of the clock should be steady and unchanging over a specified period of time. Short-term stability can be specified by requiring that the clock frequency (or its period) not be allowed to vary by more than a given percentage over a short period of time—say, a few hours. Clock signals with short-term stability can be derived from straightforward electronic circuits as shown in the following sections.

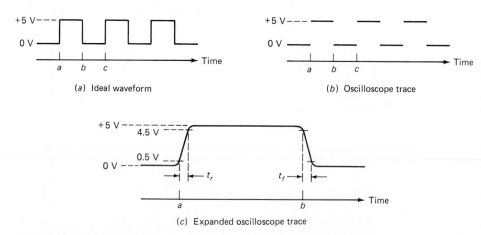

(a) Ideal waveform

(b) Oscilloscope trace

(c) Expanded oscilloscope trace

**FIG. 9–2**  Clock waveforms

Long-term stability deals with longer periods of time—perhaps days, months, or years. Clock signals that have long-term stability are generally derived from rather special circuits placed in a heated enclosure (usually called an "oven") in order to guarantee close control of temperature and hence frequency. Such circuits can provide clock frequencies having stabilities better than a few parts in $10^9$ per day.

## 9-2 TTL CLOCK

A 7404 hexadecimal inverter can be used to construct an excellent TTL-compatible clock, as shown in Fig. 9–3. This clock circuit is well known and widely used. Two inverters are used to construct a two-stage amplifier with an overall phase shift of 360° between pins 1 and 6. Then a portion of the signal at pin 6 is fed back by means of a crystal to pin 1, and the circuit oscillates at a frequency determined by the crystal. Since the feedback element is a crystal, the frequency of oscillation is very stable. Here's how the oscillator works.

Inverter 1 has a 330-$\Omega$ feedback resistor ($R_1$) connected from output (pin 2) to input (pin 1). This forms a current-to-voltage amplifier with a gain of $A_1 = v_{out}/i_{in} = -R_1$. In this case, the gain is $A_1 = -330$ V/A, where the negative sign shows 180° of phase shift. For instance, an increase of 1 mA in $i_{in}$ will cause a negative-going voltage of 1 mA × 330 = 330 mV at $v_{out}$.

Inverter 2 is connected exactly as is inverter 1. Its gain is $A_2 = -R_2$. The two amplifiers are then ac-coupled with a 0.01-$\mu$F capacitor to form an amplifier that has an overall gain of $A = A_1 \times A_2 = R_1R_2$. Notice that the overall gain has a positive sign, which shows 360° of phase shift. In this case, $A = 330 \times 330 = 1.09 \times 10^5$ V/A. For instance, an increase of 45 $\mu$A at $i_{in}$ will result in a positive-going voltage of 5.0 V at pin 6 of inverter 2. Now, if a portion of the signal at pin 6 is fed back to pin 1, it will augment $i_{in}$ (positive feedback) and the circuit will oscillate.

A series-mode crystal is used as the feedback element to return a portion of the signal at pin 6 to pin 1. The crystal acts as a series $RLC$ circuit, and at resonance it ideally appears as a low-resistance element with no phase shift. The feedback signal must there-

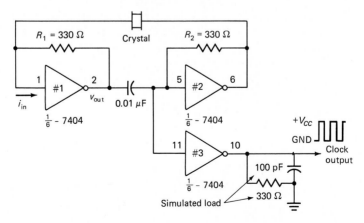

**FIG. 9–3** TTL clock circuit

fore be at resonance, and the two inverters in conjunction with the crystal form an oscillator operating at the crystal resonant frequency.

With the feedback element connected, the overall gain is sufficient to drive each inverter between saturation and cutoff, and the output signal is a periodic waveform as shown in Fig. 9–3. Typically, the output clock signal will transition between 0 and +5 V, will have rise and fall times of less than 10 ns, and will be essentially a square wave. The frequency of this clock signal is determined by the crystal, and values between 1 and 20 MHz are common.

Inverter 3 is used as an output buffer amplifier and is capable of driving a load of 330 Ω in parallel with 100 pF while still providing rise and fall times of less than 10 ns.

☐ EXAMPLE 9–3

A TTL clock circuit as shown in Fig. 9–3 is said to provide a 5-MHz clock frequency with a stability better than 5 parts per million (ppm) over a 24-h time period. What are the frequency limits of the clock?

☐ SOLUTION

A stability of 5 parts per million means that a 1-MHz clock will have a frequency of 1,000,000 plus or minus 5 Hz. So, this clock will have a frequency of 5,000,000 plus or minus 25 Hz. Over any 24-h period the clock frequency will be somewhere between 4,999,975 and 5,000,025 Hz.

## 9-3 555 TIMER—ASTABLE

The 555 timer is a TTL-compatible integrated circuit (IC) that can be used as an oscillator to provide a clock waveform. It is basically a switching circuit that has two distinct output levels. With the proper external components connected, neither of the output levels is stable. As a result, the circuit continuously switches back and forth between these two unstable states. In other words, the circuit oscillates and the output is a periodic, rectangular waveform. Since neither output state is stable, this circuit is said to be *astable* and is often referred to as a "free running," or "astable multivibrator." (Recall that a flip-flop has two stable states, is said to be *bistable,* and is referred to as a "bistable multivibrator.") The frequency of oscillation as well as the duty cycle are accurately controlled by two external resistors and a single timing capacitor.

The logic symbol for an LM555 timer connected as an oscillator is shown in Fig. 9–4. The timing capacitor $C$ is charged toward $+V_{CC}$ through resistors $R_A$ and $R_B$. The charging time $t_1$ is given as

$$t_1 = 0.693(R_A + R_B)C$$

This is the time during which the output is high as shown in Fig. 9–4.

The timing capacitor $C$ is then discharged toward GROUND (GND) through the resistor $R_B$. The discharge time $t_2$ is given as

$$t_2 = 0.693R_BC$$

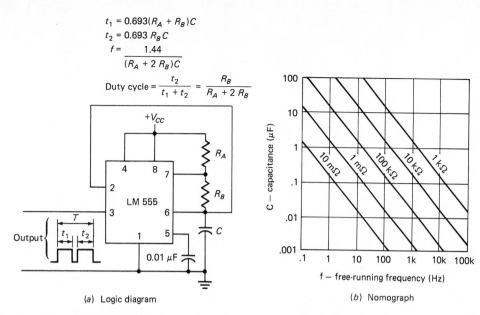

$$t_1 = 0.693(R_A + R_B)C$$
$$t_2 = 0.693\,R_B\,C$$
$$f = \frac{1.44}{(R_A + 2\,R_B)C}$$
$$\text{Duty cycle} = \frac{t_2}{t_1 + t_2} = \frac{R_B}{R_A + 2\,R_B}$$

(a)  Logic diagram            (b)  Nomograph

**FIG. 9–4**  Astable circuit

This is the time during which the output is low as shown in Fig. 9–4.

The period $T$ of the resulting clock waveform is the sum of $t_1$ and $t_2$. Thus

$$T = t_1 + t_2 = 0.693(R_A + 2R_B)C$$

The frequency of oscillation is then found as

$$f = \frac{1}{T} = \frac{1.44}{(R_A + 2R_B)C}$$

☐  EXAMPLE 9–4

Determine the frequency of oscillation for the 555 timer in Fig. 9–4, given $R_A = R_B = 1\ \text{k}\Omega$ and $C = 1000\ \text{pF}$.

☐  SOLUTION

Using the relationship given above, we obtain

$$f = \frac{1.44}{[1000 + 2(1000)] \times 10^{-9}} = 480\ \text{kHz}$$

The output of the 555 timer when connected this way is a periodic rectangular waveform, but it is not a square wave. This is because $t_1$ and $t_2$ are unequal, and the waveform is said to be asymmetrical. A measure of the asymmetry of the waveform can be stated in terms of its duty cycle. Here we define the duty cycle to be the ratio of $t_2$ to the period. Thus

$$\text{Duty cycle} = \frac{t_2}{t_1 + t_2}$$

As defined, the duty cycle is always some number between 0.0 and 1.0, although it is frequently expressed as a percent. For instance, if the duty cycle of a waveform is given as

0.45 (or 45 percent), the signal is at GND level for 45 percent of the time and at its high level for 55 percent of the time.

## □ EXAMPLE 9—5

Calculate the duty cycle for the oscillator given in Example 9–4.

## □ SOLUTION

Using the expressions given above, we obtain

$$t_1 = 0.693(1000 + 1000) \times 10^{-9} = 1.386 \ \mu s$$
$$t_2 = 0.693(1000) \times 10^{-9} = 0.693 \ \mu s$$

The period of the waveform is $T = t_1 + t_2 = 2.08 \ \mu s$. The duty cycle is then

$$\text{Duty cycle} = \frac{t_2}{t_1 + t_2} = \frac{0.693}{2.08}$$
$$= 0.333 = 33.3 \text{ percent}$$

For this particular circuit, it is generally simpler to calculate the duty cycle in terms of the two resistors. Using the above expressions for $t_1$ and $t_2$, it is a simple matter to show (see Prob. 9–14)

$$\text{Duty cycle} = \frac{t_2}{t_1 + t_2} = \frac{R_B}{R_A + 2R_B}$$

For instance, using this equation in Example 9–4 yields

$$\text{Duty cycle} = \frac{1000}{1000 + 2(1000)} = 0.333 = 33.3 \text{ percent}$$

## □ EXAMPLE 9—6

Given $R_B = 750 \ \Omega$, determine values for $R_A$ and $C$ in Fig. 9–4 to provide a 1.0-MHz clock that has a duty cycle of 25 percent.

## □ SOLUTION

A 1-MHz clock has a period of 1 $\mu s$. A duty cycle of 25 percent requires $t_1 = 0.75 \ \mu s$ and $t_2 = 0.25 \ \mu s$. Solving the expression

$$\text{Duty cycle} = R_B/(R_A + 2R_B)$$

for $R_A$ yields

$$R_A = \frac{R_B}{\text{Duty cycle}} - 2R_B = \frac{750}{0.25} - 2 \times 750 = 1500 \ \Omega$$

Solving $t_2 = 0.693R_BC$ for $C$ yields

$$C = \frac{t_2}{0.693R_B} = \frac{0.25 \times 10^{-6}}{0.693 \times 750}$$
$$= 480 \text{ pF}$$

The nomogram given in Fig. 9–4*b* can be used to estimate the free-running frequency to be achieved with various combinations of external resistors and timing capacitors. For example, the intersection of the resistance line 10 k$\Omega = (R_A + 2R_B)$ and the capacitance line 1.0 $\mu$F gives a free-running frequency of just over 100 Hz. It should be noted that there are definite constraints on timing component values and the frequency of oscillation, and you should consult the 555 data sheets.

## 9-4  555 TIMER—MONOSTABLE

With only minimal changes in wiring, the 555 timer discussed in Sec. 9–3 can be changed from a free-running oscillator (astable) into a switching circuit having one stable state and one quasistable state. The resulting monostable circuit is widely used in industry for many different timing applications. The normal mode of operation is to trigger the circuit into its quasistable state, where it will remain for a predetermined length of time. The circuit will then switch itself back (regenerate) into its stable state, where it will remain until it receives another input trigger pulse. Since it has only one stable state, the circuit is characterized by the term *monostable multivibrator,* or simply *monostable.*

The standard logic symbol for a monostable is shown in Fig. 9–5*a*. The input is labeled *TRIGGER*, and the output is $Q$. The complement of the $Q$ output may also be available at $\overline{Q}$. The input trigger circuit may be sensitive to either a positive-going or a negative-going signal. Usually the output at $Q$ is low when the circuit is in its stable state.

A typical set of waveforms showing the proper operation of a monostable circuit is shown in Fig. 9–5*b*. In this case, the circuit is sensitive to a negative-going signal at the trigger input, and the output is low when the circuit rests in its stable state. Once triggered, $Q$ goes high and remains high for a predetermined time $t$ and then switches back to its stable state until another negative-going signal appears at the trigger input.

A 555 timer wired as a monostable switching circuit (sometimes called a "one-shot") is shown in Fig. 9–6. In its stable state, the timing capacitor $C$ is completely discharged by means of an internal transistor connected to $C$ at pin 7. In this mode, the output voltage at pin 3 is at ground potential.

A negative pulse at the trigger input (pin 2) will cause the circuit to switch to its quasistable state. The output at pin 3 will go high and the discharge transistor at pin 7 will turn off, thus allowing the timing capacitor to begin charging toward $V_{CC}$.

When the voltage across $C$ reaches $\frac{2}{3} V_{CC}$, the circuit will regenerate back to its stable state. The discharge transistor will again turn on and discharge $C$ to GND, the output will go back to GND, and the circuit will remain in this state until another pulse arrives at the trigger input. A typical set of waveforms is shown in Fig. 9–6*b*.

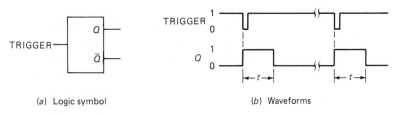

(a) Logic symbol                    (b) Waveforms

**FIG. 9–5**  Monostable circuit

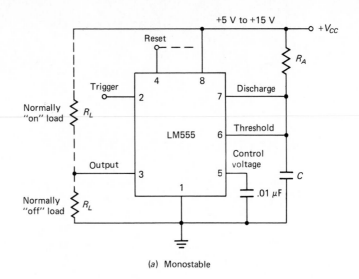

(a) Monostable

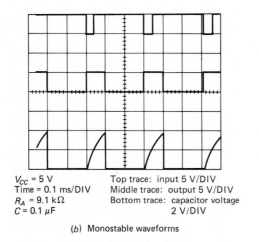

$V_{CC} = 5$ V   Top trace: input 5 V/DIV
Time = 0.1 ms/DIV   Middle trace: output 5 V/DIV
$R_A = 9.1$ kΩ   Bottom trace: capacitor voltage
$C = 0.1$ μF         2 V/DIV

(b) Monostable waveforms

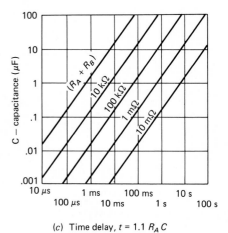

(c) Time delay, $t = 1.1 R_A C$

**FIG. 9–6** LM555 connected as a monostable circuit

The output of the monostable can be considered as a positive pulse, and the width of the pulse is given as

$$t = 1.1(R_A C)$$

Take care to note that the input voltage at the trigger input must be held at $+V_{CC}$, and that a negative pulse should then be applied when it is desired to trigger the circuit into its quasistable or timing mode.

□ EXAMPLE 9–7

Calculate the output pulse width for the timer in Fig. 9–6 given $R_A = 10$ kΩ and $C = 0.1$ μF.

□ SOLUTION

The pulse width is found as

$$t = 1.1(R_A C) = 1.1(10^4 \times 10^{-7}) = 1.1 \text{ ms}$$

□ EXAMPLE 9–8

Find the value of $C$ necessary to change the pulse width in Example 9–7 to 10 ms.

□ SOLUTION

The timing equation can be solved for $C$ as

$$C = \frac{t}{1.1 R_A} = \frac{10^{-2}}{1.1 \times 10^4}$$
$$= 0.909 \ \mu F$$

The nomograph shown in Fig. 9–6c can be used to obtain a quick, if not very accurate, idea of the sizes of $R_A$ or $C$ required for various pulse-width times. You can quickly check the validity of the results of Example 9–8 by following the $R_A = 10 \text{ k}\Omega$ line up to the $C = 0.1 \ \mu F$ line and noting pulse-width time is about 1 ms.

Once the circuit is switched into its quasistable state (the output is high), the circuit is immune to any other signals at its trigger input. That is, the timing cannot be interrupted and the circuit is said to be *nonretriggerable*. However, the timing can be interrupted by the application of a negative signal at the reset input on pin 4. A voltage level going from $+V_{CC}$ to GND at the reset input will cause the timer to immediately switch back to its stable state with the output low. As a matter of practicality, if the reset function is not used, pin 4 should be tied to $+V_{CC}$ to prevent any possibility of false triggering.

# 9-5 MONOSTABLES WITH INPUT LOGIC

The basic monostable circuit discussed in the previous section provides an output pulse of predetermined width in response to an input trigger. Logic gates have been added to the inputs of a number of commercially available monostable circuits to facilitate the use of these circuits as general-purpose delay elements. The 74121 *nonretriggerable* and the 74123 *retriggerable monostables* are two such widely used circuits.

The logic inputs on either of these circuits can be used to allow triggering of the device on either a high-to-low transition or on a low-to-high transition. Whenever the value of the input logic equation changes from false to true, the circuit will trigger. Take care to note that a transition from false to true must occur, and simply holding the input logic equation in the true state will have no effect.

The logic diagram, truth table, and typical waveforms for a 74121 are given in Fig. 9–7. The inputs to the 74121 are $\overline{A_1}$, $\overline{A_2}$, and $B$. The trigger input to the monostable appears at the output of the AND gate. Here's how the gates work:

1. If $B$ is held high, a negative transition at either $\overline{A_1}$ or $\overline{A_2}$ will trigger the circuit (see Fig. 9–7c). This corresponds to the bottom two entries in the truth table.
2. If either $\overline{A_1}$ or $\overline{A_2}$, or both are held low, a positive transition at $B$ will trigger the circuit (see Fig. 9–7d).

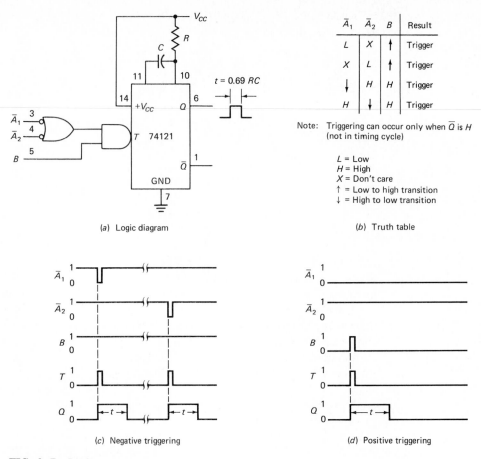

| $\bar{A}_1$ | $\bar{A}_2$ | B | Result |
|:---:|:---:|:---:|:---:|
| L | X | ↑ | Trigger |
| X | L | ↑ | Trigger |
| ↓ | H | H | Trigger |
| H | ↓ | H | Trigger |

Note: Triggering can occur only when $\bar{Q}$ is $H$ (not in timing cycle)

L = Low
H = High
X = Don't care
↑ = Low to high transition
↓ = High to low transition

(a) Logic diagram

(b) Truth table

(c) Negative triggering

(d) Positive triggering

FIG. 9–7   74121 non-retriggerable monostable

This corresponds to the top two entries in the truth table. A logic equation for the trigger input can be written as

$$T = (A_1 + A_2)B\bar{Q}$$

Note that for $T$ to be true (high), either $A_1$ or $A_2$ must be true—that is, either $\bar{A_1}$ or $\bar{A_2}$ at the gate input must be low. Also, since $\bar{Q}$ is low during the timing cycle (in the quasistable state), it is not possible for a transition to occur at T during this time. The logic equation for $T$ must be low if $\bar{Q}$ is low. In other words, once the monostable has been triggered into its quasistable state, it must time out and switch back to its stable state before it can be triggered again. This circuit is thus nonretriggerable.

The output pulse width at $Q$ is set according to the values of the timing resistor $R$ and capacitor $C$ as

$$t = 0.69RC$$

For instance, if $C = 1$ $\mu$F and $R = 10$ k$\Omega$, the output pulse width will be $t = 0.69 \times 10^4 \times 10^{-6} = 6.9$ ms.

☐ EXAMPLE 9–9

The 74121 monostable in Fig. 9–7 is connected with $R = 1$ k$\Omega$ and $C = 10,000$ pF. Pins 3 and 4 are tied to GND and a series of positive pulses are applied to pin 5. Describe the expected waveform at pin 6, assuming that the input pulses are spaced by (a) 10 $\mu$s and (b) 5 $\mu$s.

☐ SOLUTION

The circuit is connected such that positive pulses applied to pin 5 will trigger it. The output pulse width at pin 6 will be $t = 0.69 \times 10^3 \times 10^{-8} = 6.9$ $\mu$s.

(a) The monostable will trigger and time out for every input pulse appearing at $B$, as shown in Fig. 9–8a.

(b) Since the monostable is *not* retriggerable, it will trigger once and time out for every other input pulse as shown in Fig. 9–8b.

The logic diagram and truth table for a 74123 retriggerable monostable are given in Fig. 9–9. There are actually two circuits in each 16-pin DIP, and the pin numbers are given for one of them. The input logic is simpler than for the 74121. The inputs are $\overline{A}$, $B$, and $\overline{R}$, and the truth table summarizes the operation of the circuit. The first entry in the truth table shows that the circuit will trigger if $\overline{R}$ and $B$ are both high, and a high-to-low transition occurs at $\overline{A}$.

The second truth table entry states the circuit will trigger if $\overline{A}$ is held low, $\overline{R}$ is held high, and a low-to-high transition occurs at $B$.

In the third truth table entry, if $\overline{A}$ is low and $B$ is high, a low-to-high transition at $\overline{R}$ will trigger the circuit.

The last two truth table entries deal with direct reset of the circuit. Irrespective of the values of $\overline{A}$ or $B$, if the $\overline{R}$ input transitions is high to low, or is held low, the circuit will immediately reset.

The logic equation for the trigger input to the monostable can be written $T = AB\overline{R}$. Notice that the state of the output $Q$ does not appear in this equation (as it does for the 74121). This means that this circuit will trigger *every time* there is a low-to-high transition at $T$. In other words, this is a *retriggerable* monostable!

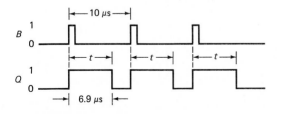

(a) Triggers on every pulse at $B$

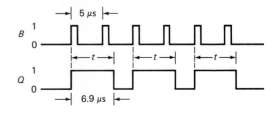

(b) Triggers on every other pulse at $B$

**FIG. 9–8**

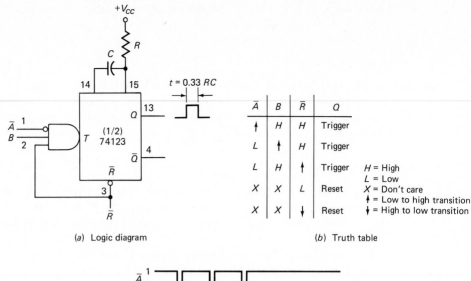

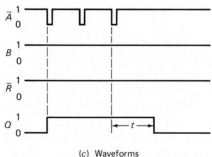

(a) Logic diagram

(b) Truth table

| $\overline{A}$ | $B$ | $\overline{R}$ | $Q$ |
|---|---|---|---|
| $\uparrow$ | $H$ | $H$ | Trigger |
| $L$ | $\uparrow$ | $H$ | Trigger |
| $L$ | $H$ | $\uparrow$ | Trigger |
| $X$ | $X$ | $L$ | Reset |
| $X$ | $X$ | $\downarrow$ | Reset |

$H$ = High
$L$ = Low
$X$ = Don't care
$\uparrow$ = Low to high transition
$\downarrow$ = High to low transition

(c) Waveforms

**FIG. 9–9** 74123

The output pulse width at $Q$ for the 74123 is set by the values of the timing resistor $R$ and the capacitor $C$. It can be approximated by the equation

$$t = 0.33RC$$

The waveforms in Fig. 9–9c show a series of negative pulses used to trigger the 74123. Notice carefully that the circuit triggers ($Q$ goes high) at the first high-to-low transition on $\overline{A}$, but that the next two negative pulses on $\overline{A}$ retrigger the circuit and the timing cycle $t$ does not begin until the very last trigger!

□ EXAMPLE 9—10

The 74123 in Fig. 9–9 is connected with $\overline{A}$ at GND, $\overline{R}$ at $+V_{CC}$, $R$ = 10 kΩ, and $C$ = 10,000 pF. Describe the expected waveform at $Q$, assuming that a series of positive pulses are applied at $B$ and the pulses are spaced at (a) 50 μs and (b) 10 μs.

□ SOLUTION

The output pulse width will be about

$$t = 0.33 \times 10^4 \times 10^{-8} = 33 \ \mu s$$

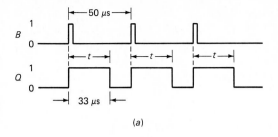

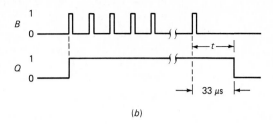

(a)                          (b)

**FIG. 9-10**

(a) The circuit will trigger and time out with every pulse as shown in Fig. 9-10a.

(b) The circuit will trigger with the first pulse and then retrigger with every following pulse. The timing cycle will be reset with every input pulse, and $Q$ will simply remain high since the circuit will never be allowed to time out (see Fig 9-10b). If the pulses at $B$ are stopped, $Q$ will be allowed to time out and will go low 33 $\mu s$ after the last pulse at $B$.

# 9-6 CONTACT BOUNCE CIRCUIT

In nearly every digital system there will be occasion to use mechanical contacts for the purpose of conveying an electrical signal; examples of this are the switches used on the keyboard of a computer system. In each case, the intent is to apply a high logic level (usually +5 V dc) or a low logic level (0 V dc). The single-pole–single-throw (SPST) switch shown in Fig. 9-11a is one such example. When the switch is open, the voltage at point $A$ is +5 V dc; when the switch is closed, the voltage at point $A$ is 0 V dc. Ideally, the voltage waveform at $A$ should appear as shown in Fig. 9-11b as the switch is moved from open to closed, or vice versa.

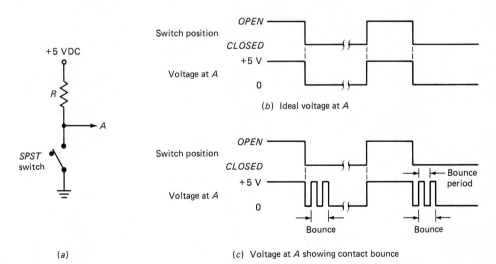

(a)

(b) Ideal voltage at $A$

(c) Voltage at $A$ showing contact bounce

**FIG. 9-11**

In actuality, the waveform at point $A$ will appear more or less as shown in Fig. 9–11c, as the result of a phenomenon known as *contact bounce*. Any mechanical switching device consists of a moving contact arm restrained by some sort of a spring system. As a result, when the arm is moved from one stable position to the other, the arm bounces, much as a hard ball bounces when dropped on a hard surface. The number of bounces that occur and the period of the bounce differ for each switching device. Notice carefully that in this particular instance, even though actual physical contact bounce occurs each time the switch is opened or closed, contact bounce appears in the voltage level at point $A$ only when the switch is closed.

If the voltage at point $A$ is applied to the input of a TTL circuit, the circuit will respond properly when the switch is opened, since no contact bounce occurs. However, when the switch is closed, the circuit will respond as if multiple signals were applied, rather than the single-switch closure intended—the undesired result of mechanical contact bounce. There is a need here for some sort of electronic circuit to eliminate the contact bounce problem, and the 74123 retriggerable monostable will do the job nicely.

The circuit shown in Fig. 9–12 uses a 74123 retriggerable monostable, two NOR gates connected to form a latch, a 3-input NOR gate, and a NOR gate used as an inverter to remove contact bounce. Here's how it works:

1. **Opening the switch.** When the switch opens, there is no bounce problem. Point $A$ immediately goes to $+V_{CC}$. Point $A$ is connected directly to one of the inputs of the 3-input NOR gate and its output at $C$ goes low. Thus the circuit output at $A'$ goes high. At the same time, $A$ sets the latch and $B$ goes high. The transition of $A$ at the monostable input has no effect, and thus its output at $Q$ remains low. The circuit output at $A'$ simply follows the switch input at $A$.
2. **Closing the switch.** Contact bounce now appears at $A$. The first high-to-low transition at the monostable input causes it to switch, and $Q$ goes high; $Q$ thus resets the latch and holds it reset, and $B$ goes low. Every subsequent high-to-low transition at $A$ restarts the timing cycle of the monostable. Thus the output of the 74123 at $Q$ remains high until allowed to time out *after the last* contact bounce. When $Q$ goes low after timing out, all inputs to the 3-input NOR gate are then low, $C$ goes high, and $A$ goes low.

You should carefully study the waveforms given in Fig. 9–12 until you fully understand the operation of the circuit. The output pulse time $t$ for the monostable is usually set just slightly longer than the bounce period—typically 5 or 10 ms.

☐ EXAMPLE 9–11

By experiment, the contact bounce period for the switch in Fig. 9–12 was found to be 750 $\mu$s. It is desired to set the delay time for the circuit at three times the bounce period. Determine the value of the capacitor $C$ if the timing resistor is set at 4.7 k$\Omega$.

☐ SOLUTION

The capacitor value can be calculated using $t = 0.33\ RC$. Thus

$$C = \frac{3 \times 750 \times 10^{-6}}{0.33 \times 4.7 \times 10^3} = 1.4\ \mu F$$

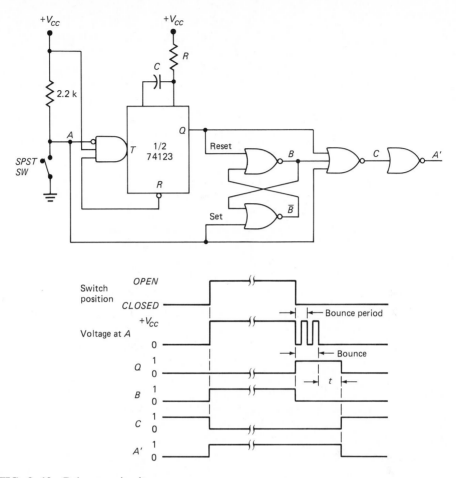

**FIG. 9–12**  Debounce circuit

## 9-7 SOME APPLICATIONS

The monostable circuits discussed in the previous sections have pulse-width times that are predictable to around 10 percent. As such, they do not represent precise timing circuits, but they do offer good short-term stability and are useful in numerous timing applications.

   One such application involves the production of a pulse that occurs after a given event with a predictable time delay. For instance, suppose that you are required to generate a 1-ms pulse exactly 2 ms after the operation of a push-button switch. Look at the waveforms in Fig. 9–13b. If the operation of the switch occurs when the waveform labeled *SWITCH* goes high, the desired pulse is shown as OUTPUT. In this case, the delay time $t_1$ will be set to 2 ms, and the time of the pulse width $t_2$ will be 1 ms.

   The two monostables in the 74123 shown in Fig. 9–13a are connected to provide a delayed pulse. The first circuit provides the delay time as $t_1 = 0.33R_1 \times C_1$, while the second circuit provides the output pulse width as $t_2 = 0.33R_2 \times C_2$. The positive transition at the INPUT triggers the first circuit into its quasistable state, and its output at $\overline{Q_1}$

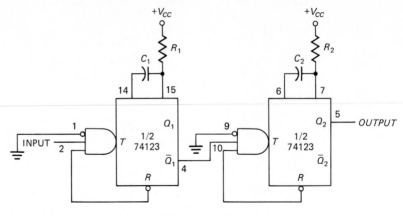

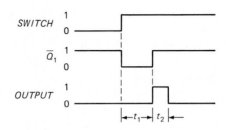

(a) A 74123 with DIP pin numbers

(b) Delayed pulse at OUTPUT

**FIG. 9–13** Delayed pulse generator

goes low. After timing out $t_1$, $\overline{Q}_1$ goes high, and this transition triggers the second circuit into its quasistable state. The OUTPUT thus goes high until the second circuit times out $t_2$, and then it returns low.

☐ EXAMPLE 9–12

The input to the circuit in Fig. 9–13a is changed to a 100-kHz square wave. It is desired to produce a 1-$\mu$s pulse 2 $\mu$s after every positive transition of the input as shown in Fig. 9–14. Find the proper timing capacitor values, given that both timing resistors are set at 500 $\Omega$.

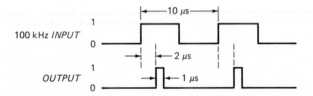

**FIG. 9–14**

## SOLUTION

The capacitor value for the pulse width is found using $t = 0.33\ RC$. Thus:

$$C = \frac{10^{-6}}{0.33 \times 500} = 6000\ \text{pF}$$

The pulse delay capacitor is twice this value, or $0.012\ \mu\text{F}$.

Whenever two or more signals at the inputs of a gate are undergoing changes at the same time, an undesired signal may appear at the gate output—this undesired signal is called a *glitch*. For example, in Fig. 9–15a, the gate output at $X$ should be low except during the time when $A = B = C = 1$ as shown. However, there is the possibility of a glitch appearing at the output at two different times. At time $T_1$, if $C$ happens to go high before $A$ and $B$ go low, a narrow positive spike will appear at the gate output—a glitch! Similarly, a glitch could occur at time $T_2$ if $B$ happens to go high before $A$ goes low.

A glitch is an unwanted signal generated usually because of different propagation delay times through different signal paths, and they generally cause random errors to occur in a digital system. They are to be avoided at all costs, and a logic circuit designer must take them into account. One method of avoiding glitches in the instance shown in Fig. 9–15a is to use a strobe pulse.

It is a simple matter to use a pulse delay circuit such as the one shown in Fig. 9–13 to generate a strobe pulse. Consider using the waveform $A$ in Fig. 9–15a as the input to the pulse delay circuit, and set the monostable times to generate a strobe pulse at the midpoint of the positive half cycle of $A$, as shown in Fig. 9–15b. If the inputs to the AND gate are

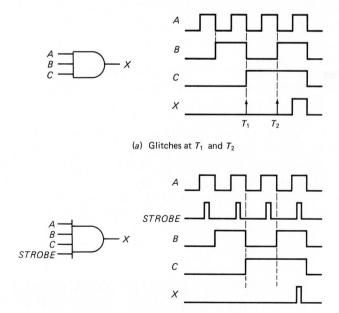

(a) Glitches at $T_1$ and $T_2$

(b) Use of *STROBE* to remove glitches

**FIG. 9–15**

CLOCKS AND TIMERS

**295**

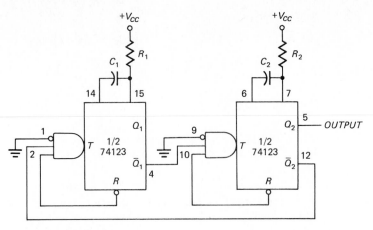

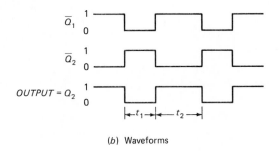

(a) Two monostable circuits connected to form an astable free-running oscillator

(b) Waveforms

**FIG. 9–16**

now $A$, $B$, $C$, and the strobe pulse, the output will occur only when $A = B = C = 1$, and a strobe pulse occurs. The glitches are completely eliminated!

An interesting variation of the pulse delay circuit in Fig. 9–13 is shown in Fig. 9–16a. Here, we have simply connected the $\overline{Q}$ output of the second circuit back to the input of the first circuit. This is a form of positive feedback. As a result, the circuit will oscillate—it becomes astable and generates a rectangular waveform as shown in Fig. 9–16. Here's how it works. The first circuit triggers into its quasistable state. When it times out $t_1$, the positive transition at $\overline{Q_1}$ will trigger the second circuit. When it times out $t_2$, the positive transition at $\overline{Q_2}$ will retrigger the first circuit and the cycle will repeat.

Independent adjustment of high and low levels of the output waveform is possible by setting the delay times of each individual monostable. Take care to note that since each circuit is edge-triggered, if a transition is missed by either circuit, oscillation will cease!

# SUMMARY

A system clock signal is a periodic waveform (usually a square wave) that has stable high and low levels, very short rise and fall times, and good frequency stability. A circuit widely used to generate a good, stable, TTL-compatible clock waveform is the crystal-controlled circuit shown in Fig. 9–3.

The 555 timer is a digital timing circuit that can be connected as either a monostable or an astable circuit. It is widely used in a number of different applications. The 74121 and 74123 monostable circuits both have logic circuits at their inputs that increase the number of possible applications.

A contact bounce circuit, a pulse delay circuit, and a free-running astable with adjustable duty cycle are only a few of the many circuits that can be constructed with the use of basic monostable circuits.

## G L O S S A R Y

*Astable* Having two output states, neither of which is stable.

*Clock* A periodic waveform (usually a square wave) that is used as a synchronizing signal in a digital system.

*Clock Cycle Time* The time period of a clock signal.

*Clock Stability* A measure of the frequency stability of a waveform; usually given in parts per million (ppm).

*Contact Bounce* Opening and closing of a set of contacts as a result of the mechanical bounce that occurs when the device is switched.

*Duty Cycle* For a periodic, rectangular waveform, the ratio of time the signal is high to the time period of the signal.

*Fall Time* The time required for a signal to transition from 90 percent of its maximum value down to 10 percent of its maximum.

*Glitch* Very narrow positive or negative pulse that appears as an unwanted signal.

*Monostable* A circuit that has two output states, only one of which is stable.

*One-Shot* Another term for a monostable circuit.

*Rise Time* The time required for a signal to transition from 10 percent of its maximum value up to 90 percent of its maximum.

*TTL Clock* A circuit that generates a clock waveform that is compatible with standard TTL logic circuits.

*10 Percent Point* A point on a rising or falling waveform that is equal to 0.1 times its highest value.

*90 Percent Point* A point on a rising or falling waveform that is equal to 0.9 times its highest value.

*555 Timer* A digital timing circuit that can be connected as either an astable or a monostable circuit.

*SECTION 9-1:*

**9–1.** Calculate the clock cycle time for a system that uses a clock that has a frequency of:

    **a**. 10 MHz    **b**. 6 MHz    **c.** 750 kHz

**9–2.** What is the clock frequency if the clock cycle time is 250 ns?

**9–3.** What is the maximum clock frequency that can be used with a master-slave flip-flop having a propagation delay of 75 ns?

**9–4.** You are selecting master-slave flip-flops that will be used in a system that has a clock frequency of 8 MHz. What is the maximum allowable propagation delay for the flip-flops?

**9–5.** What would be the 10 and 90 percent points on the waveform in Fig. 9–2c if the amplitude goes from 0 to +4.5 V?

*SECTION 9-2:*

**9–6.** Find the upper and lower frequency limits of a 5-MHz clock signal that has a stability of 10 ppm.

**9–7.** A TTL clock uses a series-mode crystal having a resonant frequency of 3.5 MHz. The circuit provides a 24-h stability of 8 ppm. Calculate the oscillator frequency limits.

**9–8.** The TTL clock shown in Fig. 9–3 uses a crystal that has frequency of 7.5 MHz. Draw the clock output waveform if $+V_{CC}$ is set at +5 V. What is the stability in ppm if the upper limit on the clock frequency is 7,499,900 Hz?

**9–9.** The flip-flop in Example 9–2 is connected with its $J$ and $K$ inputs both high, and a 15-MHz clock is applied to its clock input. Make a careful sketch of the input clock waveform and the waveform at the $Q$ output. Assume that the clock input is sensitive to negative clock transitions. (*Hint*: Be sure to consider the propagation delay time.)

*SECTION 9-3:*

**9–10.** Determine the frequency of oscillation for the 555 timer in Fig. 9–4, given $R_A = R_B = 47$ k$\Omega$ and $C = 1000$ pF. Calculate the values of $t_1$ and $t_2$, and carefully sketch the output waveform.

**9–11.** Determine the frequency of oscillation for the 555 timer in Fig. 9–4, given $R_A = 5000$ $\Omega$, $R_B = 7500$ $\Omega$, and $C = 1500$ pF. Calculate values for $t_1$ and $t_2$, and carefully sketch the output waveform.

**9–12.** Use the nomogram in Fig. 9–4b to find $(R_A + 2R_B)$, given $C = 0.1$ $\mu$F and that the desired frequency is 1 kHz. Check the results by using the formula given for the frequency.

**9–13.** Calculate the duty cycle for the circuit in Prob. 9–10. For Prob. 9–11.

**9–14.** Derive the expression

$$\text{Duty cycle} = R_B/(R_A + 2R_B).$$

**9–15.** It is desired to have a duty cycle of 25 percent for the circuit in Prob. 9–12.

Find the correct values for the two resistors.

SECTION 9-4:

**9-16.** Calculate the output pulse width for the timer in Fig. 9-6 for a 4.7-kΩ resistor and a 1.5-μF capacitor.

**9-17.** Calculate the output pulse width for the circuit in Prob. 9-16, assuming that the resistor is halved.

**9-18.** Calculate the output pulse width for the circuit in Problem 9-16, assuming that the capacitor is doubled.

**9-19.** Find the capacitor value necessary to generate a 15-ms pulse width for the monostable in Fig. 9-6, given $R_A = 100$ kΩ.

**9-20.** A 500-Hz square wave is used as the trigger input for the circuit described in Example 9-7. Make a careful sketch of the input and output waveforms (similar to those in Fig. 9-6b).

**9-21.** Repeat Prob. 9-20, assuming that the trigger input is changed to a 1-kHz square wave.

SECTION 9-5:

**9-22.** In the 74121 in Fig. 9-7, $R = 47$ kΩ and $C = 10,000$ pF. Calculate the output pulse width.

**9-23.** Redraw the 74121 logic diagram in Fig. 9-7a and show how to connect the circuit such that it will trigger on the positive transitions of a square wave. For $R = 51$ kΩ, determine a value of $C$ such that the output pulse will have a width of 750 μs.

**9-24.** Repeat Prob. 9-23, but make the circuit trigger on negative transitions of the square wave.

**9-25.** Using the circuit described in Prob. 9-23, make a careful sketch of the input and output waveforms, assuming the input square wave has a frequency of:
   **a.** 1 kHz
   **b.** 5 kHz

**9-26.** In the 74123 in Fig. 9-9, $R = 47$ kΩ and $C = 10,000$ pF. Calculate the output pulse width.

**9-27.** Redraw the 74123 logic circuit shown in Fig. 9-9 and show how to connect the circuit such that it will trigger on positive transitions of a square wave. Given $R = 51$ kΩ, determine a value of $C$ such that the output pulse will have a width of 750 μs.

**9-28.** Repeat Prob. 9-27, but make the circuit trigger on negative transitions of the square wave.

**9-29.** Using the circuit described in Prob. 9-27, make a careful sketch of the input and output waveforms if the input square wave has a frequency of:
   **a.** 1 kHz
   **b.** 5 kHz

**9–30.** Write a logic equation for the signal at point $C$ in the contact bounce circuit in Fig. 9–12 (in terms of $A$, $B$, and $Q$).

**9–31.** What would happen if the output pulse time $t$ for the circuit in Fig. 9–12 were less than the contact bounce period? Explain your response.

**9–32.** The input to the circuit in Fig. 9–13 is a 250-kHz square wave. Determine the proper timing capacitor values to generate a string of positive-going, 0.1-$\mu$s pulses, delayed by 2.0 $\mu$s from the rising edges of the input square wave. Assume $R_1 = R_2 = 1$ k$\Omega$.

**9–33.** Draw the waveforms, input and output, for the circuit in Fig. 9–13, given that both timing resistors are 470 $\Omega$, $C_1 = 0.1$ $\mu$F, $C_2 = 0.01$ $\mu$F, and the input waveform has a frequency of 20 kHz.

**9–34.** Show how to use the circuit in Fig. 9–13 to generate a 0.2-$\mu$s strobe pulse centered on the positive half cycle of a 200 kHz square wave (similar to Fig. 9–15$b$). Draw the complete circuit and calculate all timing resistor and capacitor values. Assume $R_1 = R_2 = 1$ k$\Omega$.

**9–35.** Calculate values for the timing resistors and capacitors in Fig. 9–16 to generate a clock waveform that has:
  **a.** A frequency of 100 kHz and a duty cycle of 25 percent
  **b.** A frequency of 500 kHz and a duty cycle of 50 percent

**9–36.** There is contact bounce present with the SPDT switch in Fig. 9–17 just as with the SPST switch discussed in Fig. 9–11. However, the $\overline{RS}$ latch used in Fig. 9–17 will remove all contact bounce, and $V_{out}$ will be *high* with the switch in position 1 and *low* with the switch in position 2. Explain exactly how this "debounce circuit" works. You might use waveforms as an aid. Incidentally, the 54/74279 can be used to construct four of these circuits.

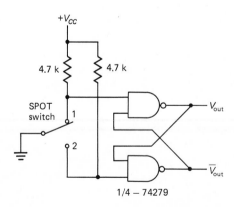

**FIG. 9–17** Debounce circuit

# 10

## SHIFT
## REGISTERS

A shift register is a very important digital building block. Registers are often used to momentarily store binary information appearing at the output of an encoding matrix. A register might be used to accept input data from an alphanumeric keyboard and then present this data at the input of a microprocessor chip. Similarly, shift registers are often used to momentarily store binary data at the output of a decoder. For instance, a register could be used to accept output data from a microprocessor chip and then present this data to the circuitry used to drive the display on a CRT screen. Thus registers form a very important link between the main digital system and the input-output channels.

A binary register also forms the basis for some very important arithmetic operations. For example, the operations of complementation, multiplication, and division are frequently implemented by means of a register. A shift register can also be connected to form a number of different types of counters. These counters offer some very distinct advantages.

The many different applications of shift registers, along with the myriad of techniques for using them, are simply too numerous to be discussed here. Our intent is to study the detailed operation of the four basic types of shift registers. With this knowledge, you will have the ability to study and understand exactly how a shift register is used in any specific application encountered.

## 10·1 TYPES OF REGISTERS

A register is simply a group of flip-flops that can be used to store a binary number. There must be one flip-flop for each bit in the binary number. For instance, a register used to store an 8-bit binary number must have eight flip-flops. Naturally the flip-flops must be connected such that the binary number can be entered (shifted) into the register and possibly shifted out. A group of flip-flops connected to provide either or both of these functions is called a *shift register*.

The bits in a binary number (let's call them the data) can be moved from one place to another in either of two ways. The first method involves shifting the data 1 bit at a time in a serial fashion, beginning with either the MSB or the LSB. This technique is referred to as *serial shifting*. The second method involves shifting all the data bits simultaneously and is referred to as *parallel shifting*.

There are two ways to shift data into a register (serial or parallel) and similarly two ways to shift the data out of the register. This leads to the construction of four basic register types as shown in Fig. 10–1—serial in–serial out, serial in–parallel out, parallel in–serial out, and parallel in–parallel out. All of these configurations are commercially available as TTL MSI/LSI circuits. For instance:

Serial in–serial out—54/74L91, 8 bits
Serial in–parallel out—54/74164, 8 bits
Parallel in–serial out—54/74165, 8 bits
Parallel in–parallel out—54/74194, 4 bits
Parallel in–parallel out—54/74198, 8 bits

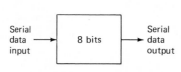

(a) Serial in/serial out

(b) Serial in/parallel out

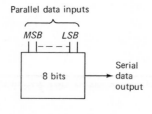

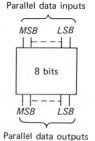

(c) Parallel in/serial out

(d) Parallel in/parallel out

**FIG. 10–1** Shift register types

We now need to consider the methods for shifting data in either a serial or parallel fashion. Data shifting techniques and methods for constructing the four different types of registers are discussed in the following sections.

## 10-2  SERIAL IN—SERIAL OUT

In this section we discuss the basic technique for shifting data in a serial fashion and then apply these ideas to construct a serial-in—serial-out shift register.

The flip-flops used to construct registers are usually either *JK* or *D* types. So, let's begin by summarizing the operation of a *JK* flip-flop.

For a *JK* flip-flop, the data bit to be shifted into the flip-flop must be present at the *J* and *K* inputs when the clock transitions (low or high). Since the data bit is either a 1 or a 0, there are two cases:

**1.** To shift a 0 into the flip-flop, $J = 0$ and $K = 1$.
**2.** To shift a 1 into the flip-flop, $J = 1$ and $K = 0$.

The important point to note is that the *J* and *K* inputs must be controlled to provide the correct input data. The *J* and *K* logic levels may be changing while the clock is high (or low), but they must be steady from just before until just after the clock transition (remember, setup time and hold time). For our discussion, we shall use *JK* master-slave flip-flops having clock inputs that are sensitive to negative clock transitions. Incidentally, this negative transition of the clock is frequently referred to as a *shift pulse*.

The waveforms in Fig. 10–2 illustrate these ideas. At time *A*, *Q* is reset low (a 0 is shifted into the flip-flop). At time *B*, *Q* does not change since the flip-flop had a 0 in it and another 0 is shifted in. At time *C*, the flip-flop is set (a 1 is shifted into it). At time *D*, another 0 is shifted into the flip-flop. In essence, we have shifted 4 data bits into this flip-flop in a time sequence: a 0 at time *A*, another 0 at time *B*, a 1 at time *C*, and a 0 at time *D*.

Now, consider adding three more flip-flops connected as shown in Fig. 10–3. Let's begin with all the flip-flops reset and then apply the exact same input signals to flip-flop *Q*

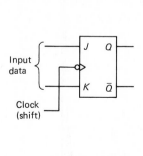

(a)

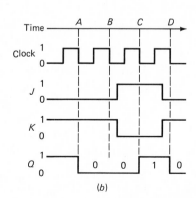

(b)

**FIG. 10–2**

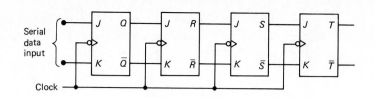

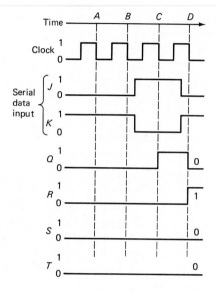

**FIG. 10–3** 4-bit serial input shift register

as we did in Fig. 10–2. Here's what happens:

**At time A:** All the flip-flops are reset, so all $J$ inputs are low and all $K$ inputs are high. Then $T$ is reset (the 0 in $S$ is shifted into $T$). Similarly, the 0 in $R$ is shifted into $S$, the 0 in $Q$ is shifted into $R$, and the 0 at the data input is shifted into $Q$. The flip-flop outputs just after time $A$ are $QRST = 0000$.

**At time B:** The flip-flops all contain 0s. Thus the 0 in $S$ is shifted into $T$, the 0 in $R$ shifts into $S$, the 0 in $Q$ shifts into $R$, and the 0 at the data input shifts into $Q$. The flip-flop outputs are $QRST = 0000$.

**At time C:** The flip-flops still all contain 0s. The 0 in $S$ shifts into $T$, the 0 in $R$ shifts into $S$, and the 0 in $Q$ shifts into $R$, but a 1 at the data input now shifts into $Q$. The flip-flop outputs are $QRST = 1000$.

**At time D:** The 0 in $S$ shifts into $T$, the 0 in $R$ shifts into $S$, the 1 in $Q$ shifts into $R$ (the $J$ input to $R$ is high and the $K$ input is low), and the 0 at the data input shifts into $Q$. The flip-flop outputs are $QRST = 0100$.

To summarize, we have shifted 4 data bits in a serial fashion into four flip-flops. These 4 data bits could represent a 4-bit binary number 0100, assuming that we began shifting with the LSB first. Notice that the LSB is in $T$ and the MSB is in $Q$. These four flip-flops

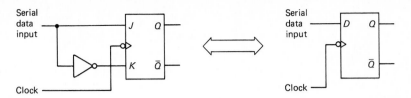

FIG. 10–4

could be defined as a 4-bit shift register; thus this is the technique used to construct a serial-input shift register.

The serial data input for the register shown in Fig. 10–3 requires two input signals— $J$ and $K$. But look carefully at the waveforms. Clearly, $K = \bar{J}$, or $J = \bar{K}$. In other words, one signal is always the complement of the other. If we were to connect an inverter between $J$ and $K$ on flip-flop $Q$ with the input at $J$, therefore, we would need to have only one data input signal—the one required for $J$! But this is precisely a $D$-type flip-flop as shown in Fig. 10–4. Remember the rules for a type $D$ flip-flop; on the negative clock transition, the data present at the $D$ input (either a 1 or a 0) will shift into the flip-flop.

Thus the 4-bit serial input shift register shown in Fig. 10–3 can be constructed by replacing the $JK$ flip-flops with type $D$ flip-flops as shown in Fig. 10–5. Besides the advantage of requiring only one data input signal (simply $D$, instead of both $J$ and $K$), the number of connections between flip-flops is reduced.

☐ EXAMPLE 10–1

Draw the waveforms to shift the number 0100 into the shift register shown in Fig. 10–5.

☐ SOLUTION

The waveforms for this register will appear exactly as in Fig. 10–3 provided the waveform labeled $K$ is eliminated and waveform $J$ is labeled $D$.

At this point, we have developed the ideas for shifting data into a register in serial form; the serial data input can be classified as either $JK$ or $D$, depending on the flip-flop type used to construct the register. Now, how about shifting data out of the register? Let's take another look at the register in Fig. 10–5, and suppose that it has the 4-bit number $QRST = 1010$ stored in it. If a clock signal is applied, the waveforms shown in

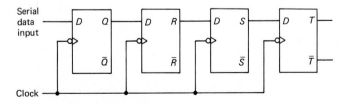

FIG. 10–5   4-bit serial input shift register

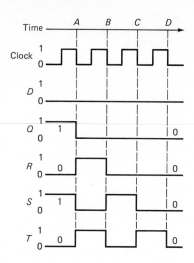

**FIG. 10–6**

Fig. 10–6 will be generated. Here's what happens:

**Before time $A$:** The register stores the number $QRST = 1010$. The LSB (a 0) appears at $T$.

**At time $A$:** The entire number is shifted one flip-flop to the right. A 0 is shifted into $Q$ and the LSB is shifted out the right end and lost. The register holds the bits $QRST = 0101$, and the second LSB (a 1) appears at $T$.

**At time $B$:** The bits are all shifted one flip-flop to the right, a 0 shifts into $Q$, and the third LSB (a 0) appears at $T$. The register holds $QRST = 0010$.

**At time $C$:** The bits are all shifted one flip-flop to the right, a 0 shifts into $Q$, and the MSB (a 1) appears at $T$. The register holds $QRST = 0001$.

**At time $D$:** The MSB is shifted out the right end and lost, a 0 shifts into $Q$, and the register holds $QRST = 0000$.

To summarize, we have caused the number stored in the register to appear at $T$ (this is the register output) 1 bit at a time, beginning with the LSB, in a serial fashion, over a time period of four clock cycles. In other words, the data stored was shifted out of the register at flip-flop $T$ in a serial fashion. Thus, not only is this a serial-input shift register, it is also a serial-output shift register. It is important to realize that the stored number is shifted out of the right end of the register and lost after four clock times. Notice that the complement of the output data stream is also available at $\overline{T}$.

The pinout and logic diagram for a 54/74L91 shift register are shown in Fig. 10–7. This is an 8-bit TTL MSI chip. There are eight $RS$ flip-flops connected to provide a serial input as well as a serial output. The clock input at each flip-flop is negative, edge-trigger-sensitive. However, since the applied clock signal is passed through an inverter, data will be shifted on the positive edges of the input clock pulses.

The $RS$ flip-flops are connected exactly as the $JK$ flip-flops in Fig. 10–3, and they behave in exactly the same way. This is true since the only two input combinations applied to the $RS$ inputs here are $R = 0$ and $S = 1$, or $R = 1$ and $S = 0$. These two

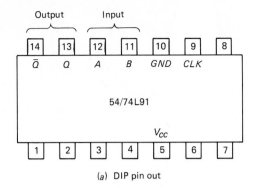

(a) DIP pin out

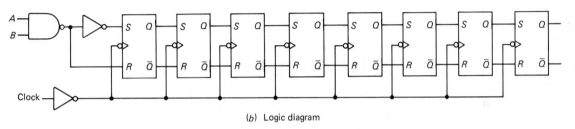

(b) Logic diagram

**FIG. 10-7** 54/74L91 8-bit shift register

combinations, of course, match those applied to the *JK* flip-flops in the shift register in Fig. 10-3.

The inverter connected between *R* and *S* on the first flip-flop means that this circuit functions as a *D*-type flip-flop. So, the input to the register is a single line on which the data to be shifted into the register appears serially. The data input is applied at either *A* (pin 12) or *B* (pin 11). Notice that a data level at *A* (or B) is complemented by the NAND gate and then applied to the *R* input of the first flip-flop. The same data level is complemented by the NAND gate and then complemented again by the inverter before it appears at the *S* input. So, a 1 at input *A* will set the first flip-flop (in other words, this 1 is shifted into the first flip-flop) on a positive clock transition.

The NAND gate with inputs *A* and *B* simply provides a gating function for the input data stream if desired. If gating is not desired, simply connect pins 11 and 12 together and apply the input data stream to this connection.

☐ EXAMPLE 10-2

Examine the logic levels at the input of a 54/74L91 and show how a 1 and then a 0 are shifted into the register.

☐ SOLUTION

The input logic and the first flip-flop are redrawn in Fig. 10-8a, and a 1 is applied at the data input A. The *R* input is 0, the *S* input is 1, and the flip-flop will clearly be set when the clock goes high. In other words, the 1 at the data input will shift into the flip-flop. In

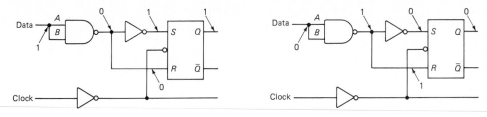

(a) Logic levels shown by arrows will set the flip-flop

(b) Logic levels shown by arrows will reset the flip-flop

**FIG. 10–8** Example 10–2

Fig. 10–8b, a 0 is applied at the data input A. The R input is 1, the S input is 0, and the flip-flop will be reset when the clock goes high. The input 0 is thus shifted into the flip-flop.

## 10-3 SERIAL IN—PARALLEL OUT

The second type of register mentioned in Sec. 10–1 is one in which data is shifted in serially, but shifted out in parallel. In order to shift the data out in parallel, it is simply necessary to have all the data bits available as outputs at the same time. This is easily accomplished by connecting the output of each flip-flop to an output pin. For instance, an 8-bit shift register would have eight output lines—one for each flip-flop in the register. The basic configuration is shown in Fig. 10–1b.

The 54/74164 is an 8-bit serial-input–parallel-output shift register. The pinout and logic diagram for this device are given in Fig. 10–9. It is constructed by using RS flip-flops having clock inputs that are sensitive to negative clock transitions. A careful examination of the logic diagram in Fig. 10–9b will reveal that this register is exactly like the 54/74L91 discussed in the previous section—with two exceptions: (1) the true side of each flip-flop is available as an output—thus all 8 bits of any number stored in the register are available simultaneously as an output (this is a parallel data output); and (2) each flip-flop has an asynchronous CLEAR input. Thus a low level at the CLEAR input to the chip (pin 9) is applied through an amplifier and will reset (clear) every flip-flop. Notice that this is an asynchronous signal and can be applied at any time, without regard to the clock waveform and also that this signal is level sensitive. As long as the CLEAR input to the chip is held low, the flip-flop outputs will all remain low. (The register will contain all zeros!)

Shifting data into the register in a serial fashion is exactly the same as the previously discussed 54/74L91. Data at the serial inputs may be changed while the clock is either low or high, but the usual setup and hold times must be observed. The data sheet for this device gives setup time as 30 ns minimum and hold time as 0.0 ns. Since data are shifted into the register on positive clock transitions, the data input line must be stable from 30 ns before the positive clock transition until the clock transition is complete.

Let's take a look at the gated serial inputs A and B. Suppose that the serial data is connected to A; then B can be used as a control line. Here's how it works:

**B is held high:** The NAND gate is enabled and the serial input data passes through the NAND gate inverted. The input data is shifted serially into the register.

**B is held low:** The NAND–gate output is forced high, the input data stream is inhibited, and the next positive clock transition will shift a 0 into the first flip-flop. Each succeeding positive clock transition will shift another 0 into the register. After eight clock pulses, the register will be full of zeros!

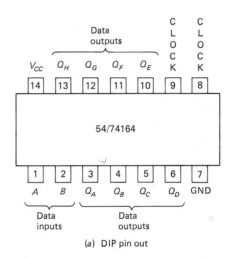

(a) DIP pin out

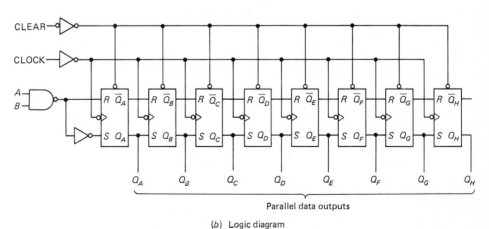

(b) Logic diagram

**FIG. 10-9**  54/74164 8-bit shift register

## ☐ EXAMPLE 10-3

How long will it take to shift an 8-bit number into a 54164 shift register if the clock is set at 10 MHz?

## ☐ SOLUTION

A minimum of eight clock periods will be required since the data is entered serially. One clock period is 100 ns, so it will require 800 ns minimum.

## ☐ EXAMPLE 10-4

For the register in Example 10-3, when must the input data be stable? When can it be changed?

## ☐ SOLUTION

The data must be stable from 30 ns before a positive clock transition until the positive transition occurs. This leaves 70 ns during which the data may be changing (see Fig. 10-10).

The waveforms shown in Fig. 10-11 show the typical response of a 54/74164. The serial data is input at $A$ (pin 1), while a gating control signal is applied at $B$ (pin 2). The first CLEAR pulse occurs at time $A$ and simply resets all flip-flops to 0.

The clock begins at time $B$, but the first positive clock does nothing since the control line is low. At time $C$ the control line goes high, and the first data bit (a 0) is shifted into the register at time $D$.

The next 7 data bits are shifted in, in order, at times $E$, $F$, $G$, $H$, $I$, $J$, and $K$. The clock remains high after time $K$, and the 8-bit number 0010 1100 now resides in the register and is available on the eight output lines. This assumes that the LSB was shifted in first and appears at $Q_H$. Notice that the clock *must be stopped* after its positive transition at time $K$; otherwise shifting will continue and the data bits will be lost.

Finally, another CLEAR pulse occurs at time $L$, the flip-flops are all reset to zero, and another shift sequence may begin. Incidentally, the register can be cleared by holding the control line at $B$ low and allowing the clock to run for eight positive clock transitions. This simply shifts eight 0s into the register.

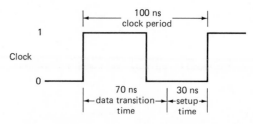

**FIG. 10-10**   Example 10-4

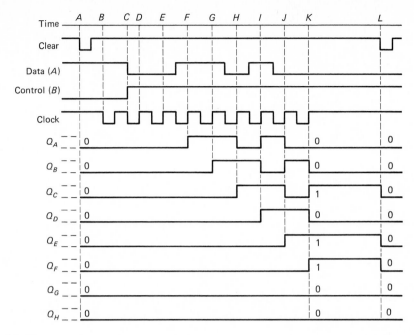

**FIG. 10–11**

# 10-4 PARALLEL IN—SERIAL OUT

In prior sections, the ideas necessary for shifting data into and out of a register in serial have been developed. We can now use these same ideas to develop methods for the parallel entry of data into a register. There are a number of different techniques for the parallel entry of data, but we shall concentrate our efforts on commercially available TTL. At first glance, the logic diagrams for some of the shift registers seem rather formidable (see, for instance, the block diagram for the 54/74166); but they aren't really. The 54/74166, for instance, is an 8-bit shift register, and the same circuit is repeated eight times. So, it's necessary to study only one of the eight circuits, and that's what we'll do here.

The pinout and logic block diagram for a 54/74166 are given in Fig. 10–12. The functional description given on the TTL data sheet says that this is an 8-bit shift register, capable of either serial or parallel data entry, and serial data output. Notice that there are eight *RS* flip-flops, each with some attached logic circuitry. Let's analyze one of these circuits by starting with the *RS* flip-flops and then adding logic blocks to accomplish our needs.

First recognize that the clocked *RS* flip-flop and the attached inverter given in Fig. 10–13a form a type *D* flip-flop. If a data bit *X* is to be clocked into the flip-flop, the complement of *X* must be present at the input. For instance, if $\overline{X} = 0$, then $R = 0$ and $S = 1$, and a 1 will be clocked into the flip-flop when the clock transitions.

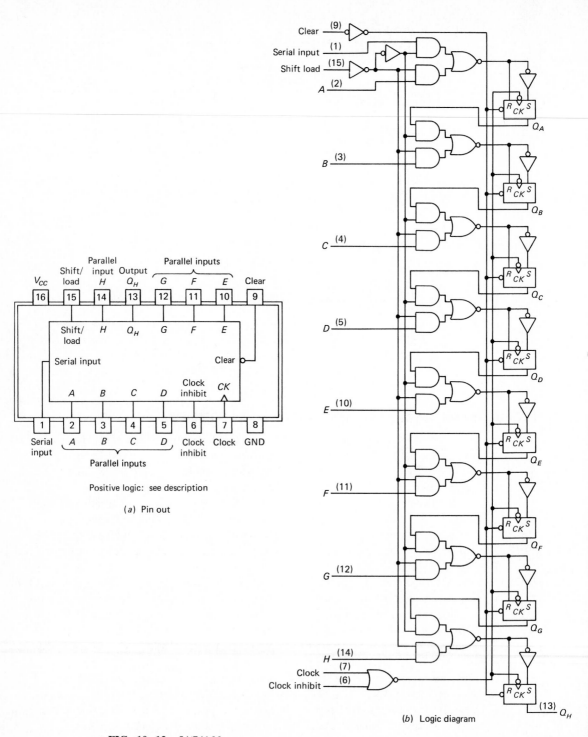

Parallel inputs
Parallel input Output
Vcc Shift/load H Qн G F E Clear

| 16 | 15 | 14 | 13 | 12 | 11 | 10 | 9 |

Shift/load H Qн G F E

Serial input Clear

Clock
A B C D inhibit CK

| 1 | 2 | 3 | 4 | 5 | 6 | 7 | 8 |

Serial A B C D Clock Clock GND
input inhibit
Parallel inputs

Positive logic: see description

(a) Pin out

(b) Logic diagram

**FIG. 10-12** 54/74166

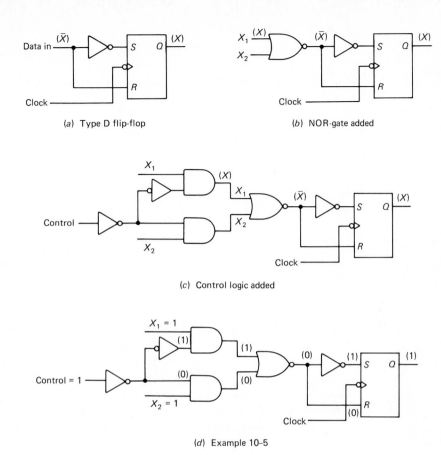

(a) Type D flip-flop

(b) NOR-gate added

(c) Control logic added

(d) Example 10-5

**FIG. 10–13**

Now, add a NOR gate as shown in Fig. 10–13$b$. If one leg of this NOR gate is at ground level, a data bit $X$ at the other leg is simply inverted by the NOR gate. For instance, if $X = 1$, then at the output of the NOR gate $\overline{X} = 0$, allowing a 1 to be clocked into the flip-flop. This NOR gate offers the option of entering data from two different sources, either $X_1$ or $X_2$. Holding $X_2$ at ground will allow the data at $X_1$ to be shifted into the flip-flop; conversely, holding $X_1$ at ground will allow data at $X_2$ to be shifted in.

The addition of two AND gates and two inverters as shown in Fig. 10–13$c$ will allow the selection of data $X_1$ or data $X_2$. If the control line is high, the upper AND gate is enabled and the lower AND gate is disabled. Thus $X_1$ will appear at the upper leg of the NOR gate while the lower leg of the NOR gate will be at ground level. On the other hand, if the control line is low, the upper AND–gate is disabled while the lower AND gate is enabled. This allows $X_2$ to appear at the lower leg of the NOR gate while the upper leg of the NOR gate is at ground level. You should now study this circuit until your understanding is crystal clear! Consider writing 0 or 1 at each gate leg in response to various inputs.

To summarize:

**CONTROL is high:** Data bit at $X_1$ will be shifted into the flip-flop at the next clock transition.

**CONTROL is low:** Data bit at $X_2$ will be shifted into the flip-flop at the next clock transition.

☐ EXAMPLE 10—5

For the circuit in Fig. 10–13c, write the logic levels present on each gate leg if CONTROL = 1, $X_1$ = 1, and $X_2$ = 1.

☐ SOLUTION

The correct levels are given in parentheses in Fig. 10–13d. The data value 1 at $X_1$ is shifted into the flip-flop when the clock transitions.

A careful examination will reveal that exactly eight of the circuits given in Fig. 10–13c are connected together to form the 54/74166 shift register shown in Fig. 10–12. The only question is: how are they connected? The answer is: they are connected to allow two different operations: (1) the parallel entry of data and (2) the operation of shifting data serially through the register from the first flip-flop $Q_A$ toward the last flip-flop $Q_H$.

If the data input labeled $X_2$ in Fig. 10–13c is brought out individually for each flip-flop, these eight inputs will serve as the parallel data entry inputs for an 8-bit number *ABCD EFGH*. These eight inputs are labeled *A, B, C, D, E, F, G,* and *H* in Fig. 10–12. The control line is labeled SHIFT/LOAD. Holding this SHIFT/LOAD control line low will enable the lower AND gate for each flip-flop, and the 8-bit number will be *LOADED* into the flip-flops with a single clock transition—PARALLEL input.

Holding the SHIFT/LOAD control line high will enable the upper AND gate for each flip-flop. If the input from this upper AND gate receives its data from the prior flip-flop in the register, each clock transition will shift a data bit from one flip-flop into the following flip-flop—proceeding in a direction from $Q_A$ toward $Q_H$. In other words, data will be shifted through the register serially! In the first flip-flop in the register, the upper AND–gate input is labeled SERIAL INPUT. Thus data can also be entered into this register in a serial fashion.

To summarize:

**SHIFT/LOAD is low:** A single clock transition loads 8 bits of data (*ABCD EFGH*) into the register in parallel.

**SHIFT/LOAD is high:** Clock transitions will shift data through the register serially, with entering data applied at the SERIAL INPUT.

Notice that the CLOCK is applied through a two-input NOR gate. When CLOCK INHIBIT is held low, the clock signal passes through the NOR gate inverted and the register flip-flops respond to "positive" transitions of the input clock signal. When CLOCK INHIBIT is high, the NOR–gate output is held low, and the clock is prevented from reaching the flip-flops. In this mode, the register can be made to stop and hold its contents.

A low level at the CLEAR input can be applied at any time without regard to the clock, and it will immediately reset all flip-flops to 0. When not in use, it should always be held high.

| Inputs | | | | | | Internal Levels | | Outputs |
|---|---|---|---|---|---|---|---|---|
| Clear | Shift/ load | Clock inhibit | Clock | Serial | Parallel A . . . H | $Q_A$ and $Q_B$ | | $Q_H$ |
| L | X | X | X | X | X | L | L | L |
| H | X | L | L | X | X | $Q_{AO}$ | $Q_{BO}$ | $Q_{HO}$ |
| H | L | L | $\uparrow$ | X | a . . . h | a | b | h |
| H | H | L | $\uparrow$ | H | X | H | $Q_{An}$ | $Q_{Gn}$ |
| H | H | L | $\uparrow$ | L | X | L | $Q_{An}$ | $Q_{Gn}$ |
| H | X | H | $\uparrow$ | X | X | $Q_{AO}$ | $Q_{BO}$ | $Q_{HO}$ |

$X$ = Irrelevant, $H$ = High level, $L$ = Low level
$\uparrow$ = Positive transition
$a$ . . . $h$ = Steady state input level at $A$ . . . $H$ respectively
$Q_{AO}$, $Q_{BO}$ = Level at $Q_A$, $Q_B$ . . . before steady state
$Q_{An}$, $Q_{Gn}$ = Level of $Q_A$ or $Q_B$ before most recent clock transition ($\uparrow$)

**FIG. 10–14** 54/74166 truth table

The truth table in Fig. 10–14 summarizes the operation of the 54/74166 8-bit shift register. You should study this table in conjunction with the logic diagram to understand clearly how the register can be used.

☐ EXAMPLE 10—6

Which entry in the truth table in Fig. 10–14 accounts for the parallel entry of data?

☐ SOLUTION

The third entry from the top; CLEAR is high, SHIFT/LOAD is low, CLOCK INHIBIT is low, a positive clock transition occurs, and the SERIAL DATA INPUT is irrelevant.

## 10-5 PARALLEL IN—PARALLEL OUT

The fourth type of register discussed in the introductory section of this chapter is designed such that data can be shifted either into or out of the register in parallel. In fact, simply adding an output line from each flip-flop in the 54/74166 discussed in the previous section would meet the parallel-in–parallel-out requirements. [It would, of course, require a larger dual in-line package (DIP)—say, a 24-pin package.]

The 54/74198 is an 8-bit TTL MSI having both parallel input and parallel output capability. The DIP pinout for this device is given in Fig. 10–15. Notice that a 24-pin package is required since 16 pins are needed just for the input and output data lines. Not only does this chip satisfy the parallel input-output requirements; it can also be used to shift data through the register in either direction—referred to as *shift right* and *shift left*. All the registers previously discussed have the ability to shift right, that is, to shift data serially from the data input flip-flop toward the right, or from a flip-flop $Q_A$ toward flip-flop $Q_B$. We now need to consider how to shift left.

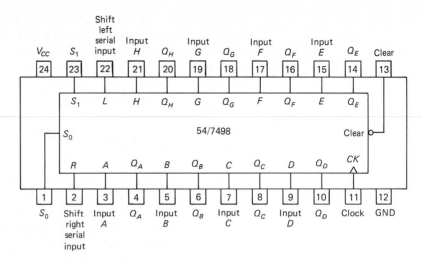

**FIG. 10–15**  54/74198, 8-bit shift register. Parallel input-parallel output

There are a number of 4-bit, parallel-input–parallel-output shift registers available since they can be conveniently packaged in a 16-pin DIP. An 8-bit register can be created by either connecting two 4-bit registers in series or by manufacturing the two 4-bit registers on a single chip and placing the chip in a 24-pin package (such as the 54/74198). Let's analyze a typical 4-bit register, say, a 54/7495A.

The data sheet for the 54/7495A describes it as a 4-bit, parallel-access shift register. It also has serial data input and can be used to shift data to the right (from $Q_A$ toward $Q_B$) and in the opposite direction—to the left. The DIP pinout and logic diagram are given in Fig. 10–16. The basic flip-flop and control logic used here is exactly the same as used in the 54/74164 as shown in Fig. 10–13c.

The parallel data outputs are simply the $Q$ sides of each of the four flip-flops in the register. In fact, note that the output $Q_D$ could be used as a serial output when data is shifted from left to right through the register (right shift).

When the MODE CONTROL line is held high, the AND gate on the right input to each NOR gate is enabled while the left AND gate is disabled. The data at inputs $A$, $B$, $C$, and $D$ will then be loaded into the register on a negative transition of the clock—this is parallel data input.

When the MODE CONTROL line is low, the AND gate on the right input to each NOR gate is disabled while the left AND gate is enabled. The data input to flip-flop $Q_A$ is now at SERIAL INPUT; the data input to $Q_B$ is $Q_A$ and so on down the line. On each negative clock transition, a data bit is entered serially into the register at the first flip-flop $Q_A$, and each stored data bit is shifted one flip-flop to the right (toward the last flip-flop $Q_D$). This is the serial input of data (at SERIAL INPUT), and also the right-shift operation.

In order to effect a shift-left operation, the input data must be connected to the $D$ DATA INPUT as shown in Fig. 10–17. It is also necessary to connect $Q_D$ to $C$, $Q_C$ to $B$, and $Q_B$ to $A$ as shown in Fig. 10–17. Now, when the MODE CONTROL line is held high,

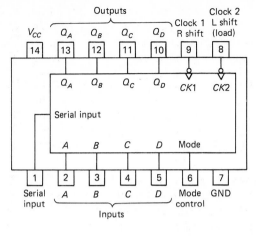

(a) Pin out

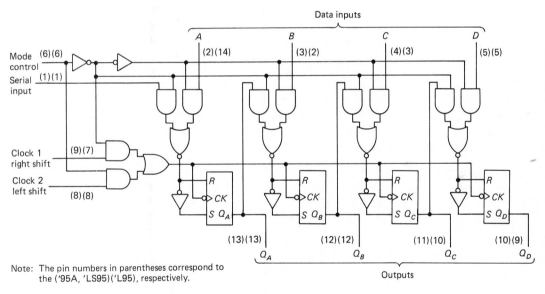

Note: The pin numbers in parentheses correspond to the ('95A, 'LS95)('L95), respectively.

(b) Logic diagram

**FIG. 10–16**  54/7495A

a data bit will be entered into flip-flop $Q_D$, and each stored data bit will be shifted one flip-flop to the left on each negative clock transition. This is also serial input of data (but at input $D$) and is the left-shift operation. Notice that the connections described here can either be hard-wired or can be made by means of logic gates.

There are two clock inputs—CLOCK 1 and CLOCK 2. This is to accommodate requirements where the clock used to shift data right is separate from the clock used to

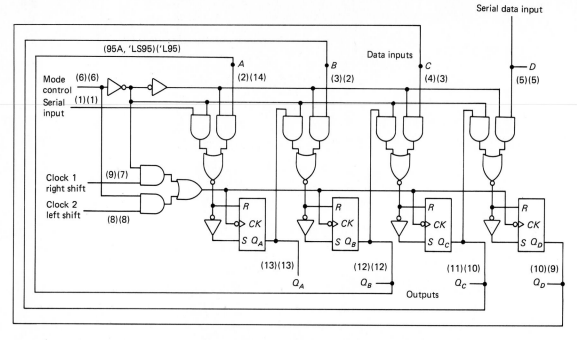

**FIG. 10-17** 54/7495A wired for ''shift left''

shift data to the left. If such a requirement is unnecessary, simply connect CLOCK 1 and CLOCK 2 together. The clock signal will then pass through the AND–OR gate combination noninverted, and the flip-flops will respond to negative clock transitions.

## ☐ EXAMPLE 10—7

Draw the waveforms you would expect if the 4-bit binary number 1010 were shifted into a 54/7495A in parallel.

## ☐ SOLUTION

The MODE CONTROL line must be high. The data input lines must be stable for more than 10 ns prior to the negative clock transition (setup time for the data sheet information). A single negative clock transition will enter the data. (The waveforms are given in Fig. 10–18.) If the clock is stopped after the transition time $T$, the levels on the input data lines may be changed. However, if the clock is not stopped, the input data line levels must be maintained.

At this point, it simply cannot be overemphasized that the input control lines to any shift register *must be controlled at all times*! Remember, the register will do something every time there is a clock transition. What it does is entirely dependent on the levels applied at the CONTROL INPUTS. If you do not account for input control levels, you simply cannot account for the behavior of the register!

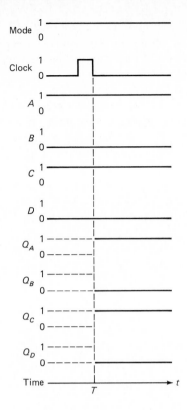

**FIG. 10–18**   Example 10–7

## 10·6 RING COUNTER

The serial shift registers discussed in the previous sections must have additional circuitry to control the *D* input to the first flip-flop (or the *RS* or the *JK* inputs) in order to ensure proper operation during shift operations. The logic circuitry that provides these control waveforms would ordinarily be derived from the control section of the system. The control section is the source of the clock and whatever other control signals are necessary. For now, let's see if any feedback techniques can be applied to a basic shift register with usable results.

Let's begin with a simple serial shift register such as the 54/74164. One of the most logical applications of feedback might be to connect the output of the last flip-flop $Q_H$ back to the *D* input of the first flip-flop *A* (Fig. 10–19a). Notice that the *A* and *B* data inputs are connected together. Now, suppose that all flip-flops are reset and the clock is allowed to run. What will happen? The answer is, nothing will happen since the *D* input to the first flip-flop is low (the input at *A* and *B*). Therefore, every time the clock goes high, the zero in each flip-flop will be shifted into the next flip-flop, while the zero in the last flip-flop *H* will travel around the feedback loop and shift into the first flip-flop *A*. In other words, all the flip-flops are in a reset state, each clock transition resets them again, and each flip-flop output simply remains low. Consider the register as a tube full of zeros (ping-pong balls) that shift round and round the register, moving ahead one flip-flop with each clock transition.

In a effort to obtain some action, suppose that $Q_A$ is high and all other flip-flops are low, and then allow the clock to run. The very first time the clock goes high, the 1 in $A$ will shift into $B$ and $A$ will be reset, since the 0 in $H$ will shift into $A$. All other flip-flops will still contain 0s. The second clock pulse will shift the 1 from $B$ to $C$, while $B$ resets. The third clock pulse will shift the 1 from $C$ to $D$, and so on. Thus this single 1 will shift down the register, traveling from one flip-flop to the next flip-flop each time the clock goes high. When it reaches flip-flop $H$, the next clock will shift it into flip-flop $A$ by means of the feedback connection. Again, consider the register as a tube full of ping-pong balls, seven "white" ones (0s) and one "black" one (a 1). The ping-pong balls simply circulate around the register in a clockwise direction, moving ahead one flip-flop with each clock transition. This configuration is frequently referred to as a *circulating register* or a *ring counter*. The waveforms present in this ring counter are given in Fig. 10–19*b*.

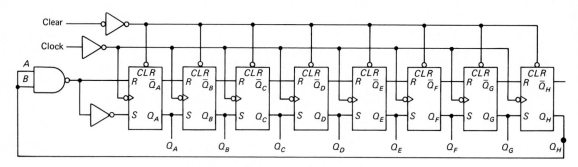

(a) 54/74164 8-bit shift register with feedback line from $Q_H$ to A–B

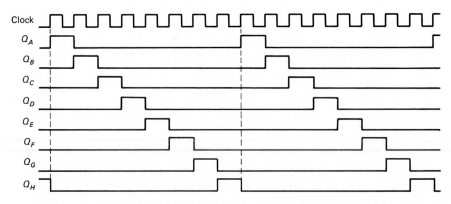

(b) Waveforms when register has a single one, and seven zeros

**FIG. 10–19** Ring counter

CHAPTER 10

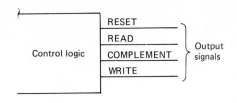 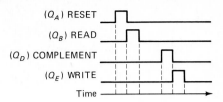

**FIG. 10–20**

Waveforms of this type are frequently used in the control section of a digital system. They are ideal for controlling events that must occur in a strict time sequence—that is, event $A$, then event $B$, then $C$, and so on. For instance, the logic diagram in Fig. 10–20 shows how to generate RESET, READ, COMPLEMENT, and WRITE (a fictitious set of control signals) as a set of control pulses that occur one after the other sequentially. The control signals are simply the outputs of flip-flops $A$, $B$, $D$, and $E$ as shown in Fig. 10–19.

There is, however, a problem with such ring counters. In order to produce the waveforms shown in Fig. 10–19, the counter should have one, and only one, 1 in it. The chances of this occurring naturally when power is first applied are very remote indeed. If the flip-flops should all happen to be in the reset state when power is first applied, it will not work at all, as we saw previously. On the other hand, if some of the flip-flops come up in the set state while the remainder come up in the reset state, a series of complex waveforms of some kind will be the result. Therefore, it is necessary to preset the counter to the desired state before it can be used.

☐ **EXAMPLE 10—8**

The register in Fig. 10–19 can easily be cleared to all 0's by using the CLEAR input. Show one method for setting a single 1 and the remaining 0's in the register.

☐ **SOLUTION**

The simple "power-on-reset" circuit in Fig. 10–21$a$ is widely used to generate the equivalent of a narrow negative pulse that occurs when power $(+V_{CC})$ is first applied to the system. Before the application of power, the voltage across the capacitor is zero. When $+V_{CC}$ is applied, the capacitor voltage charges toward $+V_{CC}$ with an $RC$ time constant, and then remains at $+V_{CC}$ as long as the system power remains, as seen by the waveform in the figure. If point $A$ is then connected to the CLEAR input of the 54/74164, all flip-flops will automatically be reset to 0's when $+V_{CC}$ is first applied.

The logic added in the feedback path in Fig. 10–21$b$ will now cause a single 1 to be set into the register. Here's how it works:

1. The "power-on-reset" pulse is inverted and used to initially set flip-flop $X$. This causes the output of the OR gate to be a 1, and the first clock transition will shift this 1 into $Q_A$.
2. When $Q_A$ goes high, this will reset flip-flop $X$. At this point, the register contains a 1 in $Q_A$, and 0's in all other flip-flops. $X$ will remain low as long as power is applied, and the data from $Q_H$ will pass through the OR gate directly to the data

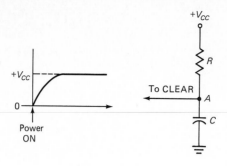

(a) "Power-on-reset" circuit.

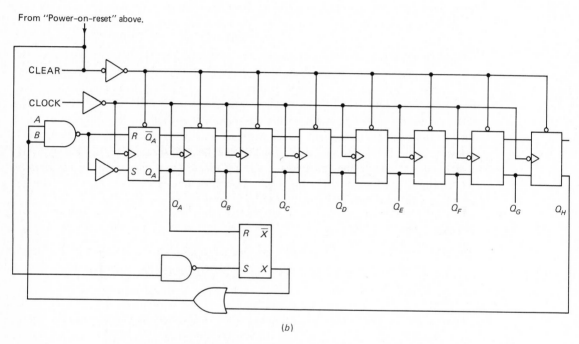

(b)

FIG. 10-21 54/74164 with logic to preset a single 1 and seven 0's

input $AB$. The single 1 and the seven 0's will now shift around the register, advancing one position with each clock transition as desired.

Since the ring counter in Fig. 10-19 can function with more than a single 1 in it, it might be desirable to operate in this fashion at some time or other. It can, for example, be used to generate a more complex control waveform. Suppose, for instance, that the waveform shown in Fig. 10-22 were needed. This waveform could easily be generated by simply presetting the counter in Fig. 10-19 with a 1 in $A$, a 1 in $C$, and all the other flip-flops reset. Notice that it is really immaterial where the two 1s are set initially. It is necessary only to ensure that they are spaced one flip-flop apart.

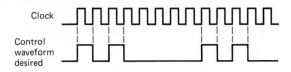

Clock

Control
waveform
desired

**FIG. 10–22**

☐ EXAMPLE 10–9

How would you preset the ring counter in Fig. 10–19 to obtain a square-wave output which is one-half the frequency of the clock? How about one-fourth the clock frequency?

☐ SOLUTION

It is necessary only to preset a 1 in every other flip-flop, while the remaining flip-flops are all reset. This will generate a waveform at each flip-flop output that is high for one clock period and then low for one clock period. The period of the output waveform is then two clock periods; therefore, the frequency is one-half the clock frequency. An output signal at one-fourth the clock frequency can be generated by presetting the shift register with two 1s, then two 0s, then two 1s, and then two 0s.

## S U M M A R Y

Shift registers are important digital building blocks that can be used to store binary data. They can accept data bits in either a serial or a parallel format and can, likewise, deliver data in either serial or parallel. There are thus four basic register types: serial input–serial output, serial input–parallel output, parallel input–serial output, and parallel input–parallel output.

In one application, a register can be used to change data from a serial format into a parallel format, or vice versa. As such, shift registers can be regarded as *data format changers*. There are a great many other shift register applications—arithmetic operations, logic operations, and counters, to name only a few. Our intent has not been to discuss all the possible applications of shift registers, but rather to consider in detail how each type of register functions. With this knowledge, one can then discover the many and varied practical applications in existing digital designs.

## G L O S S A R Y

*Parallel Shift* Data bits are shifted simultaneously with a single clock transition.

*Register Capacity* Determined by the number of flip-flops in the register. There must be one flip-flop for each binary bit; the register capacity is $2^n$, where $n$ is the number of flip-flops.

*Ring Counter* A basic shift register with direct feedback such that the contents of the register simply circulate around the register when the clock is running.

*Serial Shift* Data bits are shifted one after the other in a serial fashion with one bit shifted at each clock transition. Therefore, *n* clock transitions are needed to shift an *n*-bit binary number.

*Shift Register* A group of flip-flops connected in such a way that a binary number can be shifted into or out of the flip-flops.

## PROBLEMS

*SECTION 10-1:*

**10–1.** Determine the number of flip-flops needed to construct a shift register capable of storing:
  **a.** An 6-bit binary number
  **b.** Decimal numbers up to 32
  **c.** Hexadecimal numbers up to *F*

**10–2.** A shift register has eight flip-flops. What is the largest binary number that can be stored in it? Decimal number? Hexadecimal number?

**10–3.** Name the four basic types of shift registers, and draw a block diagram for each.

*SECTION 10-2:*

**10–4.** Draw the waveforms to shift the binary number 1010 into the register in Fig. 10–3.

**10–5.** Draw the waveforms to shift the binary number 1001 into the register in Fig. 10–5.

**10–6.** The register in Fig. 10–3 has 0100 stored in it. Draw the waveforms for four clock transitions, assuming that both *J* and *K* are low.

**10–7.** Draw the waveforms showing how the decimal number 68 is shifted into the 54/74L91 in Fig. 10–7. Show eight clock periods.

**10–8.** The hexadecimal number *AB* is stored in the 54/74L91 in Fig. 10–7. Show the waveforms at the output, assuming that the clock is allowed to run for eight cycles and that $A = B = 0$.

*SECTION 10-3:*

**10–9.** How long will it take to shift an 8-bit binary number into the 54/74164 in Fig. 10–9 if the clock is:
  **a.** 1 MHz
  **b.** 5 MHz

**10–10.** For the 54/74164 in Fig. 10–9, *B* is high, CLEAR is high, a 1-MHz clock is used to shift the decimal number 200 into the register at *A*. Draw all the waveforms (such as in Fig. 10–11).

**10–11.** On the basis of information in Example 10–4, what is the maximum frequency of the clock if the minimum data transition time is 30 ns?

**10–12.** In Fig. 10–11, if CONTROL is taken low at time $K$, will the data stored in the register remain even if the clock is allowed to run? Explain.

*SECTION 10-4:*

**10–13.** For the circuit in Fig. 10–13c, write the logic levels on each gate leg, given:
  **a.** CONTROL = 1, $X_1 = 0$, $X_2 = 1$
  **b.** CONTROL = 0, $X_1 = 0$, $X_2 = 1$

**10–14.** Redraw the 54/74166 in Fig. 10–12 showing only those gates used to shift data into the register in parallel. If a gate is disabled, don't draw it.

**10–15.** Redraw the 54/74166 in Fig. 10–12 showing only those gates used to shift data into the register in serial. If a gate is disabled, don't draw it.

**10–16.** Explain the operation of the 54/74166 for each of the six truth table entries in Fig. 10–14.

**10–17.** Draw all the input and output waveforms for the 54/74166 in Fig. 10–12, assuming that the decimal number 190 is shifted into the register in:
  **a.** Parallel
  **b.** Serial

*SECTION 10-5:*

**10–18.** Redraw the 54/7495A shift register in Fig. 10–16 showing only those gates used to shift data into the register in parallel. If a gate is disabled, don't draw it.

**10–19.** Repeat Prob. 10–18, assuming that the data is shifted in serially.

**10–20.** Draw the waveforms necessary to enter, and shift to the right a single 1 through the shift register in Fig. 10–16.

**10–21.** Repeat Prob. 10–20, but do a left shift. (See Fig. 10–17.)

*SECTION 10-6:*

**10–22.** Draw the waveforms that would result if the circulating register (ring counter) in Fig. 10–19 had alternate 1s and 0s stored in it and a 1-MHz clock were applied.

**10–23.** The register in Fig. 10–19 can easily be cleared to all 0s by using the CLEAR input. See if you can design logic circuitry to set the register with alternating 1s and 0s.

**10–24.** Explain the operation of the 54/74165 shift register. Redraw one of the eight flip-flops along with its two NAND gates, and analyze:
  **a.** Parallel data entry
  **b.** Shift right
  **c.** Serial data entry

The logic diagram is given in Fig. 10–23.

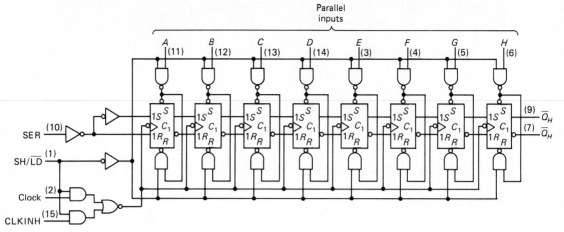

Parallel inputs

Pin numbers shown are for *J* and *N* packages.

(a) Logic diagram (positive logic)

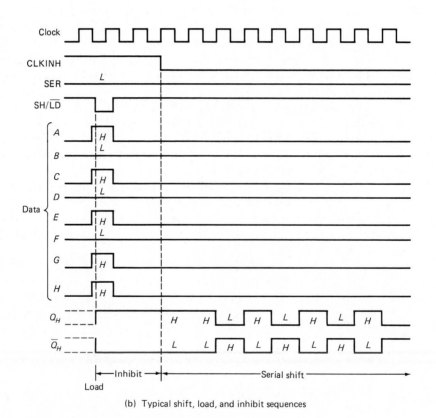

(b) Typical shift, load, and inhibit sequences

**FIG. 10–23**  54/74165, 8-bit shift register

# 11

## COUNTERS

A counter is probably one of the most useful and versatile subsystems in a digital system. A counter driven by a clock can be used to count the number of clock cycles. Since the clock pulses occur at known intervals, the counter can be used as an instrument for measuring time and therefore period or frequency. There are basically two different types of counters—synchronous and asynchronous.

The ripple counter is simple and straightforward in operation and construction and usually requires a minimum of hardware. It does, however, have a speed limitation. Each flip-flop is triggered by the previous flip-flop, and thus the counter has a cumulative settling time. Counters such as these are called *serial* or *asynchronous*.

An increase in speed of operation can be achieved by use of a *parallel* or *synchronous* counter. Here, every flip-flop is triggered by the clock (in synchronism), and thus settling time is simply equal to the delay time of a single flip-flop. The increase in speed is usually obtained at the price of increased hardware.

Serial and parallel counters are used in combination to compromise between speed of operation and hardware count. Serial, parallel, or combination counters can be designed such that each clock pulse advances the contents of the counter by one; it is then operating in a count-up mode. The opposite is also possible; the counter then operates in the count-down mode. Furthermore, many counters can be either "cleared" so that every flip-flop contains a zero, or preset such that the contents of the flip-flops represent any desired binary number.

Now, let's take a look at some of the techniques used to construct counters.

## 11·1 ASYNCHRONOUS COUNTERS

A binary ripple counter can be constructed by use of clocked *JK* flip-flops. Figure 11–1 shows three master-slave, *JK* flip-flops connected in cascade. The system clock, a square wave, drives flip-flop *A*. The output of *A* drives *B*, and the output of *B* drives flip-flop *C*. All the *J* and *K* inputs are tied to $+V_{CC}$. This means that each flip-flop will change state (toggle) with a negative transition at its clock input.

When the output of a flip-flop is used as the clock input for the next flip-flop, we call the counter a *ripple counter*, or *asynchronous counter*. The *A* flip-flop must change states before it can trigger the *B* flip-flop, and the *B* flip-flop has to change states before it can trigger the *C* flip-flop. The triggers move through the flip-flops like a ripple in water. Because of this, the overall propagation delay time is the sum of the individual delays. For instance, if each flip-flop in this three-flip-flop counter has a propagation delay time of 10 ns, the overall propagation delay time for the counter is 30 ns.

The waveforms given in Fig. 11–1*b* show the action of the counter as the clock runs. Let's assume that the flip-flops are all initially reset to produce 0 outputs. If we consider *A* to be the least-significant bit (LSB) and *C* the most-significant bit (MSB), we can say the contents of the counter is *CBA* = 000.

Every time there is a negative clock transition, flip-flop *A* will change states. Thus at point *a* on the time line, *A* goes high, at point *b* it goes back low, at *c* it goes back high, and so on. Notice that the waveform at the output of flip-flop *A* is one-half the clock frequency.

Since *A* acts as the clock for *B*, each time the waveform at *A* goes low, flip-flop *B* will toggle. Thus at point *b* on the time line, *B* goes high; it then goes low at point *d* and toggles back high again at point *f*. Notice that the waveform at the output of flip-flop *B* is one-half the frequency of *A* and one-fourth the clock frequency.

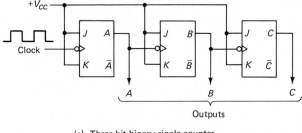

(a) Three-bit binary ripple counter

(b) Waveforms

| Clock transitions | C | B | A |
|---|---|---|---|
| 0 | 0 | 0 | 0 |
| 1 | 0 | 0 | 1 |
| 2 | 0 | 1 | 0 |
| 3 | 0 | 1 | 1 |
| 4 | 1 | 0 | 0 |
| 5 | 1 | 0 | 1 |
| 6 | 1 | 1 | 0 |
| 7 | 1 | 1 | 1 |
| 0 | 0 | 0 | 0 |

(c) Truth table

**FIG. 11–1**

Since $B$ acts as the clock for $C$, each time the waveform at $B$ goes low, flip-flop $C$ will toggle. Thus $C$ goes high at point $d$ on the time line and goes back low again at point $h$. The frequency of the waveform at $C$ is one-half that at $B$, but it is only one-eighth the clock frequency.

□ EXAMPLE 11—1

What is the clock frequency in Fig. 11–1 if the period of the waveform at $C$ is 24 $\mu$s?

□ SOLUTION

Since there are eight clock cycles in one cycle of $C$, the period of the clock must be $24/8 = 3$ $\mu$s. The clock frequency must then be $1/(3 \times 10^{-6}) = 333$ kHz.

Notice that the output condition of the flip-flops is a binary number equivalent to the number of negative clock transitions that have occurred. Prior to point $a$ on the time line the output condition is $CBA = 000$. At point $a$ on the time line the output condition changes to $CBA = 001$, at point $b$ it changes to $CBA = 010$, and so on. In fact, a careful examination of the waveforms will reveal that the counter content advances one count with each negative clock transition in a "straight binary progression" that is summarized in the truth table in Fig. 11–1$c$.

Because each output condition shown in the truth table is the binary equivalent of the number of clock transitions, the three cascaded flip-flops in Fig. 11–1 comprise a 3-bit binary ripple counter. This counter can be used to count the number of clock transitions up to a maximum of seven. The counter begins at count 000 and advances one count for each clock transition until it reaches count 111. At this point it resets back to 000 and begins the count cycle all over again. We can say that this ripple counter is operating in a count-up mode.

Since a binary ripple counter counts in a straight binary sequence, it is easy to see that a counter having $n$ flip-flops will have $2^n$ output conditions. For instance, the three-flip-flop counter just discussed has $2^3 = 8$ output conditions (000 through 111). Five flip-flops would have $2^5 = 32$ output conditions (00000 through 11111), and so on. The largest binary number that can be represented by $n$ cascaded flip-flops has a decimal equivalent of $2^n - 1$. For example, the three-flip-flop counter reaches a maximum decimal number of $2^3 - 1 = 7$. The maximum decimal number for five flip-flops is $2^5 - 1 = 31$, while six flip-flops have a maximum count of 63.

A three-flip-flop counter is often referred to as a modulus-8 (or mod-8) counter since it has eight states. Similarly, a four-flip-flop counter is a mod-16 counter, and a six-flip-flop counter is a mod-64 counter. The *modulus* of a counter is the total number of states through which the counter can progress.

□ EXAMPLE 11—2

How many flip-flops are required to construct a mod-128 counter? A mod-32? What is the largest decimal number that can be stored in a mod-64 counter?

□ SOLUTION

A mod-128 counter must have seven flip-flops, since $2^7 = 128$. Five flip-flops are needed to construct a mod-32 counter. The largest decimal number that can be stored in a six-flip-

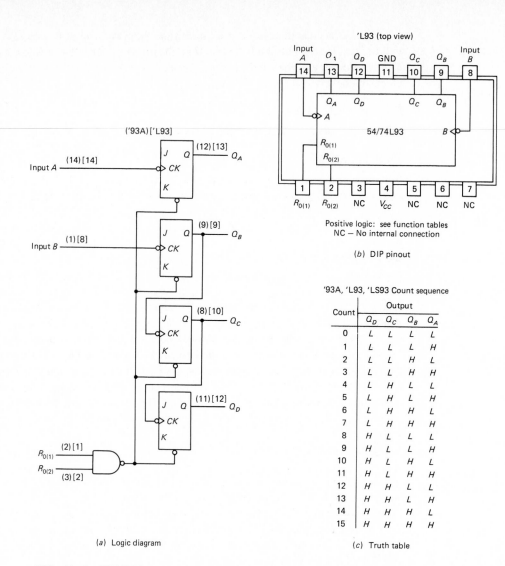

'L93 (top view)

| Input A | $O_1$ | $Q_D$ | GND | $Q_C$ | $Q_B$ | Input B |
|---|---|---|---|---|---|---|
| 14 | 13 | 12 | 11 | 10 | 9 | 8 |

$Q_A$ $Q_D$ $Q_C$ $Q_B$

A

54/74L93

B

$R_{0(1)}$
$R_{0(2)}$

| 1 | 2 | 3 | 4 | 5 | 6 | 7 |
|---|---|---|---|---|---|---|
| $R_{0(1)}$ | $R_{0(2)}$ | NC | $V_{CC}$ | NC | NC | NC |

Positive logic: see function tables
NC — No internal connection

(b) DIP pinout

('93A) ['L93]

(14) [14]
Input A ——— CK
J Q (12) [13] $Q_A$
K

(1) [8]
Input B ——— CK
J Q (9) [9] $Q_B$
K

CK
J Q (8) [10] $Q_C$
K

CK
J Q (11) [12] $Q_D$
K

$R_{0(1)}$ (2) [1]
$R_{0(2)}$ (3) [2]

(a) Logic diagram

'93A, 'L93, 'LS93 Count sequence

| Count | Output | | | |
|---|---|---|---|---|
| | $Q_D$ | $Q_C$ | $Q_B$ | $Q_A$ |
| 0 | L | L | L | L |
| 1 | L | L | L | H |
| 2 | L | L | H | L |
| 3 | L | L | H | H |
| 4 | L | H | L | L |
| 5 | L | H | L | H |
| 6 | L | H | H | L |
| 7 | L | H | H | H |
| 8 | H | L | L | L |
| 9 | H | L | L | H |
| 10 | H | L | H | L |
| 11 | H | L | H | H |
| 12 | H | H | L | L |
| 13 | H | H | L | H |
| 14 | H | H | H | L |
| 15 | H | H | H | H |

(c) Truth table

**FIG. 11–2**   54/74L93

flop counter (mod-64) is $111111 = 63$. Note carefully the difference between the modulus (total number of states) and the maximum decimal number.

The logic diagram, DIP pinout, and truth table for a 54/74L93 are given in Fig. 11–2. This TTL MSI circuit is a 4-bit binary counter that can be used in either a mod-8 or a mod-16 configuration. If the clock is applied at input $B$, the outputs will appear at $Q_B$, $Q_C$, and $Q_D$, and this is a mod-8 binary ripple counter exactly like that in Fig. 11–1. In this case, flip-flop $Q_A$ is simply unused.

On the other hand, if the clock is applied at input $A$ and flip-flop $Q_A$ is connected to input $B$, we have a mod-16, 4-bit, binary ripple counter. The outputs are $Q_A$, $Q_B$, $Q_C$, and $Q_D$. The proper truth table for this connection is given in Fig. 11–2c.

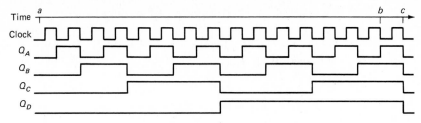

FIG. 11–3   Waveforms for a mod-16, 74L93

All the flip-flops in the 54/74L93 have direct reset inputs that are active low. Thus a high level at both reset inputs of the NAND gate, $R_{0(1)}$ and $R_{0(2)}$, is needed to reset all flip-flops simultaneously. Notice that this reset operation will occur without regard to the clock.

☐ EXAMPLE 11–3

Draw the correct output waveforms for a 74L93 connected as a mod-16 counter.

☐ SOLUTION

The correct waveforms are shown in Fig. 11–3. The contents of the counter is 0000 at point $a$ on the time line. With each negative clock transition, the counter is advanced by one until the counter contents are 1111 at point $b$ on the time line. At point $c$, the counter resets to 0000, and the counting sequence repeats. Clearly, this is a mod-16 counter, since there are 16 states (0000 through 1111), and the maximum decimal number that can be stored in the flip-flops is decimal 15 (1111).

An interesting and useful variation of the 3-bit ripple counter in Fig. 11–1 is shown in Fig. 11–4. The system clock is still used at the clock input to flip-flop $A$, but the comple-

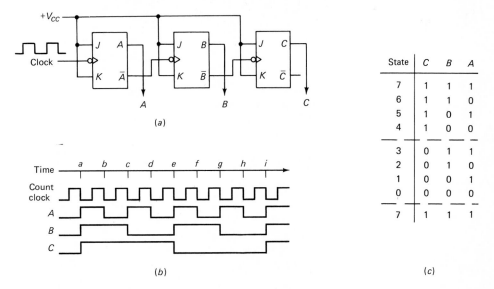

FIG. 11–4   A down counter

ment of $A$, $\overline{A}$, is used to drive flip-flop $B$; likewise, $\overline{B}$ is used to drive flip-flop $C$. Take a look at the resulting waveforms.

Flip-flop $A$ simply toggles with each negative clock transition as before. But flip-flop $B$ will toggle each time $A$ *goes high!* Notice that each time $A$ goes high, $\overline{A}$ goes low, and it is this negative transition on $\overline{A}$ that triggers $B$. On the time line, $B$ toggles at points $a$, $c$, $e$, $g$, and $i$.

Similarly, flip-flop $C$ is triggered by $\overline{B}$ and so $C$ will toggle each time $B$ goes high. Thus $C$ toggles high at point $a$ on the time line, toggles back low at point $e$, and goes back high again at point $i$.

The counter contents become $ABC = 111$ at point $a$ on the time line, change to 110 at point $b$, and change to 101 at point $c$. Notice that the counter contents are reduced by one count with each clock transition! In other words, the counter is operating in a count-down mode. The results are summarized in the truth table in Fig. 11–4c. This is still a mod-8 counter, since it has eight discrete states, but it is connected as a *down-counter*.

A 3-bit, asynchronous, *up-down* counter that counts in a straight binary sequence is shown in Fig. 11–5. It is simply a combination of the two counters discussed previously. For this counter to progress through a count-up sequence, it is necessary to trigger each flip-flop with the true side of the previous flip-flop (as opposed to the complement side). If the count-down control line is low and the count-up control line high, this will be the case, and the counter will have count-up waveforms such as those shown in Fig. 11–1.

On the other hand, if count-down is high and count-up is low, each flip-flop will be triggered from the complement side of the previous flip-flop. The counter will then be in a count-down mode and will progress through the waveforms as shown in Fig. 11–4.

This process can be continued to other flip-flops down the line to form an up-down counter of larger moduli. It should be noticed, however, that the gates introduce additional delays that must be taken into account when determining the maximum rate at which the counter can operate.

## 11·2 DECODING GATES

A decoding gate can be connected to the outputs of a counter in such a way that the output of the gate will be high (or low) only when the counter contents are equal to a given state. For instance, the decoding gate connected to the 3-bit ripple counter in Fig. 11–6a will decode state 7 ($CBA = 111$). Thus the gate output will be high only when $A = 1$, $B = 1$, and $C = 1$ and the waveform appearing at the output of the gate is labeled 7. The boolean

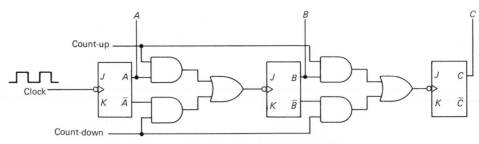

Note: The $J$ and $K$ inputs are all tied to $+V_{CC}$.
The counter outputs are A, B, and C.

**FIG. 11–5** 3-bit binary up-down counter

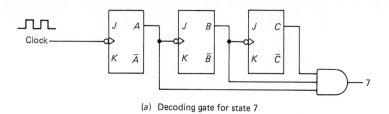

(a) Decoding gate for state 7

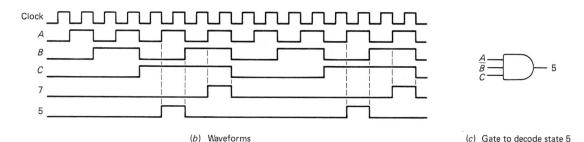

(b) Waveforms

(c) Gate to decode state 5

**FIG. 11–6**

expression for this gate can be written $7 = CBA$. A comparison with the truth table for this counter (in Fig. 11–1) will reveal that the condition $CBA = 111$ is true only for state 7.

The other seven states of the counter can be decoded in a similar fashion. It is only necessary to examine the truth table for the counter, and then the proper boolean expression for each gate can be written. For instance, to decode state 5, the truth table reveals that $CBA = 101$ is the unique state. For the gate output to be high during this time, we must use $C$, $\bar{B}$, and $A$ at the gate inputs. Notice carefully that if $B = 0$, then $\bar{B} = 1$! The correct boolean expression is then $5 = C\bar{B}A$, and the desired gate is that given in Fig. 11–6c. The waveform is again that given in Fig. 11–6b and is labeled 5.

All eight gates necessary to decode the eight states of the 3-bit counter in Fig. 11–1 are shown in Fig. 11–7a. The gate outputs are shown in Fig. 11–7b. These decoded waveforms are a series of positive pulses that occur in a strict time sequence and are very useful as control signals throughout a digital system. If we consider state 0 as the first event, then state 1 will be the second, state 2 the third, and so on, up to state 7. Clearly the counter is counting upward in decimal notation from 0 to 7 and then beginning over again at 0.

If these eight gates are connected to the up-down counter shown in Fig. 11–5, the decoded waveforms will appear exactly as shown in Fig. 11–7b, provided the counter is operating in the count-up mode. If the counter is operated in the count-down mode, the decoded waveforms will appear as in Fig. 11–7c. In this case, if state 0 is considered the first event, then state 7 is the second event, then state 6, and so on, down to state 1. Clearly the counter is counting downward in decimal notation from 7 to 0 and then beginning again at 7.

□ EXAMPLE 11—4

Show how to use a 54LS11, triple 3-input AND gate to decode states 1, 4, and 6 of the counter in Fig. 11–5.

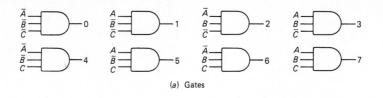

(a) Gates

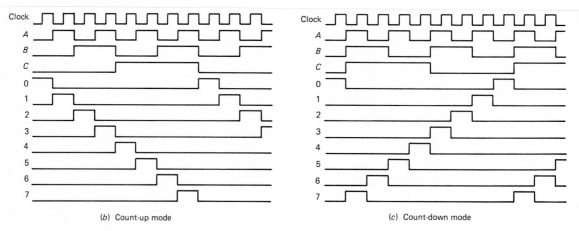

(b) Count-up mode           (c) Count-down mode

**FIG. 11–7**   Decoding gates for a 3-bit binary ripple counter

☐ SOLUTION

The logic diagram and pinout for a 54LS11 is given in Fig. 11–8. The correct boolean expressions for the desired states are $1 = \overline{C}\,\overline{B}\,A$, $4 = C\,\overline{B}\,\overline{A}$, and $6 = CB\overline{A}$. Wiring from the counter flip-flop outputs to the chip is given in Fig. 11–8.

Let's take a more careful look at the waveforms generated by the counter in Fig. 11–5 as it operates in the count-up mode. The clock and each flip-flop output are redrawn in

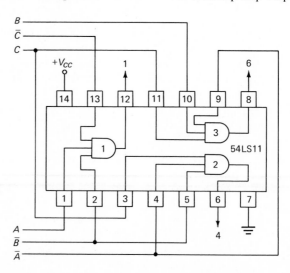

**FIG. 11–8**

CHAPTER 11

Fig. 11–9, and the propagation delay time of each flip-flop is taken into account. Notice carefully that the clock is the trigger for flip-flop $A$, and the $A$ waveform is thus delayed by $t_p$ from the negative clock transition. For reference purposes, the complement of $A$, $\overline{A}$, is also shown. Naturally it is the exact mirror image of $A$.

Since $A$ acts as the trigger for $B$, the $B$ waveform is delayed by one flip-flop delay time from the negative transition of $A$. Similarly, the $C$ waveform is delayed by $t_p$ from each negative transition of $B$.

At first glance, these delay times would seem to offer no more serious problem than a speed limitation for the counter, but a closer examination reveals a much more serious problem. When the decoding gates in Fig. 11–7 are connected to this counter (or, indeed, when decoding gates are connected to any ripple counter), glitches may appear at the outputs of one or more of the gates. Consider, for instance, the gate used to decode state 6. The proper boolean expression is $6 = CB\overline{A}$. So, in Fig. 11–9 the correct output waveform for this gate is high only when $C = 1$, $B = 1$, and $\overline{A} = 1$.

But look at the glitch that occurs when the counter progresses from state 7 to state 0. On the time line, $A$ goes low ($\overline{A}$ goes high) at point $a$. Because of flip-flop delay time, however, $B$ does not go low until point $b$ on the time line! Thus between points $a$ and $b$ on the time line we have the condition $C = 1$, $B = 1$, and $\overline{A} = 1$—therefore, the gate output is high, and we have a glitch! Look at the waveform $6 = CB\overline{A}$.

Depending on how the decoder gate outputs are used, the glitches (or unwanted pulses) may or may not be a problem. Admittedly the glitches are only a few nanoseconds wide and may even be very difficult to observe on an oscilloscope. But TTL is *very fast*, and TTL circuits will respond to even the smallest-appearing glitches—usually when you least expect it, and always at unwanted times! Therefore, you must beware to avoid this condition. There are at least two solutions to the glitch problem. One method involves strobing the gates; we discuss that technique here. A second method is to use synchronous counters; we consider that topic in the next section.

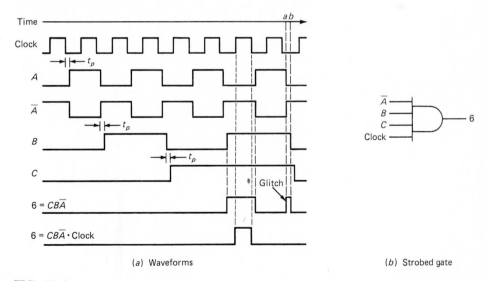

(a) Waveforms                    (b) Strobed gate

**FIG. 11–9**

Consider using a 4-input AND gate to decode state 6 as shown in Fig. 11–9b, where the clock is now used as a strobe. An examination of the waveforms in this figure clearly reveals that the clock is low between points $a$ and $b$ on the time line. Since the clock must be high for the gate output to be high, the glitch cannot possibly occur! On the other hand, the clock is high when $C = 1$, $B = 1$, and $\overline{A} = 1$, and the waveform for 6 appears exactly as it should. Notice that the width of the positive pulse at 6 is exactly the width of the positive portion of the clock. Look at the waveform $6 = CB\overline{A} \cdot \text{clock}$. This technique can be applied to the other seven decoding gates for this counter (or for any other counter), and the decoded output waveforms will be glitch-free.

## 11-3 SYNCHRONOUS COUNTERS

The ripple counter is the simplest to build, but there is a limit to its highest operating frequency. As previously discussed, each flip-flop has a delay time. In a ripple counter these delay times are additive, and the total "settling" time for the counter is approximately the delay time times the total number of flip-flops. Furthermore, there is the possibility of glitches occurring at the output of decoding gates used with a ripple counter. Both of these problems can be overcome by the use of a synchronous or parallel counter. The main difference here is that every flip-flop is triggered in synchronism with the clock.

The construction of one type of parallel binary counter is shown in Fig. 11–10, along with the truth table and the waveforms for the natural count sequence. Since each state corresponds to an equivalent binary number (or count), we refer to each state as a count from now on. The basic idea here is to keep the $J$ and $K$ inputs of each flip-flop high, such that the flip-flop will toggle with any negative clock transition at its clock input. We then

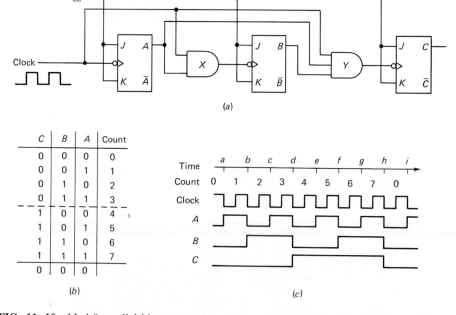

(a)

(b)

(c)

**FIG. 11–10** Mod-8 parallel binary counter

use AND gates to gate every second clock to flip-flop $B$, every fourth clock to flip-flop $C$, and so on. This logic configuration is often referred to as "steering logic" since the clock pulses are gated or steered to each individual flip-flop.

The clock is applied directly to flip-flop $A$. Since the $JK$ flip-flop used responds to a negative transition at the clock input and toggles when both the $J$ and $K$ inputs are high, flip-flop $A$ will change state with each negative clock transition.

Whenever $A$ is high, AND gate $X$ is enabled and a clock pulse is passed through the gate to the clock input of flip-flop $B$. Thus $B$ changes state with every other negative clock transition at points $b$, $d$, $f$, and $h$ on the time line.

Since AND gate $Y$ is enabled and will transmit the clock to flip-flop $C$ only when both $A$ and $B$ are high, flip-flop $C$ changes state with every fourth negative clock transition at points $d$ and $h$ on the time line.

Examination of the waveforms and the truth table reveals that this counter progresses upward in a natural binary sequence from count 000 up to count 111, advancing one count with each negative clock transition. This is a mod-8, parallel or synchronous, binary counter operating in the count-up mode.

Let's see if this counter configuration has cured the glitch problem discussed previously. The waveforms for this counter are expanded and redrawn in Fig. 11–11, and we have accounted for the individual flip-flop propagation times. Study these waveforms carefully and note the following:

1. The negative clock transition is the mechanism that toggles each flip-flop.
2. Therefore, whenever a flip-flop changes state, it toggles at exactly the same time as all the other flip-flops—in other words, all the flip-flops change states in synchronism!
3. As a result of the synchronous changes of state, it is not possible to produce a glitch at the output of a decoding gate, such as the gate for 6 shown in Fig. 11–11. Therefore, the decoding gates need not be strobed. All the decoding gates in Fig. 11–7 can be used with this counter without fear of glitches!

You should take time to compare these waveforms with those generated by the ripple counter as shown in Fig. 11–9.

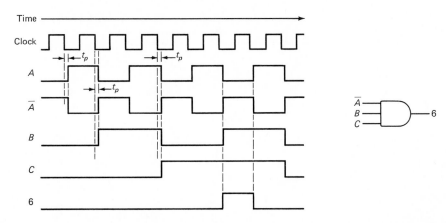

FIG. 11–11

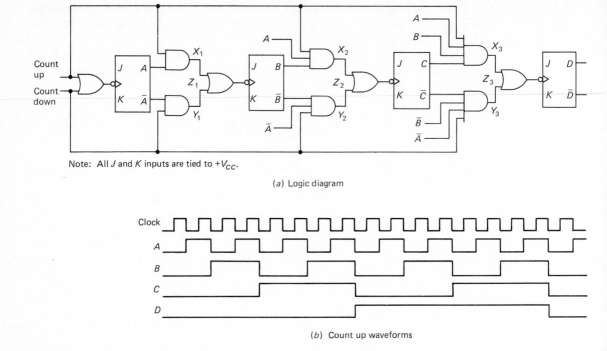

Note: All $J$ and $K$ inputs are tied to $+V_{CC}$.

(a) Logic diagram

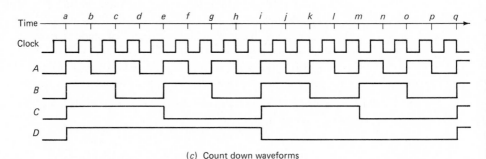

(b) Count up waveforms

(c) Count down waveforms

**FIG. 11–12** Synchronous, 4-bit up-down counter

A parallel, up-down counter can be constructed in a fashion similar to that shown in Fig. 11–12. In any parallel counter, the time at which any flip-flop changes state is determined by the states of all previous flip-flops in the counter. In the count-up mode, a flip-flop must toggle every time all previous flip-flops are in a 1 state, and the clock makes a transition. In the count-down mode, flip-flop toggles must occur when all prior flip-flops are in a 0 state.

The counter in Fig. 11–12 is a synchronous, 4-bit, up-down counter. To operate in the count-up mode, the system clock is applied at the count-up input, while the count-down input is held low. To operate in the count-down mode, the system clock is applied at the count-down input while holding the count-up input low.

Holding the count-down input low (at ground) will disable AND gates $Y_1$, $Y_2$, and $Y_3$. The clock applied at count-up will then go directly into flip-flop $A$ and will be steered into the other flip-flops by AND gates $X_1$, $X_2$, and $X_3$. This counter will then function exactly as the previously discussed parallel counter shown in Fig. 11–10. The only difference here is that this is a mod-16 counter that advances one count with each negative clock transition, beginning with 0000 and ending with 1111. The correct waveforms are shown in Fig. 11–12*b*.

If the count-up line is held low, the upper AND gates $X_1$, $X_2$, and $X_3$ are disabled. The clock applied at input count-down will go directly into flip-flop $A$ and be steered into the following flip-flops by AND gates $Y_1$, $Y_2$, and $Y_3$.

Flip-flop $A$ will toggle each time there is a negative clock transition as shown in Fig. 11–12*c*. Each time $\overline{A}$ is high, AND gate $Y_1$ will be enabled and the clock transition will toggle flip-flop $B$ at points $a$, $c$, $e$, $g$, and so on. Whenever both $\overline{A}$ and $\overline{B}$ are high, AND gate $Y_2$ is enabled, and thus a clock will be steered into flip-flop $C$ at points $a$, $e$, $i$, $m$, and $q$. Similarly, AND gate $Y_3$ will steer a clock into flip-flop $D$ only when $\overline{A}$, $\overline{B}$, and $\overline{C}$ are all high. Thus flip-flop $D$ will toggle at points $a$ and $i$ on the time line. The waveforms in Fig. 11–12*c* clearly show that the counter is operating in a count-down mode, progressing one count at a time from 1111 to 0000.

If you examine the logic diagram for the 54/74193 TTL circuit shown in Fig. 11–13, you will see that it uses steering logic just like the counter in Fig. 11–12. This MSI circuit is a synchronous, 4-bit, up-down counter that can also be cleared and preset to any desired count—attributes that we discuss later. For now, you should carefully examine the steering logic for each flip-flop and study the OR gate and the two AND gates at the input of the OR gate used to provide the clock to each flip-flop.

The waveforms for the 54/74193 are exactly the same as those shown in Fig. 11–12, except that the flip-flop outputs change states when the clock makes a low-to-high transition. Note carefully that the external clock (applied at either the count-up or the count-

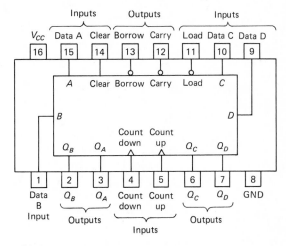

'192, '193 . . . J, N, or W package
'L192, 'L193 . . . J or N package
'LS192, 'LS193 . . . J, N, or W package
(Top view)

Logic: low input to load sets $Q_A$ = A,
$Q_B$ = B, $Q_C$ = C, and $Q_D$ = D.

(*a*) Pin out

**FIG. 11–13**   54/74193

'193, 'L193, 'LS193

(13) Borrow output

(12) Carry output

Data input A (15)

Count down (4)

Count up (5)

(3) Output $Q_A$

Data input B (1)

(2) Output $Q_B$

Data input C (10)

(6) Output $Q_C$

Data input D (9)

Clear (14)

(7) Output $Q_D$

Load (11)

(b) Logic

**FIG. 11–13** (continued)

**Typical clear, load, and count sequences**

Illustrated below is the following sequence.
1. Clear outputs to zero.
2. Load (preset) to binary thirteen.
3. Count up to fourteen, fifteen, carry, zero, one, and two.
4. Count down to one, zero, borrow, fifteen, fourteen, and thirteen.

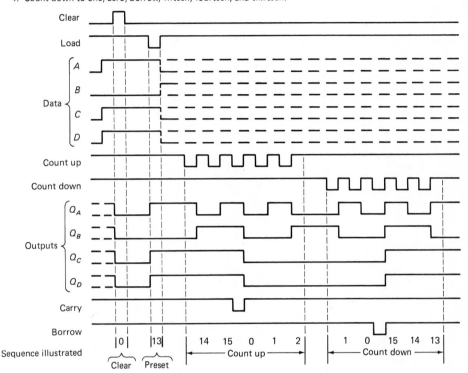

Notes: A. Clear overrides load, data, and count inputs.
      B. When counting up, count-down input must be high; when counting down, count-up input must be high.

(*c*) Waveforms for 54/74193

**FIG. 11–13** (continued)

down input) passes through an inverter before being applied to the AND–OR–gate logic of each flip-flop clock input.

☐ EXAMPLE 11—5

Write a boolean expression for the AND gate connected to the lower leg of the OR gate that drives the clock input to flip-flop $Q_D$ in the 54/74193.

☐ SOLUTION

The correct expression is

$$x = \overline{(\text{count-up clock})}(Q_A)(Q_B)(Q_C)$$

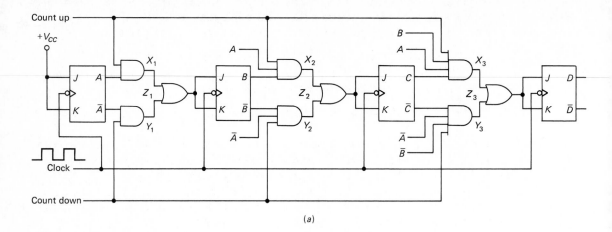

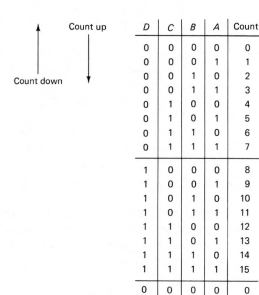

(a)

(b)

| D | C | B | A | Count |
|---|---|---|---|-------|
| 0 | 0 | 0 | 0 | 0 |
| 0 | 0 | 0 | 1 | 1 |
| 0 | 0 | 1 | 0 | 2 |
| 0 | 0 | 1 | 1 | 3 |
| 0 | 1 | 0 | 0 | 4 |
| 0 | 1 | 0 | 1 | 5 |
| 0 | 1 | 1 | 0 | 6 |
| 0 | 1 | 1 | 1 | 7 |
| 1 | 0 | 0 | 0 | 8 |
| 1 | 0 | 0 | 1 | 9 |
| 1 | 0 | 1 | 0 | 10 |
| 1 | 0 | 1 | 1 | 11 |
| 1 | 1 | 0 | 0 | 12 |
| 1 | 1 | 0 | 1 | 13 |
| 1 | 1 | 1 | 0 | 14 |
| 1 | 1 | 1 | 1 | 15 |
| 0 | 0 | 0 | 0 | 0 |

**FIG. 11–14** Parallel up-down counter

A parallel, up-down counter can be formed by using a slightly different logic scheme, as shown in Fig. 11–14. Remember that in a parallel counter, the time at which any flip-flop changes state is determined by the states of all previous flip-flops in the counter. In the count-up mode, a flip-flop must toggle every time all previous flip-flops are in a 1 state, and the clock makes a transition. In the count-down mode, flip-flop toggles must occur when all prior flip-flops are in a 0 state.

This particular counter works in an inhibit mode, since each flip-flop changes state on a negative clock transition provided its $J$ and $K$ inputs are both high; a change of state will not occur when the $J$ and $K$ inputs are low. We might consider this as "look-ahead logic", since the mode of operation occurs in a strict time sequence as follows:

1. Establish a level on the $J$ and $K$ inputs (low or high)
2. Let the clock transition high to low
3. Look at the flip-flop output to determine whether it toggled

To understand the logic used to implement this counter, refer to the truth table shown in Fig. 11–14$b$.

$A$ is required to change state each time the clock goes, and flip-flop $A$ therefore has both its $J$ and $K$ inputs held in a high state. This is true in both the count-up and count-down modes, and thus no other logic is necessary for this flip-flop.

In the count-up mode, $B$ is required to change state each time $A$ is high and the clock goes low. Whenever the count-up line and $A$ are both high, the output of gate $X_1$ is high. Whenever either input to $Z_1$ is high, the output is high. Therefore, the $J$ and $K$ inputs to flip-flop $B$ are high whenever both count-up and $A$ are high. Then, in the count-up mode, a negative clock transition will toggle $B$ each time $A$ is high, such as in going from count 1 to count 2, or 3 to 4, and so on.

In the count-down mode, $B$ must change state each time $\overline{A}$ is high and the clock goes low. The output of gate $Y_1$ is high, and thus the $J$ and $K$ inputs to flip-flop $B$ are high whenever $\overline{A}$ and count-down are high. Thus, in the count-down mode, $B$ changes state every time $\overline{A}$ is high and the clock goes low—going from 0 to 15, or from 14 to 13, and so forth.

In the count-up mode, a negative clock transition must toggle $C$ every time both $A$ and $B$ are high (transitions 3 to 4, 7 to 8, 11 to 12, and 15 to 0). The output of gate $X_2$ is high whenever both $A$ and $B$ are high and the count-up line is high. Therefore, the $J$ and $K$ inputs to flip-flop $C$ are high during these times and $C$ changes state during the desired transitions.

In the count-down mode, $C$ is required to change state whenever both $\overline{A}$ and $\overline{B}$ are high. The output of gate $Y_2$ is high any time both $\overline{A}$ and $\overline{B}$ are high, and the count-down line is high. Thus the $J$ and $K$ inputs to flip-flop $C$ are high during these times, and $C$ then changes state during the required transitions—that is, 0 to 15, 12 to 11, 8 to 7, and 4 to 3.

In the count-up mode, $D$ must toggle every time $A$, $B$, and $C$ are all high. The output of gate $X_3$ is high, and thus the $J$ and $K$ inputs to flip-flop $D$ are high whenever $A$, $B$, and $C$ and count-up are all high. Thus $D$ changes state during the transitions from 7 to 8 and from 15 to 0.

In the count-down mode, a negative clock transition must toggle $D$ whenever $\overline{A}$, $\overline{B}$, and $\overline{C}$ are all high. The output of gate $Y_3$ is high, and thus the $J$ and $K$ inputs to flip-flop $D$ are high whenever $\overline{A}$, $\overline{B}$, and $\overline{C}$ and count-down are all high. Thus D changes state during the transitions from 0 to 15 and from 8 to 7. The count-up and count-down waveforms for this counter are exactly like those shown in Fig. 11–12.

Take a look at the logic diagram for the 54/74191 TTL MSI circuit shown in Fig. 11–15. This is a synchronous up-down counter. A careful examination of the AND–OR–gate logic used to precondition the $J$ and $K$ inputs to each flip-flop will reveal that this counter uses look-ahead logic exactly like the counter in Fig. 11–14. Additional logic

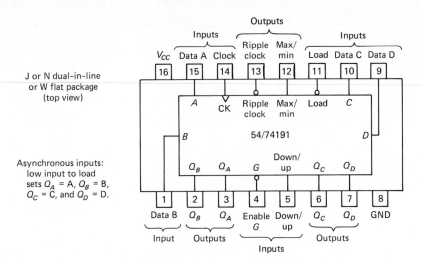

Inputs: $V_{CC}$, Data A, Clock
Outputs: Ripple clock, Max/min
Inputs: Load, Data C, Data D

16 | 15 | 14 | 13 | 12 | 11 | 10 | 9

J or N dual-in-line
or W flat package
(top view)

A    CK    Ripple    Max/    Load    C
          clock     min

B         54/74191                    D

Asynchronous inputs:
low input to load
sets $Q_A$ = A, $Q_B$ = B,
$Q_C$ = C, and $Q_D$ = D.

$Q_B$   $Q_A$   G   Down/   $Q_C$   $Q_D$
                    up

1 | 2 | 3 | 4 | 5 | 6 | 7 | 8

Data B    $Q_B$    $Q_A$    Enable    Down/    $Q_C$    $Q_D$    GND
                            G        up

Input    Outputs    Inputs    Outputs

**FIG. 11–15**   54/74191

allows one to clear or preset this counter to any desired count, and we study these functions later. For now, carefully compare the logic diagram with the counter in Fig. 11–14 to be certain you understand its operation.

Notice carefully that the clock input passes through an inverter before it is fed to the individual flip-flops. Thus the outputs of the four master-slave flip-flops will change states only on low-to-high transitions of the input clock. Typical waveforms are given in Fig. 11–15. Incidentally, these are precisely the same waveforms one would expect when using the 54/74193 discussed previously.

☐ EXAMPLE 11–6

Write a boolean expression for the 4-input AND gate connected to the lower leg of the OR gate that conditions the J and K inputs to the $Q_D$ flip-flop in a 54/74191.

☐ SOLUTION

The correct logic expression is

$$x = (\overline{\text{down-up}})\,(Q_A)\,(Q_B)\,(Q_C)\,(\overline{\text{Enable}})$$

## 11-4 MOD-3 COUNTER

At this point, we have discussed asynchronous (ripple) counters and two different types of synchronous (parallel) counters, all of which have the ability to operate in either a count-up or count-down mode. All of these counters progress one count at a time in a strict binary progression, and they all have a modulus given by $2^n$, where $n$ indicates the number of flip-flops. Such counters are said to have a "natural count" of $2^n$.

A mod-2 counter consists of a single flip-flop; a mod-4 counter requires two flip-flops, and it counts through four discrete states. Three flip-flops form a mod-8 counter, while four flip-flops form a mod-16 counter. Thus we can construct counters that have a natural count of 2, 4, 8, 16, 32, and so on by using the proper number of flip-flops.

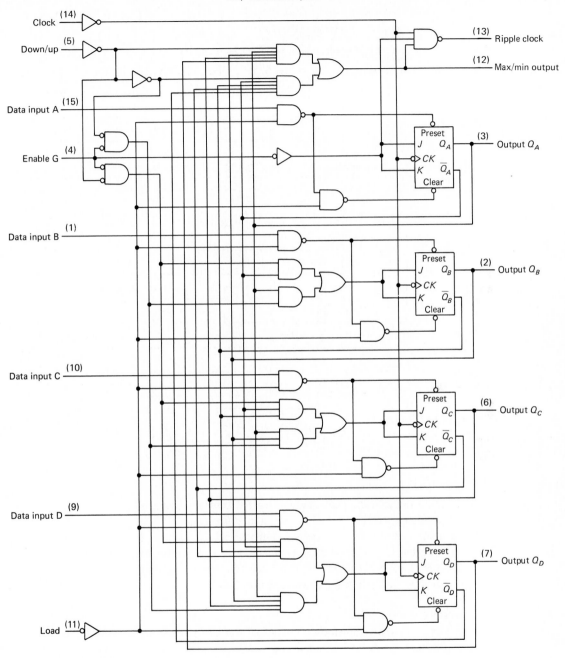

**FIG. 11–15** (continued)

It is often desirable to construct counters having a modulus other than 2, 4, 8, and so on. For example, a counter having a modulus of 3, or 5, would be useful. A smaller

**Typical load, count, and inhibit sequences**

Illustrated below is the following sequence.
1. Load (preset) to binary thirteen.
2. Count up to fourteen, fifteen (maximum), zero, one, and two.
3. Inhibit.
4. Count down to one, zero (minimum), fifteen, fourteen, and thirteen.

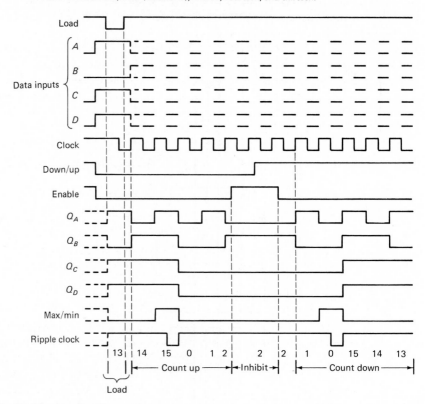

**FIG. 11–15** (continued)

modulus counter can always be constructed from a larger modulus counter by skipping states. Such counters are said to have a *modified count*. It is first necessary to determine the number of flip-flops required. The correct number of flip-flops is determined by choosing the lowest natural count that is greater than the desired modified count. For example, a mod-7 counter requires three flip-flops, since 8 is the lowest natural count greater than the desired modified count of 7.

☐ EXAMPLE 11—7

Indicate how many flip-flops are required to construct each of the following counters: (*a*) mod-3, (*b*) mod-6, and (*c*) mod-9.

☐ SOLUTION

(*a*) The lowest natural count greater than 3 is 4. Two flip-flops provide a natural count of 4. Therefore, it requires at least two flip-flops to construct a mod-3 counter.

(*b*) Construction of a mod-6 counter requires at least three flip-flops, since 8 is the lowest natural count greater than 6.

(*c*) A mod-9 counter requires at least four flip-flops, since 16 is the lowest natural count greater than 9.

A single flip-flop has a natural count of 2; thus we could use a single flip-flop to construct a mod-2 counter, and that's all. However, a two flip-flop counter has a natural count of 4. Skipping one count will lead to a mod-3 counter. So, two flip-flops can be used to construct either a mod-4 or mod-3 counter.

Similarly, a three-flip-flop counter has a natural count of 8, but by skipping counts we can use three flip-flops to construct a counter having a modulus of 8, 7, 6, or 5. Note that counters having a modulus of 4 or 3 could also be constructed, but these two counters can be constructed by using only two flip-flops.

## ☐ EXAMPLE 11−8

What modulus counters can be constructed with the use of four flip-flops?

## ☐ SOLUTION

A four-flip-flop counter has a natural count of 16. We can thus construct any counter that has a modulus between 16 and 2, inclusive. We might choose to use four flip-flops only for counters having a modulus between 16 and 9, since only three flip-flops are required for a modulus of less than 8, and only two are required for a modulus of less than 4.

There are a great many different methods for constructing a counter having a modified count. A counter can be synchronous, asynchronous, or a combination of these two types; furthermore, there is the decision of which count to skip. For instance, if a mod-6 counter using three flip-flops is to be constructed, which two of the eight discrete states should be skipped? Our purpose here is not to consider all possible counter configurations and how to design them; rather, we devote our efforts to one or two designs widely used in TTL MSI. A mod-3 counter is considered in this section and a mod-5 in the next section, and then we consider the use of presettable counters to achieve any desired modulus.

The two flip-flops in Fig. 11–16 have been connected to provide a mod-3 counter. Since two flip-flops have a natural count of 4, this counter skips one state. The waveforms and the truth table in Fig. 11–16 show that this counter progresses through the count sequence 00, 01, 10, and then back to 00. It clearly skips count 11. Here's how it works:

1. Prior to point *a* on the time line, $A = 0$ and $B = 0$. A negative clock transition at *a* will cause:
   **a.** $A$ to toggle to a 1, since its $J$ and $K$ inputs are high
   **b.** $B$ to reset to 0 (it's already a 0), since its $J$ input is low and its $K$ input is high
2. Prior to point *b* on the time line, $A = 1$, and $B = 0$. A negative clock transition at *b* will cause:
   **a.** $A$ to toggle to a 0, since its $J$ and $K$ inputs are high
   **b.** $B$ to toggle to a 1, since its $J$ and $K$ inputs are high
3. Prior to point *c* on the time line, $A = 0$ and $B = 1$. A negative clock transition at *c* will cause:

**a.** *A* to reset to 0 (it's already 0), since its *J* input is low and its *K* input is high

**b.** *B* to reset to 0 since its *J* input is low and its *K* input is high

4. The counter has now progressed through all three of its states, advancing one count with each negative clock transition.

This two-flip-flop mod-3 counter can be considered as a logic building block as shown in Fig. 11–16*d*. It has a clock input and outputs at *A* and *B*. It can be considered as a divide-by-3 block, since the output waveform at *B* (or at A) has a period equal to three times that of the clock—in other words, this counter divides the clock frequency by 3. Notice that this is a synchronous counter since both flip-flops change state in synchronism with the clock.

If we consider a basic flip-flop to be a mod-2 counter, we see that a mod-4 counter (two flip-flops in series) is simply two mod-2 counters in series. Similarly, a mod-8 counter is simply a $2 \times 2 \times 2$ connection, and so on. Thus a great number of higher-modulus counters can be formed by using the product of any number of lower-modulus counters. For instance, suppose that we connect a flip-flop at the B output of the mod-3 counter in Fig. 11–16. The result is a ($3 \times 2 = 6$) mod-6 counter as shown in Fig. 11–17. The output of the single flip-flop is labeled *C*. Notice that it is a symmetrical waveform, and it also has a frequency of one-sixth that of the input clock. Notice that this can no longer be considered a synchronous counter since flip-flop *C* is triggered by flip-flop *B*; that is, the flip-flops do not all change states in synchronism with the clock.

☐ EXAMPLE 11–9

Draw the waveforms you would expect from the mod-6 counter by connecting a single flip-flop in front of the mod-3 counter in Fig. 11–16.

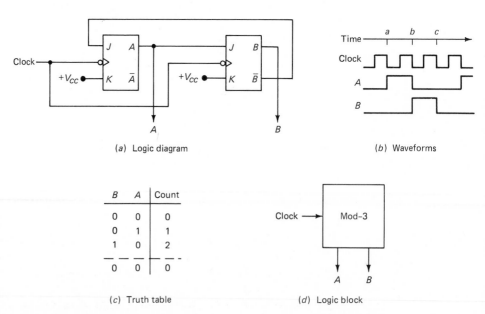

(a) Logic diagram

(b) Waveforms

| B | A | Count |
|---|---|-------|
| 0 | 0 | 0 |
| 0 | 1 | 1 |
| 1 | 0 | 2 |
| 0 | 0 | 0 |

(c) Truth table

(d) Logic block

**FIG. 11–16**  Mod-3 counter

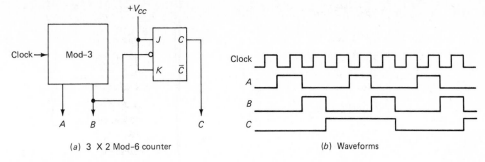

(a) 3 X 2 Mod-6 counter  (b) Waveforms

**FIG. 11–17**

☐ SOLUTION

The resulting counter is a $2 \times 3$ = mod-6 counter that has the waveforms shown in Fig. 11–18. Notice that $B$ now has a period equal to six clock periods, but it is not symmetrical.

The 54/7492A in Fig. 11–19 is a TTL, MSI divide-by-12 counter. A careful examination of the logic diagram will reveal that flip-flops $Q_B$, $Q_C$, and $Q_D$ are exactly the same as the $3 \times 2$ counter in Fig. 11–17. Thus if the clock is applied to input $B$ of the '92A and the outputs are taken at $Q_B$, $Q_C$, and $Q_D$, this is a mod-6 counter ('92A is a popular shorthand notation for 54/7492A).

On the other hand, if the clock is applied at input $A$ and $Q_A$ is connected to input $B$, we have a $2 \times 3 \times 2$ mod-12 counter. The proper truth table for the mod-12 configuration is given in Fig. 11–19b. Again, this must be considered as an asynchronous counter since

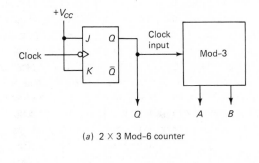

(a) 2 X 3 Mod-6 counter

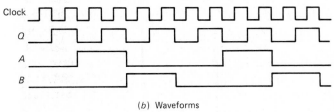

(b) Waveforms

**FIG. 11–18**

'92A, 'LS92 Count sequence
(See Note C)

| Count | Output | | | |
|---|---|---|---|---|
| | $Q_D$ | $Q_C$ | $Q_B$ | $Q_A$ |
| 0 | L | L | L | L |
| 1 | L | L | L | H |
| 2 | L | L | H | L |
| 3 | L | L | H | H |
| 4 | L | H | L | L |
| 5 | L | H | L | H |
| 6 | H | L | L | L |
| 7 | H | L | L | H |
| 8 | H | L | H | L |
| 9 | H | L | H | H |
| 10 | H | H | L | L |
| 11 | H | H | L | H |

(b) Truth table

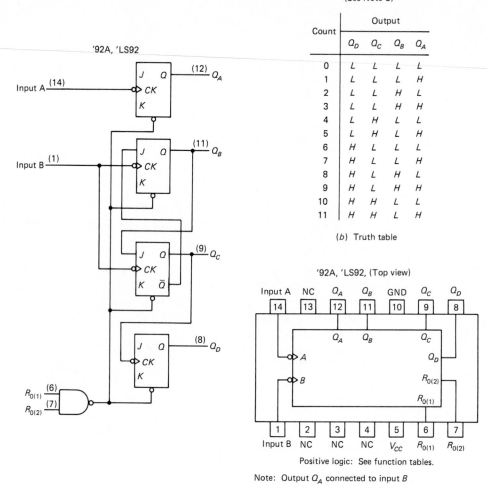

'92A, 'LS92

Input A (14)

Input B (1)

(12) $Q_A$

(11) $Q_B$

(9) $Q_C$

(8) $Q_D$

$R_{0(1)}$ (6)

$R_{0(2)}$ (7)

'92A, 'LS92, (Top view)

| Input A | NC | $Q_A$ | $Q_B$ | GND | $Q_C$ | $Q_D$ |
|---|---|---|---|---|---|---|
| 14 | 13 | 12 | 11 | 10 | 9 | 8 |

$Q_A$  $Q_B$  $Q_C$

A  B  $Q_D$  $R_{0(2)}$  $R_{0(1)}$

| 1 | 2 | 3 | 4 | 5 | 6 | 7 |
|---|---|---|---|---|---|---|
| Input B | NC | NC | NC | $V_{CC}$ | $R_{0(1)}$ | $R_{0(2)}$ |

Positive logic: See function tables.

Note: Output $Q_A$ connected to input $B$

(a) Logic

(c) Pin out

**FIG. 11–19**  54/7492A

all flip-flops do not change states at the same time. Thus there is the possibility of glitches occurring at the outputs of any decoding gates used with the counter.

☐ EXAMPLE 11–10

Use the truth table for the '92A to write a boolean expression for a gate to decode count 8.

☐ SOLUTION

The correct expression is "8" = $Q_D \overline{Q_C} Q_B \overline{Q_A}$.

At this point, we can construct counters that have any natural count (2, 4, 8, 16, etc.) and, in addition, a mod-3 counter. Furthermore, we can cascade these counters in any combination, such as $2 \times 2$, $2 \times 3$, $3 \times 4$, and so on. So far we can construct counters having a modulus of 2, 3, 4, 6, 8, 9, 12, and so on. Let's consider next a mod-5 counter.

## 11·5 A MOD·5 COUNTER

The three-flip-flop counter shown in Fig. 11–20 has a natural count of 8, but it is connected in such a way that it will skip over three counts. It will, in fact, advance one count at a time, through a strict binary sequence, beginning with 000 and ending with 100; therefore, it is a mod-5 counter. Let's see how it works.

The waveforms show that flip-flop $A$ changes state each time the clock goes negative, except during the transition from count 4 to count 0. Thus flip-flop $A$ should be triggered by the clock and must have an inhibit during count 4—that is, some signal must be provided during the transition from count 4 to count 0. Notice that $\overline{C}$ is high during all counts except count 4. If $\overline{C}$ is connected to the $J$ input of flip-flop $A$, therefore, we have the desired inhibit signal. This is true since the $J$ and $K$ inputs to flip-flop $A$ are both true for all counts except count 4; thus the flip-flop triggers each time the clock goes negative. However, during count 4, the $J$ side is low, and the next time the clock goes negative the flip-flop will be prevented from being set. The connections which cause flip-flop $A$ to progress through the desired sequence are shown in Fig. 11–20.

The desired waveforms in Fig. 11–20$b$ show that flip-flop $B$ must change state each time $A$ goes negative. Thus the clock input of flip-flop $B$ will be driven by $A$ as shown in Fig. 11–20$c$.

If flip-flop $C$ is triggered by the clock while the $J$ input is held low and the $K$ input high, every clock pulse will reset it. Now, if the $J$ input is high only during count 3, $C$

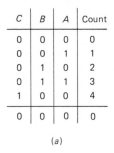

| C | B | A | Count |
|---|---|---|-------|
| 0 | 0 | 0 | 0 |
| 0 | 0 | 1 | 1 |
| 0 | 1 | 0 | 2 |
| 0 | 1 | 1 | 3 |
| 1 | 0 | 0 | 4 |
| 0 | 0 | 0 | 0 |

($a$)

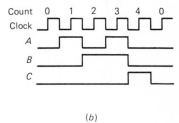

($b$)

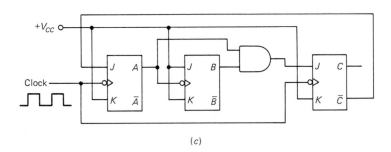

($c$)

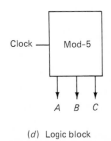

($d$) Logic block

**FIG. 11–20** Mod-5 binary counter

will be high during count 4 and low during all other counts. The necessary levels for the $J$ input can be obtained by ANDing flip-flops $A$ and $B$. Since $A$ and $B$ are both high only during count 3, the $J$ input to flip-flop $C$ is high only during count 3. Thus, when the clock goes negative during the transition from count 3 to count 4, flip-flop $C$ will be set. At all other times, the $J$ input to flip-flop $C$ is low and is held in the reset state. The complete mod-5 counter is shown in Fig. 11–20.

In constructing a counter of this type, it is always necessary to examine the omitted states to make sure that the counter will not malfunction. This counter omits states 5, 6, and 7 during its normal operating sequence. There is, however, a very real possibility that the counter may set up in one of these omitted (illegal) states when power is first applied to the system. It is necessary to check the operation of the counter when starting from each of the three illegal states to ensure that it progresses into the normal count sequence and does not become inoperative.

Begin by assuming that the counter is in state 5 ($CBA = 101$). When the next clock pulse goes low, the following events occur:

1. Since $\overline{C}$ is low, flip-flop $A$ resets. Thus $A$ changes from a 1 to a 0.
2. When $A$ changes from a 1 to a 0, flip-flop $B$ triggers and $B$ changes from a 0 to a 1.
3. Since the $J$ input to flip-flop $C$ is low, flip-flop $C$ is reset and $C$ changes from a 1 to a 0.
4. Thus the counter progresses from the illegal state 5 to the legal state 2 ($CBA = 010$) after one clock.

Now, assume that the counter starts in the illegal state 6 ($CBA = 110$). On the next negative clock transition, the following events occur:

1. Since $\overline{C}$ is low, flip-flop $A$ is reset. Since $A$ is already a 0, it simply remains a 0.
2. Since $A$ does not change, flip-flop $B$ does not change and $B$ remains a 1.
3. Since the $J$ input to flip-flop $C$ is low, flip-flop $C$ is reset and $C$ changes from a 1 to a 0.
4. Thus the counter progresses from the illegal state 6 to the legal state 2 after one clock transition.

Finally, assume that the counter begins in the illegal state 7 ($CBA = 111$). On the next negative clock transition, the following events occur:

1. Since $\overline{C}$ is low, flip-flop $A$ is reset and $A$ changes from a 1 to a 0.
2. Since $A$ changes from a 1 to a 0, flip-flop $B$ triggers and $B$ changes from a 1 to a 0.
3. The $J$ input to flip-flop $C$ is high; therefore, flip-flop $C$ toggles from a 1 to a 0.
4. Thus the counter progresses from the illegal count 7 to the legal count 0 after one clock transition.

None of the three illegal states will cause the counter to malfunction, and it will automatically work itself out of any illegal state after only one clock transition.

This mod-5 counter configuration can be considered as a logic block as shown in Fig. 11–20$d$ and can be used in cascade to construct higher-modulus counters. For instance, a $2 \times 5$ or a $5 \times 2$ will form a mod-10, or "decade" counter.

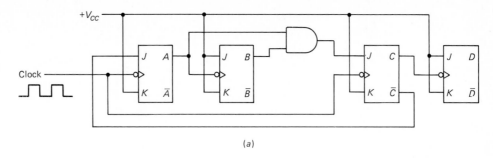

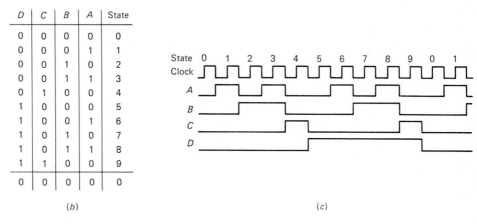

| D | C | B | A | State |
|---|---|---|---|-------|
| 0 | 0 | 0 | 0 | 0 |
| 0 | 0 | 0 | 1 | 1 |
| 0 | 0 | 1 | 0 | 2 |
| 0 | 0 | 1 | 1 | 3 |
| 0 | 1 | 0 | 0 | 4 |
| 1 | 0 | 0 | 0 | 5 |
| 1 | 0 | 0 | 1 | 6 |
| 1 | 0 | 1 | 0 | 7 |
| 1 | 0 | 1 | 1 | 8 |
| 1 | 1 | 0 | 0 | 9 |
| 0 | 0 | 0 | 0 | 0 |

(b)

(c)

FIG. 11–21   A decade counter

□ EXAMPLE 11–11

Show a method for constructing a $5 \times 2$ mod-10 decade counter.

□ SOLUTION

A decade counter can be constructed by using the mod-5 counter in Fig. 11–20 and adding an additional flip-flop, labeled $D$, as shown in Fig. 11–21. The appropriate waveforms and truth table are included. Notice that the counter progresses through a *biquinary* count sequence and does not count in a straight binary sequence.

A decade counter could be formed just as easily by using the mod-5 counter in Fig. 11–20 in conjunction with a flip-flop, but connected in a $2 \times 5$ configuration as shown in Fig. 11–22. The truth table for this configuration, and the resulting waveforms are shown in the Fig. 11–20. This is still a mod-10 (decade) counter since it still has 10 discrete states. Notice that this counter counts in a straight binary sequence from 0000 up to 1001, and then back to 0000.

The 54/7490A is a TTL MSI decade counter. Its logic diagram, truth table, and pinout are given in Fig. 11–23. A careful examination will reveal that flip-flops $Q_B$, $Q_C$, and $Q_D$ form a mod-5 counter exactly like the one in Fig. 11–20. Notice, however, that flip-flop $Q_D$ in the '90A is an $RS$ flip-flop that has a direct connection from its $Q$ output

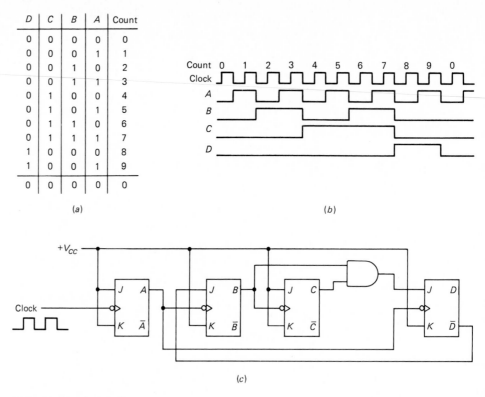

| D | C | B | A | Count |
|---|---|---|---|-------|
| 0 | 0 | 0 | 0 | 0 |
| 0 | 0 | 0 | 1 | 1 |
| 0 | 0 | 1 | 0 | 2 |
| 0 | 0 | 1 | 1 | 3 |
| 0 | 1 | 0 | 0 | 4 |
| 0 | 1 | 0 | 1 | 5 |
| 0 | 1 | 1 | 0 | 6 |
| 0 | 1 | 1 | 1 | 7 |
| 1 | 0 | 0 | 0 | 8 |
| 1 | 0 | 0 | 1 | 9 |
| 0 | 0 | 0 | 0 | 0 |

(a)

(b)

(c)

**FIG. 11–22** A decade counter

back to its $R$ input. The net result in this case is that $Q_D$ behaves exactly like a $JK$ flip-flop.

If the system clock is applied at input $A$ and $Q_A$ is connected to input $B$, we have a true binary decade counter exactly as in Fig. 11–22. On the other hand, if the system clock is applied at input $B$ and $Q_D$ is connected to input $A$, we have the biquinary counter as discussed in Example 11–11. Take time to study the logic diagram and the truth table for the '90A; it is widely used in industry, and the time spent will be well worth your while.

An interesting application using three 54/7490A decade counters is shown in Fig. 11–24. The three '90A's are connected in series such that the first one (on the right) counts the number of input pulses at its clock input. We call it a *units counter*.

The middle '90A will advance one count each time the units counter counts 10 input pulses, because $D$ from the units counter will have a single negative transition as that counter progresses from count 9 to 0. This middle block is then called *tens counter*.

The left '90A will advance one count each time the tens counter progresses from count 9 to 0. This will occur once for every 100 input pulses. Thus this block is called the *hundreds counter*.

Now the operation should be clear. This logic circuit is capable of counting input pulses from one up to 999. The procedure is to reset all the '90A's and then count the

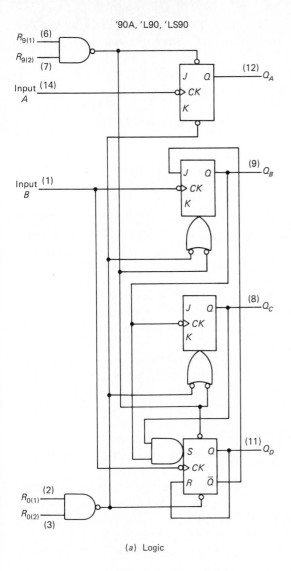

'90A, 'L90, 'LS90

| '90A, 'L90, 'LS90 BCD count sequence (See note A) | | | | | '90A, 'L90, 'LS90 Bi-quinary (5-2) (See note B) | | | | |
|---|---|---|---|---|---|---|---|---|---|---|
| Count | Output | | | | | Count | Output | | | |
| | $Q_D$ | $Q_C$ | $Q_B$ | $Q_A$ | | | $Q_A$ | $Q_D$ | $Q_C$ | $Q_B$ |
| 0 | L | L | L | L | | 0 | L | L | L | L |
| 1 | L | L | L | H | | 1 | L | L | L | H |
| 2 | L | L | H | L | | 2 | L | L | H | L |
| 3 | L | L | H | H | | 3 | L | L | H | H |
| 4 | L | H | L | L | | 4 | L | H | L | L |
| 5 | L | H | L | H | | 5 | H | L | L | L |
| 6 | L | H | H | L | | 6 | H | L | L | H |
| 7 | L | H | H | H | | 7 | H | L | H | L |
| 8 | H | L | L | L | | 8 | H | L | H | H |
| 9 | H | L | L | H | | 9 | H | H | L | L |

(*b*) Truth table

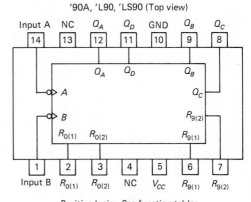

'90A, 'L90, 'LS90 (Top view)

Positive logic: See function tables

Note: A Output $Q_A$ connected to input *B*
B Output $Q_D$ connected to input *A*

(*a*) Logic

(*c*) Pin out

**FIG. 11–23**  54/7490A

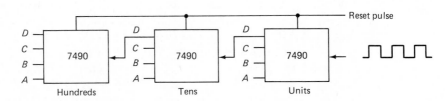

**FIG. 11–24**  Cascaded 7490's can count to 999

number of pulses at the input to the units counter. This cascaded arrangement is widely used in digital voltmeters, frequency counters, and many other applications where a decimal count is required.

It should be pointed out that the 54/7490A is only one of a number of TTL MSI decade counters. In particular, the 54/74176 is another popular asynchronous decade counter, and the 54/74160, 54/74162, 54/74190, and 54/74192 are all popular synchronous decade counters. Each has particular attributes that you should consider, and a study of their individual data sheets would be worth while.

## 11-6 PRESETTABLE COUNTERS

Up to this point we have discussed the operation of counters that progress through a natural binary count sequence in either a count-up or count-down mode and have studied two counters that have a modified count—a mod-3 and a mod-5. With these basic configurations, and with cascaded combinations of these basic units, it is possible to construct counters having moduli of 2, 3, 4, 5, 6, 8, 9, 10, and so on. The ability to quickly and easily construct a counter having any desired modulus is so important that the semiconductor industry has provided a number of TTL MSI circuits for this purpose. The presettable counter is the basic building block that can be used to implement a counter that has any modulus.

Nearly all the presettable counters available as TTL MSI are constructed by using four flip-flops (usually master-slave *JK* types), and they are generally referred to as *4-bit counters*. They may be either synchronous or asynchronous. When connected such that the count advances in a natural binary sequence from 0000 to 1111, it is simply referred to as a *binary counter*. For instance, the 54/74161 and the 54/74163 are both synchronous binary counters that operate in a count-up mode. The 54/74191 and the 54/74193 are also synchronous binary counters, but they can operate in either a count-down or count-up mode.

Since the decade counter is a very important and useful configuration, many of the basic 4-bit counters are internally connected to provide a modified count of 10—a mod-10 or decade counter. For instance, the 54/74160 and the 54/74162 are synchronous decade counters that operate in the count-up mode. The 54/74190 and the 54/74192 are also synchronous decade counters, but they can operate in either a count-up or count-down mode.

The counters mentioned above are all TTL MSI circuits, and as such we have little control over the internal logic used to implement each counter. Our concern is directed at how each unit can be used in a digital system. Thus we consider each of these counters as a logic block, and our efforts are concentrated on inputs, outputs, and control signals. Even so, the logic block diagram is given for each counter, since a knowledge of the internal logic gives a depth of understanding that is invaluable in practical applications.

The pinout and logic diagram for a 54/74163 synchronous 4-bit counter are given in Fig. 11–25. The pinout contains a logic block diagram for this unit. The power requirements are $+V_{CC}$ and GROUND on pins 16 and 8, respectively. The "clock" is applied on pin 2, and you will notice from the diagram that the outputs change states on positive clock transitions.

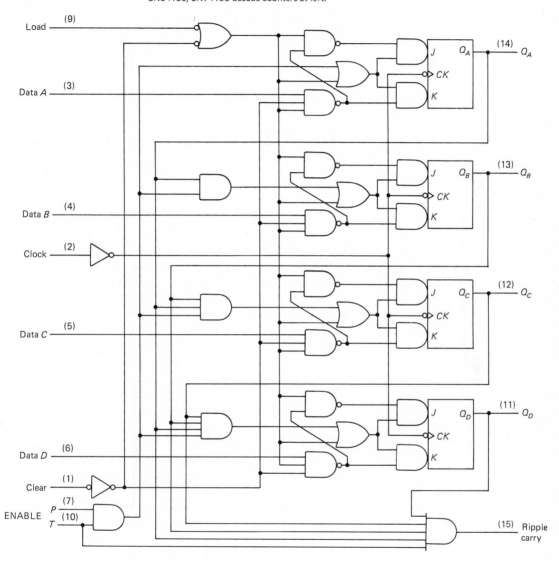

SN54163, SN74163 Synchronous binary counters

SN54161, SN74161 Synchronous binary counters are similar; however, the CLEAR is asynchronous as shown for the SN54160, SN74160 decade counters at left.

**FIG. 11-25**   54/74161 and 54/74163

The four flip-flop outputs are $Q_A$, $Q_B$, $Q_C$, and $Q_D$, while the CARRY output on pin 15 can be used to enable successive counter stages (e.g., in a units, tens, hundreds application).

The two ENABLE inputs ($P$ on pin 7 and $T$ on pin 10) are used to control the counter. If either ENABLE input is low, the counter will cease to advance; both of these inputs must be high for the counter to count.

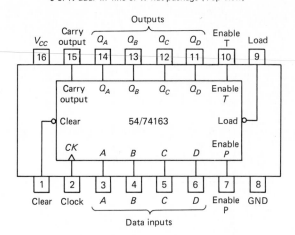

Positive logic: See description.

**FIG. 11–25**   (continued)

A low level on the CLEAR input will reset all flip-flop outputs low at the very next clock transition, regardless of the levels on the ENABLE inputs. This is called a *synchronous reset* since it occurs at a positive clock transition. On the other hand, note that the 54/74161 has an asynchronous clear, since it occurs immediately when the CLEAR input goes low, regardless of the levels on the CLOCK, ENABLE, or LOAD inputs.

When a low level is applied to the LOAD input, the counter is disabled, and the very next positive clock transition will set the flip-flops to agree with the levels present on the four data inputs ($D$, $C$, $B$, and $A$). For instance, suppose that the data inputs are $DCBA = 1101$, and the LOAD input is taken low. The very next positive clock transition will load these data into the counter and the outputs will become $Q_D Q_C Q_B Q_A = 1101$. This is a very useful function when it is desired to have the counter begin counting from a predetermined count.

For the counter to count upward in its normal binary count sequence, it is necessary to hold the ENABLE inputs ($P$ and $T$), the LOAD input, and the CLEAR input all high. Under these conditions, the counter will advance one count for each positive clock transition, progressing from count 0000 up to count 1111 and then repeating the sequence. Since the flip-flops are clocked synchronously, the outputs change states simultaneously and there are no counting spikes or glitches associated with the counter outputs. The state diagram given in Fig. 11–26*a* show the normal count sequence, where each box corresponds to one count (or state), and the arrows show how the counter progresses from one state to the next.

The count length can be very easily modified by making use of the synchronous CLEAR input. It is a simple matter to use a NAND gate to decode the maximum count desired, and use the output of this NAND gate to clear the counter synchronously to count 0000. The counter will then count from 0000 up to the maximum desired count and then clear back to 0000. This is the technique that can be used to construct a counter that has any desired modulus.

For instance, if a maximum count of 9 is desired, we connect the inputs of the NAND

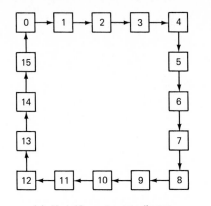

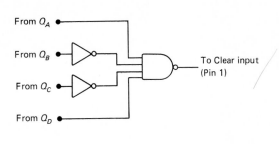

(a) Mod–16 counter state diagram

(b) Gate to decode count 9 (1001)

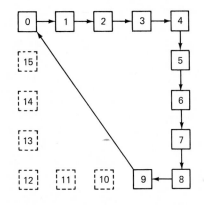

(c) Modified state diagram for Mod–10 counter

FIG. 11–26

gate to decode count $9 = DCBA = 1001$. We then have a mod-10 counter, since the count sequence is from 0000 up to 1001. The NAND gate used to decode count 9 along with the modified state diagram are shown in Fig. 11–26$b$ and $c$, respectively. Notice that it was necessary to use two inverters to obtain $\overline{Q_B}$ and $\overline{Q_C}$. The modified state diagram has solid boxes for states in the modified, mod-10 counter, and dashed boxes for omitted states.

☐ EXAMPLE 11—12

What are the NAND–gate inputs in Fig. 11–26$b$ if this figure is to be used to construct a mod-12 counter?

☐ SOLUTION

The counter must progress from 0000 up to 1011 (decimal 11); the NAND–gate inputs must then be $Q_D$, $\overline{Q_C}$, $Q_B$, and $Q_A$.

A set of typical waveforms showing clear, preset, count, and inhibit operation for a 54/74163 (and 54/74161) are given in Fig. 11–27. You should take time to study them carefully until you understand exactly how these four operations are controlled.

**Typical clear, preset, count, and inhibit sequences**

Illustrated below is the following sequence.

1. Clear outputs to zero.
2. Preset to binary twelve.
3. Count to thirteen, fourteen, fifteen, zero, one, and two.
4. Inhibit.

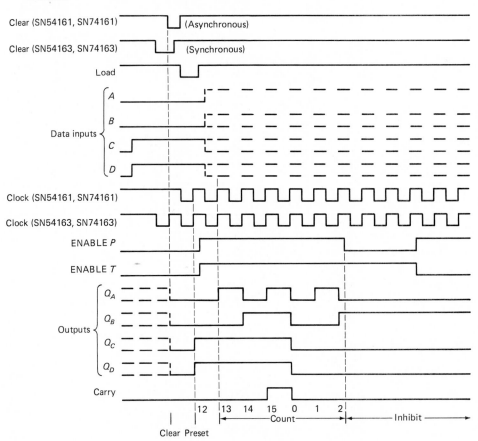

**FIG. 11–27**

The logic diagram and a typical set of waveforms for the 54/74160 and the 54/74162 are given in Fig. 11–28. (The pinout is identical for the previously given 54/74163.) These two counters have been modified internally and are decade counters. Other than that, the input, output, and control lines for these two counters are identical with the previously discussed 54/74163 and 54/74161. These counters advance one count with each positive clock transition, progressing from 0000 to 1001 and back to 0000. The state diagram for these two units would appear exactly as shown in Fig. 11–26c; this is the state diagram for a mod-10 or decade counter.

The 54/74193 is a 4-bit synchronous up-down binary counter. It has a master reset input and can be reset to any desired count with the parallel load inputs. The logic symbol

CHAPTER 11

SN54162, SN74162 Synchronous decade counters are similar; however, the clear is synchronous as shown for the SN54163, SN74163 binary counters at right.

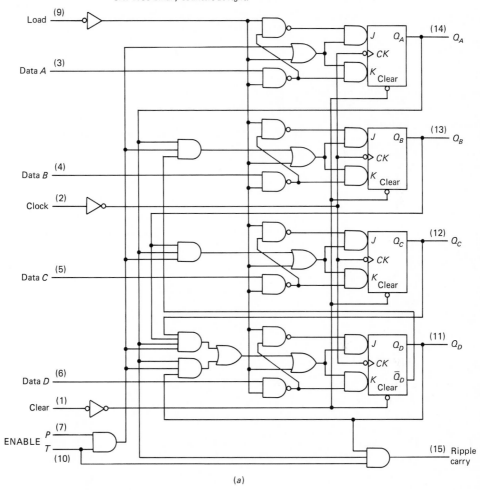

(a)

**FIG. 11-28** 54/74160

for this TTL MSI is shown in Fig. 11-29a. Pin $\overline{PL}$ is a control input for loading data into pins $P_A$, $P_B$, $P_C$, and $P_D$. When the device is used as a counter, these four pins are left open and $\overline{PL}$ must be held high. Pin $MR$ is the master reset, and it is normally held low. (A high level on $MR$ will reset all flip-flops.)

Outputs $TC_U$ and $TC_D$ are to be used to drive the following units, such as in a cascade arrangement. The clock inputs are $CP_U$ and $CP_D$. Placing the clock on $CP_U$ will cause the counter to count up, and placing the clock on $CP_D$ will cause the counter to count down. Notice that the clock should be connected to either $CP_U$ or $CP_D$, but not both, and the unused input should be held high. The outputs of the counter are $Q_A$, $Q_B$, $Q_C$, and $Q_D$.

**Typical clear, preset, count, and inhibit sequences**

Illustrated below is the following sequence.

1. Clear outputs to zero.
2. Preset to BCD seven.
3. Count to eight, nine, zero, one, two, and three.
4. Inhibit

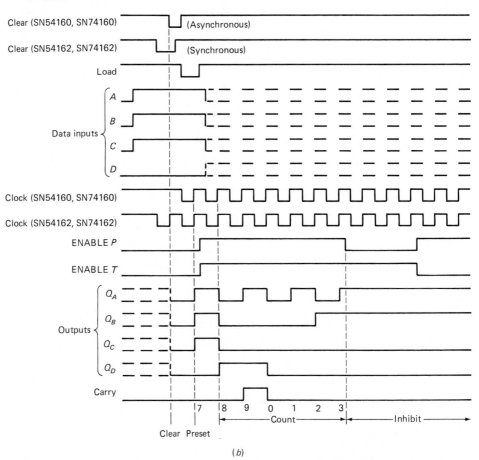

(b)

**FIG. 11–28** (continued)

A state diagram is a simple drawing which shows the stable states of the counter, as well as how the counter progresses from one count to the next. The state diagram for the 54/74193 is shown in Fig. 11–29b. Each box represents a stable state, and the arrows indicate the count sequence for both count-up and count-down operations. This is a 4-bit counter, and clearly there are 16 stable states, numbered 0, 1, 2, . . ., 15.

The 54/74193 has a parallel-data-entry capability which permits the counter to be preset to the number present on the parallel-data-entry inputs ($P_A$, $P_B$, $P_C$, and $P_D$). Whenever the parallel load input ($\overline{PL}$) is low, the data present at these four inputs is shifted into the counter; that is, the counter is preset to the number held by $P_D P_C P_B P_A$.

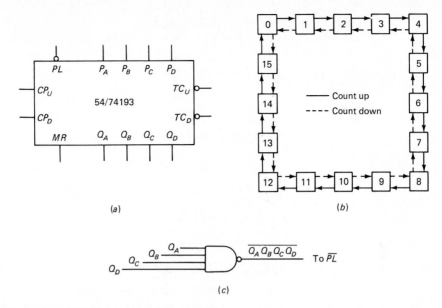

(a)

(b)

$Q_A Q_B Q_C Q_D$ ... $\overline{Q_A Q_B Q_C Q_D}$ To $\overline{PL}$

(c)

**FIG. 11–29**  4-bit binary counter (presetable)

Now, here is another technique for modifying the count. Simply use a NAND gate to detect any of the stable states, say, state 15 (1111), and use this gate output to take $\overline{PL}$ low. The only time $\overline{PL}$ will be low is when $Q_D$, $Q_C$, $Q_B$, and $Q_A$ are all high, or state 15 (1111). At this time, the counter will be preset to the data $P_D P_C P_B P_A$.

For example, suppose that $P_D P_C P_B P_A = 1001$ (the number 9). When the clock is applied, the counter will progress naturally to count 15 (1111). At this time, $\overline{PL}$ will go low and the number 9 (1001) will be shifted into the counter. The counter will then progress through states 9, 10, 11, 12, 13, and 14, and at count 15 it will again be preset to 9.

The count sequence is easily shown by the state diagram in Fig. 11–30. Notice that count 15 (1111) is no longer a stable state; it is the short time during which the counter is

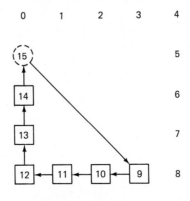

**FIG. 11–30**

preset. The stable states in this example are 9, 10, 11, 12, 13, and 14. This is, then, a mod-6 counter. Notice that this technique is asynchronous since the preset action is not in synchronism with the clock. Therefore, you should be aware that counting spikes or glitches may be associated with the outputs of this presetting arrangement.

□ EXAMPLE 11–13

Suppose that the counter just discussed is still preset to 1001 (the number 9) but the clock is applied to count down rather than count up. What are the counting states? What is the modulus?

□ SOLUTION

The counter will count down to 15, then preset back to 9, and repeat. The resulting state diagram is given in Fig. 11–31. The modulus is clearly 10.

## 11-7 SHIFT COUNTERS

In Chap. 10, the output of the last flip-flop in a shift register was connected back to the control input of the first flip-flop in the register. This arrangement resulted in a ring counter, as shown in Fig. 10–19, and the technique is referred to as *direct feedback*. If the outputs of the last flip-flop are first crossed over and then connected back to the control inputs of the first flip-flop, the technique is called *inverse feedback*. This connection results in a counter that has some very unique characteristics—the *Johnson*, or *shift*, *counter*.

The three master-slave *JK* flip-flops in Fig. 11–32 are connected in a standard shift register configuration. In addition, the outputs of the last flip-flop are crossed over and then connected back to the inputs of the first flip-flop. Specifically, $C$ is connected to the $K$ input of $A$, and $\overline{C}$ is connected to the $J$ input of $A$.

Now assume that all flip-flops are in a reset condition and the clock is allowed to run. Since $\overline{C}$ is high and $C$ is low, a 1 is set in flip-flop $A$ during the first cycle of the clock. At the same time, $B$ and $C$ remain low since their $J$ inputs are low and their $K$ inputs are high.

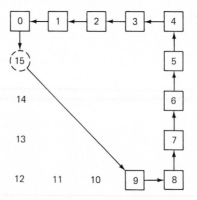

**FIG. 11–31**

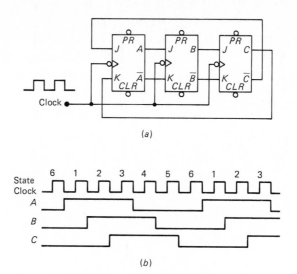

(a)

(b)

**FIG. 11–32**   Three stage shift register using inverse feedback to form a shift counter

During the second cycle of the clock, $A$ remains high since $\overline{C}$ is still high and $C$ is still low. At the same time, $B$ is set high since $A$ is now high and $\overline{A}$ is low; $C$ remains unchanged since $B$ is low during this period.

During the third clock period, $A$ and $B$ remain high and $C$ is set high since $B$ is now high. Thus after three cycles of the clock, all three flip-flops have been changed from the low state to the high state.

During the fourth clock period, $\overline{C}$ is low and $C$ is high; therefore, $A$ is reset to the low state, and $B$ and $C$ remain high.

During the fifth cycle of the clock, $A$ remains low, $B$ is reset low (since $A$ is now low and $\overline{A}$ is high), and $C$ remains high.

The sixth clock period returns the counter to the initial starting point since $C$ is reset low while $A$ and $B$ both remain low. Thus this shift register with inverse feedback has progressed through a complete cycle of counts in six clock periods.

Examination of Fig. 11–32b reveals that the waveform of each flip-flop is a square wave that has a period six times that of the clock. Moreover, all three flip-flop outputs are identical except that they are shifted with respect to one another by one clock period. This square wave apparently shifts through the flip-flops, advancing one flip-flop each time the clock goes low. Since the operation is cyclic and the waveforms shift through the flip-flops, this configuration is commonly called a shift or Johnson counter (as mentioned in the first paragraph of this section).

In order to investigate this counter more carefully, let us make a truth table showing the states through which the counter progresses. This truth table, constructed with the aid of the counter waveforms, is shown in Fig. 11–33a. For easy comparison, a straight binary truth table for three flip-flops is also shown. Notice that the three flip-flop shift counter counts through six discrete states. The six ordered states through which the shift

| C | B | A | State | Equivalent binary count |
|---|---|---|---|---|
| 0 | 0 | 1 | 1 | 1 |
| 0 | 1 | 1 | 2 | 3 |
| 1 | 1 | 1 | 3 | 7 |
| 1 | 1 | 0 | 4 | 6 |
| 1 | 0 | 0 | 5 | 4 |
| 0 | 0 | 0 | 6 | 0 |
| 0 | 0 | 1 | 1 | 1 |

(a)

| C | B | A | Count |
|---|---|---|---|
| 0 | 0 | 0 | 0 |
| 0 | 0 | 1 | 1 |
| 0 | 1 | 0 | 2 |
| 0 | 1 | 1 | 3 |
| 1 | 0 | 0 | 4 |
| 1 | 0 | 1 | 5 |
| 1 | 1 | 0 | 6 |
| 1 | 1 | 1 | 7 |
| 0 | 0 | 0 | 0 |

(b)

FIG. 11–33 Three flip-flop shift counter truth table

counter progresses correspond to the binary counts of 1-3-7-6-4-0. Thus the six states of the shift counter are, indeed, discrete and can be decoded.

Notice, however, that the shift counter omits binary counts 2 (010) and 5 (101). Therefore, the shift counter must be examined to determine whether it will work its way out of either of these two states, since it is possible that one of them may occur when power is first applied to the system. In fact, one of these two illegal states could occur during normal operation because of noise or some other malfunction.

First, suppose that the counter is in binary count 2 (010). The next time the clock goes negative, $A$ goes high, $B$ goes low, and $C$ goes high. Thus the counter will advance to binary count 5 (101), which is the second illegal state.

During the second clock cycle, $A$ goes low, $B$ goes high, and $C$ goes low. Thus the second clock period will advance the counter right back into the first illegal state, binary count 2 (010). Therefore, the counter will simply oscillate between the two illegal states and will not function as a mod-6 counter.

To avoid this situation, it should be ensured that the counter cannot remain in one of the illegal states. One method of accomplishing this is to use the NAND gate shown in Fig. 11–34. When $\overline{A}$, $B$, and $\overline{C}$ are all high (corresponding to the illegal state 010), the output of the NAND gate is low. If the NAND–gate output is applied to the PRESET (PR) of flip-flop $A$, then $A$ will be set high whenever this condition occurs. Thus the counter will immediately advance from binary count 2 (010) to binary count 3 (011). This is the second state in the normal counting sequence, and the counter will thus operate as desired.

☐ EXAMPLE 11—14

Draw the diagram for a four-flip-flop shift counter. Using these waveforms, make a truth table showing the desired count sequence. List the illegal states, and examine them to determine whether the counter will get into an undesired mode of operation. If the counter does malfunction because of illegal states, show a method to cure the problem.

FIG. 11–34 Preset gate for the counter in Fig. 11–32

The desired counter and waveforms are shown in Fig. 11–35. The truth table, derived from the desired waveforms, is also given. Now, since this counter is constructed using four flip-flops, there are 16 possible states. In the desired mode of operation, the counter sequences through only eight states; therefore, eight illegal states must be examined. The eight illegal states correspond to the binary counts 2, 4, 5, 6, 9, 10, 11, and 13, and they are shown in the table in Fig. 11–36.

The two columns to the right of the double vertical line in the table show the new state and the new binary equivalent count, to which the counter will progress after one cycle of the clock. For example, after one cycle of the clock, the counter will advance from binary count 2 (0010) to binary count 5 (0101). The table shows that if the counter comes up in any one of the illegal states, it will count through the binary sequence 2-5-11-6-13-10-4-9-2. Thus the counter will enter an undesired mode of operation and remain there. Notice that the counter still divides by 8, which was part of the original intent. However, to decode the counter it is necessary to have some means of forcing the counter back into the desired count, since the output waveforms in this secondary mode are quite different from the desired mode.

The waveforms for the illegal count sequence are shown in Fig. 11–36b. One method of forcing the counter into the desired mode can be found by observing that the condition $\overline{A}B\overline{C}$ is true only twice during the undesired count sequence (counts 2 and 10). This condition is never true during the desired count sequence, so the NAND gate shown in Fig. 11–34 can be used to correct the problem. It is necessary, however, to connect the

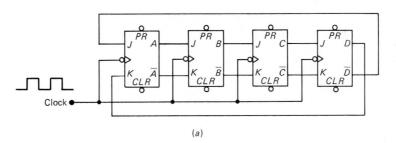

(a)

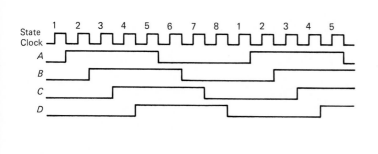

(b)

| D | C | B | A | State | Binary equivalent |
|---|---|---|---|-------|-------------------|
| 0 | 0 | 0 | 0 | 1 | 0 |
| 0 | 0 | 0 | 1 | 2 | 1 |
| 0 | 0 | 1 | 1 | 3 | 3 |
| 0 | 1 | 1 | 1 | 4 | 7 |
| 1 | 1 | 1 | 1 | 5 | 15 |
| 1 | 1 | 1 | 0 | 6 | 14 |
| 1 | 1 | 0 | 0 | 7 | 12 |
| 1 | 0 | 0 | 0 | 8 | 8 |
| 0 | 0 | 0 | 0 | 1 | 0 |

(c)

**FIG. 11–35** Four stage shift counter

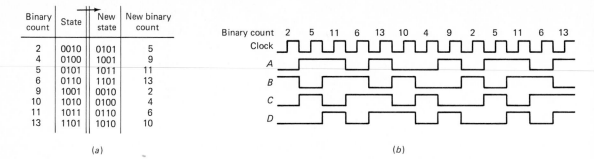

| Binary count | State | New state | New binary count |
|:---:|:---:|:---:|:---:|
| 2 | 0010 | 0101 | 5 |
| 4 | 0100 | 1001 | 9 |
| 5 | 0101 | 1011 | 11 |
| 6 | 0110 | 1101 | 13 |
| 9 | 1001 | 0010 | 2 |
| 10 | 1010 | 0100 | 4 |
| 11 | 1011 | 0110 | 6 |
| 13 | 1101 | 1010 | 10 |

(a)

(b)

**FIG. 11–36** (a) The eight illegal states for the counter shown in Fig. 11–35, and the states to which the counter will advance after one clock period (b) Illegal counting sequence waveforms

output of the NAND gate to PR on both flip-flops $C$ and $D$. This forces the counter to state 6 whenever the output of the gate goes low (i.e., whenever $\overline{AB}\overline{C}$ is high), and the counter will then be back into the desired count sequence.

Notice that $n$ flip-flops can be used to divide the clock by $2n$. That is, two flip-flops divide the clock by 4, three flip-flops divide the clock by 6, and so on. Alternatively, we can say that $n$ flip-flops connected in the shift counter configuration provide $2n$ discrete states through which the counter will progress. Thus it is possible to construct a counter of any even modulus by simply choosing the proper number of flip-flops and connecting them in the shift counter configuration.

Construction of other modulus counters from the basic shift-counter configuration is amazingly simple. The waveforms in Fig. 11–32 show that if $A$ is reset low one clock period earlier, $B$ is also reset one clock period earlier, as is $C$. This means that $A$, $B$, and $C$ are all high for only two clock periods but will still be low for three periods. Therefore, the counter is now dividing by 5, instead of 6, and there are five discrete states of a mod-5 counter.

The method of accomplishing this is shown in Fig. 11–37, along with the appropriate waveforms and truth table. Since it is necessary to cause $A$ to go low only one clock period earlier than before, this means that the $K$ input to flip-flop $A$ must go low one period earlier. Therefore, it is necessary only to obtain the $K$ input to flip-flop $A$ from $B$ instead of from $C$. Thus any odd modulus counter of count $m$ can be constructed from an even counter of modulus $m + 1$ by simply removing the $K$ input of the first flip-flop from the true side of the last flip-flop and connecting it to the true output of the next-to-last flip-flop.

This counter still has the two illegal counts, 2 (010) and 5 (101), and also the illegal count 7 (111). The additional count 7 will not cause any problem since the counter will advance naturally from 7 (111) to 6 (110), and this is state 4 in the desired counting sequence. Furthermore, it can be seen that the counter will advance from 2 (010) to 5 (101) and then to 3 (011), which is state 2 in the desired count sequence. Thus this counter has no permanent illegal modes.

CHAPTER 11

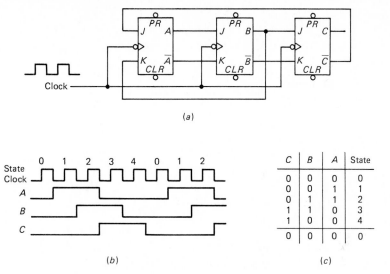

FIG. 11–37   Mod-5 shift counter

# 11·8 A MOD·10 SHIFT COUNTER WITH DECODING

A five-flip-flop shift counter can be constructed as shown in Fig. 11–38. This is a very useful configuration since it divides by 10 and can be used as a decade counter. The desired waveforms and corresponding truth table are also shown in Fig. 11–38. Since there are five flip-flops, there are a possible 32 states in which the counter can exist. The desired count sequence uses only 10 of these possible states; thus there are 22 illegal states.

It can be shown that (see Prob. 11–37) if the counter comes up in any one of these illegal states it will do one of two things. First, it will continue to divide the clock by 10, but it will advance through one of the two following illegal count sequences: 2-5-11-23-14-29-26-20-8-17-2, or 4-9-19-6-13-27-22-12-25-18-4. Second, if the counter comes up in either count 10 or 21, it will simply alternate back and forth between these two counts. Again, the cure to ensure that the counter operates in the proper sequence is the NAND gate shown in Fig. 11–34. The output of the gate must be connected to PR of flip-flops $C$, $D$, and $E$ (see Prob. 11–38).

The decade shift counter in Fig. 11–38 has 10 discrete states; therefore, the counter can be decoded to form 10 timing waveforms similar to those obtained for the mod-8 counter in Fig. 11–7. In decoding the mod-8 ripple counter, we needed eight 3-input AND gates since there were three flip-flops in the counter. One might expect to use ten 5-input AND gates to decode this mod-10 shift counter since there are five flip-flops. However, examination of the waveforms in Fig. 11–38 reveals that a much simpler arrangement is possible.

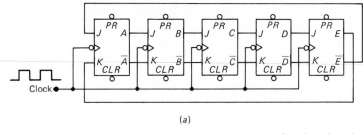

(a)

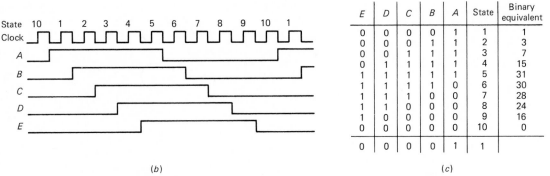

| E | D | C | B | A | State | Binary equivalent |
|---|---|---|---|---|---|---|
| 0 | 0 | 0 | 0 | 1 | 1 | 1 |
| 0 | 0 | 0 | 1 | 1 | 2 | 3 |
| 0 | 0 | 1 | 1 | 1 | 3 | 7 |
| 0 | 1 | 1 | 1 | 1 | 4 | 15 |
| 1 | 1 | 1 | 1 | 1 | 5 | 31 |
| 1 | 1 | 1 | 1 | 0 | 6 | 30 |
| 1 | 1 | 1 | 0 | 0 | 7 | 28 |
| 1 | 1 | 0 | 0 | 0 | 8 | 24 |
| 1 | 0 | 0 | 0 | 0 | 9 | 16 |
| 0 | 0 | 0 | 0 | 0 | 10 | 0 |
| 0 | 0 | 0 | 0 | 1 | 1 |  |

(b)  (c)

**FIG. 11–38**  Five flip-flop shift counter (decade counter)

Notice that state 1 can be uniquely determined as the time when $A$ is high and $B$ is low. Thus only a 2-input AND gate is required to decode state 1. The inputs to this AND gate are $A$ and $\bar{B}$, and the boolean expression is $x_1 = A\bar{B}$.

Similarly, state 2 is the only time when $B$ is high and $C$ is low. The inputs to the gate to decode 2 are $B$ and $\bar{C}$, and the appropriate boolean expression is $x_2 = B\bar{C}$. The logic equations for the remaining states are found in a similar manner. These equations, along with appropriate decoding gates, are given in Fig. 11–39.

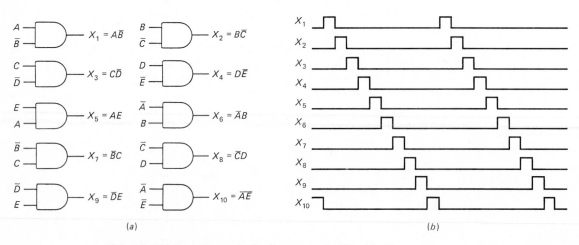

(a)  (b)

**FIG. 11–39**  Decoding gates for the counter in Fig. 11–38

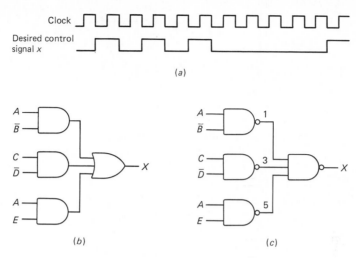

FIG. 11–40

Thus, another of the advantages of using a shift counter is the fact that it requires only 2-input gates to decode any of the individual counter states.

☐ EXAMPLE 11–15

The decade counter in Fig. 11–38 has been constructed, and its outputs are available. It is necessary to develop the waveform shown in Fig. 11–40 to be used as a control signal. Show the logic necessary to provide this signal.

☐ SOLUTION

The desired control signal corresponds to a high level during states 1, 3, and 5 and is low at all other times. The most obvious solution is to use the three AND gates shown in Fig. 11–39 for decoding gates. If the outputs of these three gates are used as the inputs to an OR gate, as shown in Fig. 11–40b, the control signal $X$ will be produced at the output of the OR gate.

A slightly different method for developing the desired waveform $X$ is shown in Fig. 11–40c. The outputs of NAND gates 1, 3, and 5 will be low only during states 1, 3, and 5, respectively. Whenever 1, 3, or 5 is low, the output $X$ must be high, and thus the desired control signal is developed.

# 11·9 A DIGITAL CLOCK

A very interesting application of counters and decoding arises in the design of a digital clock. Suppose that we want to construct an ordinary clock which will display hours, minutes, and seconds. The power supply for this system is the usual 60-Hz, 120-V ac commercial power. Since the 60-Hz frequency of most power systems is very closely controlled, it is possible to use this signal as the basic clock frequency for our system.

In order to obtain pulses occurring at a rate of one cycle per second $(1\text{-s}^{-1})$, it is necessary to divide the 60-Hz power source by 60. If the resulting 1-Hz waveform is again

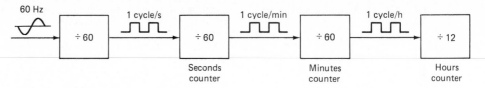

**FIG. 11–41** Digital clock block diagram

divided by 60, a 1-min$^{-1}$ waveform is the result. Dividing this signal by 60 then provides a 1-h$^{-1}$ waveform. This, then, is the basic idea to be used in forming a digital clock.

A block diagram showing the functions to be performed is given in Fig. 11–41. The first divide-by-60 counter simply divides the 60-Hz power signal down to a 1-Hz square wave. The second divide-by-60 counter changes state once each second and has 60 discrete states and can, therefore, be decoded to provide signals to display seconds. This counter is then referred to as the *seconds counter*.

The third divide-by-60 counter changes state once each minute and has 60 discrete states. It can thus be decoded to provide the necessary signals to display minutes. This counter is then the minutes counter.

The last counter changes state once each 60 minutes (once each hour). Thus, if it is a divide-by-12 counter, it will have 12 states that can be decoded to provide signals to display the correct hour. This, then, is the hours counter.

As you know, there are a number of ways to implement a counter. What is desired here is to design the counters in such a way as to minimize the hardware required. The first counter must divide by 60, and it need not be decoded. Therefore, it should be constructed in the easiest manner with the minimum number of flip-flops.

For instance, the divide-by-60 counter could be implemented by cascading counters ($12 \times 5 = 60$, or $10 \times 6 = 60$, etc.). The TTL MSI 7490 decade counter can be used as a divide-by-10 counter, and the TTL MSI 7492 can be used as a divide-by-6 counter. Cascading these two will provide a divide-by-60 counter as shown in Fig. 11–42. The amplifier at the input provides a 60-Hz square wave of the proper amplitude to drive the 7490. The 7492 is connected as a divide-by-12 counter, but only outputs $Q_A$, $Q_B$, and $Q_C$ are used. In this fashion, the 7492 operates essentially as a divide-by-6 counter.

The seconds counter in the system also divides by 60 and could be implemented in the same way. However, the seconds counter must be decoded. We are interested in decoding this counter to represent each of the 60 s in 1 min. This can most easily be accomplished by constructing a mod-10 counter in series with a mod-6 counter for the divide-by-60 counter. The mod-10 counter can then be decoded to represent the units digit of seconds, and the mod-6 counter can be decoded to represent the tens digits of seconds.

Since both the 7490 and the 7492 count in straight 8421 binary, a 7447 decoder-driver can be used with each to drive two 7-segment indicators, as shown in Fig. 11–43.

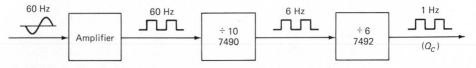

**FIG. 11–42**

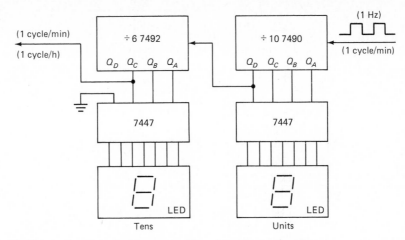

**FIG. 11–43** A 10 × 6 mod-60 counter with units and tens decoding

Notice that the 7492 is connected as a divide-by-12 counter, but only outputs $Q_A$, $Q_B$, and $Q_C$ are used to drive the 7447 decoder-driver.

The minutes counter is exactly the same as the seconds counter, except that it is driven by the 1-min$^{-1}$ square wave from the output of the seconds counter, and its output is a 1-h$^{-1}$ square wave, as shown in Fig. 11–43.

The divide-by-12 hours counter must be decoded into 12 states to display hours. This can be accomplished by connecting a mod-10 (54/74160) decade counter in series with a single flip-flop $E$ as shown in Fig. 11–44. This forms a divide-by-20 (10 × 2 = 20) counter. Feedback is then used to form a mod-12 counter.

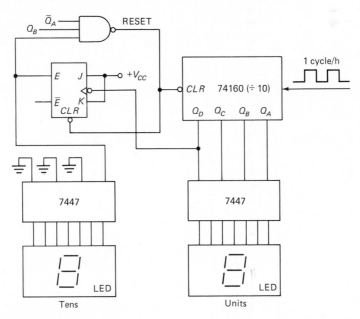

**FIG. 11–44** Mod-12 hours counter

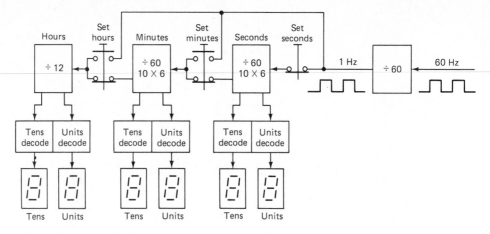

**FIG. 11–45**  Digital clock

The hours counter must count through states 00, 01, 02, $\cdots$, 11, and then back to 00. The NAND gate in Fig. 11–44 will go low as the counter progresses from count 11 to count 12, and this will immediately clear the 74160 to 0000 and reset the flip-flop $E$ to 0. The counter actually skips from count 11 to count 00 omitting the eight counts in between. This is the mod-12 hours counter; the 74160 will provide the units of hours while the flip-flop will provide the tens of hours. Notice that the 74160 is reset asynchronously and there might then be glitches at the outputs of the decoding gates. However, this is one case where these glitches will have no effect, since they are too narrow to cause a visible indication on the light-emitting diodes (LEDs).

Finally, some means must be found to set the clock because the flip-flops will assume random states when the power is turned off and then turned back on again. Setting the clock can be quite easily accomplished by means of the SET push-buttons shown in Fig. 11–45. Depressing the SET HOURS button causes the hours counter to advance at a $1\text{-s}^{-1}$ rate, and thus this counter can be set to the desired hour. The minutes counter can be similarly set by depression of the SET MINUTES button. Depression of the SET SECONDS button removes the signal from the seconds counter, and the clock can thus be brought into synchronization.

By means of large-scale integration (LSI), it is possible to construct a digital clock entirely on one semiconductor chip. Such units are commercially available, and they perform essentially the function shown in the logic diagram in Fig. 11–45 (the seven-segment indicators are, of course, separate). The National Semiconductor 5318 is one such commercially available LSI digital clock. It is available in a 24-pin dual in-line (DIP) package measuring $0.54 \times 1.25$ in.

## S U M M A R Y

A counter has a natural count of $2^n$, where $n$ is the number of flip-flops in the counter. Counters of any modulus can be constructed by incorporating logic which causes certain states to be skipped over or omitted. One technique for skipping counts is to steer the clock pulses to certain flip-flops at the proper time—this is called steering logic. A second technique is to precondition the logic inputs to each flip-flop in order to omit certain states. This is called look-ahead logic.

Logic can be included such that the counter can operate in either a count-up or count-down mode. Furthermore, logic gates can be designed to uniquely decode each state of a counter.

Higher-modulus counters can be easily constructed by using combinations of lower-modulus counters. Such configurations represent a compromise between speed and hardware count.

Cross-coupled feedback around a basic shift register leads to the formation of the shift counter. Shift counters of any modulus can be formed by taking the feedback from the proper flip-flops. The shift counter has the great advantage of simplified decoding. This type of counter does, however, have undesired states, and these must be provided for in the counter design.

The digital clock is an interesting application that illustrates some of the methods employing counters and decoders.

# GLOSSARY

*Decoding Gate* A logic gate whose output is high (or low) only during one of the unique states of a counter.

*Glitch* An undesired positive or negative pulse appearing at the output of a logic gate.

*Modulus* Defines the number of states through which a counter can progress.

*Natural Count* The maximum number of states through which a counter can progress. Given by $2^n$, where $n$ is the number of flip-flops in the counter.

*Parallel Counter* A synchronous counter in which all flip-flops change states simultaneously since all clock inputs are driven by the same clock.

*Presettable Counter* A counter incorporating logic such that it can be preset to any desired state.

*Ripple Counter* An asynchronous counter in which each flip-flop is triggered by the output of the previous flip-flop.

*Shift Counter* A basic shift register with inverse feedback such that a cyclic counter is formed.

*Up-Down Counter* A basic counter, synchronous or asynchronous, that is capable of counting in either an upward or a downward direction.

# PROBLEMS

*SECTION 11-1:*

**11-1.** Draw the logic diagram, truth table, and waveforms for a two-flip-flop ripple counter similar to that in Fig. 11-1.

**11–2.** Draw the logic diagram, truth table, and waveforms for a three-flip-flop ripple counter that uses $JK$ master-slave flip-flops sensitive to a positive clock transition.

**11–3.** What is the clock frequency if the period of $B$ in Fig. 11–1 is 1000 ns?

**11–4.** Determine the number of possible states in a counter composed of the following number of flip-flops:

   **a.** 7       **c.** 8

   **b.** 10

**11–5.** See if you can draw the waveforms for a 10-flip-flop ripple counter. What difficulties do you encounter?

**11–6.** What is the largest decimal number that can be stored in each counter in Prob. 11–4?

**11–7.** Draw the waveforms at $Q_B$, $Q_C$, and $Q_D$ for a 74L93, assuming that a 1-MHz clock is applied at input $B$.

**11–8.** Draw the logic diagram, truth table, and waveforms for a two-flip-flop ripple counter operating in the count-down mode.

*SECTION 11-2:*

**11–9.** Draw the gates necessary to decode the 16 states of a 74L93 operating as in Fig. 11–3.

**11–10.** Assume that the clock for the ripple counter in Fig. 11–1 is a 1-MHz square wave and each flip-flop has a delay time of 0.25 $\mu$s. Carefully draw the waveforms for the clock and each flip-flop and the output decoded signals. Do you see any sources of difficulty?

**11–11.** Use the waveforms in Fig. 11–9 and study the remaining seven decoding gates in Fig. 11–7. Show whether glitches will appear by drawing the decoded waveform for each gate.

*SECTION 11-3:*

**11–12.** Draw the logic diagram, truth table, and waveforms for the synchronous counter in Fig. 11–13 in the count-up mode.

**11–13.** Repeat Prob. 11–12, but in the count-down mode.

**11–14.** Write a boolean expression for the AND gate connected to the upper leg of the OR gate that drives the clock input to flip-flop $Q_D$ in a 74193.

**11–15.** Draw a complete set of waveforms for the 74191 in Fig. 11–15 operating in the count-up mode.

**11–16.** Repeat Prob. 11–15, but operating in the count-down mode.

*SECTION 11-4:*

**11–17.** Determine the number of flip-flops that would be required to build the following counters:

   **a.** Mod-6       **d.** Mod-19

   **b.** Mod-11      **e.** Mod-31

   **c.** Mod-15

**11–18.** Draw decoding gates and all waveforms for the mod-3 counter in Fig. 11–16.

**11–19.** Draw decoding gates and all waveforms for the mod-6 counter in Fig. 11–17.

**11–20.** Draw decoding gates and all waveforms for the counter in Fig. 11–18.

**11–21.** Draw the logic diagram, truth table, and waveforms for a mod-9 counter using two mod-3 counters connected in series.

*SECTION 11-5*

**11–22.** Draw decoding gates and all waveforms for the decade counter in Fig. 11–21.

**11–23.** Draw decoding gates and all waveforms for the counter in Fig. 11–22.

**11–24.** Draw waveforms for $Q_B$, $Q_C$, and $Q_D$, assuming that the clock is applied to input $B$ of a 7490A.

**11–25.** Show how an AND gate might be used in Fig. 11–24 to count an unknown number of pulses that occur during a known time interval. This is the basic idea used in a frequency counter.

*SECTION 11-6:*

**11–26.** Draw the logic block for a 74163 and show how to construct a mod-13 counter. Use the same technique as in Fig. 11–26. Draw the state diagram.

**11–27.** Repeat Prob. 11–26 for a mod-11 counter and then a mod-7 counter.

**11–28.** Draw the waveforms expected in Prob. 11–26.

**11–29.** Draw the logic block for a 74162 and show how to construct a mod-7 counter. Use the same technique as in Fig. 11–26. Draw the state diagram.

**11–30.** Use a 74193 presettable counter to implement a mod-8 counter. List the omitted states and normal count sequence; draw a complete logic diagram. Draw the set of waveforms you would expect, showing the clock and the four outputs. Remember that the output transitions occur on positive clock transitions.

*SECTION 11-7:*

**11–31.** Estimate the number of flip-flops required to construct shift counters of the following moduli:

    **a.** 5        **d.** 10
    **b.** 8        **e.** 21
    **c.** 9

**11–32.** How many inputs would be required on each gate used to decode a seven-flip-flop shift counter? Draw the gate to decode state 000 0000.

**11–33.** Draw a two-flip-flop shift counter and its waveforms. Make a truth table and check for any illegal states.

**11–34.** Will the counter in Fig. 11–35 be forced into the proper mode if $\overline{A}$ is omitted from the NAND gate in Fig. 11–34?

**11–35.** In the worst case, how many clock periods will be required for the counter in Fig. 11–35 to reenter the proper count sequence?

**11–36.** Will the counter in Fig. 11–35 be forced into the proper mode if $\overline{C}$ is omitted from the NAND gate in Fig. 11–34?

*SECTION 11-8:*

**11–37.** List the 22 illegal states for the decade counter in Fig. 11–38, and verify the two illegal count sequences. Also verify the results of the counter existing in either count 10 or count 21.

**11–38.** Verify that the NAND gate in Fig. 11–34 will force the decade counter in Fig. 11–38 into the desired mode of operation. On what illegal counts will the counter be corrected? How many clock pulses are required in the worst case to get the counter back into the proper sequence? Is the $\overline{C}$ input to the NAND gate necessary?

**11–39.** Draw the logic diagram, waveforms, and truth table to form a mod-9 counter out of the decade counter shown in Fig. 11–38.

**11–40.** Do the complete design for a mod-7 shift counter. Draw the waveforms and truth table; list the illegal states. Check the operation of the counter if it were to appear in any of the illegal states. Design a cure to place the counter back into the proper count sequence if one is needed.

**11–41.** Demonstrate that the mod-5 counter in Fig. 11–37 will always operate in the desired mode. Do the same for a mod-3 shift counter.

**11–42.** Draw a 2-input AND gate to decode state 3 of the counter in Fig. 11–37.

**11–43.** Draw the 2-input AND gates to decode states 1, 3, and 5 of the counter in Prob. 11–40.

# 12

# SEMICONDUCTOR MEMORIES

Semiconductor MSI and LSI, along with magnetic recording on tapes and disks, are the most widely used methods for storing digital information. The semiconductor memory has replaced older devices such as magnetic cores and is so important in digital systems that we devote this entire chapter to the different types of semiconductor memories and their operating characteristics. The smaller memory chips are discussed first because the basics are easiest to understand with simpler configurations. These fundamental ideas are then applied to larger and more complicated memory devices.

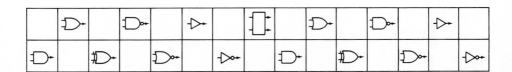

# 12·1  BASICS

Recent advances in semiconductor technology have provided a number of reliable and economical MSI and LSI memory circuits. The typical semiconductor memory consists of a rectangular array of memory cells, fabricated on a silicon wafer, and housed in a convenient package, such as a DIP. The basic memory cell is typically a transistor flip-flop or a circuit capable of storing charge and is used to store 1 bit of information. Memories are usually classified as either bipolar, metal oxide semiconductor (MOS), or complementary metal oxide semiconductor (CMOS) according to the type of transistor used to construct the individual memory cells. The total number of cells in a memory determine its capacity. For instance, a 1024 bipolar memory chip is a semiconductor memory that has 1024 memory cells, each cell consisting of a flip-flop constructed with the use of bipolar transistors. *Chip* is a term commonly used to refer to a semiconductor memory device.

In general, faster operation is obtained with a bipolar memory chip, but greater packing density and thus reduced size and cost, as well as lower power requirements, are characteristics of MOS and CMOS memory chips. Memories can generally be divided into the following categories—random-access, read-write memories (RAM), and read-only memories (ROM).

An application in which data changes frequently calls for the use of a RAM. The logic circuitry associated with a RAM will allow a single bit of information to be stored in any of the memory cells—this is the write operation. There is also logic circuitry that will detect whether a 0 or a 1 is stored in any particular cell—this is the read operation. The fact that a bit can be written (stored) in any cell or read (detected) from any cell suggests the description *random-access*. A control signal, usually called *WRITE ENABLE*, defines the mode of operation (read or write). In the read mode, data from the selected memory cell is made available at the output. In the write mode, information at the data input is written into (stored in) the selected cell. Since each cell is a flip-flop, a loss of power means a loss of data—a RAM that has this type of memory cell is said to provide ''volatile'' storage.

An application in which the data does not change dictates the use of a ROM. For instance, a ''lookup table'' that stores the values of mathematical constants such as trigonometric functions or a fixed program such as that used to find the square root of a number could be stored in a ROM. The content of a ROM is fixed during manufacturing, perhaps by metallization or by the presence or absence of a working transistor in a memory cell, by opening or shorting of the gate structure, or by the oxide-layer thickness. A ROM is still random-access, since there is logic circuitry to select any desired cell in the memory. In the read mode, data from the selected cell is made available at the output. There is, of course, no write mode. Since data is permanently stored in each cell, a loss of power does not cause a loss of data, and thus a ROM provides ''nonvolatile'' data storage.

An application in which the data does not change but the required data will not be available until a later time suggests the use of a *programmable* ROM (PROM). Operation is essentially the same as that of a ROM, but the stored data can be set in the memory by writing into the PROM at the user's convenience. However, it can be programmed only once!

An application in which the data may change from time to time might call for the use of an *erasable* PROM (EPROM). The data can be programmed into the EPROM and can then be erased and reprogrammed if desired.

☐ EXAMPLE 12–1

State the most likely type of semiconductor memory for each application: (*a*) main memory in a hand calculator; (*b*) storing values of logarithms; (*c*) storing prices of vegetable produce; (*d*) emergency stop procedures for an industrial mill now in the design stage.

☐ SOLUTION

(*a*) RAM; (*b*) ROM; (*c*) EPROM; (*d*) PROM.

# 12-2 MEMORY ADDRESSING

Addressing is the process of selecting one of the cells in a memory to be written into or to be read from. In order to facilitate selection, memories are generally arranged by placing cells in a rectangular arrangement of rows and columns as shown in Fig. 12–1*a*. In this particular case, there are $m$ rows and $n$ columns, for a total of $n \times m$ cells in the memory.

The control circuitry that accompanies the basic memory array is designed such that if one and only one row line is activated and one and only one column line is activated, the memory cell at the intersection of these two lines is selected. For instance, in Fig. 12–1*b*, if row $A$ is activated and column $B$ is activated, the cell at the intersection of this row and column is selected—that is, it can be read from or written into. For convenience, this cell is then called $AB$, corresponding to the row and the column selected. This designation is defined as the ''address'' of the cell. The activation of a line (row or column) is achieved by placing a logic 1 (or perhaps a logic 0) on it. Exactly how this action selects a cell is discussed in the last section of this chapter.

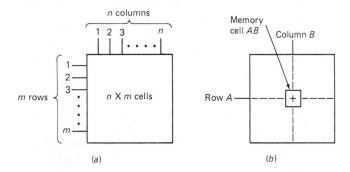

FIG. 12–1 (*a*) A rectangular array of $m \times n$ cells (*b*) Selecting the cell at memory address $AB$

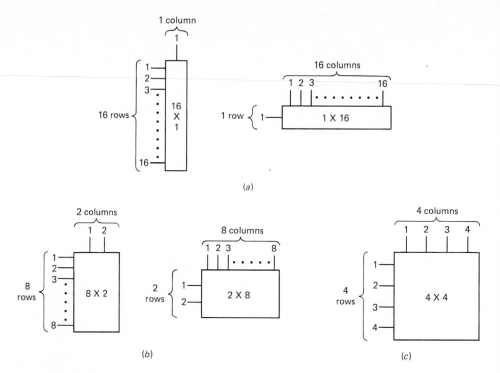

FIG. 12–2

Let's take a little time to consider the various possible configurations for a rectangular array of memory cells. The different rectangular arrays of 16 cells are shown in Fig. 12–2. In each of the five cases given, there are exactly 16 cells. The 16 × 1 and the 1 × 16 arrangements in Fig. 12–2a are really equivalent; likewise, the 8 × 2 and the 2 × 8 are essentially the same. So, there are really only three different configurations, each of which contain the exact same number of cells.

For any of the three configurations, the selection of a single cell still requires a single row and a single column to define a unique address. In Fig. 12–2a, a total of 17 address lines must be used—16 rows and 1 column, or 1 row and 16 columns. The minimum requirement in either case is really only 16 lines. However, either arrangement in Fig. 12–2b requires only 10 address lines—8 rows and 2 columns, or 2 rows and 8 columns. Clearly the best arrangement is given in Fig. 12–2c, since this configuration only requires 8 address lines—4 rows and 4 columns!

In general, the arrangement that requires the fewest address lines is a square array of $n$ rows and $n$ columns for a total memory capacity of $n \times n = n^2$ cells. It is exactly for this reason that the square configuration is so widely used in industry. This arrangement of $n$ rows and $n$ columns is frequently referred to as *matrix addressing*. In contrast, a single column that has $n$ rows (such as the 16 × 1 array of cells) is frequently called *linear addressing*, since selection of a cell simply means selection of the corresponding row, and the column is always used.

For instance, a 54/7481A is a 16-bit bipolar RAM, arranged in a 4 × 4 array as shown in Fig. 12–3. There are four X-address lines (rows $X_1$, $X_2$, $X_3$, and $X_4$), and four

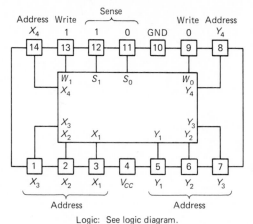

J or N dual-in-line or W flat package (top view)

SN5481A/SN7481A Circuits

SN5484A/SN7484A Circuits

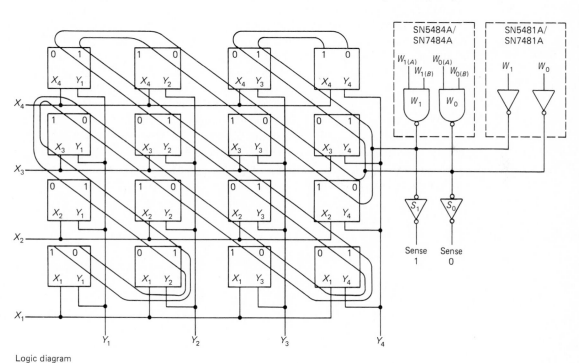

Logic diagram

**FIG. 12–3** Texas Instruments SN5481A

Y-address lines (columns $Y_1$, $Y_2$, $Y_3$, and $Y_4$). Selection of a cell is accomplished by placing a high level on one X line and on one Y line. For instance, $X_1 = 1$ and $Y_4 = 1$ will select the cell in the bottom row and the right-hand column. (All other address lines must be low.) The address of this cell is $X_1Y_4$.

Take another look at the $4 \times 4$ memory in Fig. 12–2c. To select a single cell, we must activate one and only one row, and one and only one column. This suggests the use of two 1 of 4 binary to decimal decoders as shown in Fig. 12–4. Consider the selection of the cell at address 43 (row 4 and column 3). If $A_4 = 1$ and $A_3 = 1$, the decoder will hold the row 4 line high while all other row lines will be low. Similarly, if $A_2 = 1$ and $A_1 = 0$, the decoder will hold column 3 high and all other column lines low. Thus an input $A_4A_3A_2A_1 = 1110$ will select cell 43. We can consider $A_4A_3$ as a row address of 2 bits and $A_2A_1$ as a column address of 2 bits. Taken together, any cell in the array can be uniquely specified by the 4-bit address $A_4A_3A_2A_1$. As another example, the address $A_4A_3A_2A_1 = 0110$ selects the cell at row 2 and column 3 (address 23).

The address decoders shown in Fig. 12–4 further reduce the number of address lines needed to uniquely locate a memory cell, and they are almost always included on the memory chip. Recall that a binary-to-decimal decoder having $n$ binary inputs will select one of $2^n$ output lines. For instance, a decoder that has 3 binary inputs will have $2^3 = 8$ outputs, or a decoder having 4 inputs will have 16 outputs, and so on.

In general, an address of $B$ bits can be used to define a square memory of $2^B$ cells, where there are $B/2$ bits for the rows and $B/2$ bits for the columns, as shown in Fig. 12–5. Notice that the total number of address bits $B$ must be an even integer (2, 4, 6, 8, . . . ). Since the input to each decoder is $B/2$ bits, the output of each decoder must be $2^{B/2}$ lines. So the capacity of the memory must be $2^{B/2} \times 2^{B/2} = 2^B$. For instance, an address of 12 bits can be used in this way for a memory that has $2^{12} = 4096$ bits. There will be 6 address bits providing $2^6 = 64$ rows and likewise 6 address bits providing 64 columns. The memory will then be arranged as a square array of $64 \times 64 = 4096$ memory cells.

You may have noticed that most commercially available memories have capacities like 1024, 2048, 4096, 16,384, and so on. The reason for this is now clear—all of these numbers are clearly integer powers of 2! Incidentally, a memory having 1024 bits is usually referred to as a 1K memory (1000 bits) simply for convenience. Similarly, a memory advertised as 16K really has 16,384 bits, 4K is really 4096, and so on.

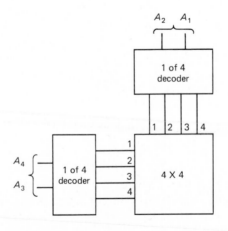

**FIG. 12–4**

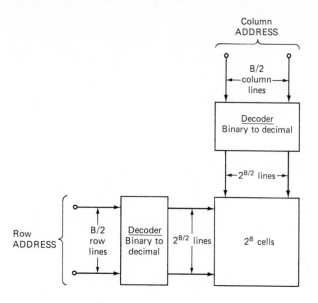

Column
ADDRESS

B/2
←column→
lines

Decoder
Binary to decimal

←$2^{B/2}$ lines→

Row
ADDRESS

B/2
row
lines

Decoder
Binary to
decimal

$2^{B/2}$ lines

$2^B$ cells

**FIG. 12–5**

☐ EXAMPLE 12–2

What would be the structure of the binary address for a memory system having a capacity of 1024 bits?

☐ SOLUTION

Since $2^{10} = 1024$, there would have to be 10 bits in the address word. The first 5 bits could be used to designate one of the required 32 rows, and the second 5 bits could be used to designate one of the required 32 columns. Notice that $32 \times 32 = 1024$.

☐ EXAMPLE 12–3

For the memory system described in the previous example, what is the decimal address for the binary address 10110 01101? What is the address in hexadecimal?

☐ SOLUTION

The first 5 bits are the row address. Thus row = 10110 = 22. The second 5 bits are the column address. So, column = 13. The decimal address is thus 22 13. In hexadecimal, this same address is 16 0D.

So far, we have only discussed memories that provide access to a single cell or bit at a time. It is often advantageous to access groups of bits—particularly groups of 4 bits (a nibble) and groups of 8 bits (a byte). It is not difficult to extend our discussion here to accommodate such requirements. There are at least two popular methods. The first simply accesses groups of cells on the same memory chip, and we discuss this idea next. The second connects memory chips in parallel, and we consider this technique in a following section.

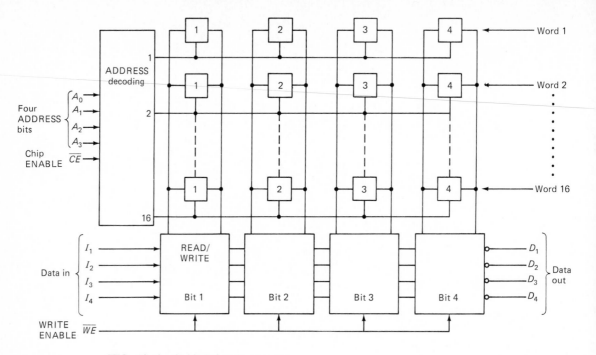

**FIG. 12–6**   64-bit (16 × 4) memory

The logic diagram for a 64-bit (16 × 4) bipolar memory is given in Fig. 12–6. There are 16 rows of cells with four cells in each row; thus the description (16 × 4). Each cell is a bipolar junction transistor flip-flop. The address decoder has 4 address bits and thus 16 select lines—one for each row. In this case, each select line is connected to all four of the cells in a row. So, each select line will now select four cells at a time. Therefore, each select line will select a 4-bit word (a nibble), rather than a single cell.

You might think of this arrangement as a "stack" of sixteen 4-bit registers. This is really a form of linear addressing, since the 4 address bits, when decoded, select one of the sixteen 4-bit registers. In any case, when data is read from this memory it appears at the four data output lines $D_1$, $D_2$, $D_3$, and $D_4$ as a 4-bit data word. Similarly, data is presented to the memory for storage as a 4-bit data word at input lines $I_1$, $I_2$, $I_3$, and $I_4$. The 54/74S89 and the 54/74S189 both are 64-bit (16 × 4) bipolar scratch pad memories arranged in exactly this configuration (look ahead in Fig. 12–13). The idea is easily extended to memories that access a word of 8 bits (a byte) at a time—for instance, the 54/7488A ROM discussed in the next section.

## 12-3  ROMS, PROMS, AND EPROMS

Having gained a clear understanding of memory addressing, let's turn our attention to the operation of a ROM. The 54/7488A is a 256-bit (32 × 8) ROM arranged as a stack of thirty-two 8-bit words. As shown in Fig. 12–7, the 5 row address bits are labeled *ABCDE*, and the 8 output bits in a word are labeled $Y_1Y_2Y_3Y_4Y_5Y_6Y_7Y_8$.

Types SN5488A, SN7488A 256-bit read-only memories

Functional block diagram and word selection

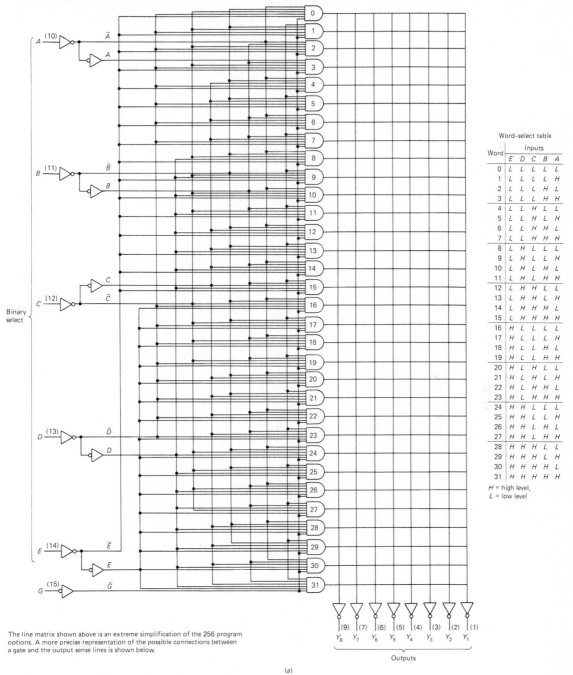

(a)

**FIG. 12-7** 5488A, 256-bit ROM (32 × 8)

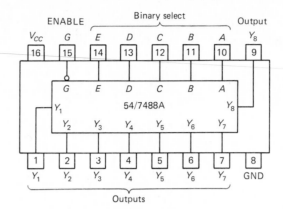

J or N dual-in-line or W flat package (top view)

Positive logic: See description.

(b)

**FIG. 12–7** (continued)

Input G is used to enable or disable the entire set of 32 input decoding gates. When G is high, all the address decoding gates are inhibited and the memory chip is disabled, causing the eight output data bit lines to be high. When G is low, the data at the outputs will correspond to the 8-bit word in memory selected by the input address. On most memory chips there is an input line that performs the same function as G, and it is usually referred to as the *CHIP-ENABLE* or *CHIP-SELECT* line.

Any pattern of 1s and 0s can be stored in the 8 bits of each of the 32 words. The desired pattern must be decided on before purchase and exact specifications given to the manufacturer. The manufacturer will then program the memory before delivery to the customer. Once the chip is programmed, its contents cannot be changed. Complete details for specifying memory content and ordering a chip are given on each individual manufacturer's data sheets. When the programming of the chip is to be done by the manufacturer, the chip is defined as being *mask-programmable*.

Using the 54/7488A ROM is relatively simple. First, since the logic circuits are TTL, a supply voltage and ground connections must be made. The data sheet calls for a nominal supply voltage of $+V_{CC} = 5.0$ V dc on pin 16, with ground connected to pin 8. The inputs and outputs are all TTL-compatible.

The eight data outputs are *open-collector* transistors, so a pull-up resistor is required at each data output—pins 1 through 7 and pin 9. Typically a resistance between 5.1 k$\Omega$ and 510 $\Omega$ is connected from each output pin up to $+V_{CC}$.

Now, all that is required is to apply the correct input address to read a desired 8-bit word and then take the G input line (select line) low. The 54/7488A data sheet states an access time $t_p$ of 25 ns (typical) and 45 ns (maximum). So, an 8-bit data word will be available at the outputs $Y_1 \cdots Y_8$ within 45 ns after the falling edge of G, as shown in Fig. 12–8a. The address lines should, of course, be stable during the time data is being read out of the memory. There are two output lines in Fig. 12–8a showing that a data line may

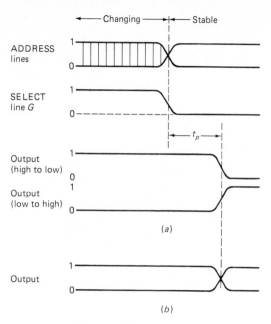

*Changing* ← → *Stable*

ADDRESS
lines

SELECT
line $G$

$t_p$

Output
(high to low)

Output
(low to high)

(a)

Output

(b)

FIG. 12–8   Access time, $t_p$, for a 54/7488A ROM

transition low to high, or high to low. To save time and space, this idea is usually conveyed in a single waveform as seen in Fig. 12–8b—this single waveform is the equivalent of the two output waveforms above it in Fig. 12–8a.

The logic diagram for a 54/74186 is shown in Fig. 12–9. This chip is a TTL LSI (large-scale integrated circuit) and is a 512-bit PROM organized as 64 words of 8 bits each. The 6 bits of address (*ABCDEF*) are decoded to uniquely select one of the 64, 8-bit words in a linear addressing scheme similar to that of the 54/7488A. The unique thing about this memory chip is that it is *field-programmable* by the user.

As it comes from the manufacturer, the chip is stored full of 0s—that is to say, the contents of any 8-bit word is 0000 0000. *Field programming* the chip means to determine the contents desired for each word in the memory and then storing a 1 in those positions requiring a 1. This is done by altering that bit position electrically. For instance, if the desired contents of word 27 is 0101 1100, bit positions 3, 4, 5, and 7 will be altered electrically to change them from a permanent low state (a 0) to a permanent high state (a 1). Once the field programming is done, it is permanent!

Basically, the programming is done by applying a current pulse to each output terminal where a logic 1 must appear. So, in Fig. 12–9b, the chip is programmed by:

1. Applying the correct address (*ABCDEF*) for the word to be programmed. For word 27, the address would be 0001 1011. A 1 is an open switch, and a 0 is a closed switch.
2. Applying a current pulse to each bit that must store a 1. To store 0101 1100, requires current pulses, one at a time, at data positions $Y_3$, $Y_4$, $Y_5$, and $Y_7$.
3. Repeating the above steps for all words to be stored in the memory.

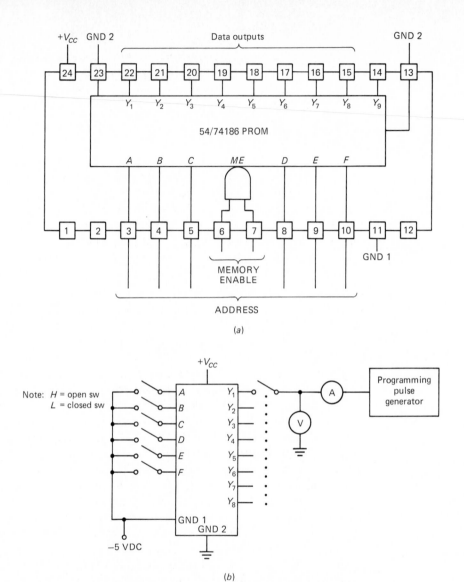

FIG. 12-9 (a) Logic diagram and pin out (b) Programming the 54/74186

Specific information for chip programming must be taken from the manufacturer's data sheets—current pulse limits, voltage levels, and so on.

Once programmed, the chip is used in basically the same fashion as that in the previously discussed 54/7488A. Here there are two memory-enable lines, either of which will inhibit the chip if held low. In other words, to enable the chip, both memory-enable lines must be high. The data sheet advertises typical access times of 50 ns and full TTL compatibility.

## EXAMPLE 12—4

The TTL LSI 54/74187 is advertised as a 1024-bit ROM. Since $2^{10} = 1024$, it would seem to require 10 address bits, but the data sheet shows only 8 bits of address. Can you explain how the memory on this chip must be organized by looking at the data sheet in Fig. 12–10?

## SOLUTION

There are 4 bits appearing at the output of the chip, so it must be organized as 256 words of 4 bits each ($256 \times 4 = 1024$). The 1024 memory cells are arranged in a square consisting of 32 rows and 32 columns. Five of the address bits (*DEFGH*) are used to select one of the 32 rows ($2^5 = 32$). The 32 columns are divided into eight groups of 4 bits each. So, it is only necessary to select one of the eight groups, and this can be done with three address lines (*ABC*), since $2^3 = 8$. As an example, the address *HGFEDCBA* = 10110 110 will select row 10110 = 22 and the 4 bits (four columns) in group 110 = 6.

One disadvantage of a PROM is that once it is programmed, the contents are stored in that memory chip permanently—it can't be changed; a mistake in programming the chip can't be corrected. The EPROM overcomes this difficulty.

The EPROM has a structure and addressing scheme similar to those of the previously discussed PROM, but it is constructed using MOS devices rather than bipolar devices. Many MOS EPROMs are TTL-compatible, and even the technique used to program the chip is similar to that used with a bipolar memory. The only difference is really the mechanism for permanently storing a 1 or a 0 in an MOS memory cell.

The current pulse used to store a 1 when programming a bipolar PROM is used to destroy ("burn out") a connection on the chip. The same technique is used to program an MOS-type EPROM, but the current pulse is now applied for a period of time (usually a few milliseconds) in order to store a fixed charge in that particular memory cell. This stored charge will cause the cell to store a logic 1.

The interesting thing about this phenomenon is that the charge can be removed (or erased), and the cell will now contain a logic 0! Furthermore, the process can be repeated. The memory cells are "erased" by shining an ultraviolet light through a quartz window onto the top of the chip. The light bleeds off the charge and all cells will now contain 0s. The requirements for programming and erasing an EPROM vary widely from chip to chip, and data sheet information must be consulted in each individual case.

An example of an MOS EPROM is the widely used Intel 8708. This chip is an 8192-bit erasable PROM (8192 bits arranged as 1024 words, 8 bits each). From the logic diagram in Fig. 12–11, it is clear that the memory is arranged as a rectangular array of 8192 cells having 128 rows and 64 columns. The X decoder uses 7 address bits to select one of the 128 rows, while the Y decoder uses 3 address bits to select one of the eight groups of 8 bits from the 64 columns. The 8 data bits are available at the outputs through three-state buffer amplifiers.

Access time is given as 450 ns, and this is the maximum address to output delay time, $t_{ACC}$, as shown in the waveforms. The maximum chip select-to-output delay time $t_{\overline{CO}}$ is also given on the waveforms, and it has a maximum value of 120 ns. So, to read a data

Functional block diagram and schematics of inputs and outputs

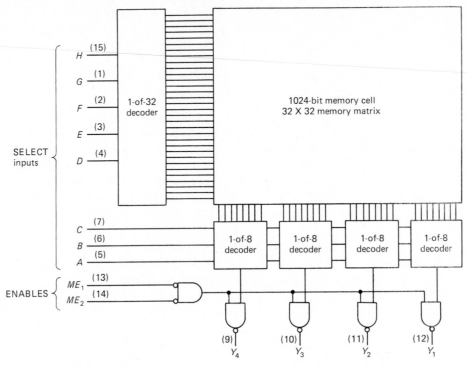

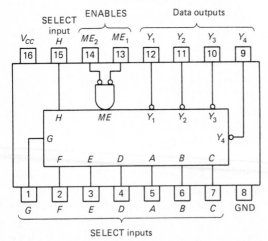

SN54187 . . . J or W package
SN74187 . . . J, N, or W package
(Top view)

Positive logic: See description.

Word selection

Word selection is accomplished in a conventional 8-bit positive-logic binary code with the A SELECT input being the least-significant bit progressing alphabetically through the select inputs to H which is the most-significant bit.

Word–select table

| Word | Inputs | | | | | | | |
|------|--------|---|---|---|---|---|---|---|
|      | H | G | F | E | D | C | B | A |
| 0 | L | L | L | L | L | L | L | L |
| 1 | L | L | L | L | L | L | L | H |
| 2 | L | L | L | L | L | L | H | L |
| 3 | L | L | L | L | L | L | H | H |
| 4 | L | L | L | L | L | H | L | L |
| 5 | L | L | L | L | L | H | L | H |
| 6 | L | L | L | L | L | H | H | L |
| 7 | L | L | L | L | L | H | H | H |
| 8 | L | L | L | L | H | L | L | L |

Words 9 through 250 omitted

| 251 | H | H | H | H | H | L | H | H |
| 252 | H | H | H | H | H | H | L | L |
| 253 | H | H | H | H | H | H | L | H |
| 254 | H | H | H | H | H | H | H | L |
| 255 | H | H | H | H | H | H | H | H |

**FIG. 12–10**   54/74187, 1024 bit ROM

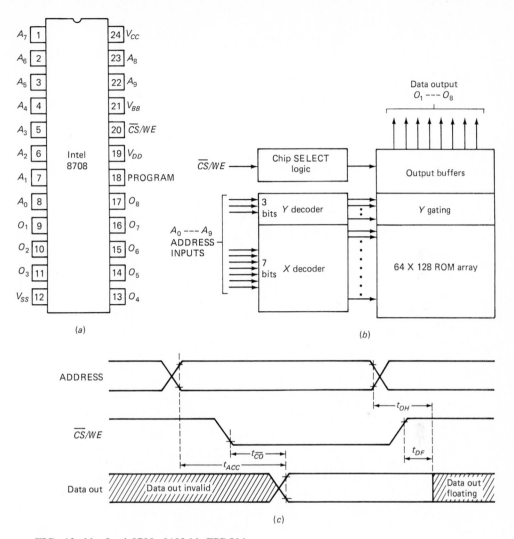

FIG. 12–11   Intel 8708, 8192-bit EPROM

word from the chip, the address lines must stabilize, and then within 120 ns after the CHIP-SELECT line ($\overline{CS}$) goes low, the data will appear at the outputs. A disadvantage with this chip is that it requires +12 Vdc and both +5 Vdc and −5 Vdc power supplies. The Intel 2716, 2732, and 2764 overcome this disadvantage since they require only +5 Vdc.

The Intel 2716 is a 16K (2K × 8) EPROM. There are actually 2048 words, 8 bits each, for a total storage capacity of 16,384 bits. The chip is completely TTL compatible and has an access time of less than 450 ns. The 11 address bits will uniquely select one of the two thousand, forty-eight, 8-bit words ($2^{11} = 2048$), and the selected word will appear at the data output lines if ($\overline{CE}$) and OUTPUT ENABLE ($\overline{OE}$) are low. The Intel 2732

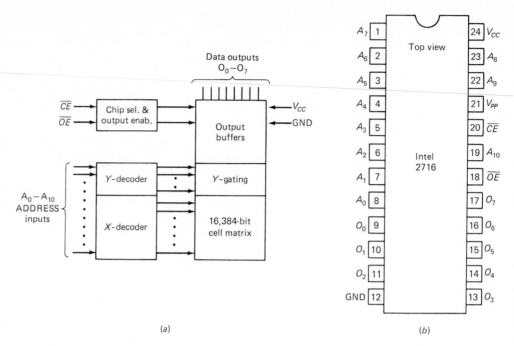

Data outputs $O_0$–$O_7$

$\overline{CE}$ → Chip sel. &
$\overline{OE}$ → output enab.

Output buffers ← $V_{CC}$
← GND

$A_0$–$A_{10}$ ADDRESS inputs

Y-decoder

X-decoder

Y-gating

16,384-bit cell matrix

(a)

| $A_7$ | 1 | Top view | 24 | $V_{CC}$ |
| $A_6$ | 2 | | 23 | $A_8$ |
| $A_5$ | 3 | | 22 | $A_9$ |
| $A_4$ | 4 | | 21 | $V_{PP}$ |
| $A_3$ | 5 | | 20 | $\overline{CE}$ |
| $A_2$ | 6 | Intel 2716 | 19 | $A_{10}$ |
| $A_1$ | 7 | | 18 | $\overline{OE}$ |
| $A_0$ | 8 | | 17 | $O_7$ |
| $O_0$ | 9 | | 16 | $O_6$ |
| $O_1$ | 10 | | 15 | $O_5$ |
| $O_2$ | 11 | | 14 | $O_4$ |
| GND | 12 | | 13 | $O_3$ |

(b)

**FIG. 12–12**   Intel 2716 EPROM

is a 32K (4K × 8) EPROM that is pin-compatible with the 2716—it simply has twice the memory storage. Likewise, the 2764 is a 64K (8K × 8) EPROM. The logic diagram and pinout for the 2716 are given in Fig. 12–12.

As a matter of fact, the logic diagram in Fig. 12–12a is virtually the same for any ROM, PROM, or EPROM. Essentially the only variation is the total number of address inputs to accommodate the number of bits in the cell matrix. As an example, Advanced Micro Devices (AMD) offers a CMOS EPROM with essentially the same logic diagram; it is the AM27512. But this chip has 16 address input lines and a 534,288-bit cell matrix, organized as 65,536 words, each 8 bits in length. This is most impressive when you consider that there are over a half-million bits of information stored in a 28-pin package that measures less than 1.5 in in length and about 0.5 in in width!

## 12·4  RAMS

The basic difference between a RAM and a ROM is that data can be written into (stored in) a RAM at any address as often as desired. Naturally data can be read from any address in either a RAM or a ROM, and the addressing and read cycles for both devices are similar. The characteristics of both bipolar and MOS "static" RAM's are discussed in this section.

A static RAM uses a flip-flop as the basic memory cell (either bipolar or MOS), and once a bit is stored in a flip-flop, it will remain there as long as power is available to the chip—essentially forever—thus the term "static." On the other hand, the basic memory cell in a "dynamic" RAM utilizes stored charge in conjunction with an MOS device to store a bit of information. Since this stored charge will not remain for long periods of time, it must periodically be recharged (refreshed), and thus the term "dynamic" RAM. Both static and dynamic RAMs are "volatile" memory storage devices, since a loss of power supply voltages means a loss of stored data. The dynamic RAM is the topic of the following section.

The 7489 shown in Fig. 12–13 is a TTL, LSI, 64-bit RAM, arranged as 16 words of 4 bits each. Holding the MEMORY-ENABLE (ME) input low will enable the chip for either a read or a write operation, and the four data address lines will select which one of the sixteen 4-bit word positions to read from or write into. Then, if the write-enable (WE) is held low, the 4 bits present at the data inputs ($D_1$, $D_2$, $D_3$, $D_4$) will be stored in the selected address. Conversely, if WE is high, the data currently stored in the memory address will be presented to the four data output lines ($S_1$, $S_2$, $S_3$, $S_4$). Incidentally, the outputs are open-collector transistors, and a pull-up resistor from each output up to $+V_{CC}$ is normally required. The operations for this chip are summarized in the truth table in Fig. 12–13.

The read operation is no different from that for a ROM. For this chip, simply hold ME low and WE high, and select the desired address. The 4-bit data word then appears at the "sense" outputs. The timing for a read operation is shown by the waveforms in Fig. 12–13d. The propagation delay time $t_{PHL}$ is that period of time from the fall of ME until stable data appears at the outputs—the data sheet gives a maximum value of 50 ns, with 33 ns typical. Naturally the address input lines must be stable during the entire read operation, beginning with the fall of ME. Notice carefully that the data appearing at the four sense outputs will be the *complement* of the stored data word!

You will notice from the truth table that when the chip is *deselected*, that is, when ME is high, the outputs all go to a high level, provided we are in a read mode (WE is high). So, in the read operation waveforms, the time $t_{PLH}$ is the delay time from the rise of ME until the outputs assume the high state. The data sheet gives 50 ns maximum and 26 ns typical for this delay time.

During a write operation the 4 bits present at the data inputs will be stored in the selected memory address by holding the ME input low (selecting the chip) and holding the WE low. At the same time, the complement of the data present at the four input lines will appear at the four output lines. Timing waveforms for the write operation are also shown in Fig. 12–13d.

Let's look carefully at the timing requirements for the write cycle. First, the WE must be held low for a minimum period of time in order to store information in the memory cells—this is given as time $t_W$ on the waveforms, and the data sheet calls for 40 ns minimum. Memory-enable selects the chip when low, and is allowed to go low coincident with or before a write operation is called for by WE going low.

Next, the data to be written into memory must be stable at the data inputs for a minimum period of time before WE and, also for a minimum period of time after WE. The time period prior to WE is called the *data-setup time* $t_2$. This time is measured from the end of the write-enable signal back to the point where the data must be stable. The data sheet calls for 40 ns, and in this case it is the same as $t_W$. Also, the data inputs must be held stable for a period of time after WE rises—this is called the *data-hold time* $t_3$, and the data sheet calls for 5 ns minimum.

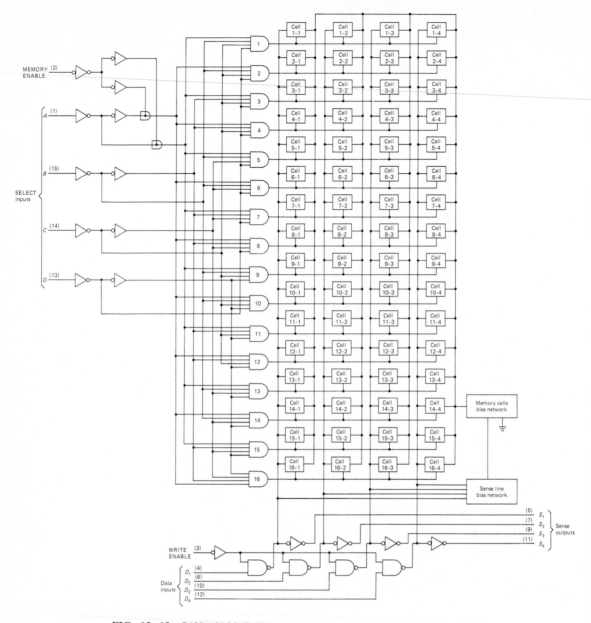

**FIG. 12–13** 7489, 64-bit RAM

The address lines must also be stable for a period of time before as well as after the WE signal. The time period before WE is called the *address-setup* or *select-setup* time $t_4$. This time is measured from the fall of WE back to the point where the input address lines must be stable; the data sheet calls for 0.0 ns minimum. In other words, the address lines are allowed to become stable coincident with or before WE goes low. The address lines

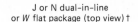

SELECT inputs    Data   Sense   Data   Sense
input output input output

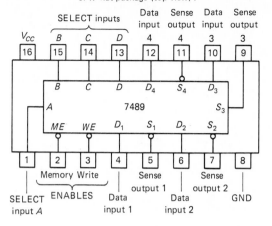

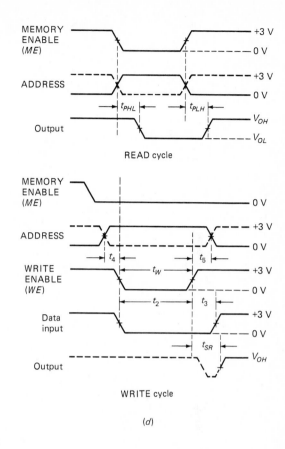

Positive logic: See description

†Pin assignments for these circuits are the same for all packages.

(b)

| ME | WE | Operation | Condition of outputs |
|----|----|-----------|----------------------|
| L | L | Write | Complement of data inputs |
| L | H | Read | Complement of selected word |
| H | L | Inhibit storage | Complement of data inputs |
| H | H | Do nothing | High |

(c)

(d)

**FIG. 12–13** (continued)

must also be stable for a period of time after the rise of WE; this is called the *address-hold* or *select-hold* time $t_5$, and the data sheet calls for 5 ns minimum.

Finally, after a write operation, if the chip is deselected (ME goes high), the outputs will return to a high state. The maximum time for this to occur is the sense-recovery time $t_{SR}$, given as 70 ns maximum on the data sheet.

The operation of a 7489 is straightforward and easy to understand; therefore it is a good chip to study in elementary discussions of RAMs. It can be used to construct memories having larger capacities by connecting chips in parallel, but it's not too practical when we wish to consider memories of 16K, 32K, . . . , 256K, 512K, and so on. Nevertheless, the time spent studying this chip is well invested since the fundamentals of addressing and the read and write operations are essentially the same for all static RAMs. So, with these fundamentals in mind, let's take a look at some chips that have more memory capacity.

The block diagram in Fig. 12–14 can be used to describe the operation of most static RAMs. Most of these are constructed with $n$ address lines that will uniquely select only one of the $2^n$ cells in the memory array—that is, selection is 1 bit at a time. There will be

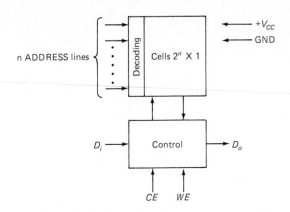

**FIG. 12–14**  Generalized block diagram for a static RAM

a chip-enable control (CE), a write-enable (WE), and a provision for a single input data bit ($D_i$) and a single output data bit ($D_o$).

For instance, the 54/74200 in Fig. 12–15 is a 256-bit RAM, organized as 256 words, each 1 bit in length. The 256 cells are arranged in a square array of 16 rows and 16 columns. The 8 address bits ($2^8 = 256$) are divided into 4 bits that are decoded to select one of the 16 rows and 4 bits that are decoded to select one of the 16 columns. There is a single input data bit, a single output data bit, and a WRITE-ENABLE line (WE). There are three memory enable inputs ($M_1$, $M_2$, $M_3$), and all three of them must be low to select or enable the chip.

The truth table shows that if the chip is enabled, a write cycle is initiated by holding WE low, or a read cycle can be initiated by holding WE high. If any or all of the MEMORY-ENABLE inputs are high, the chip is inhibited and the output goes to a high-impedance state. Naturally the proper timing must be observed as defined by the timing waveforms; you will see that the timing requirements are very similar to the previously described 7489.

### ☐ EXAMPLE 12–5

Using the information in Fig. 12–15, determine how long the address lines for the 54/74200 must be held before WE goes low and after WE goes high.

### ☐ SOLUTION

The setup time, address-to-write-enable is 0.0 ns. The hold time, address-from-write-enable is 10 ns. Therefore, the address lines must be stable from the fall of WE until at least 10 ns after WE rises.

Now that we understand the operation of the 54/74200 (abbreviated as '200), it is a simple matter to use multiple '200 chips to construct larger memories. For instance, we can connect four '200 chips in parallel as shown in Fig. 12–16 to construct a RAM organized as 256 words, each 4 bits in length. Connecting eight '200 chips in parallel will form a memory having 8-bit words, and so on. The nice thing about connecting chips in

Types SN54S200, SN74S200 256-bit READ/WRITE memories with 3-state outputs

Switching characteristics over recommended operating ranges of $T_A$ and $V_{CC}$ (unless otherwise noted)

| Parameter | | Test conditions | SN54S200 | | SN74S200 | | Unit |
|---|---|---|---|---|---|---|---|
| | | | Typ‡ | Max | Typ‡ | Max | |
| $t_{PLH}$ Propagation delay time, low-to-high-level output | Access times from ADDRESS | | 33 | 70 | 33 | 50 | ns |
| $t_{PHL}$ Propagation delay time, high-to-low-level output | | $C_L$ = 15 pF, | 29 | 70 | 29 | 50 | |
| $t_{ZH}$ Output enable time to high level | Access times from memory ENABLE | $R_L$ = 560 Ω (SN54S'), | 21 | 45 | 21 | 35 | ns |
| $t_{ZL}$ Output enable time to low level | | 400 Ω (SN74S'), | 20 | 45 | 20 | 35 | |
| $t_{ZH}$ Output enable time to high level | Sense recovery times from write ENABLE | See figure 1 | 19 | 50 | 19 | 40 | ns |
| $t_{ZL}$ Output enable time to low level | | | 17 | 50 | 17 | 40 | |
| $t_{HZ}$ Output disable time from high level | Disable times from memory ENABLE | $C_L$ = 5 pF, | 7 | 30 | 7 | 20 | ns |
| $t_{LZ}$ Output disable time from low level | | $R_L$ = 560 Ω (SN54S'), | 9 | 30 | 9 | 20 | |
| $t_{HZ}$ Output disable time from high level | Disable times from write ENABLE | 400 Ω (SN74S'), | 13 | 40 | 13 | 30 | ns |
| $t_{LZ}$ Output disable time from low level | | See figure 1 | 16 | 40 | 16 | 30 | |

‡ All typical values are at $V_{CC}$ = 5 V, $T_A$ = 25°C.

## Parameter measurement information

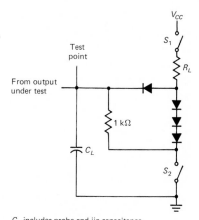

$C_L$ includes probe and jig capacitance.
All diodes are 1N3064.

Load circuit

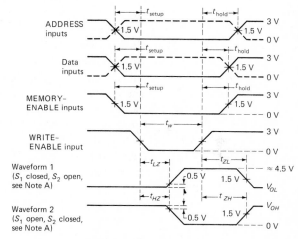

Write–cycle voltage waveforms

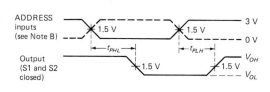

Access time from ADDRESS inputs
voltage waveforms

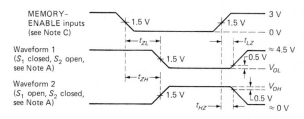

Access (ENABLE) time and disable time from memory ENABLE
voltage waveforms

**FIG. 12–15**  74S200, 256-bit RAM

Recommended operating conditions

| | SN54S200 | | | SN74S200 | | | Unit |
|---|---|---|---|---|---|---|---|
| | Min | Nom | Max | Min | Nom | Max | |
| Supply voltage, $V_{CC}$ | 4.5 | 5 | 5.5 | 4.75 | 5 | 5.25 | V |
| High-level output current, $I_{OH}$ | | | −5.2 | | | −10.3 | mA |
| Low-level output current, $I_{OL}$ | | | 8 | | | 12 | mA |
| Width of WRITE-ENABLE pulse, $t_w$ | 50 | | | 40 | | | ns |
| Setup time, $t_{setup}$ — ADDRESS to WRITE-ENABLE | 0 | | | 0 | | | |
| Setup time, $t_{setup}$ — Data to WRITE-ENABLE | 0 | | | 0 | | | ns |
| Setup time, $t_{setup}$ — MEMORY-ENABLE to WRITE-ENABLE | 0 | | | 0 | | | |
| Hold time, $t_{hold}$ — ADDRESS from WRITE-ENABLE | 10 | | | 10 | | | |
| Hold time, $t_{hold}$ — Data from WRITE-ENABLE | 10 | | | 10 | | | ns |
| Hold time, $t_{hold}$ — MEMORY-ENABLE from WRITE-ENABLE | 0 | | | 0 | | | |
| Operating free-air temperature, $T_A$ | −55 | | 125 | 0 | | 70 | °C |

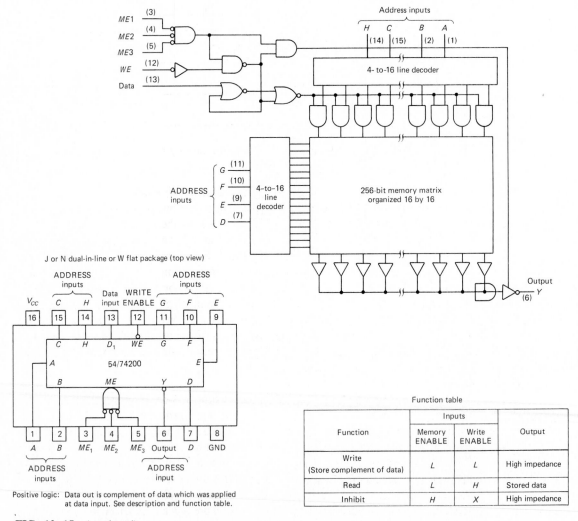

J or N dual-in-line or W flat package (top view)

Positive logic: Data out is complement of data which was applied at data input. See description and function table.

Function table

| Function | Inputs | | Output |
|---|---|---|---|
| | Memory ENABLE | Write ENABLE | |
| Write (Store complement of data) | L | L | High impedance |
| Read | L | H | Stored data |
| Inhibit | H | X | High impedance |

**FIG. 12–15** (continued)

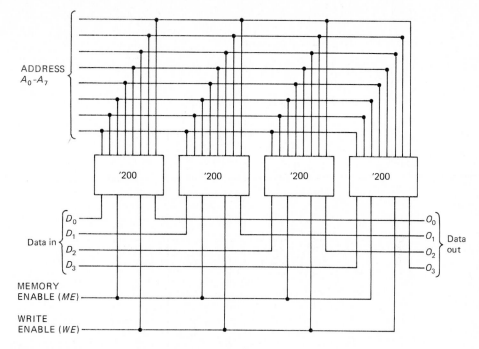

**FIG. 12–16**  Four 54/74200's arranged as a 1024-bit memory having 256 4-bit words

parallel like this is that the control and timing are exactly the same as if there were only a single chip. The only difference is that there are 4 data bits in and 4 data bits out (or 8 in and 8 out), all of which are in parallel with one another.

As a matter of fact, even larger memories can be constructed by connecting basic chips such as the '200 in both series and parallel. For instance, thirty-two '200 chips are connected in a $4 \times 8$ matrix in Fig. 12–17 to form a memory having one thousand, twenty-four 8-bit words. This configuration requires a 10-bit address: 2 bits can be used to select one of the four rows of eight '200s, and the remaining 8 bits will be wired in parallel to all the chips; they will work exactly as for the two-hundred fifty-six 8-bit word memory in Fig. 12–16. This concept can be continued, of course, but it becomes somewhat impractical with larger memory requirements, especially since there are MOS chips readily available with greater memory capacity.

A very popular and widely used MOS memory chip is the INTEL 2114. This is a static RAM having 4096 bits arranged as 1024 words of 4-bits each (selected information from the data sheet is given in Fig. 12–18). The organization of this chip is quite similar to that of the 7489 shown in Fig. 12–13, but notice that the 2114 is sixteen times larger!

The basic memory is arranged as 64 rows and 64 columns for a total of $64 \times 64 = 4096$ bits. Six address bits ($A_3$ through $A_8$) are used to select one of the 64 rows ($2^6 = 64$). The 64 columns are divided into 16 groups of 4-bit words, and four address bits ($A_0$, $A_1$, $A_2$ and $A_9$) are used to select one of these 16 groups. A 10-bit address will then select a single 4-bit word from 64 rows and 16 columns, to provide a memory of $64 \times 16 = 1024$, 4-bit words.

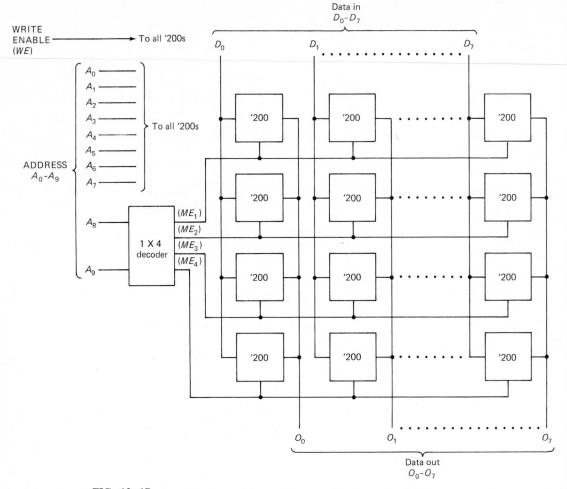

**FIG. 12–17**

$I/O_1$, $I/O_2$, $I/O_3$ and $I/O_4$ are the input lines for a 4-bit data word during a WRITE operation, and a 4-bit data word will also appear on these lines during a READ operation. Holding CHIP SELECT low will enable the chip for either a READ or a WRITE operation. WRITE ENABLE must be held high for a READ cycle, and low for a WRITE cycle. The READ and WRITE cycle waveforms are given in Fig. 12–18, and you will notice that they are quite similar to those required for the 7489.

Nearly all static RAMs larger than 1024 bits are MOS types, but the generalized block diagram in Fig. 12–14 still holds. The only variation is that the address bits are sometimes multiplexed (applied to the memory chip in two groups) in order to keep the required number of pins on the DIP lower. For instance, a memory that has 16 address bits might be designed with only eight address pins. The address is then applied to the chip, in two groups of eight, first bits 1 through 8 and then bits 9 through 16. The bits are simply stored in a register on the memory chip and then applied to the cell matrix all at once to

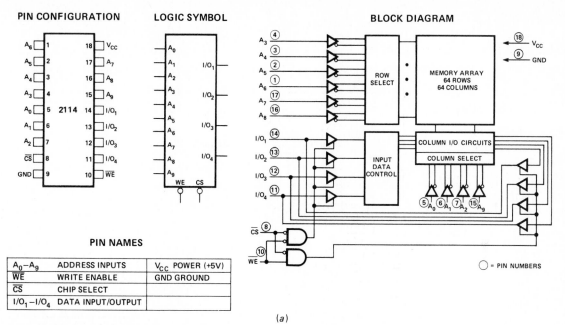

PIN CONFIGURATION    LOGIC SYMBOL

PIN NAMES

| $A_0 - A_9$ | ADDRESS INPUTS | $V_{CC}$ POWER (+5V) |
|---|---|---|
| $\overline{WE}$ | WRITE ENABLE | GND GROUND |
| $\overline{CS}$ | CHIP SELECT | |
| $I/O_1 - I/O_4$ | DATA INPUT/OUTPUT | |

BLOCK DIAGRAM

$\bigcirc$ = PIN NUMBERS

*(a)*

## A.C. CHARACTERISTICS  $T_A = 0°C$ to $70°C$, $V_{CC} = 5V \pm 5\%$, unless otherwise noted.

### READ CYCLE [1]

| SYMBOL | PARAMETER | 2114-2 | | 2114-3, 2114L3 | | 2114, 2114L | | UNIT |
|---|---|---|---|---|---|---|---|---|
| | | Min. | Max. | Min. | Max. | Min. | Max. | |
| $t_{RC}$ | Read Cycle Time | 200 | | 300 | | 450 | | ns |
| $t_A$ | Access Time | | 200 | | 300 | | 450 | ns |
| $t_{CO}$ | Chip Selection to Output Valid | | 70 | | 100 | | 100 | ns |
| $t_{CX}$ | Chip Selection to Output Active | 0 | | 0 | | 0 | | ns |
| $t_{OTD}$ | Output 3-state from Deselection | 0 | 40 | 0 | 80 | 0 | 100 | ns |
| $t_{OHA}$ | Output Hold from Address Change | 10 | | 10 | | 10 | | ns |

### WRITE CYCLE [2]

| SYMBOL | PARAMETER | 2114-2 | | 2114-3, 2114L3 | | 2114, 2114L | | UNIT |
|---|---|---|---|---|---|---|---|---|
| | | Min. | Max. | Min. | Max. | Min. | Max. | |
| $t_{WC}$ | Write Cycle Time | 200 | | 300 | | 450 | | ns |
| $t_W$ | Write Time | 100 | | 150 | | 200 | | ns |
| $t_{WR}$ | Write Release Time | 20 | | 0 | | 0 | | ns |
| $t_{OTW}$ | Output 3-state from Write | 0 | 40 | 0 | 80 | 0 | 100 | ns |
| $t_{DW}$ | Data to Write Time Overlap | 100 | | 150 | | 200 | | ns |
| $t_{DH}$ | Data Hold From Write Time | 0 | | 0 | | 0 | | ns |

NOTES:  1. A Read occurs during the overlap of a low $\overline{CS}$ and a high $\overline{WE}$.
          2. A Write occurs during the overlap of a low $\overline{CS}$ and a low $\overline{WE}$.

*(b)*

**FIG. 12–18**   AMD, AM99C88, 8k $\times$ 8 static CMOS RAM

## WAVEFORMS
### READ CYCLE ①

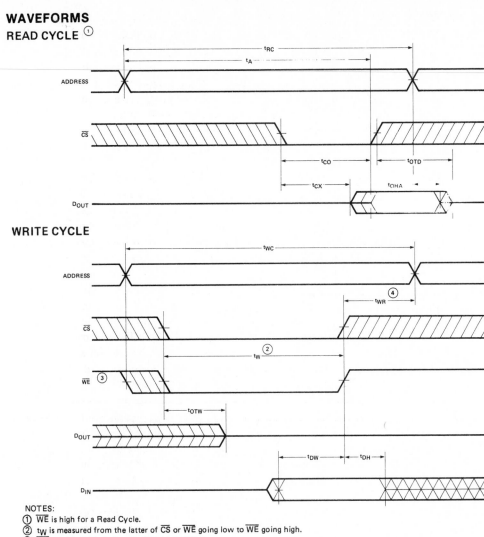

### WRITE CYCLE

NOTES:
① $\overline{WE}$ is high for a Read Cycle.
② $t_W$ is measured from the latter of $\overline{CS}$ or $\overline{WE}$ going low to $\overline{WE}$ going high.
③ $\overline{WE}$ must be high during all address transitions.
④ $t_{WR}$ is referenced to the high transition of $\overline{WE}$.

(c)

**FIG. 12–18** (continued)

decode the desired 16-bit address. A good example of this type of address multiplexing is the 4116 ''dynamic'' RAM to be discussed in the next section.

## 12-5 DRAMS—DYNAMIC RAMS

As stated previously, the basic difference between a static and a dynamic RAM is the memory cell. The dynamic memory cell utilizes an MOS transistor circuit to store charge and, therefore, store 1 bit. This dynamic cell requires a much smaller surface area on a silicon chip; thus a dynamic memory chip will have a much greater memory capacity than will a static RAM chip of the same overall dimensions. However, since the stored charge dissipates with time, it must be periodically restored or ''refreshed,'' and the dynamic RAM thus has an additional operational requirement—the refresh cycle.

A typical dynamic RAM is essentially the same as the previously discussed static RAM chip, with the exception of the required refresh cycle. The 4116 is a widely used 16K (16,384 × 1) dynamic RAM available from a number of different sources, such as Mostek (MK4116), Motorola (MCM4116), and Texas Instruments TMS4116. The Advanced Micro Devices Am9016 is their latest replacement for a 4116, and the Intel 2116 is their equivalent.

The Texas Instruments TMS4116 is shown in Fig. 12–19. There is a single data input line $D$, a single data output line $Q$ and a write-enable control, $\overline{W}$. The 16,384 memory cells are arranged in a square array having 128 rows and 128 columns, with a requirement for 14 address lines ($2^{14} = 16,384$). Notice that the pinout provides for only seven address lines; this means that the 14 address bits must be multiplexed into the chip in two groups of 7 bits each. This is the function of the other two control inputs, $\overline{RAS}$ and $\overline{CAS}$.

Let's define the 14 required address bits as $A_{13}, A_{12}, \ldots, A_0$. Designate the seven LSBs ($A_6, A_5, \ldots, A_0$) as the row address and the seven MSBs ($A_{13}, A_{12}, \ldots, A_7$) as the column address. The selection of a single bit in the memory requires addressing the chip in two steps. First, the 7-bit row address is applied to the seven address pins on the chip; a negative pulse on the row address strobe ($\overline{RAS}$) control line will strobe these 7 bits into the row latches on the chip. Next, the 7-bit column address is applied to the seven address pins on the chip; a negative pulse on the column address strobe ($\overline{CAS}$) control line will strobe these 7 bits into the column latches on the chip. The 14 bits of address are now stored on the chip and decoded to select a single memory cell. If both $\overline{RAS}$ and $\overline{CAS}$ are held low, the decoded address will remain for a period of time, and the selected memory cell contents can be read out, or a new data bit can be stored in the selected cell.

If the write enable ($\overline{W}$) is held high, the contents of the selected bit will appear at the data output, $Q$—thus forming a read cycle. On the other hand, if $\overline{W}$ is held low, the data bit present at the data input $D$ will be stored in the selected bit position—thus forming a write cycle.

Naturally there are strict timing requirements that must be met. These time requirements are specified in the cycle timing waveforms included in Fig. 12–19. For instance, the row address inputs must be stable for a minimum period of time before $\overline{RAS}$ goes low—this is the address setup time for rows, $t_{ASR}$, or $t_{su(RA)}$. Look carefully at the read cycle timing waveforms for this time period—in the data sheets, this time is given as 0.0 ns minimum. In other words, the row addresses must be stable any time after $\overline{RAS}$ goes low. You should take the time to study these waveforms carefully—a clear understanding of the timing requirements is necessary if you are to understand how and where this chip is used.

☐ EXAMPLE 12–6

According to the information in Fig. 12–19, what is the maximum memory access time—in other words, the time from the fall of $\overline{RAS}$ until data is available at the output?

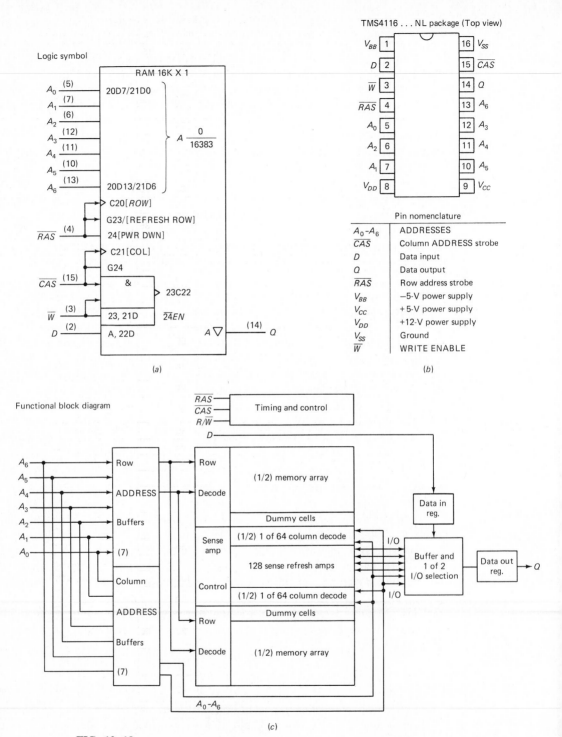

Logic symbol

RAM 16K X 1

$A_0$ (5)    20D7/21D0
$A_1$ (7)
$A_2$ (6)
$A_3$ (12)                $A \dfrac{0}{16383}$
$A_4$ (11)
$A_5$ (10)
$A_6$ (13)    20D13/21D6

C20[ROW]
G23/[REFRESH ROW]
$\overline{RAS}$ (4)    24[PWR DWN]
C21[COL]
G24
$\overline{CAS}$ (15)    &
                         23C22
$\overline{W}$ (3)    23, 21D    $\overline{24EN}$
D (2)    A, 22D    $A \triangledown$    (14) Q

(a)

| | TMS4116 . . . NL | |
|---|---|---|
| $V_{BB}$ | 1 | 16 $V_{SS}$ |
| D | 2 | 15 $\overline{CAS}$ |
| $\overline{W}$ | 3 | 14 Q |
| $\overline{RAS}$ | 4 | 13 $A_6$ |
| $A_0$ | 5 | 12 $A_3$ |
| $A_2$ | 6 | 11 $A_4$ |
| $A_1$ | 7 | 10 $A_5$ |
| $V_{DD}$ | 8 | 9 $V_{CC}$ |

Pin nomenclature

| | |
|---|---|
| $A_0$-$A_6$ | ADDRESSES |
| $\overline{CAS}$ | Column ADDRESS strobe |
| D | Data input |
| Q | Data output |
| $\overline{RAS}$ | Row address strobe |
| $V_{BB}$ | −5-V power supply |
| $V_{CC}$ | +5-V power supply |
| $V_{DD}$ | +12-V power supply |
| $V_{SS}$ | Ground |
| $\overline{W}$ | WRITE ENABLE |

(b)

Functional block diagram

$\overline{RAS}$
$\overline{CAS}$    Timing and control
$R/\overline{W}$
D

$A_6$
$A_5$    Row
$A_4$    ADDRESS
$A_3$
$A_2$    Buffers
$A_1$
$A_0$    (7)

Row    Row
ADDRESS    Decode    (1/2) memory array
Buffers    Dummy cells
(7)    Sense    (1/2) 1 of 64 column decode    I/O
Column    amp    128 sense refresh amps
ADDRESS    Control    (1/2) 1 of 64 column decode    I/O
Buffers    Dummy cells
(7)    Row
         Decode    (1/2) memory array

Data in reg.

Buffer and 1 of 2 I/O selection    Data out reg.    Q

$A_0$-$A_6$

(c)

**FIG. 12–19**

**Timing requirements over recommended supply voltage range and operating free-air temperature range**

| | Parameter | Alt. symbol | TMS4116-15 Min | TMS4116-15 Max | TMS4116-20 Min | TMS4116-20 Max | TMS4116-25 Min | TMS4116-25 Max | Unit |
|---|---|---|---|---|---|---|---|---|---|
| $t_{c(P)}$ | Page-mode cycle time | $t_{PC}$ | 170 | | 225 | | 275 | | ns |
| $t_{c(rd)}$ | Read cycle time | $t_{RC}$ | 375 | | 375 | | 410 | | ns |
| $t_{c(W)}$ | Write cycle time | $t_{WC}$ | 375 | | 375 | | 410 | | ns |
| $t_{c(rdW)}$ | Read, modify-write cycle time | $t_{RWC}$ | 375 | | 375 | | 515 | | ns |
| $t_{w(CH)}$ | Pulse width, $\overline{CAS}$ high (precharge time) | $t_{CP}$ | 60 | | 80 | | 100 | | ns |
| $t_{w(CL)}$ | Pulse width, $\overline{CAS}$ low | $t_{CAS}$ | 100 | 10,000 | 135 | 10,000 | 165 | 10,000 | ns |
| $t_{w(RH)}$ | Pulse width, $\overline{RAS}$ high (precharge time) | $t_{RP}$ | 100 | | 120 | | 150 | | ns |
| $t_{w(RL)}$ | Pulse width, $\overline{RAS}$ low | $t_{RAS}$ | 150 | 10,000 | 200 | 10,000 | 250 | 10,000 | ns |
| $t_{w(W)}$ | Write pulse width | $t_{WP}$ | 45 | | 55 | | 75 | | ns |
| $t_t$ | Transition times (rise and fall) for $\overline{RAS}$ and $\overline{CAS}$ | $t_T$ | 3 | 35 | 3 | 50 | 3 | 50 | ns |
| $t_{su(CA)}$ | Column address setup time | $t_{ASC}$ | −10 | | −10 | | −10 | | ns |
| $t_{su(RA)}$ | Row address setup time | $t_{ASR}$ | 0 | | 0 | | 0 | | ns |
| $t_{su(D)}$ | Data setup time | $t_{DS}$ | 0 | | 0 | | 0 | | ns |
| $t_{su(rd)}$ | Read command setup time | $t_{RCS}$ | 0 | | 0 | | 0 | | ns |
| $t_{su(WCH)}$ | Write command setup time before $\overline{CAS}$ high | $t_{CWL}$ | 60 | | 80 | | 100 | | ns |
| $t_{su(WRH)}$ | Write command setup time before $\overline{RAS}$ high | $t_{RWL}$ | 60 | | 80 | | 100 | | ns |
| $t_{h(CLCA)}$ | Column address hold time after $\overline{CAS}$ low | $t_{CAH}$ | 45 | | 55 | | 75 | | ns |
| $t_{h(RA)}$ | Row address hold time | $t_{RAH}$ | 20 | | 25 | | 35 | | ns |
| $t_{h(RLCA)}$ | Column address hold time after $\overline{RAS}$ low | $t_{AR}$ | 95 | | 120 | | 160 | | ns |
| $t_{h(CLD)}$ | Data hold time after $\overline{CAS}$ low | $t_{DH}$ | 45 | | 55 | | 75 | | ns |
| $t_{h(RLD)}$ | Data hold time after $\overline{RAS}$ low | $t_{DHR}$ | 95 | | 120 | | 160 | | ns |
| $t_{h(WLD)}$ | Data hold time after $\overline{W}$ low | $t_{DH}$ | 45 | | 55 | | 75 | | ns |
| $t_{h(rd)}$ | Read command hold time | $t_{RCH}$ | 0 | | 0 | | 0 | | ns |
| $t_{h(CLW)}$ | Write command hold time after $\overline{CAS}$ low | $t_{WCH}$ | 45 | | 55 | | 75 | | ns |
| $t_{h(RLW)}$ | Write command hold time after $\overline{RAS}$ low | $t_{WCR}$ | 95 | | 120 | | 160 | | ns |
| $t_{RLCH}$ | Delay time, $\overline{RAS}$ low to $\overline{CAS}$ high | $t_{CSH}$ | 150 | | 200 | | 250 | | ns |
| $t_{CHRL}$ | Delay time, $\overline{CAS}$ high to $\overline{RAS}$ low | $t_{CRP}$ | −20 | | −20 | | −20 | | ns |
| $t_{CLRH}$ | Delay time, $\overline{CAS}$ low to $\overline{RAS}$ high | $t_{RSH}$ | 100 | | 135 | | 165 | | ns |
| $t_{CLWL}$ | Delay time, $\overline{CAS}$ low to $\overline{W}$ low (read, modify–write-cycle only) | $t_{CWD}$ | 70 | | 95 | | 125 | | ns |
| $t_{RLCL}$ | Delay line, $\overline{RAS}$ low to $\overline{CAS}$ low (maximum value specified only to guarantee access time) | $t_{RCD}$ | 20 | 50 | 25 | 65 | 35 | 85 | ns |
| $t_{RLWL}$ | Delay time, $\overline{RAS}$ low to $\overline{W}$ low (read, modify–write-cycle only) | $t_{RWD}$ | 120 | | 160 | | 200 | | ns |
| $t_{WLCL}$ | Delay time, $\overline{W}$ low to $\overline{CAS}$ low (early write cycle) | $t_{WCS}$ | −20 | | −20 | | −20 | | ns |
| $t_{rf}$ | Refresh time interval | $t_{REF}$ | | 2 | | 2 | | 2 | ms |

**FIG. 12–19** (continued)

| Parameter | | Test conditions | Alt. symbol | TMS4116-15 | | TMS4116-20 | | TMS4116-25 | | Unit |
|---|---|---|---|---|---|---|---|---|---|---|
| | | | | Min | Max | Min | Max | Min | Max | |
| $t_{a(C)}$ | Access time from $\overline{CAS}$ | $C_L$ = 100 pF, Load = 2 series, 74 TTL gates | $t_{CAC}$ | | 100 | | 135 | | 165 | ns |
| $t_{a(R)}$ | Access time from $\overline{RAS}$ | $t_{RLCL}$ = MAX, $C_L$ = 100 pF, Load = 2 series, 74 TTL gates | $t_{RAC}$ | | 150 | | 200 | | 250 | ns |
| $t_{dis(CH)}$ | Output disable time after $\overline{CAS}$ high | $C_L$ = 100 pF, Load = 2 series, 74 TTL gates | $t_{OFF}$ | 0 | 40 | 0 | 50 | 0 | 60 | ns |

TMS4116 16,384-bit dynamic random-access memory

**Read cycle timing**

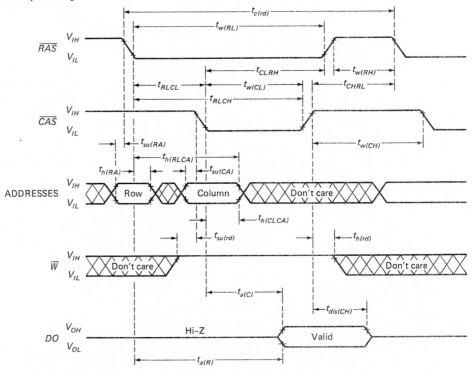

**FIG. 12–19** (continued)

☐ SOLUTION

On the read cycle waveforms, this time is given as $t_{RAC}$, or $t_{a(R)}$, and the data sheet states 250 ns maximum.

The data sheet for a 4116 states that each bit location must be refreshed (recharged) at least once every 2 ms. The design of this chip is such that when the seven LSBs used as the row address bits are strobed into the row latches, the contents of all 128 of the cells in the decoded row are read into the sense-refresh amplifiers and then stored back in the

TMS4116 16,384-bit dynamic random-access memory

Write cycle timing

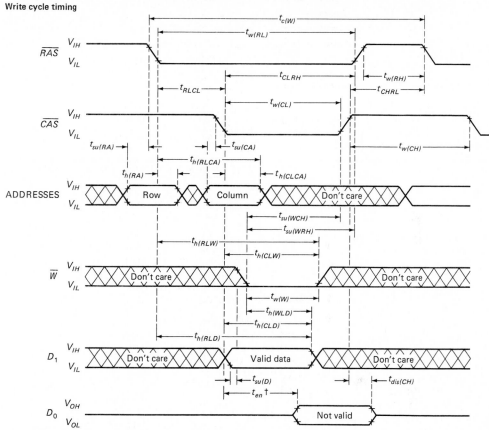

†The enable time ($t_{en}$) for a WRITE cycle is equal in duration to the access time from $\overline{CAS}$ ($t_{a(C)}$) in a READ cycle; but the active levels at the output are invalid.

$\overline{RAS}$-only refresh timing

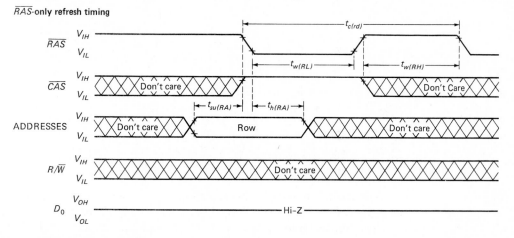

**FIG. 12–19** (continued)

TMS4500A . . . NL package
(Top view)

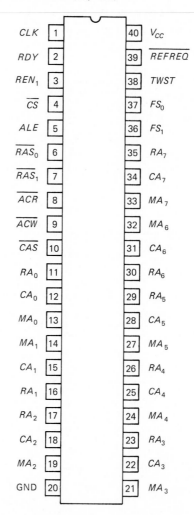

| | | |
|---|---|---|
| CLK | 1 | 40 | $V_{CC}$ |
| RDY | 2 | 39 | $\overline{REFREQ}$ |
| $REN_1$ | 3 | 38 | TWST |
| $\overline{CS}$ | 4 | 37 | $FS_0$ |
| ALE | 5 | 36 | $FS_1$ |
| $\overline{RAS_0}$ | 6 | 35 | $RA_7$ |
| $\overline{RAS_1}$ | 7 | 34 | $CA_7$ |
| $\overline{ACR}$ | 8 | 33 | $MA_7$ |
| $\overline{ACW}$ | 9 | 32 | $MA_6$ |
| $\overline{CAS}$ | 10 | 31 | $CA_6$ |
| $RA_0$ | 11 | 30 | $RA_6$ |
| $CA_0$ | 12 | 29 | $RA_5$ |
| $MA_0$ | 13 | 28 | $CA_5$ |
| $MA_1$ | 14 | 27 | $MA_5$ |
| $CA_1$ | 15 | 26 | $RA_4$ |
| $RA_1$ | 16 | 25 | $CA_4$ |
| $RA_2$ | 17 | 24 | $MA_4$ |
| $CA_2$ | 18 | 23 | $RA_3$ |
| $MA_2$ | 19 | 22 | $CA_3$ |
| GND | 20 | 21 | $MA_3$ |

**FIG. 12–20**  Texas Instruments TMS 4500A, DRAM controller

same cells. As a result, all 128 of the cells in a selected row are refreshed during any read or write cycle. Furthermore, simply taking $\overline{RAS}$ low will accomplish the same thing. So, the $\overline{RAS}$-only cycle given in the waveforms specifies the control signals necessary to refresh a single row of 128 cells. However, all 128 rows must be refreshed during every 2-ms time period.

Refreshing can be accomplished by using a 7-bit refresh control counter to supply address bits for the selection of one row at a time, sequentially. An $\overline{RAS}$ only is used at each address position to refresh that row. Assuming a cycle time, $t_{c(W)}$ or $t_{c(rd)}$ of 375 ns, it will require $0.375 \times 128 = 48$ $\mu$s to refresh all 128 rows. Thus 48 $\mu$s out of every 2 ms

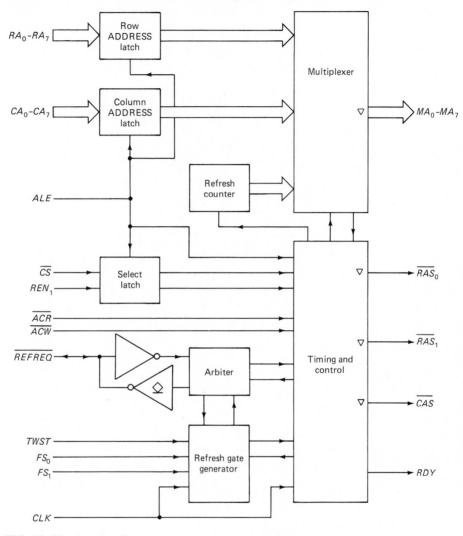

TMS4500A dynamic RAM controller
Block diagram

**FIG. 12–20**   (continued)

must be used for refresh time, but this amounts only to $48/2000 = 2.4$ percent of the available memory time. Refreshing can be done to all 128 rows at once, in a "burst," or by refreshing a single row every 2 ms, with equal time spacing between rows in a "periodic" mode.

An interesting and useful feature of the TMS4116 is that it can be addressed in a *page address mode*. In this mode, a row address is stored in the input decoders by $\overline{RAS}$, and then sequential columns are selected in the same row using $\overline{CAS}$. The great advantage here is that the read or write cycle time $t_{c(P)}$ is reduced by more than half—from 375 down to 170 ns!

TMS4500A dynamic RAM controller

Access cycle timing

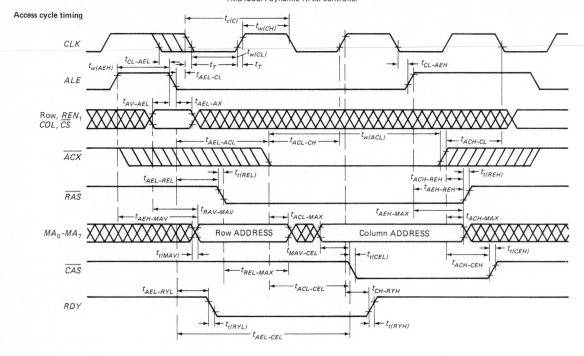

Refresh request timing

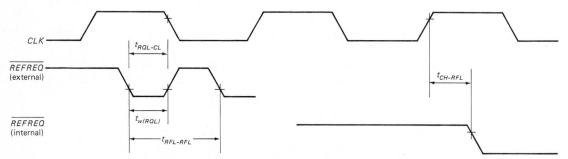

**FIG. 12–20**   (continued)

Using a dynamic RAM is slightly more complicated than using a static RAM, since the address bits must be multiplexed, the memory must be refreshed periodically, and storing data in the memory or reading data from the memory must be kept separate from the refresh cycles. All of these functions can be accomplished with external control logic, of course, but the problem is somewhat simplified with a dynamic RAM controller. The Texas Instruments TMS 4500A is one of a number of dynamic RAM controller circuits available on a single chip. The 4500A can be used with any 4116, as well as with 8K, 32K and 64K DRAMs.

The TMS4500A is used to interface dynamic RAMs to a microprocessor system and contains an address multiplexer, refresh control, timing control, and arbitration logic to

Typical access/refresh/access cycle
(three cycle, TWST = 0)

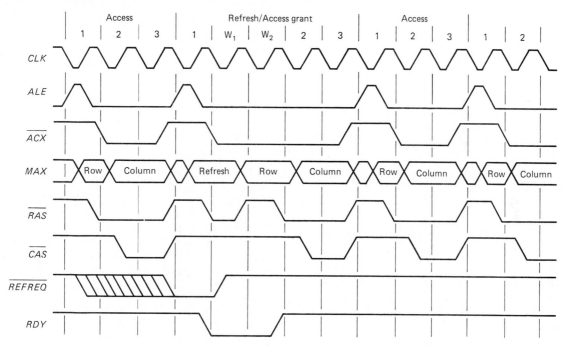

**FIG. 12–20**   (continued)

resolve memory access requests and refresh requests. The outputs $MA_0$ through $MA_7$ are 8 bits of memory address used to drive the DRAMs. The format is 8 bits of row address followed by 8 bits of column address, with 8 bits of refresh address at desired refresh times, as shown in Fig. 12–20d. There are two row address strobe outputs, $\overline{RAS_0}$ and $\overline{RAS_1}$, used to strobe two different banks of DRAMs; $\overline{CAS}$ is the column address strobe output to the DRAMs.

The address multiplexer will multiplex the 16 bits of input address to the eight output address pins. The input address latch enable (ALE) will latch the 16 address inputs and will also begin an access cycle provided CHIP SELECT $(\overline{CS})$ is low.

The input $REN_1$ is used to select one of two possible banks of DRAMs being controlled by means of the two RAS outputs. When high, $\overline{RAS_1}$ is selected, and $\overline{RAS_0}$ is selected when it is low.

A low level on either access control read $(\overline{ACR})$ or access control write $(\overline{ACW})$ will cause the column address to appear on the eight output address lines $MA_0$ through $MA_7$. However, when both these inputs are low, all the outputs assume a high-impedance state.

The input $\overline{REFREQ}$ initiates a refresh cycle. The CLOCK input, labeled CLK, is the system clock input, and $FS_0$, $FS_1$, and TWST are used to determine refresh rate and timing.

We conclude this section by noting that the organization and operation of the 4116 is typical for a dynamic RAM (DRAM). There is one other widely used DRAM, the 4164, organized as a $64,536 \times 1$ chip. The operation of the chip is quite similar to the 4116, except that it requires only a single 5 Vdc power supply. Note that these, or comparable, chips are available from a number of different manufacturers, but the part numbers may differ even though the chips are compatible. For instance, the 64K DRAM ($65,536 \times 1$) is available under the following part numbers: Advanced Micro Devices, AM9064; Fairchild, F4164; Intel, 2164; Mostek, MK4564; Motorola, MCM6665; and Texas Instruments, TMS4164.

## 12-6  MEMORY CELLS

A memory cell is the basic unit for storing a single bit of information in a memory. The cells used to construct semiconductor memories are usually flip-flops designed by using either bipolar or MOS transistors or a charge storage circuit that uses MOS transistors. The intent here is not to study these memory cells with the detail required to design these circuits. (Such subjects are more appropriately covered elsewhere.) Rather, we are interested primarily in understanding at least one example that illustrates exactly how a memory cell operates. Secondly, such an in-depth understanding leads to a clear concept of the circuit requirements, as well as an appreciation for the operating limitations imposed on a memory chip. It is not possible to discuss all the different circuit designs, but the basic circuits covered here illustrate the principles used in most memory cells. For each circuit discussed, we assume the signal voltage levels to be high = $+V_{CC}$ and low = 0.0 Vdc.

A memory cell used in a TTL static RAM, along with a sense amplifier and a write amplifier, are shown in Fig. 12–21. Two bipolar junction transistors ($Q_1$ and $Q_2$), cross-coupled to form a simple latch and capable of storing 1 bit of information, serve as the memory element. The only unusual aspect is that each transistor has three emitter leads. Power supply connections necessary for the circuits are $+V_{CC}$ and ground (GND).

If either the ROW or COLUMN select line is low (GND), the cell is disabled (deselected). In order to select a cell, the ROW and COLUMN lines must both be high ($+V_{cc}$). When selected, a data bit can be stored in the cell (write), or the contents of the cell can be sensed (read out). Here's how it works.

When the ROW and COLUMN select lines are high, the emitters connected to these lines are reverse-biased, and the cell behaves as a simple latch. So, one of the transistors is on, and one is off. The transistor that is on conducts current out through its emitter, down through a sense resistor $R_s$ and into the base of the sense amplifier transistor, turning it on. In the cell, there is no current coming out of the emitter of the transistor that is off, and thus the sense amplifier transistor on that side of the cell is off. The net result is that the sense amplifier transistors "mimic" the transistors in the memory cell. Therefore, when selected, the contents of a cell is immediately available at the sense amplifier data outputs. For instance, if a 1 is stored in the cell, $Q_1$ is on and $Q_2$ is off. When selected, DATA OUT will be high (a 1), and its complement, $\overline{\text{DATA OUT}}$, will be low (a 0).

The WRITE amplifier is used to store information in the cell when its W input is held high. When W is low, the amplifier is disabled and its outputs are disconnected from the sense resistors $R_s$. There are numerous configurations, but the basic requirements are that WRITE 1 goes high and WRITE 0 goes low whenever DATA IN is high; conversely, WRITE 0 goes high and WRITE 1 goes low whenever DATA IN is low. To store a bit in the cell, it is first selected by taking both ROW and COLUMN high. Then a high at the

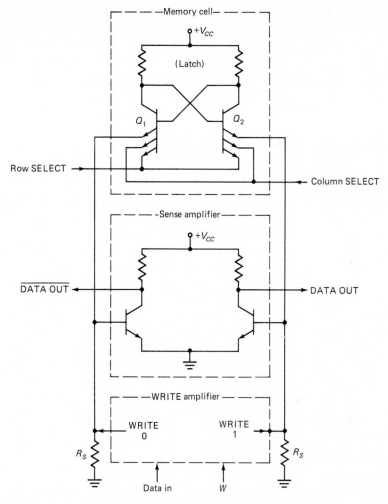

**FIG. 12–21**

DATA IN of the write amplifier will cause WRITE 1 to go high and WRITE 0 to go low. Thus all three of the emitters of $Q_2$ will be high, forcing $Q_2$ off and $Q_1$ on, and the latch will store a 1. Conversely, a 0 at the DATA IN will take all three of the emitters of $Q_1$ high, forcing $Q_1$ OFF and $Q_2$ on, storing a 0 in the latch.

In a static $n$-type-channel metal oxide semiconductor (NMOS) memory, the memory cell is constructed by using two NMOS inverters, cross-coupled to form a simple latch. In a static CMOS memory, two CMOS inverters are cross-coupled to form a latch. In either case, a data bit is coupled into (stored in) the cell, or the contents are sensed (read out) by means of NMOS transmission gates. Let's first consider how a transmission gate functions.

Figure 12–22 shows an NMOS, enhancement-mode transistor connected as a transmission gate. The transistor is assumed to be completely bilateral—that is, the source and

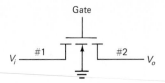

**FIG. 12–22** NMOS transmission gate

the drain leads may be interchanged with no difference in performance. Since this is the case, let's simply call these two leads 1 and 2. Let lead 1 be the input and lead 2 the output. The GATE lead is the control. The voltage signals are either 0.0 Vdc or $+V_{CC}$. Here's how it works.

If the GATE is held low (at 0.0 Vdc), the transistor is off, and the input and output terminals are isolated from one another. There is no transmission of signal through the gate.

If the GATE is high ($+V_{CC}$), and the voltage at lead 1 is low, the transistor is ON, and current will flow through the transistor until $V_o = V_i = 0.0$ Vdc.

If the GATE is high and the voltage at lead 1 is also high, the transistor will be on if $V_o$ is low, and current will flow through the transistor until $V_o = V_i = +V_{CC}$. Or, if $V_o$ is also high, the transistor will remain off since $V_o = V_i = +V_{CC}$.

To summarize, the transmission gate is disabled whenever the GATE is low, and the input is isolated from the output; thus it acts like an open switch. The transmission gate is enabled whenever the GATE is high, and the output will equal the input, $V_o = V_i$; thus it acts like a closed switch.

The circuit in Fig. 12–23 is typical of the memory cell and sense amplifier used in an NMOS static RAM. The memory cell consists of two NMOS inverters, cross-coupled to form a latch. There are four transmission gates, $T_1, T_2, T_3$, and $T_4$, used to select the cell. When both ROW and COLUMN are high, the cell is selected and its contents can be read out by the sense amplifier, or a data bit can be stored in the cell by using the WRITE gate (W).

Let's assume the cell is selected by holding ROW and COLUMN both high. Then the two transmission gates $T_1$ and $T_2$ are both enabled. (They act like closed switches.) If W is held high, the WRITE transmission gate is also enabled and a data bit at the DATA IN (either a high or a low) will be connected directly to the latch at $D$. If DATA IN is high, a 1 will be stored, since the latch will stabilize with $D$ high; conversely, a low at DATA IN will store a 0, since the latch will stabilize with $D$ low. This is the WRITE cycle.

Again, let's assume that the cell is selected with both ROW and COLUMN high. The two transmission gates $T_3$ and $T_4$ are enabled. (They act like closed switches.) Whatever data bit is stored in the latch at $D$, its complement appears at $\overline{D}$ and is coupled by the transmission gates $T_3$ and $T_4$ out to the sense amplifier. If the READ gate is enabled ($R$ is high), the data bit stored in the latch $D$ will appear at the DATA OUT. This is a READ cycle.

The first two transistors in the sense amplifier form an inverter whose output is $D$. The two transistors at the sense amplifier output ($Q_1$ and $Q_2$) have input signals of $D$ and $\overline{D}$ with the result that one of them is always on and the other one is always off. For instance, if the data bit stored in the latch is a 1, then $D$ = high. So, $Q_1$ will be on and $Q_2$

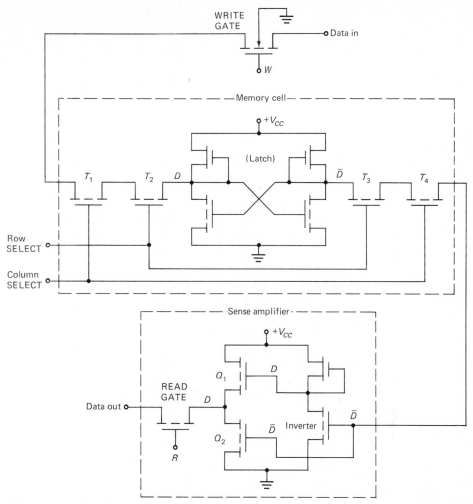

Note: All transistors are NMOS with their substrate leads connected to ground (substrate leads not shown)

**FIG. 12–23** Memory cell for an NMOS static RAM

will be off, and the DATA OUT will be high (1). Conversely, if a 0 is stored in the latch, $D$ = low. So, $Q_1$ will be off, $Q_2$ will be on, and the DATA OUT will be low (0).

The circuit in Fig. 12–24 is typical of the memory cell and sense amplifier used in a CMOS static RAM. The memory cell consists of two CMOS inverters, cross-coupled to form a simple latch. There are four transmission gates, $T_1$, $T_2$, $T_3$, and $T_4$, used to select the cell. When both ROW and COLUMN are high, the cell is selected and its contents can be read by using the sense amplifier, or a data bit can be stored in the cell by use of the WRITE gate (W).

The WRITE cycle for this circuit is exactly like the WRITE cycle for the NMOS circuit described previously. The cell is selected by holding both ROW and COLUMN

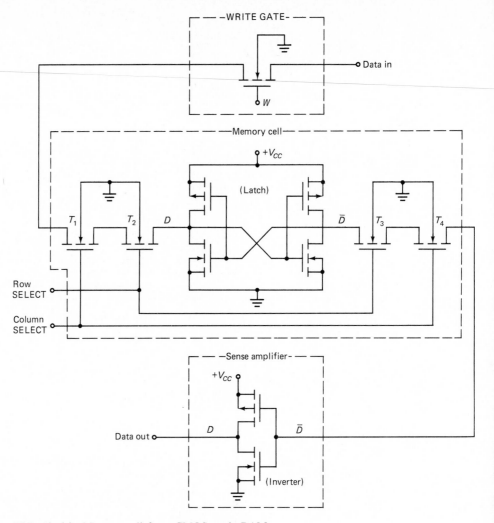

**FIG. 12–24** Memory cell for a CMOS static RAM

high, and then the bit present at DATA IN will be stored in the latch if W is high (the WRITE GATE is enabled).

The READ cycle is also exactly the same as for the previously discussed NMOS circuit. The cell is selected with both ROW and COLUMN high, and the cell contents appears at the DATA OUT after passing through the sense amplifier inverter.

The circuit used for a dynamic MOS memory cell consists of one or more MOS transistors. (There are numerous different configurations using one, two, and three devices.) Consequently, this cell requires less surface area on a chip and usually dissipates less power than do the cells used in either bipolar or MOS static memories. Because of this, most RAMs larger than 16K are dynamic rather than static.

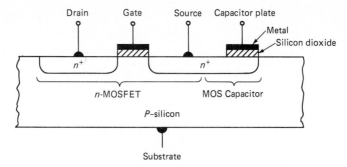

(a) Fabrication of a one-transistor dynamic memory cell

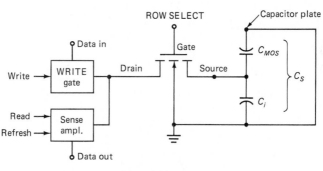

(b) Circuit diagram

**FIG. 12–25** One transistor dynamic memory cell

A data bit stored in a dynamic memory cell is in the form of a charge stored on a capacitor. Since this charge will leak off with time, the capacitor must periodically be recharged. Furthermore, the process of sensing, or reading out, the contents of the dynamic cell may dissipate the stored charge, resulting in a loss of the stored data bit. This is referred to as *destructive readout*, a problem not encountered with static memory cells. Even though dynamic memories require additional circuitry and timing signals to deal with refreshing as well as the destructive readout problem, the advantage of greater memory capacity on a single chip is generally worth the effort.

Let's take a look at a one-transistor MOS dynamic memory cell as shown in Fig. 12–25. The cell consists of an NMOS transistor and a MOS capacitor fabricated on a chip as shown. If the capacitor PLATE lead and the substrate are connected to ground, the storage capacitance $C_s$ used to store the charge for a bit consists of the MOS capacitor $C_{mos}$ in parallel with the transistor junction capacitance $C_i$. A 1 is stored in the cell if the capacitor $C_s$ is charged, and a 0 is stored if $C_s$ is discharged.

The transistor itself behaves essentially like a transmission gate. The ROW select will enable all the transistors in an entire row in the memory. Then on a READ cycle, the sense amplifier must detect the contents of the cell and at the same time recharge the storage capacitance $C_s$. Notice that the sense amplifier thus solves the destructive readout problem and at the same time will be used for the refresh operation. On a WRITE cycle, the WRITE gate is used to store a 1 or a 0 in the cell—charge or discharge $C_s$, respectively.

Let's conclude this section by again noting that the circuit treatment here is not intended to be exhaustive, but is typical of the circuits used in industry. The reader interested in pursuing this topic is referred to texts on bipolar and MOS IC design.

## S U M M A R Y

This entire chapter has been devoted to the study of semiconductor memories. We considered first the various rectangular arrays of memory cells on a chip and found that a square array containing the same number of rows and columns requires the fewest number of address lines.

Programmable, erasable-programmable, and plain read-only memories (PROMs, EPROMs, and ROMs) are used to store data in applications where the data changes not at all, or only infrequently. These memory chips are available as either bipolar or MOS, but the MOS devices offer much greater capacity per chip.

Random-access memories (RAMs) are also available as either bipolar or MOS devices and are used to store data that must be readily available and may be changed frequently. The dynamic random-access memory (DRAM) offers the greater advantage of more storage capacity on a chip but has the disadvantage of requiring refreshing. Careful attention to timing requirements is absolutely essential with the use of any memory chip.

The basic memory cell on a bipolar chip is a simple latch using cross-coupled bipolar junction transistors. The same is true for a static MOS or CMOS memory chip, except that the transistors used are MOS or CMOS, respectively. A dynamic memory, on the other hand, uses a capacitor and one or more MOS transistors to store charge and, therefore, a single bit.

We have not undertaken an exhaustive study of all the memory chips available, but the chips discussed in detail are representative of the most popular ones in present use.

## G L O S S A R Y

*Access time* In general, the delay time measured from chip enable (or address) until valid data appears at the output.

*Address* Selection of a cell in a memory array for a read or a write operation.

*Capacity* The total number of bits that can be stored in a memory.

*Chip* A semiconductor circuit on a single silicon die.

*Dynamic Memory* A memory whose contents must be restored periodically.

*EPROM* An eraseable-programmable read-only memory.

*Field-Programmable* A PROM that can be programmed by the user.

*Mask-Programmable* A PROM that can be programmed only by the manufacturer.

*Matrix Addressing* Selection of a single cell in a rectangular array of cells by choosing the intersection of a single row and a single column.

*Memory Cell* The circuit used to store a single bit of information in a semiconductor memory chip.

*Nonvolatile Storage* A method whereby a loss of power will not result in a loss of stored data.

*PROM* Programmable read-only memory.

*RAM* Random-access memory.

*Read Operation* The act of detecting the contents of a memory.

*ROM* Read-only memory.

*Static Memory* A memory capable of storing data indefinitely, provided there is no loss of power.

*Volatile Storage* A method of storing information whereby a loss of power will result in a loss of the data stored.

*Write Operation* The act of storing information in a memory.


# PROBLEMS

## SECTION 12-1:

**12–1.** State the most appropriate memory type to use for each of the following:
   **a.** The working memory in a small computer
   **b.** The memory used to store permanent programs in a small computer
   **c.** A memory used to store development programs in a small computer

**12–2.** Explain the difference between an EPROM and a PROM.

**12–3.** Explain the term volatile memory.

**12–4.** Explain why an EPROM is or is not a volatile memory.

## SECTION 12-2:

**12–5.** Show the different possible rectangular arrangements for a memory that contains 32 memory cells. How many rows and columns for each case?

**12–6.** How many address lines are required for each case in Prob. 12–5?

**12–7.** What is the required address $A_4A_3A_2A_1$ to select cell 21 in Fig. 12–2c?

**12–8.** Determine how many address bits are required for a memory that has the following number of bits:
   **a.** 1024      **c.** 256
   **b.** 4098      **d.** 16,384

**12–9.** A memory chip available from Advanced Micro Devices is the Am9016, advertised as a 16K memory. How many bits of storage are there? How many address lines are required to access one bit at a time?

**12–10.** What address must be applied to the 54/74S89 in Fig. 12–13 to select the 4-bit word stored in row 14? Give the address in both binary and hexadecimal.

SECTION 12-3:

**12–11.** What is the required address in both binary and hexadecimal to select the 8-bit word in row 27 of the 54/7488A in Fig. 12–7?

**12–12.** Show a method for scanning the contents of a 5488A beginning with word 1, then word 2, and so on up to word 32, and then repeating. (*Hint*: Try using a five-flip-flop binary counter for the address *ABCDE*, or use a mod-5 counter with decoding gates, or maybe a shift counter, or . . . )

**12–13.** Write a boolean expression for address row 15 in the 7488A in Fig. 12–7.

**12–14.** Define the term mask-programmable.

**12–15.** Draw a set of timing waveforms for a 7488A similar to Fig. 12–8, assuming an access time of 35 ns.

**12–16.** Redraw Fig. 12–9*b* and show exactly how to set the switches to program the 8-bit word at address 110 101. Explain exactly what must be done to program the word 1010 0011 at this address.

**12–17.** Explain the function of $\overline{CS}$/WE for the Intel 8708 in Fig. 12–11.

**12–18.** Draw a set of control waveforms for the Intel 2716 similar to Fig. 12–11*c*. Account for address, $\overline{CE}$, $\overline{OE}$, data out, and access time.

SECTION 12-4:

**12–19.** Show how to connect 7489s in series to construct a memory that has thirty-two 4-bit words.

**12–20.** Show how to connect 7489s in parallel to construct a memory that contains sixteen 8-bit words.

**12–21.** Design the logic circuits to provide a read and a write cycle for a 7489.

**12–22.** Refer to the 74200 information in Fig. 12–15 and determine the following:
   **a.** Minimum write-enable pulse width
   **b.** Setup time, address-to-write enable
   **c.** Hold time, data from write-enable
   **d.** Maximum access time from address

**12–23.** Draw the logic diagram for a two-hundred fifty-six word 8-bit memory using '200s.

SECTION 12-5:

**12–24.** Explain the meaning of the terms RAS and CAS.

**12–25.** For the Texas Instruments TMS4116 in Fig. 12–19, what are the minimum and maximum pulse widths for $\overline{RAS}$ and for $\overline{CAS}$?

**12–26.** For the Texas Instruments TMS4116 in Fig. 12–19, what is the minimum read or write cycle time?

**12–27.** What connections must be made between the Texas Instruments TMS4500A controller in Fig. 12–20 and the TMS4116 in Fig. 12–19?

**12–28.** Explain the difference between burst and distributed refresh in a DRAM.

*SECTION 12-6:*

**12–29.** Redraw the circuit in Fig. 12–21 and show all voltage levels, assuming that the cell is not selected and storing a 1.

**12–30.** Repeat Prob. 12–29, but with the cell storing a 0.

**12–31.** Redraw the circuit in Fig. 12–21, showing the voltage levels if a 1 is stored in the cell and it is selected, assuming that W is low.

**12–32.** Repeat Prob. 12–31 with a 0 stored in the cell.

**12–33.** Draw the transmission gate in Fig. 12–22 and show the voltage levels for the following conditions:
    **a.** GATE high and $V_i$ high
    **b.** GATE high and $V_i$ low
    **c.** GATE low and $V_i$ high

**12–34.** Explain in a step-by-step fashion the voltage levels that must be applied and the order of application in order to store a 1 in the circuit in Fig. 12–23.

**12–35.** Repeat Prob. 12–34, but read a stored 0 from the cell.

**12–36.** Explain in a step-by-step fashion the voltage levels that must be applied and the order of application in order to store a 0 in the circuit in Fig. 12–24.

**12–37.** Repeat Prob. 12–36, but read a stored 1 from the cell.

# 13

## D/A AND
## A/D CONVERSION

Digital-to-analog (D/A) and analog-to-digital (A/D) conversion form two very important aspects of digital data processing. Digital-to-analog conversion involves translation of digital information into equivalent analog information. As an example, the output of a digital system might be changed to analog form for the purpose of driving a pen recorder. Similarly, an analog signal might be required for the servomotors which drive the cursor arms of a plotter. In this respect, a D/A converter (DAC) is sometimes considered a decoding device.

The process of changing an analog signal to an equivalent digital signal is accomplished by the use of an A/D converter. For example, an A/D converter is used to change the analog output signals from transducers (measuring temperature, pressure, vibration, etc.) into equivalent digital signals. These signals would then be in a form suitable for entry into a digital system. An A/D converter (ADC) is often referred to as an *encoding device* since it is used to encode signals for entry into a digital system.

Digital-to-analog conversion is a straightforward process and is considerably easier than A/D conversion. In fact, a D/A converter is usually an integral part of any A/D converter. For this reason, we consider the D/A conversion process first.

# 13-1 VARIABLE-RESISTOR NETWORK

The basic problem in converting a digital signal into an equivalent analog signal is to change the $n$ digital voltage levels into one equivalent analog voltage. This can be most easily accomplished by designing a resistive network that will change each digital level into an equivalent binary weighted voltage (or current).

As an example of what is meant by *equivalent binary weight*, consider the truth table for the 3-bit binary signal shown in Fig. 13–1. Suppose that we want to change the eight possible digital signals in this figure into equivalent analog voltages. The smallest number represented is 000; let us make this equal to 0 V. The largest number is 111; let us make this equal to +7 V. This then establishes the range of the analog signal to be developed. (There is nothing special about the voltage levels chosen; they were simply selected for convenience.)

Now, notice that between 000 and 111 there are seven discrete levels to be defined. Therefore, it will be convenient to divide the analog signal into seven levels. The smallest incremental change in the digital signal is represented by the least-significant bit (LSB), $2^0$. Thus we would like to have this bit cause a change in the analog output that is equal to one-seventh of the full-scale analog output voltage. The resistive divider will then be designed such that a 1 in the $2^0$ position will cause $+7 \times \frac{1}{7} = +1$ V at the output.

Since $2^1 = 2$ and $2^0 = 1$, it can be clearly seen that the $2^1$ bit represents a number that is twice the size of the $2^0$ bit. Therefore, a 1 in the $2^1$ bit position must cause a change in the analog output voltage that is twice the size of the LSB. The resistive divider must then be constructed such that a 1 in the $2^1$ bit position will cause a change of $+7 \times \frac{2}{7} = +2$ V in the analog output voltage.

Similarly, $2^2 = 4 = 2 \times 2^1 = 4 \times 2^0$, and thus the $2^2$ bit must cause a change in the output voltage equal to four times that of the LSB. The $2^2$ bit must then cause an output voltage change of $+7 \times \frac{4}{7} = +4$ V.

The process can be continued, and it will be seen that each successive bit must have a value twice that of the preceding bit. Thus the LSB is given a binary equivalent weight of $\frac{1}{7}$ or 1 part in 7. The next LSB is given a weight of $\frac{2}{7}$, which is twice the LSB, or 2 parts in 7. The MSB (in the case of this 3-bit system) is given a weight of $\frac{4}{7}$, which is 4 times the LSB or 4 parts in 7. Notice that the sum of the weights must equal 1. Thus $\frac{1}{7} + \frac{2}{7} + \frac{4}{7} = \frac{7}{7} = 1$. In general, the binary equivalent weight assigned to the LSB is $1/(2n - 1)$, where $n$ is the number of bits. The remaining weights are found by multiplying by 2, 4, 8, and so on.

| $2^2$ | $2^1$ | $2^0$ |
|---|---|---|
| 0 | 0 | 0 |
| 0 | 0 | 1 |
| 0 | 1 | 0 |
| 0 | 1 | 1 |
| 1 | 0 | 0 |
| 1 | 0 | 1 |
| 1 | 1 | 0 |
| 1 | 1 | 1 |

**FIG. 13–1**

| Bit | Weight |
|---|---|
| $2^0$ | 1/7 |
| $2^1$ | 2/7 |
| $2^2$ | 4/7 |
| Sum | 7/7 |

| Bit | Weight |
|---|---|
| $2^0$ | 1/15 |
| $2^1$ | 2/15 |
| $2^2$ | 4/15 |
| $2^3$ | 8/15 |
| Sum | 15/15 |

(a)                    (b)

**FIG. 13–2**  Binary equivalent weights

## EXAMPLE 13—1

Find the binary equivalent weight of each bit in a 4-bit system.

## SOLUTION

The LSB has a weight of $1/(2^4 - 1) = 1/(16 - 1) = \frac{1}{15}$, or 1 part in 15. The second LSB has a weight of $2 \times \frac{1}{15} = \frac{2}{15}$. The third LSB has a weight of $4 \times \frac{1}{15} = \frac{4}{15}$, and the MSB has a weight of $8 \times \frac{1}{15} = \frac{8}{15}$. As a check, the sum of the weights must equal 1. Thus $\frac{1}{15} + \frac{2}{15} + \frac{4}{15} + \frac{8}{15} = \frac{15}{15} = 1$. The binary equivalent weights for 3-bit and 4-bit system are summarized in Fig. 13–2.

What is now desired is a resistive divider that has three digital inputs and one analog output as shown in Fig. 13–3a. Assume that the digital input levels are $0 = 0$ V and $1 = +7$ V. Now, for an input of 001, the output will be $+1$ V. Similarly, an input of 010 will provide an output of $+2$ V, and an input of 100 will provide an output of $+4$ V. The digital input 011 is seen to be a combination of the signals 001 and 010. If the $+1$ V from the $2^0$ bit is added to the $+2$ V from the $2^1$ bit, the desired $+3$ V output for the 011 input is achieved. The other desired voltage levels are shown in Fig. 13–3b; they, too, are additive combinations of voltages.

Thus the resistive divider must do two things in order to change the digital input into an equivalent analog output voltage:

1.  The $2^0$ bit must be changed to $+1$ V, and $2^1$ bit must be changed to $+2$ V, and $2^2$ bit must be changed to $+4$ V.
2.  These three voltages representing the digital bits must be summed together to form the analog output voltage.

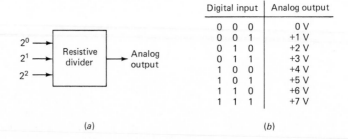

| Digital input | | | Analog output |
|---|---|---|---|
| 0 | 0 | 0 | 0 V |
| 0 | 0 | 1 | +1 V |
| 0 | 1 | 0 | +2 V |
| 0 | 1 | 1 | +3 V |
| 1 | 0 | 0 | +4 V |
| 1 | 0 | 1 | +5 V |
| 1 | 1 | 0 | +6 V |
| 1 | 1 | 1 | +7 V |

(a)                    (b)

**FIG. 13–3**

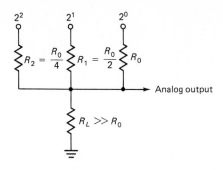

**FIG. 13–4**  Resistive ladder

A resistive divider that performs these functions is shown in Fig. 13–4. Resistors $R_0$, $R_1$, and $R_2$ form the divider network. Resistance $R_L$ represents the load to which the divider is connected and is considered to be large enough that it does not load the divider network.

Assume that the digital input signal 001 is applied to this network. Recalling that $0 = 0$ V and $1 = +7$ V, you can draw the equivalent circuit shown in Fig. 13–5. Resistance $R_L$ is considered large and is neglected. The analog output voltage $V_A$ can be most easily found by use of Millman's theorem. Millman's theorem states that the voltage appearing at any node in a resistive network is equal to the summation of the currents entering the node (found by assuming that the node voltage is zero) divided by the summation of the conductances connected to the node. In equation form, Millman's theorem is

$$V = \frac{V_1/R_1 + V_2/R_2 + V_3/R_3 + \cdots}{1/R_1 + 1/R_2 + 1/R_3 + \cdots}$$

Applying Millman's theorem to Fig. 13–5, we obtain

$$V_A = \frac{V_0/R_0 + V_1/(R_0/2) + V_2/(R_0/4)}{1/R_0 + 1/(R_0/2) + 1/(R_0/4)}$$

$$= \frac{7/R_0}{1/R_0 + 2/R_0 + 4/R_0} = \frac{7}{7} = +1 \ V$$

Drawing the equivalent circuits for the other 7-input combinations and applying Millman's theorem will lead to the table of voltages shown in Fig. 13–3 (see Prob. 13–3).

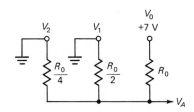

**FIG. 13–5**

☐ EXAMPLE 13—2

For a 4-input resistive divider (0 = 0 V, 1 = +10 V), find (a) the full-scale output voltage; (b) the output voltage change due to the LSB; (c) the analog output voltage for a digital input of 1011.

☐ SOLUTION

(a) The maximum output voltage occurs when all the inputs are at +10 V. If all four inputs are at +10 V, the output must also be at +10 V (ignoring the effects of $R_L$).

(b) For a 4-bit digital number, there are 16 possible states. There are 15 steps between these 16 states, and the LSB must be equal to $\frac{1}{15}$ of the full-scale output voltage. Therefore, the change in output voltage due to the LSB is $+10 \times \frac{1}{15} = +\frac{2}{3}$ V.

(c) According to Millman's theorem, the output voltage for a digital input of 1011 is

$$V_A = \frac{10/R_0 + 10/(R_0/2) + 0/(R_0/4) + 10/(R_0/8)}{1/R_0 + 1(R_0/2) + 1/(R_0/4) + 1(R_0/8)}$$

$$= \frac{110}{15} = \frac{22}{3} = +7\frac{1}{3} \text{ V}$$

To summarize, a resistive divider can be built to change a digital voltage into an equivalent analog voltage. The following criteria can be applied to this divider:

1. There must be one input resistor for each digital bit.
2. Beginning with the LSB, each following resistor value is one-half the size of the previous resistor.
3. The full-scale output voltage is equal to the positive voltage of the digital input signal. (The divider would work equally well with input voltages of 0 and $-V$.)
4. The LSB has a weight of $1/(2^n - 1)$, where $n$ is the number of input bits.
5. The change in output voltage due to a change in the LSB is equal to $V/(2^n - 1)$, where $V$ is the digital input voltage level.
6. The output voltage $V_A$ can be found for any digital input signal by using the following modified form of Millman's theorem:

$$V_A = \frac{V_0 2^0 + V_1 2^1 + V_2 2^2 + V_3 2^3 + \cdots + V_{n-1} 2^{n-1}}{2^n - 1} \tag{13-1}$$

where $V_0, V_1, V_2, V_3, \ldots, V_{n-1}$ are the digital input voltage levels (0 or V) and $n$ is the number of input bits.

☐ EXAMPLE 13—3

For a 5-bit resistive divider, determine the following: (a) the weight assigned to the LSB; (b) the weight assigned to the second and third LSB; (c) the change in output voltage due to a change in the LSB, the second LSB, and the third LSB; (d) the output voltage for a digital input of 10101. Assume 0 = 0 V and 1 = +10 V.

☐ SOLUTION

(a) The LSB weight is $1/(2^5 - 1) = 1/31$.
(b) The second LSB weight is 2/31, and the third LSB weight is 4/31.

(*c*) The LSB causes a change in the output voltage of 10/31 V. The second LSB causes an output voltage change of 20/31 V, and the third LSB causes an output voltage change of 40/31 V.

(*d*) The output voltage for a digital input of 10101 is

$$V_A = \frac{10 \times 2^0 + 0 \times 2^1 + 10 \times 2^2 + 0 \times 2^3 + 10 \times 2^4}{2^5 - 1}$$

$$= \frac{10(1 + 4 + 16)}{32 - 1} = \frac{210}{31} = +6.77 \text{ V}$$

This resistive divider has two serious drawbacks. The first is the fact that each resistor in the network has a different value. Since these dividers are usually constructed by using precision resistors, the added expense becomes unattractive. Moreover, the resistor used for the MSB is required to handle a much greater current than that used for the LSB resistor. For example, in a 10-bit system, the current through the MSB resistor is approximately 500 times as large as the current through the LSB resistor (see Prob. 13–5). For these reasons, a second type of resistive network, called a "ladder," has been developed.

## 13-2  BINARY LADDER

The binary ladder is a resistive network whose output voltage is a properly weighted sum of the digital inputs. Such a ladder, designed for 4 bits, is shown in Fig. 13–6. It is constructed of resistors that have only two values and thus overcomes one of the objections to the resistive divider previously discussed. The left end of the ladder is terminated in a resistance of $2R$, and we shall assume for the moment that the right end of the ladder (the output) is open-circuited.

Let us now examine the resistive properties of the network, assuming that all the digital inputs are at ground. Beginning at node $A$, the total resistance looking into the terminating resistor is $2R$. The total resistance looking out toward the $2^0$ input is also $2R$. These two resistors can be combined to form an equivalent resistor of value $R$ as shown in Fig. 13–7a.

Now, moving to node $B$, we see that the total resistance looking into the branch toward node $A$ is $2R$, as is the total resistance looking out toward the $2^1$ input. These resistors can be combined to simplify the network as shown in Fig. 13–7b.

From Fig. 13–7b, it can be seen that the total resistance looking from node $C$ down the branch toward node $B$ or out the branch toward the $2^2$ input is still $2R$. The circuit in Fig. 13–7b can then be reduced to the equivalent as shown in Fig. 13–7c.

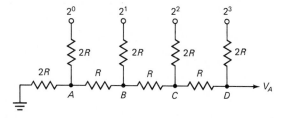

**FIG. 13–6**  Binary ladder

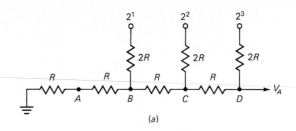

(a)

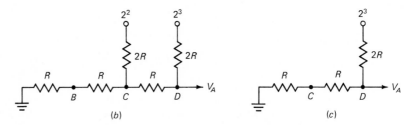

(b)                    (c)

**FIG. 13–7**

From this equivalent circuit, it is clear that the resistance looking back toward node $C$ is $2R$, as is the resistance looking out toward the $2^3$ input.

From the preceding discussion, we can conclude that the total resistance looking from any node back toward the terminating resistor or out toward the digital input is $2R$. Notice that this is true regardless of whether the digital inputs are at ground or $+V$. The justification for this statement is the fact that the internal impedance of an ideal voltage source is $0\ \Omega$, and we are assuming that the digital inputs are ideal voltage sources.

We can use the resistance characteristics of the ladder to determine the output voltages for the various digital inputs. First, assume that the digital input signal is 1000. With this input signal, the binary ladder can be drawn as shown in Fig. 13–8a. Since there are no voltage sources to the left of node $D$, the entire network to the left of this node can be replaced by a resistance of $2R$ to form the equivalent circuit shown in Fig. 13–8b. From this equivalent circuit, it can be easily seen that the output voltage is

$$V_A = V \times \frac{2R}{2R + 2R} = \frac{+V}{2}$$

Thus a 1 in the MSB position will provide an output voltage of $+V/2$.

To determine the output voltage due to the second MSB, assume a digital input signal of 0100. This can be represented by the circuit shown in Fig. 13–9a. Since there are no voltage sources to the left of node $C$, the entire network to the left of this node can be replaced by a resistance of $2R$, as shown in Fig. 13–9b. Let us now replace the network to the left of node $C$ with its Thévenin equivalent by cutting the circuit on the jagged line shown in Fig. 13–9b. The Thévenin equivalent is clearly a resistance $R$ in series with a voltage source $+V/2$. The final equivalent circuit with the Thévenin equivalent included is shown in Fig. 13–9c. From this circuit, the output voltage is clearly

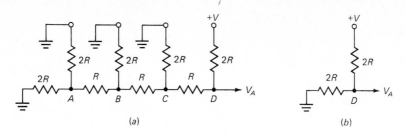

**FIG. 13–8** (*a*) Binary ladder with a digital input of 1000 (*b*) Equivalent circuit for a digital input of 1000

$$V_A = \frac{+V}{2} \times \frac{2R}{R + R + 2R} = \frac{+V}{4}$$

Thus the second MSB provides an output voltage of $+V/4$.

This process can be continued, and it can be shown that the third MSB provides an output voltage of $+V/8$, the fourth MSB provides an output voltage of $+V/16$, and so on. The output voltages for the binary ladder are summarized in Fig. 13–10; notice that each digital input is transformed into a properly weighted binary output voltage.

☐ EXAMPLE 13–4

What are the output voltages caused by each bit in a 5-bit ladder if the input levels are $0 = 0$ V and $1 = +10$ V?

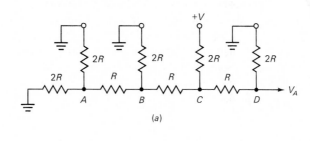

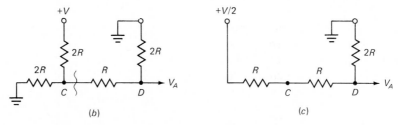

**FIG. 13–9** (*a*) Binary ladder with a digital input of 0100 (*b*) Partially reduced equivalent circuit (*c*) Final equivalent circuit using Thévenin's theorem

The output voltages can be easily calculated by using Fig. 13–10. They are

$$\text{First MSB} \quad V_A = \frac{V}{2} = \frac{+10}{2} = +5 \text{ V}$$

$$\text{Second MSB} \quad V_A = \frac{V}{4} = \frac{+10}{4} = +2.5 \text{ V}$$

$$\text{Third MSB} \quad V_A = \frac{V}{8} = \frac{+10}{8} = +1.25 \text{ V}$$

$$\text{Fourth MSB} \quad V_A = \frac{V}{16} = \frac{+10}{16} = +0.625 \text{ V}$$

$$\text{LSB} = \text{fifth MSB} \quad V_A = \frac{V}{32} = \frac{+10}{32} = +0.3125 \text{ V}$$

Since this ladder is composed of linear resistors, it is a linear network and the principle of superposition can be used. This means that the total output voltage due to a combination of input digital levels can be found by simply taking the sum of the output levels caused by each digital input individually.

In equation form, the output voltage is given by

$$V_A = \frac{V}{2} + \frac{V}{4} + \frac{V}{8} + \frac{V}{16} + \cdots + \frac{V}{2^n} \qquad (13\text{–}2)$$

where $n$ is the total number of bits at the input.

This equation can be simplified somewhat by factoring and collecting terms. The output voltage can then be given in the form

$$V_A = \frac{V_0 2^0 + V_1 2^1 + V_2 2^2 + V_3 2^3 + \cdots + V_{n-1} 2^{n-1}}{2^n} \qquad (13\text{–}3)$$

where $V_0, V_1, V_2, V_3, \ldots, V_{n-1}$ are the digital input voltage levels. Equation (13–3) can be used to find the output voltage from the ladder for any digital input signal.

| Bit position | Binary weight | Output voltage |
|---|---|---|
| MSB | 1/2 | $V/2$ |
| 2d MSB | 1/4 | $V/4$ |
| 3d MSB | 1/8 | $V/8$ |
| 4th MSB | 1/16 | $V/16$ |
| 5th MSB | 1/32 | $V/32$ |
| 6th MSB | 1/64 | $V/64$ |
| 7th MSB | 1/128 | $V/128$ |
| ⋮ | ⋮ | ⋮ |
| $N$th MSB | $1/2^N$ | $V/2^N$ |

**FIG. 13–10**  Binary ladder output voltages

Find the output voltage from a 5-bit ladder that has a digital input of 11010. Assume that $0 = 0$ V and $1 = +10$ V.

□ SOLUTION

By Eq. (13–3):

$$V_A = \frac{0 \times 2^0 + 10 \times 2^1 + 0 \times 2^2 + 10 \times 2^3 + 10 \times 2^4}{2^5}$$

$$= \frac{10(2 + 8 + 16)}{32} = \frac{10 \times 26}{32} = +8.125 \text{ V}$$

This solution can be checked by adding the individual bit contributions calculated in Example 13–4.

Notice that Eq. (13–3) is very similar to Eq. (13–1), which was developed for the resistive divider. They are, in fact, identical with the exception of the denominators. This is a subtle but very important difference. Recall that the full-scale voltage for the resistive divider is equal to the voltage level of the digital input 1. On the other hand, examination of Eq. (13–2) reveals that the full-scale voltage for the ladder is given by

$$V_A = V\left(\frac{1}{2} + \frac{1}{4} + \frac{1}{8} + \frac{1}{16} + \cdots + \frac{1}{2^n}\right)$$

The terms inside the brackets form a geometric series whose sum approaches 1, given a sufficient number of terms. However, it never quite reaches 1. Therefore, the full-scale output voltage of the ladder approaches $V$ in the limit, but never quite reaches it.

□ EXAMPLE 13—6

What is the full-scale output voltage of the 5-bit ladder in Example 13–4?

□ SOLUTION

The full-scale voltage is simply the sum of the individual bit voltages. Thus

$$V = 5 + 2.5 + 1.25 + 0.625 + 0.3125 = +9.6875 \text{ V}$$

To keep the ladder in perfect balance and to maintain symmetry, the output of the ladder should be terminated in a resistance of $2R$. This will result in a lowering of the output voltage, but if the $2R$ load is maintained constant, the output voltages will still be a properly weighted sum of the binary input bits. If the load is varied, the output voltage will not be a properly weighted sum, and care must be exercised to ensure that the load resistance is constant.

Terminating the output of the ladder with a load of $2R$ also ensures that the input resistance to the ladder seen by each of the digital voltage sources is constant. With the ladder balanced in this manner, the resistance looking into any branch from any node has a value of $2R$. Thus the input resistance seen by any input digital source is $3R$. This is a definite advantage over the resistive divider, since the digital voltage sources can now all be designed for the same load.

## EXAMPLE 13–7

Suppose that the value of $R$ for the 5-bit ladder described in Example 13–4 is 1000 Ω. Determine the current that each input digital voltage source must be capable of supplying. Also determine the full-scale output voltage, assuming that the ladder is terminated with a load resistance of 2000 Ω.

## SOLUTION

The input resistance into the ladder seen by each of the digital sources is $3R = 3000$ Ω. Thus, for a voltage level of $+10$ V, each source must be capable of supplying $I = 10/(3 \times 10^3) = 3\frac{1}{3}$ mA (without the $2R$ load resistor, the resistance looking into the MSB terminal is actually $4R$). The no-load output voltage of the ladder has already been determined in Example 13–6. This open-circuit output voltage along with the open-circuit output resistance can be used to form a Thévenin equivalent circuit for the output of the ladder. The resistance looking back into the ladder is clearly $R = 1000$ Ω. Thus the Thévenin equivalent is as shown in Fig. 13–11. From this figure, the output voltage is

$$V_A = +9.6875 \times \frac{2R}{2R + R} = +6.4583 \text{ V}$$

The operational amplifier (OA) shown in Fig. 13–12$a$ is connected as a unity-gain noninverting amplifier. It has a very high input impedance, and the output voltage is equal to the input voltage. It is thus a good buffer amplifier for connection to the output of a resistive ladder. It will not load down the ladder and thus will not disturb the ladder output voltage $V_A$; $V_A$ will then appear at the output of the OA.

Connecting an OA with a feedback resistor $R$ as shown in Fig. 13–12$b$ results in an amplifier that acts as an inverting current-to-voltage amplifier. That is, the output voltage $V_A$ is equal to the negative of the input current $I$ multiplied by $R$. The input impedance to this amplifier is essentially 0 Ω; thus, when it is connected an $R$-$2R$ ladder, the connecting point is virtually at ground potential. In this configuration, the $R$-$2R$ ladder will produce a current output $I$ that is a binary weighted sum of the input digital levels. For instance, the MSB produces a current of $V/2R$. The second MSB produces a current of $V/4V$, and so on. But the OA multiplies these currents by $-R$, and thus $V_A$ is

$$V_A = (-R)(\frac{V}{2R} + \frac{V}{4R} + \cdots) = -\frac{V}{2} - \frac{V}{4} - \cdots$$

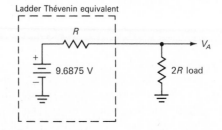

Ladder Thévenin equivalent

9.6875 V  ·  $V_A$  ·  $2R$ load

**FIG. 13–11**  Example 11–7

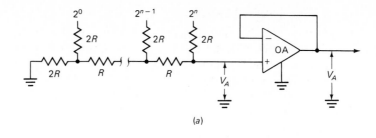

(a)

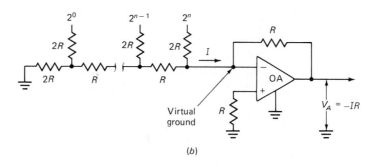

(b)

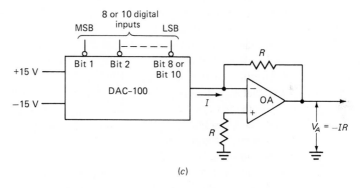

(c)

**FIG. 13–12**

This is exactly the same expression given in Eqs. (13–2) and (13–3) except for the sign. Thus the D/A converter in Fig. 13–12a and b will provide the same output voltage $V_A$ except for sign. In Fig. 13–12a, the R-2R ladder and OA are said to operate in a voltage mode, while the connection in Fig. 13–12b is said to operate in a current mode.

The Precision Monolithics DAC-100 is a precision 8- or 10-bit D/A converter which is available in a 16-pin dual in-line package (DIP). It incorporates transistors as current-source switches in conjunction with a thin-film precision R-2R ladder network. A block diagram is shown in Fig. 13–12c. Complementary logic is used with the DAC-100. That is, all 0s at the input will produce full-scale output at $V_A$ (either +10 or +5 V), while all 1s at the input will produce zero output voltage. For the DAC-100, $I$ has a range of 0 to 2 mA, and thus this connection is operating in a current mode.

## 13-3 D/A CONVERTER

Either the resistive divider or the ladder can be used as the basis for a digital-to-analog converter (DAC). It is in the resistive network that the actual translation from a digital signal to an analog voltage takes place. There is, however, the need for additional circuitry to complete the design of the D/A converter.

As an integral part of the D/A converter there must be a register that can be used to store the digital information. This register could be any one of the many types discussed in previous chapters. The simplest register is formed by use of $RS$ flip-flops, with one flip-flop per bit. There must also be level amplifiers between the register and the resistive network to ensure that the digital signals presented to the network are all of the same level and are constant. Finally, there must be some form of gating on the input of the register such that the flip-flops can be set with the proper information from the digital system. A complete D/A converter in block-diagram form is shown in Fig. 13–13a.

Let us expand on the block diagram shown in this Fig. 13–13a by drawing the complete schematic for a 4-bit D/A converter as shown in Fig. 13–13b. You will recognize that the resistor network used is of the ladder type.

The level amplifiers each have two inputs: one input is the $+10$ V from the precision voltage source, and the other is from a flip-flop. The amplifiers work in such a way that when the input from a flip-flop is high, the output of the amplifier is at $+10$ V. When the input from the flip-flop is low, the output is 0 V.

The four flip-flops form the register necessary for storing the digital information. The flip-flop on the right represents the MSB, and the flip-flop on the left represents the LSB. Each flip-flop is a simple $RS$ latch and requires a positive level at the $R$ or $S$ input to reset or set it. The gating scheme for entering information into the register is straightforward and should be easy to understand. With this particular gating scheme, the flip-flops need not be reset (or set) each time new information is entered. When the READ IN line goes high, only one of the two gate outputs connected to each flip-flop is high, and the flip-flop is set or reset accordingly. Thus data are entered into the register each time the READ IN (strobe) pulse occurs.

Quite often it is necessary to decode more than one signal—for example, the $X$ and $Y$ coordinates for a plotting board. In this event, there are two ways in which to decode the signals.

The first and most obvious method is simply to use one D/A converter for each signal. This method, shown in Fig. 13–14a, has the advantage that each signal to be decoded is held in its register and the analog output voltage is then held fixed. The digital input lines are connected in parallel to each converter. The proper converter is then selected for decoding by the select lines.

The second method involves the use of only one D/A converter and switching its output. This is called *multiplexing*, and such a system is shown in Fig. 13–14b. The disadvantage here is that the analog output signal must be held between sampling periods, and the outputs must therefore be equipped with sample-and-hold amplifiers.

An operational amplifier (OA) connected as in Fig. 13–15a is a unity-gain, noninverting voltage amplifier—that is, $V_{out} = V_{in}$. Two such OAs are used with a capacitor in Fig. 13–15b to form a sample-and-hold amplifier. When the switch is closed, the capacitor charges to the D/A converter output voltage. When the switch is opened, the capacitor

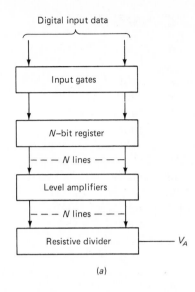

(a)

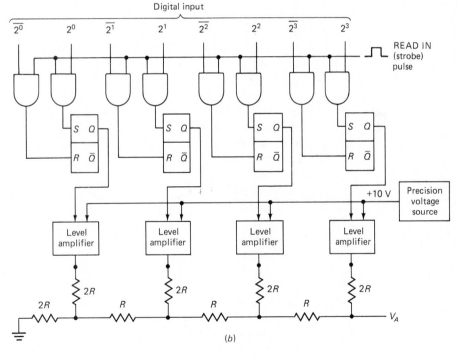

(b)

**FIG. 13–13**   4-bit D/A converter

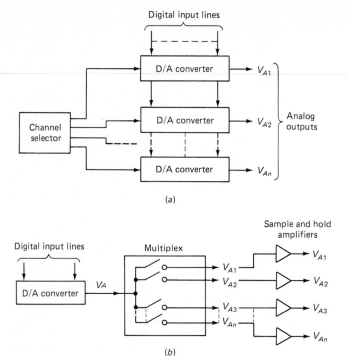

(a)

(b)

**FIG. 13–14** Decoding a number of signals (a) Channel selection method (b) Multiplex method

holds the voltage level until the next sampling time. The operational amplifier provides a large input impedance so as not to discharge the capacitor appreciably and at the same time offers gain to drive external circuits.

When the D/A converter is used in conjunction with a multiplexer, the maximum rate at which the converter can operate must be considered. Each time data is shifted into the register, transients appear at the output of the converter. This is due mainly to the fact that each flip-flop has different rise and fall times. Thus a settling time must be allowed between the time data is shifted into the register and the time the analog voltage is read out. This settling time is the main factor in determining the maximum rate of multiplexing the output. The worst case is when all bits change (e.g., from 1000 to 0111).

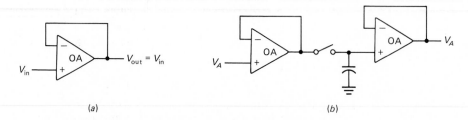

(a)                                                                      (b)

**FIG. 13–15** (a) Unity gain amplifier (b) Sample and hold circuit

Naturally, the capacitors on the sample-and-hold amplifiers are not capable of holding a voltage indefinitely; therefore, the sampling rate must be sufficient to ensure that these voltages do not decay appreciably between samples. The sampling rate is a function of the capacitors as well as the frequency of the analog signal which is expected at the output of the converter.

At this point, you might be curious to know just how fast a signal must be sampled in order to preserve its integrity. Common sense leads to the conclusion that the more often the signal is sampled, the less the sample degrades between samples. On the other hand, if too few samples are taken, the signal degrades too much (the sample and hold capacitors discharge too much), and the signal information is lost. We would like to reduce the sampling rate to the minimum necessary to extract all the necessary information from the signal. The solution to this problem involves more that we have time for here, but the results are easy enough to apply.

First, if the signal in question is sinusoidal, it is necessary to sample at only *twice* the signal frequency. For instance, if the signal is a 5-kHz sine wave, it must be sampled at a rate greater than or equal to 10 kHz. In other words, a sample must be taken every $\frac{1}{10000}$ s = 100 $\mu$s. What if the waveform is not sinusoidal? Any waveform that is periodic can be represented by a summation of sine and cosine terms, with each succeeding term having a higher frequency. In this case, it will be necessary to sample at a rate equal to twice the highest frequency of interest.

Two simple but important tests that can be performed to check the proper operation of the D/A converter are the *steady-state accuracy test* and the *monotonicity test*.

The steady-state accuracy test involves setting a known digital number in the input register, measuring the analog output with an accurate meter, and comparing with the theoretical value.

Checking for monotonicity means checking that the output voltage increases regularly as the input digital signal increases. This can be accomplished by using a counter as the digital input signal and observing the analog output on an oscilloscope. For proper monotonicity, the output waveform should be a perfect staircase waveform, as shown in Fig. 13–16. The steps on the staircase waveform must be equally spaced and of the exact same amplitude. Missing steps, steps of different amplitude, or steps in a downward fashion indicate malfunctions.

The monotonicity test does not check the system for accuracy, but if the system passes the test, it is relatively certain that the converter error is less than 1 LSB. Converter accuracy and resolution are the subjects of the next section.

A DAC can be regarded as a logic block having numerous digital inputs and a single analog output as seen in Fig. 13–16*b*. It is interesting to compare this logic block with the potentiometer shown in Fig. 13–16*c*. The analog output voltage of the DAC is controlled by the digital input signals, while the analog output voltage of the potentiometer is controlled by mechanical rotation of the potentiometer shaft. Considered in this fashion, it is easy to see how a DAC could be used to generate a voltage waveform (sawtooth, triangular, sinusoidal, etc.). It is, in effect, a digitally controlled voltage generator!

□ EXAMPLE 13—8

Suppose that in the course of a monotonicity check on the 4-bit converter in Fig. 13–13, the waveform shown in Fig. 13–17 is observed. What is the probable malfunction in the converter?

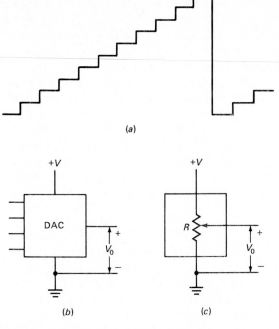

(a)

(b)                    (c)

**FIG. 13–16**   Correct output voltage waveform for monotonicity test

☐   SOLUTION

There is obviously some malfunction since the actual output waveform is not continuously increasing as it should be. The actual digital inputs are shown directly below the waveform. Notice that the converter functions correctly up to count 3. At count 4, however, the output should be 4 units in amplitude. Instead, it drops to 0. It remains 4 units below the

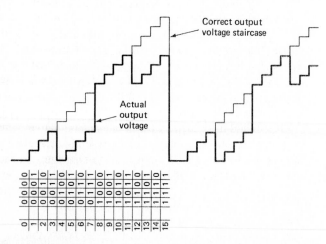

**FIG. 13–17**   Irregular output voltage for Example 13–8

correct level until it reaches count 8. Then, from count 8 to 11, the output level is correct. But again at count 12 the output falls 4 units below the correct level and remains there for the next four levels. If you examine the waveform carefully, you will note that the output is 4 units below normal during the time when the $2^2$ bit is supposed to be high. This then suggests that the $2^2$ bit is being dropped (i.e., the $2^2$ input to the ladder is not being held high). This means that the $2^2$-level amplifier is malfunctioning or the $2^2$ AND gate is not operating properly. In any case, the monotonicity check has clearly shown that the second MSB is not being used and that the converter is not operating properly.

Digital-to-analog converters, as well as sample-and-hold amplifiers, are readily obtainable commercial products. Each unit is constructed in a single package; general-purpose economy units are available with 6-, 8-, 10-, and 12-bit resolution, and high-resolution units with up to 16-bit resolution are available.[1]

# 13-4 D/A ACCURACY AND RESOLUTION

Two very important aspects of the D/A converter are the resolution and the accuracy of the conversion. There is a definite distinction between the two, and you should clearly understand the differences.

The accuracy of the D/A converter is primarily a function of the accuracy of the precision resistors used in the ladder and the precision of the reference voltage supply used. Accuracy is a measure of how close the actual output voltage is to the theoretical output value.

For example, suppose that the theoretical output voltage for a particular input should be +10 V. An accuracy of 10 percent means that the actual output voltage must be somewhere between +9 and +11 V. Similarly, if the actual output voltage were somewhere between +9.9 and +10.1 V, this would imply an accuracy of 1 percent.

Resolution, on the other hand, defines the smallest increment in voltage that can be discerned. Resolution is primarily a function of the number of bits in the digital input signal; that is, the smallest increment in output voltage is determined by the LSB.

In a 4-bit system using a ladder, for example, the LSB has a weight of $\frac{1}{16}$. This means that the smallest increment in output voltage is $\frac{1}{16}$ of the input voltage. To make the arithmetic easy, let us assume that this 4-bit system has input voltage levels of +16 V. Since the LSB has a weight of $\frac{1}{16}$, a change in the LSB results in a change of 1 V in the output. Thus the output voltage changes in steps (or increments) of 1 V. The output voltage of this converter is then the staircase shown in Fig. 13–16 and ranges from 0 to +15 V in 1-V increments. This converter can be used to represent analog voltages from 0 to +15 V, but it cannot resolve voltages into increments smaller than 1 V. If we desired to produce +4.2 V using this converter, therefore, the actual output voltage would be +4.0 V. Similarly, if we desired a voltage of +7.8 V, the actual output voltage would be +8.0 V. It is clear that this converter is not capable of distinguishing voltages finer than 1 V, which is the resolution of the converter.

[1]The manufacturers of these units are Datel Systems, Inc., 1020 Turnpike St., Canton, MA; Analog Devices, Inc., Norwood, MA; and Precision Monolithics, Inc., 1500 Space Park Drive, Santa Clara, CA.

If we wanted to represent voltages to a finer resolution, we would have to use a converter with more input bits. As an example, the LSB of a 10-bit converter has a weight of $\frac{1}{1024}$. Thus the smallest incremental change in the output of this converter is approximately $\frac{1}{1000}$ of the full-scale voltage. If this converter has a $+10$ V full-scale output, the resolution is approximately $+10 \times \frac{1}{1000} = 10$ mV. This converter is then capable of representing voltages to within 10 mV.

☐ EXAMPLE 13—9

What is the resolution of a 9-bit D/A converter which uses a ladder network? What is this resolution expressed as a percent? If the full-scale output voltage of this converter is $+5$ V, what is the resolution in volts?

☐ SOLUTION

The LSB in a 9-bit system has a weight of $\frac{1}{512}$. Thus this converter has a resolution of 1 part in 512. The resolution expressed as a percentage is $\frac{1}{512} \times 100$ percent $\cong 0.2$ percent. The voltage resolution is obtained by multiplying the weight of the LSB by the full-scale output voltage. Thus the resolution in volts is $\frac{1}{512} \times 5 \cong 10$ mV.

☐ EXAMPLE 13—10

How many bits are required at the input of a converter if it is necessary to resolve voltages to 5 mV and the ladder has $+10$ V full scale?

☐ SOLUTION

The LSB of an 11-bit system has a resolution of $\frac{1}{2048}$. This would provide a resolution at the output of $\frac{1}{2048} \times +10 \cong +5$ mV.

It is important to realize that resolution and accuracy in a system should be compatible. For example, in the 4-bit system previously discussed, the resolution was found to be 1 V. Clearly it would be unjustifiable to construct such a system to an accuracy of 0.1 percent. This would mean that the system would be accurate to 16 mV but would be capable of distinguishing only to the nearest 1 V.

Similarly, it would be wasteful to construct the 11-bit system described in Example 13–10 to an accuracy of only 1 percent. This would mean that the output voltage would be accurate only to 100 mV, whereas it is capable of distinguishing to the nearest 5 mV.

## 13-5 A/D CONVERTER—SIMULTANEOUS CONVERSION

The process of converting an analog voltage into an equivalent digital signal is known as *analog-to-digital conversion* (ADC). This operation is somewhat more complicated than the converse operation of D/A conversion (DAC). A number of different methods have been developed, the simplest of which is probably the simultaneous method.

The simultaneous method of A/D conversion is based on the use of a number of comparator circuits. One such system using three comparator circuits is shown in Fig. 13–18. The analog signal to be digitized serves as one of the inputs to each comparator. The second input is a standard reference voltage. The reference voltages used are $+V/4$,

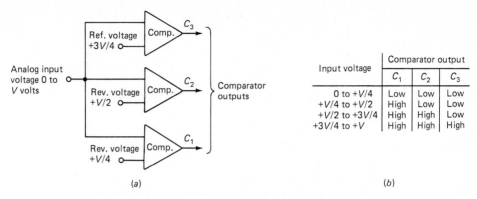

**FIG. 13–18** Simultaneous A/D conversion (*a*) Logic diagram (*b*) Comparator outputs for input voltage ranges

$+V/2$, and $+3V/4$. The system is then capable of accepting an analog input voltage between 0 and $+V$.

If the analog input signal exceeds the reference voltage to any comparator, that comparator turns on. (Let's assume that this means that the output of the comparator goes high.) Now, if all the comparators are off, the analog input signal must be between 0 and $+V/4$. If $C_1$ is high (comparator $C_1$ is on) and $C_2$ and $C_3$ are low, the input must be between $+V/4$ and $+V/2$ V. If $C_1$ and $C_2$ are high while $C_3$ is low, the input must be between $+V/2$ and $+3V/4$. Finally, if all comparator outputs are high, the input signal must be between $+3V/4$ and $+V$. The comparator output levels for the various ranges of input voltages are summarized in Fig. 13–18*b*.

Examination of Fig. 13–18 reveals that there are four voltage ranges that can be detected by this converter. Four ranges can be effectively discerned by two binary digits (bits). The three comparator outputs can then be fed into a coding network to provide 2 bits which are equivalent to the input analog voltage. The bits of the coding network can then be entered into a flip-flop register for storage. The complete block diagram for such an A/D converter is shown in Fig. 13–19.

In order to gain a clear understanding of the operation of the simultaneous A/D converter, let us investigate the 3-bit converter shown in Fig. 13–20*a*. Notice that in order to convert the input signal to a digital signal having 3 bits, it is necessary to have seven comparators (this allows a division of the input into eight ranges). For the 2-bit converter, remember that three comparators were necessary for defining four ranges. In general, it can be said that $2^n - 1$ comparators are required to convert to a digital signal that has $n$ bits. Some of the comparators have inverters at their outputs since both $C$ and $\overline{C}$ are needed for the encoding matrix.

The encoding matrix must accept seven input levels and encode them into a 3-bit binary number (having eight possible states). Operation of the encoding matrix can be most easily understood by examination of the table of outputs in Fig. 13–21.

The $2^2$ bit is easiest to determine since it must be high (the $2^2$ flip-flop must be set) whenever $C_4$ is high.

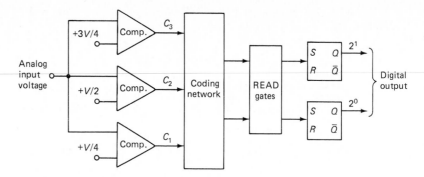

**FIG. 13–19** 2-bit simultaneous A/D converter

The $2^1$ line must be high whenever $C_2$ is high and $\overline{C_4}$ is high, or whenever $C_6$ is high. In equation form, we can write $2^1 = C_2\overline{C_4} + C_6$.

The logic equation for the $2^0$ bit can be found in a similar manner; it is

$$2^0 = C_1\overline{C_2} + C_3\overline{C_4} + C_5\overline{C_6} + C_7$$

The transfer of data from the encoding matrix into the register must be carried out in two steps. First, a positive reset pulse must appear on the RESET line to reset all the flip-flops low. Then, a positive READ pulse allows the proper READ gates to go high and thus transfer the digital information into the flip-flops.

Interestingly, a convenient application for a 9318 priority encoder is to use it to replace all the digital logic as shown in Fig. 13–20b. Of course, the inputs $C_1, C_2, \ldots, C_7$ must be TTL-compatible. In essence, the output of the 9318 is a digital number that reflects the highest-order zero input; this corresponds to the lowest reference voltage that still exceeds the input analog voltage.

The construction of a simultaneous A/D converter is quite straightforward and relatively easy to understand. However, as the number of bits in the desired digital number increases, the number of comparators increases very rapidly ($2^n - 1$), and the problem soon becomes unmanageable. Even though this method is simple and is capable of extremely fast conversion rates, there are preferable methods for digitizing numbers having more than 3 or 4 bits. Incidentally, because it is so fast, this type of converter is frequently called a "flash" converter.

# 13-6 A/D CONVERTER-COUNTER METHOD

A higher-resolution A/D converter using only one comparator could be constructed if a variable reference voltage were available. This reference voltage could then be applied to

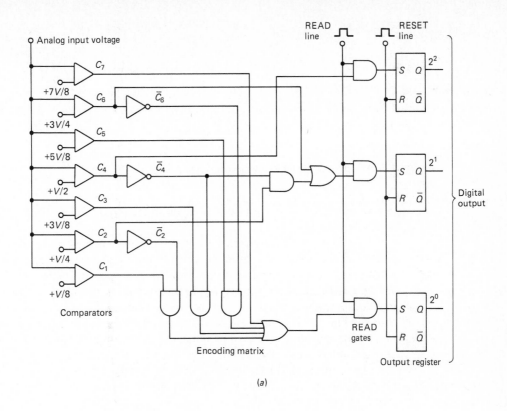

(a)

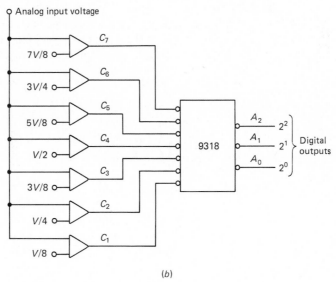

(b)

**FIG. 13–20**  3-bit simultaneous A/D converter (a) Logic diagram (b) Using a 9318 priority encoder

| Input voltage | Comparator for level | | | | | | | Binary output | | |
|---|---|---|---|---|---|---|---|---|---|---|
| | $C_1$ | $C_2$ | $C_3$ | $C_4$ | $C_5$ | $C_6$ | $C_7$ | $2^2$ | $2^1$ | $2^0$ |
| 0 to $V/8$ | Low | Low | Low | Low | Low | Low | Low | 0 | 0 | 0 |
| $V/8$ to $V/4$ | High | Low | Low | Low | Low | Low | Low | 0 | 0 | 1 |
| $V/4$ to $3V/8$ | High | High | Low | Low | Low | Low | Low | 0 | 1 | 0 |
| $3V/8$ to $V/2$ | High | High | High | Low | Low | Low | Low | 0 | 1 | 1 |
| $V/2$ to $5V/8$ | High | High | High | High | Low | Low | Low | 1 | 0 | 0 |
| $5V/8$ to $3V/4$ | High | High | High | High | High | Low | Low | 1 | 0 | 1 |
| $3V/4$ to $7V/8$ | High | High | High | High | High | High | Low | 1 | 1 | 0 |
| $7V/8$ to $V$ | High | High | High | High | High | High | High | 1 | 1 | 1 |

**FIG. 13–21**  Logic table for the converter in Fig. 13–20

the comparator, and when it became equal to the input analog voltage, the conversion would be complete.

To construct such a converter, let us begin with a simple binary counter. The digital output signals will be taken from this counter, and thus we want it to be an $n$-bit counter, where $n$ is the desired number of bits. Now let us connect the output of this counter to a standard binary ladder to form a simple D/A converter. If a clock is now applied to the input of the counter, the output of the binary ladder is the familiar staircase waveform shown in Fig. 13–16. This waveform is exactly the reference voltage signal we would like to have for the comparator! With a minimum of gating and control circuitry, this simple D/A converter can be changed into the desired A/D converter.

Figure 13–22 shows the block diagram for a counter-type A/D converter. The operation of the counter is as follows. First, the counter is reset to all 0s. Then, when a convert signal appears on the START line, the gate opens and clock pulses are allowed to pass through to the input of the counter. The counter advances through its normal binary count sequence, and the staircase waveform is generated at the output of the ladder. This waveform is applied to one side of the comparator, and the analog input voltage is applied to the other side. When the reference voltage equals (or exceeds) the input analog voltage,

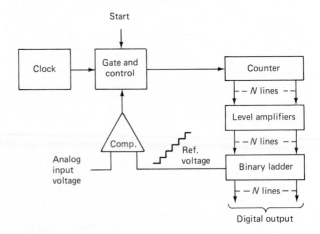

**FIG. 13–22**  Counter type A/D converter

the gate is closed, the counter stops, and the conversion is complete. The number stored in the counter is now the digital equivalent of the analog input voltage.

Notice that this converter is composed of a D/A converter (the counter, level amplifiers, and the binary ladder), one comparator, a clock, and the gate and control circuitry. This can really be considered as a closed-loop control system. An error signal is generated at the output of the comparator by taking the difference between the analog input signal and the feedback signal (staircase reference voltage). The error is detected by the control circuit, and the clock is allowed to advance the counter. The counter advances in such a way as to reduce the error signal by increasing the feedback voltage. When the error is reduced to zero, the feedback voltage is equal to the analog input signal, the control circuitry stops the clock from advancing the counter, and the system comes to rest.

The counter-type A/D converter provides a very good method for digitizing to a high resolution. This method is much simpler than the simultaneous method for high resolution, but the conversion time required is longer. Since the counter always begins at zero and counts through its normal binary sequence, as many as $2^n$ counts may be necessary before conversion is complete. The average conversion time is, of course, $2^n/2$ or $2^{n-1}$ counts.

The counter advances one count for each cycle of the clock, and the clock therefore determines the conversion rate. Suppose, for example, that we have a 10-bit converter. It requires 1024 clock cycles for a full-scale count. If we are using a 1-MHz clock, the counter advances 1 count every microsecond. Thus, to count full-scale requires $1024 \times 10^{-6} = 1.024$ ms. The converter reaches one-half full scale in half this time, or in 0.512 ms. The time required to reach one-half full scale can be considered the *average* conversion time for a large number of conversions.

☐ EXAMPLE 13—11

Suppose that the converter shown in Fig. 13–22 is an 8-bit converter driven by a 500-kHz clock. Find (*a*) the maximum conversion time; (*b*) the average conversion time; (*c*) the maximum conversion rate.

☐ SOLUTION

(*a*) An 8-bit converter has a maximum of $2^8 = 256$ counts. With a 500-kHz clock, the counter advances at the rate of 1 count each 2 $\mu$s. To advance 256 counts requires $256 \times 2 \times 10^{-6} = 512 \times 10^{-6} = 512$ $\mu$s.

(*b*) The average conversion time is one-half the maximum conversion time. Thus it is $1/2 \times 0.512 \times 10^{-3} = 0.256$ ms.

(*c*) The maximum conversion rate is determined by the longest conversion time. Since the converter has a maximum conversion time of 0.512 ms, it is capable of making at least $1/(0.512 \times 10^{-3}) \cong 1953$ conversions per second.

Figure 13–23 shows one method of implementing the control circuitry for the converter shown in Fig. 13–22. The waveforms for one conversion are also shown. A conversion is initiated by the receipt of a START signal. The positive edge of the START pulse is used to reset all the flip-flops in the counter and to trigger the one-shot. The output of the one-shot sets the control flip-flop, which makes the AND gate true and allows clock pulses to advance the counter.

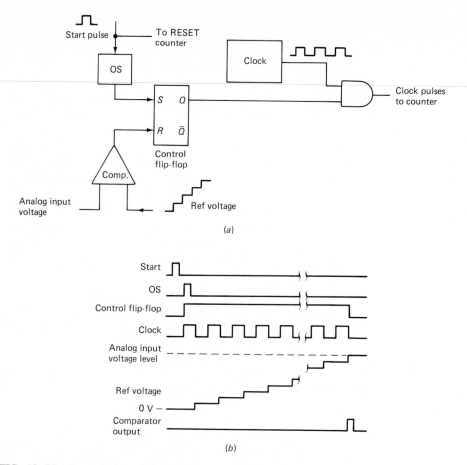

**FIG. 13–23** Control of the A/D converter in Fig. 13–22

The delay between the RESET pulse to the flip-flops and the beginning of the clock pulses (ensured by the one-shot) is to ensure that all flip-flops are reset before counting begins. This is a definite attempt to avoid any racing problems.

With the control flip-flop set, the counter advances through its normal count sequence until the staircase voltage from the ladder is equal to the analog input voltage. At this time, the comparator output changes state, generating a positive pulse which resets the control flip-flop. Thus the AND gate is closed and counting ceases. The counter now holds a digital number which is equivalent to the analog input voltage. The converter remains in this state until another conversion signal is received.

If a new start signal is generated immediately after each conversion is completed, the converter will operate at its maximum rate. The converter could then be used to digitize a signal as shown in Fig. 13–24a. Notice that the conversion times in digitizing this signal are not constant but depend on the amplitude of the input signal. The analog input signal can be reconstructed from the digital information by drawing straight lines from each digitized point to the next. Such a reconstruction is shown in Fig. 13–24b; it is, indeed, a

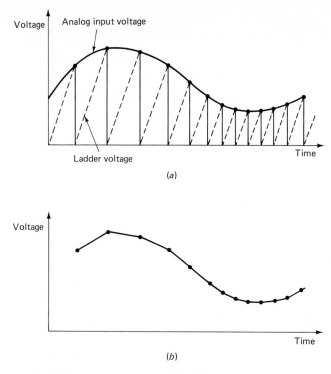

Voltage

Analog input voltage

Ladder voltage

Time

(a)

Voltage

Time

(b)

FIG. 13–24  (a) Digitizing an analog voltage (b) Reconstructed signal from the digital data

reasonable representation of the original input signal. In this case, it is important to note that the conversion times are smaller than the transient time of the input waveform.

On the other hand, if the transient time of the input waveform approaches the conversion time, the reconstructed output signal is not quite so accurate. Such a situation is shown in Fig. 13–25a and b. In this case, the input waveform changes at a rate faster than the converter is capable of recognizing. Thus the need for reducing conversion time is apparent.

## 13·7  CONTINUOUS A/D CONVERSION

An obvious method for speeding up the conversion of the signal as shown in Fig. 13–25 is to eliminate the need for resetting the counter each time a conversion is made. If this were done, the counter would not begin at zero each time, but instead would begin at the value of the last converted point. This means that the counter would have to be capable of counting either up or down. This is no problem; we are already familiar with the operation of up-down counters (Chap. 11).

There is, however, the need for additional logic circuitry, since we must decide whether to count up or down by examining the output of the comparator. An A/D converter which uses an up-down counter is shown in Fig. 13–26. This method is known as *continuous conversion*, and thus the converter is called a *continuous-type A/D converter*.

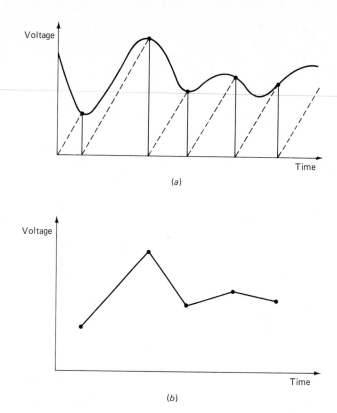

Voltage

Time

(a)

Voltage

Time

(b)

**FIG. 13–25**  (a) Digitizing an analog voltage (b) Reconstructed signal from the digital data

The D/A portion of this converter is the same as those previously discussed, with the exception of the counter. It is an up-down counter and has the up and down count control lines in addition to the advance line at its input.

The output of the ladder is fed into a comparator which has two outputs instead of one as before. When the analog voltage is more positive than the ladder output, the *up* output of the comparator is high. When the analog voltage is more negative than the ladder output, the *down* output is high.

If the *up* output of the comparator is high, the AND gate at the input of the *up* flip-flop is open, and the first time the clock goes positive, the *up* flip-flop is set. If we assume for the moment that the *down* flip-flop is reset, the AND gate which controls the *count-up* line of the counter will be true and the counter will advance one count. The counter can advance only one count since the output of the one-shot resets both the *up* and the *down* flip-flops just after the clock goes low. This can then be considered as one count-up conversion cycle.

Notice that the AND gate which controls the *count-up* line has inputs of *up* and $\overline{down}$. Similarly, the count-down line AND gate has inputs of *down* and $\overline{up}$. This could be considered an exclusive–OR arrangement and ensures that the count-down and count-up lines cannot both be high at the same time.

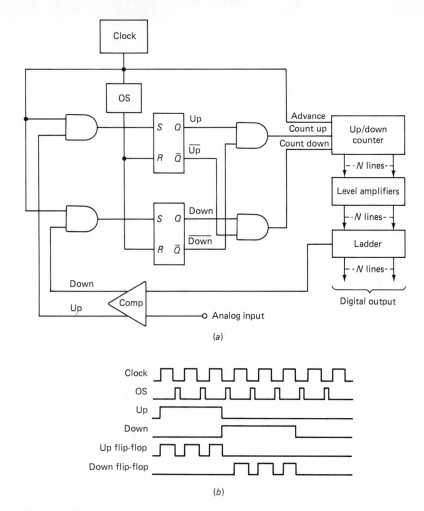

FIG. 13-26 Continuous A/D converter

As long as the *up* line out of the comparator is high, the converter continues to operate one conversion cycle at a time. At the point where the ladder voltage becomes more positive than the analog input voltage, the *up* line of the comparator goes low and the *down* line goes high. The converter then goes through a count-down conversion cycle. At this point, the ladder voltage is within 1 LSB of the analog voltage, and the converter oscillates about this point. This is not desirable since we want the converter to cease operation and not jump around the final value. The trick here is to adjust the comparator such that its outputs do not change at the same time.

We can accomplish this by adjusting the comparator such that the *up* output will not go high unless the ladder voltage is more than $\frac{1}{2}$ LSB below the analog voltage. Similarly, the *down* output will not go high unless the ladder voltage is more than $\frac{1}{2}$ LSB above the analog voltage. This is called *centering on the LSB* and provides a digital output which is within $\frac{1}{2}$ LSB.

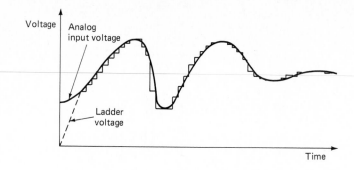

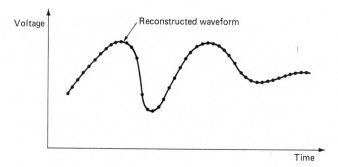

**FIG. 13–27**   Continuous A/D conversion

A waveform typical of this type of converter is shown in Fig. 13–27. You can see that this converter is capable of following input voltages that change at a much faster rate.

☐ EXAMPLE 13—12

Quite often, additional circuitry is added to a continuous converter to ensure that it cannot count off scale in either direction. For example, if the counter contained all 1s, it would be undesirable to allow it to progress through a count-up cycle, since the next count would advance it to all 0s. We would like to design the logic necessary to prevent this.

☐ SOLUTION

The two limit points which must be detected are all 1s and all 0s in the counter. Suppose that we construct an AND–gate having the 1 sides of all the counter flip-flops as its inputs. The output of this gate will be true whenever the counter contains all 1s. If the gate is then connected to the reset side of the *up* flip-flop, the counter will be unable to count beyond all 1s.

Similarly, we might construct an AND gate in which the inputs are the 0 sides of all the counter flip-flops. The output of this gate can be connected to the reset side of the *down* flip-flop, and the counter will then be unable to count beyond all 0s. The gates are shown in Fig. 13–28.

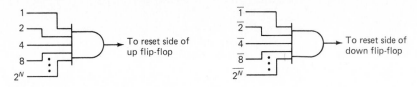

**FIG. 13–28**  Count limiting gates for the converter in Fig. 13–26

# 13-8 A/D TECHNIQUES

There are a variety of other methods for digitizing analog signals—too many to discuss in detail. Nevertheless, we shall take the time to examine two more techniques and the reasons for their importance.

Probably the most important single reason for investigating other methods of conversion is to determine ways to reduce the conversion time. Recall that the simultaneous converter has a very fast conversion time but becomes unwieldy for more than a few bits of information. The counter converter is simple logically but has a relatively long conversion time. The continuous converter has a very fast conversion time once it is locked on the signal but loses this advantage when multiplexing inputs.

If multiplexing is required, the successive-approximation converter is most useful. The block diagram for this type of converter is shown in Fig. 13–29a. The converter operates by successively dividing the voltage ranges in half. The counter is first reset to all 0s, and the MSB is then set. The MSB is then left in or taken out (by resetting the MSB flip-flop) depending on the output of the comparator. Then the second MSB is set in, and a comparison is made to determine whether to reset the second MSB flip-flop. The process is repeated down to the LSB, and at this time the desired number is in the counter. Since the conversion involves operating on one flip-flop at a time, beginning with the MSB, a ring counter may be used for flip-flop selection.

The successive-approximation method thus is the process of approximating the analog voltage by trying 1 bit at a time beginning with the MSB. The operation is shown in diagram form in Fig. 13–29b. It can be seen from this diagram that each conversion takes the same time and requires one conversion cycle for each bit. Thus the total conversion time is equal to the number of bits, $n$, times the time required for one conversion cycle. One conversion cycle normally requires one cycle of the clock. As an example, a 10-bit converter operating with a 1-MHz clock has a conversion time of $10 \times 10^{-6} = 10^{-5} = 10 \ \mu s$.

When dealing with conversion times this short, it is usually necessary to take into account the other delays in the system (e.g., switching time of the multiplexer, settling time of the ladder network, comparator delay, and settling time).

All the logic blocks inside the dashed line in Fig. 13–29a, or some equivalent arrangement, are frequently constructed on a single MSI chip; this chip is called a *successive-approximation register* (SAR). For instance, the Motorola MC14549 is an 8-bit SAR that can be used with the Motorola MC1408 DAC to construct a successive-approximation ADC as shown in Fig. 13–29c.

Another method for reducing the total conversion time of a simple counter converter is to divide the counter into sections. Such a configuration is called a *section counter*. To

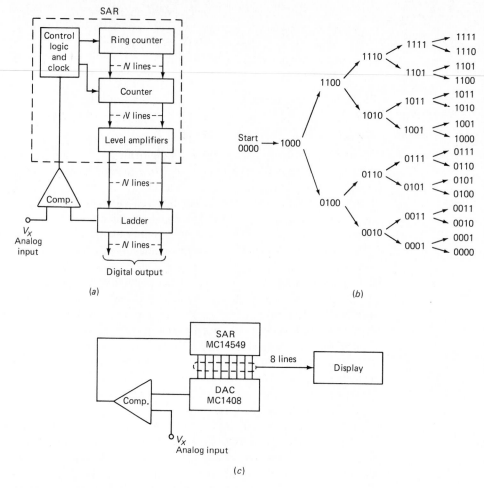

(a)

(b)

(c)

**FIG. 13–29** Successive approximation converter

determine how the total conversion time might be reduced by this method, assume that we have a standard 8-bit counter. If this counter is divided into two equal counters of 4 bits each, we have a section converter. The converter operates by setting the section containing the four LSBs to all 1s and then advancing the other sections until the ladder voltage exceeds the input voltage. At this point the four LSBs are all reset, and this section of the counter is then advanced until the ladder voltage equals the input voltage.

Notice that a maximum of $2^4 = 16$ counts is required for each section to count full scale. Thus this method requires only $2 \times 2^4 = 2^5 = 32$ counts to reach full scale. This is a considerable reduction over the $2^8 = 256$ counts required for the straight 8-bit counter. There is, of course, some extra time required to set the counters initially and to switch from counter to counter during the conversion. This logical operation time is very small, however, compared with the total time saved by this method.

This type of converter is quite often used for digital voltmeters, since it is very convenient to divide the counters by counts of 10. Each counter is then used to represent one of the digits of the decimal number appearing at the output of the voltmeter. We discuss this subject in detail in Chap. 14.

# 13-9 DUAL-SLOPE A/D CONVERSION

Up to this point, our interest in different methods of A/D conversion has centered on reducing the actual conversion time. If a very short conversion time is not a requirement, there are other methods of A/D conversion that are simpler to implement and much more economical. Basically, these techniques involve comparison of the unknown input voltage with a reference voltage that begins at zero and increases linearly with time. The time required for the reference voltage to increase to the value of the unknown voltage is directly proportional to the magnitude of the unknown voltage, and this time period is measured with a digital counter. This is referred to as a "single-ramp" method, since the reference voltage is sloped like a ramp. A variation on this method involves using an operational amplifier integrating circuit in a dual-ramp configuration. The dual-ramp method is very popular, and widely used in digital voltmeters and digital panel meters. It offers good accuracy, good linearity, and very good noise-rejection characteristics.

Let's take a look at the single ramp A/D converter in Fig. 13–30. The heart of this converter is the *ramp generator*. This is a circuit that produces an output voltage ramp as shown in Fig. 13–31a. The output voltage begins at zero and increases linearly up to a maximum voltage $V_m$. It is important that this voltage be a straight line—that is, it must have a constant slope. For instance, if $V_m = 1.0$ Vdc, and it takes 1.0 ms for the ramp to move from 0.0 up to 1.0 V, the slope is 1 V/ms, or 1000 V/s.

This ramp generator can be constructed in a number of different ways. One way might be to use a DAC driven by a simple binary counter. This would generate the staircase waveform previously discussed and shown in Fig. 13–16a. A second method is to use an operational amplifier (OA) connected as an integrator as shown in Fig. 13–31b. For this circuit, if $V_i$ is a constant, the output voltage is given by the relationship $V_o = (V_i/RC)t$. Since $V_i$, $R$, and $C$ are all constants, this is the equation of a straight line that has a slope $(V_i/RC)$ as shown in Fig. 31–31a. Now that we have a way to generate a voltage ramp and we understand its characteristics, let's return to the converter in Fig. 13–30.

We assume that the clock is running continuously and that any input voltage $V_x$ that we wish to digitize is positive. If it is not, there are circuits that we can use to adjust for negative input signals. The three decade counters are connected in cascade, and their outputs can be strobed into three 4-flip-flop latch circuits. The latches are then decoded by seven-segment decoders to drive the LED displays as units, tens, and hundreds of counts. We can begin a conversion cycle by depressing the MANUAL RESET switch.

Refer carefully to the logic diagram and the waveforms in Fig. 13–30. MANUAL RESET generates a RESET pulse that clears all the decade counters to 0s and resets the ramp voltage to zero. Since $V_x$ is positive and RAMP begins at zero, the output of the comparator OA, $V_c$, must be high. This voltage enables the CLOCK gate allowing the clock, CLK, to be applied to the decade counter. The counter begins counting upward, and the RAMP continues upward until the ramp voltage is equal to the unknown input $V_x$.

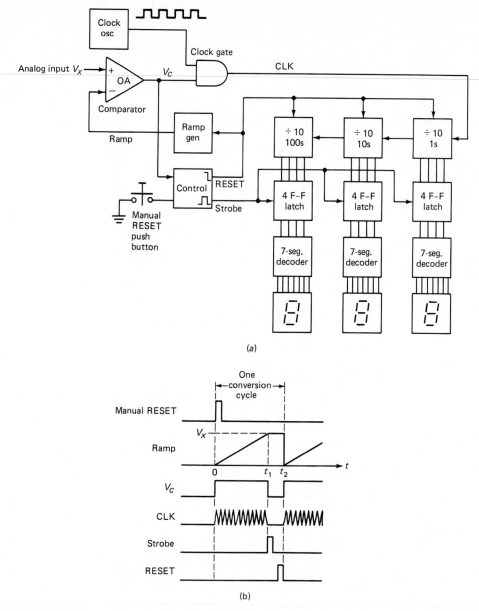

(a)

(b)

**FIG. 13–30**  Single slope A/D converter

At this point, time $t_1$, the output of the comparator $V_c$ goes low, thus disabling the CLOCK gate and the counters cease to advance. Simultaneously, this negative transition on $V_c$ generates a STROBE signal in the CONTROL box that shifts the contents of the three decade counters into the three 4-flip-flop latch circuits. Shortly thereafter, a reset pulse is generated by the CONTROL box that resets the RAMP and clears the decade

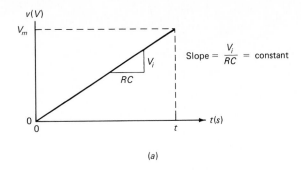

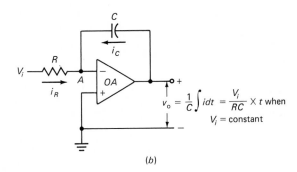

**FIG. 13–31** An integrating circuit

counters to 0s, and another conversion cycle begins. In the meantime, the contents of the previous conversion are contained in the latches and are displayed on the seven-segment LEDs.

As a specific example, suppose that the clock in Fig. 13–30 is set at 1.0 MHz and the ramp voltage slope is 1.0 V/ms. Note that the decade counters have the ability to store and display any decimal number from 000 up to 999. From the beginning of a conversion cycle, it will require 999 clock pulses (999 microseconds) for the counters to advance full scale. During this same time period, the ramp voltage will have increased from 0.0 V up to 999 mV. So, this circuit as it stands will display the value of any input voltage between 0.0 V and 999 mV.

In effect, we have a digital voltmeter! For instance, if $V_x = 345$ mV, it will require 345 clock pulses for the counter to advance from 000 to 345, and during the same time period the ramp will have increased to 345 mV. So, at the end of the conversion cycle, the display output will read 345—we supply the units of millivolts.

One weakness of the single-slope ADC is its dependency on an extremely accurate ramp voltage. This in turn is strongly dependent on the values of $R$ and $C$ and variations of these values with time and temperature. The dual-slope ADC overcomes these problems.

The logic diagram for a basic dual-slope ADC is given in Fig. 13–32. With the exception of the ramp generator and the comparator, the circuit is similar to the single-slope ADC in Fig. 13–30. In this case, the integrator forms the desired ramp—in fact, two different ramps—as the input is switched first to the unknown input voltage $V_x$ and then to a known reference voltage $V_r$. Here's how it works.

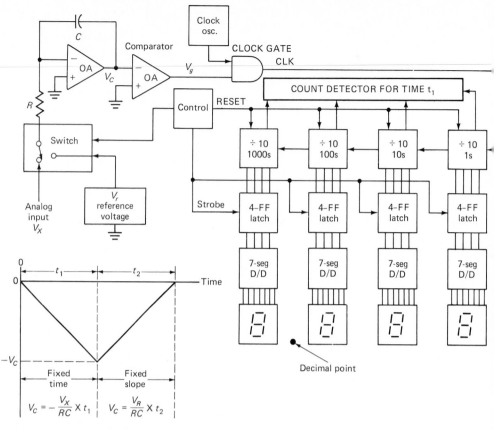

**FIG. 13-32** Dual slope A/D converter

We begin with the assumptions that the clock is running, and that the input voltage $V_x$ is positive. A conversion cycle begins with the decade counters cleared to all 0s, the ramp reset to 0.0 V, and the input switched to the unknown input voltage $V_x$. Since $V_x$ is positive, the integrator output $V_c$ will be a negative ramp. The comparator output $V_g$ is thus positive and the clock is allowed to pass through the CLOCK GATE to the counters. We allow the ramp to proceed for a fixed time period $t_1$, determined by the count detector for time $t_1$. The actual voltage $V_c$ at the end of the fixed time period $t_1$ will depend on the unknown input $V_x$, since we know that $V_c = -(V_x/RC) \times t_1$ for an integrator.

When the counter reaches the fixed count at time $t_1$, the CONTROL unit generates a pulse to clear the decade counters to all 0s and switch the integrator input to the negative reference voltage $V_r$. The integrator will now begin to generate a ramp beginning at $-V_c$ and increasing steadily upward until it reaches 0.0 V. All this time, the counter is counting, and the conversion cycle ends when $V_c = 0.0$ V since the CLOCK GATE is now disabled. The equation for this positive ramp is $V_c = (V_r/RC) \times t_2$. In this case, the slope of this ramp $(V_r/RC)$ is constant, but the time period $t_2$ is variable.

In fact, since the integrator output voltage begins at 0.0 V, integrates down to $-V_c$, and then integrates back up to 0.0 V, we can equate the two equations given for $V_c$. That is:

$$\frac{V_x}{RC} \times t_1 = \frac{V_r}{RC} \times t_2$$

The value $RC$ will cancel from both sides, leaving

$$V_x = V_r \times \frac{t_2}{t_1}$$

Since $V_r$ is a known reference voltage and $t_1$ is a predetermined time, clearly the unknown input voltage is directly proportional to the variable time period $t_2$. However, this time period is exactly the contents of the decade counters at the end of a conversion cycle! The obvious advantage here is that the $RC$ terms cancel from both sides of the equation above—in other words, this technique is free from the absolute values of either $R$ or $C$ and also from variations in either value.

As a concrete example, let's suppose that the clock in Fig. 13–32 is 1.0 MHz, the reference voltage is $-1.0$ Vdc, the fixed time period $t_1$ is 1000 $\mu$s, and the $RC$ time constant of the integrator is set at $RC = 1.0$ ms. During the time period $t_1$, the integrator voltage $V_c$ will ramp down to $-1.0$ Vdc if $V_x = 1.0$ V. Then, during time $t_2$, $V_c$ will ramp all the way back up to 0.0 V, and this will require a time of 1000 $\mu$s, since the slope of this ramp is fixed at 1.0 V/ms. The output display will now read 1000, and with placement of a decimal as shown, this reads 1.000 V.

Another way of expressing the operation of this ADC is to solve the equation $V_x = V_r(t_2/t_1)$ for $t_2$, since $t_2$ is the digital readout. Thus $t_2 = (V_x/V_r)t_1$. If the same values as given above are applied, an unknown input voltage $V_x = 2.75$ V will be digitized and the readout will be $t_2 = (2.75/1.0)1000 = 2750$, or 2.75 V, using the decimal point on the display. Notice that we have used $t_1 = 1000$, the number of clock pulses that occur during the time period $t_1$. *Likewise,* $t_2$ is the number of clock pulses that occur during the time period $t_2$.

## 13·10 A/D ACCURACY AND RESOLUTION

Since the A/D converter is a closed-loop system involving both analog and digital systems, the overall accuracy must include errors from both the analog and digital positions. In determining the overall accuracy it is easiest to separate the two sources of error.

If we assume that all components are operating properly, the source of the digital error is simply determined by the resolution of the system. In digitizing an analog voltage, we are trying to represent a continuous analog voltage by an equivalent set of digital numbers. When the digital levels are converted back into analog form by the ladder, the output is the familiar staircase waveform. This waveform is a representation of the input voltage but is certainly not a continuous signal. It is, in fact, a discontinuous signal composed of a number of discrete steps. In trying to reproduce the analog input signal, the best we can do is to get on the step which most nearly equals the input voltage in amplitude.

The simple fact that the ladder voltage has steps in it leads to the digital error in the system. The smallest digital step, or quantum, is due to the LSB and can be made smaller only by increasing the number of bits in the counter. This inherent error is often called the *quantization error* and is commonly $\pm$ 1 bit. If the comparator is centered, as with the continuous converter, the quantization error can be made $\pm \frac{1}{2}$ LSB.

The main source of analog error in the A/D converter is probably the comparator. Other sources of error are the resistors in the ladder, the reference-voltage supply ripple, and noise. These can, however, usually be made secondary to the sources of error in the comparator.

The sources of error in the comparator are centered around variations in the dc switching point. The dc switching point is the difference between the input voltage levels that cause the output to change state. Variations in switching are due primarily to offset, gain, and linearity of the amplifier used in the comparator. These parameters usually vary slightly with input voltage levels and quite often with temperature. It is these changes which give rise to the analog error in the system.

An important measure of converter performance is given by the differential linearity. *Differential linearity* is a measure of the variation in voltage-step size that causes the converter to change from one state to the next. It is usually expressed as a percent of the average step size. This performance characteristic is also a function of the conversion method and is best for the converters having counters that count continuously. The counter-type and continuous-type converters usually have better differential linearity than do the successive-approximation-type converters. This is true since the ladder voltage is always approaching the analog voltage from the same direction in the one case. In the other case, the ladder voltage is first on one side of the analog voltage and then on the other. The comparator is then being used in both directions, and the net analog error from the comparator is thus greater.

The next logical question that might be asked is: what should be the relative order of magnitudes of the analog and digital errors? As mentioned previously, it would be difficult to justify construction of a 15-bit converter that has an overall error of 1 percent. In general, it is considered good practice to construct converters having analog and digital errors of approximately the same magnitudes. There are many arguments for and against this, and any final argument would have to depend on the situation. As an example, an 8-bit converter would have a quantization error of $\frac{1}{256} \cong 0.4$ percent. It would then seem reasonable to construct this converter to an accuracy of 0.5 percent in an effort to achieve an overall accuracy of 1.0 percent. This might mean constructing the ladder to an accuracy of 0.1 percent, the comparator to an accuracy of 0.2 percent, and so on, since these errors are all accumulative.

☐ EXAMPLE 13—13

What overall accuracy could one reasonably expect from the construction of a 10-bit A/D converter?

☐ SOLUTION

A 10-bit converter has a quantization error of $\frac{1}{1024} \cong 0.1$ percent. If the analog portion can be constructed to an accuracy of 0.1 percent, it would seem reasonable to strive for an overall accuracy of 0.2 percent.

# SUMMARY

Digital-to-analog conversion, the process of converting digital input levels into an equivalent analog output voltage, is most easily accomplished by the use of resistance networks. The binary ladder has been found to have definite advantages over the resistance divider. The complete D/A converter consists of a binary ladder (usually) and a flip-flop register to hold the digital input information.

The simultaneous method for A/D conversion is very fast but becomes cumbersome for more than a few bits of resolution. The counter-type A/D converter is somewhat slower but represents a much more reasonable solution for digitizing high-resolution signals. The continuous-converter method, the successive-approximation method, and the section-counter method are all variations of the basic counter-type A/D converter which lead to a much faster conversion time. A dual-slope A/D converter is somewhat slower than the previously discussed methods but offers excellent accuracy in a relatively inexpensive circuit. Dual-slope ADCs are widely used in digital voltmeters.

The DAC and ADC logic circuits given in this chapter are all drawn in logic block diagram form and can all be constructed by simply connecting these commercially available logic blocks. For instance, a DAC can be constructed by connecting resistors that have values of $R$ and $2R$, or an ADC can be constructed by connecting the various inverters, gates, flip-flops, and so on; however, you must realize that these units are now readily available as MSI circuits. The only really practical and economical way to build DACs or ADCs is to make use of these commercially available circuits; this is exactly the subject pursued in Chap. 14!

# GLOSSARY

*ADC* Analog-to-digital converter.

*A/D Conversion* The process of converting an analog input voltage to a number of equivalent digital output levels.

*DAC* Digital-to-analog converter.

*D/A Conversion* The process of converting a number of digital input signals to one equivalent analog output voltage.

*Differential Linearity* A measure of the variation in size of the input voltage to an A/D converter which causes the converter to change from one state to the next.

*Equivalent Binary Weight* The value assigned to each bit in a digital number, expressed as a fraction of the total. The values are assigned in binary fashion according to the sequence 1, 2, 4, 8, . . . , $2^n$, where $n$ is the total number of bits.

*Millman's Theorem* A theorem from network analysis which states that the voltage at any node in a resistive network is equal to the sum of the currents entering the node divided by the sum of the conductances connected to the node, all determined by assuming that the voltage at the node is zero.

*Quantization Error* The error inherent in any digital system due to the size of the LSB.

*SAR* Sequential approximation register, used in a sequential ADC.

## PROBLEMS

*SECTION 13-1:*

**13–1.** What is the binary equivalent weight of each bit in a 6-bit resistive divider?

**13–2.** Draw the schematic for a 6-bit resistive divider.

**13–3.** Verify the voltage output levels for the network in Fig. 13–4, using Millman's theorem. Draw the equivalent circuits.

**13–4.** Assume that the divider in Prob. 13–2 has +10 V full-scale output, and find the following:
  **a.** The change in output voltage due to a change in the LSB
  **b.** The output voltage for an input of 110110

**13–5.** A 10-bit resistive divider is constructed such that the current through the LSB resistor is 100 $\mu$A. Determine the maximum current that will flow through the MSB resistor.

*SECTIONS 13-2 THROUGH 13-4:*

**13–6.** What is the full-scale output voltage of a 6-bit binary ladder if 0 = 0 V and 1 = +10 V? Of an 8-bit ladder?

**13–7.** Find the output voltage of a 6-bit binary ladder with the following inputs:
  **a.** 101001      **c.** 110001
  **b.** 111011

**13–8.** Check the results of Prob. 13–7 by adding the individual bit contributions.

**13–9.** What is the resolution of a 12-bit D/A converter which uses a binary ladder? If the full-scale output is +10 V, what is the resolution in volts?

**13–10.** How many bits are required in a binary ladder to achieve a resolution of 1mV if full scale is +5 V?

*SECTION 13-5:*

**13–11.** How many comparators are required to build a 5-bit simultaneous A/D converter?

**13–12.** Redesign the encoding matrix and READ gates in Fig. 13–20, using NAND gates.

**13–13.** Assuming that the input reference voltage is $V$ = 10.0 Vdc, determine the digital output of the ADC in Fig. 13–20 for an input voltage of:
  **a.** 1.25 V      **c.** 8.05 V
  **b.** 3.33 V

*SECTION 13-6:*

**13–14.** Find the following for a 12-bit counter-type A/D converter using a 1-MHz clock:
   **a.** Maximum conversion time
   **b.** Average conversion time
   **c.** Maximum conversion rate

**13–15.** What clock frequency must be used with a 10-bit counter-type A/D converter if it must be capable of making at least 7000 conversions per second?

**13–16.** Design additional control circuitry for Fig. 13–23 such that the A/D converter in Fig. 13–22 will continue to make conversions after an initial START pulse is applied.

*SECTIONS 13-7 AND 13-8:*

**13–17.** What is the conversion time of a 12-bit successive-approximation-type A/D converter using a 1-MHz clock?

**13–18.** What is the conversion time of a 12-bit section-counter-type A/D converter using a 1-MHz clock? The counter is divided into three equal sections.

*SECTION 13-9:*

**13–19.** For the integrator in Fig. 13–31a, show that the output voltage is given by $V_o = (V_i/RC)t$, assuming that the input voltage $V_i$ is a constant. [*Hint*: Using Kirchhoff's current law at node $A$, the resistor current $i_R$ is equal to the capacitor current $i_C$, but $i_R = V_i/R$ and $i_C = q/t = (V_oC)/t$.

**13–20.** Design the control logic for the CONTROL box in Fig. 13–30 to generate the proper control signals shown in that figure.

**13–21.** Calculate a value for $C$ in Fig. 13–31 to obtain a fixed slope $V_i/(RC) = 1000$ V/s, given $V_i = 1.0$ V dc and $R = 100$ k$\Omega$.

**13–22.** Can you design an amplifier such that the output is always positive and is equal to the magnitude of the input voltage? In other words, the input can be either $+V_i$ or $-V_i$, but in either case, the output will be $+V_i$.

**13–23.** Design the CONTROL logic for the converter in Fig. 13–32.

*SECTION 13-10:*

**13–24.** What overall accuracy could you reasonably expect from a 12-bit A/D converter?

**13–25.** Discuss the overall acceptable accuracy of a 10-bit ADC in terms of quantization error, ladder accuracy, comparator accuracy, converter accuracy, and other factors.

# 14

# S O M E
# A P P L I C A T I O N S

This chapter is intended to tie together many of the fundamental ideas presented previously by considering some of the more common digital circuit designs encountered in industry. The multiplexing of digital LED displays is considered first since it requires the use of a number of different TTL circuits studied in detail in prior chapters. Digital instruments that can be used to measure time and frequency are considered next, and the concept of display multiplexing is applied here. A microprocessor-compatible, integrating-type ADC is studied, and then a similar ADC used as a digital voltmeter. In most of the applications considered, specific TTL part numbers have been specified, but in the interest of clarity, detailed designs including pin numbers have not been provided. However, it is a simple matter to consult the appropriate data sheets for this information.

In some cases, a specific part number has not been assigned; an example of this is the 1-MHz clock oscillator shown in Fig. 14–14, or a divide-by-10 counter in the same figure. In such cases, it is left to you to select any one of a number of divide-by-10 circuits, or to choose an oscillator circuit such as discussed in a previous chapter, on the basis of availability, cost, ease of use, compatibility with the overall system, and other factors.[1]

[1]The detailed design of these or similar applications can be found in the following laboratory manual: Donald P. Leach, *Experiments in Digital Principles*, 3d ed., McGraw-Hill, New York, 1985.

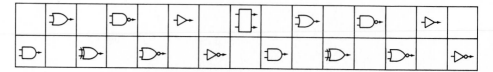

# 14·1 MULTIPLEXING DISPLAYS

The decimal outputs of digital instruments such as digital voltmeters (DVMs) and frequency counters are often displayed using seven-segment indicators. Such indicators are constructed by using a fluorescent bar, a liquid crystal bar, or a LED bar for each segment. Light-emitting-diode-type indicators are convenient because they are directly compatible with TTL circuits, do not require the higher voltages used with fluorescents, and are generally brighter than liquid crystals. On the other hand, LEDs do generally require more power than either of the other two types, and *multiplexing* is a technique used to reduce indicator power requirements.

The circuit in Fig. 14–1a is a common-anode, LED-type, seven-segment indicator used to display a single decimal digit. The 7447 BCD-to-seven-segment decoder is used to

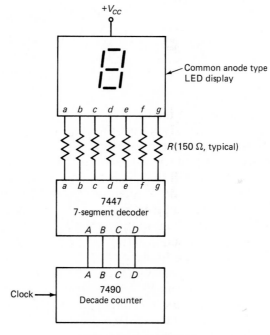

(a) Single decimal-digit display

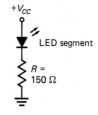

(b) Equivalent circuit for an illuminated segment

**FIG. 14–1**

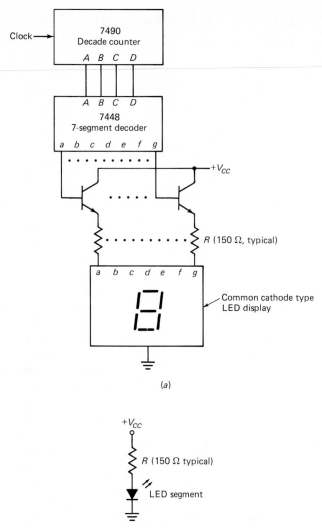

Clock → | 7490
Decade counter
A B C D

A B C D
7448
7-segment decoder
a b c d e f g

+$V_{CC}$

R (150 Ω, typical)

a b c d e f g

Common cathode type
LED display

(a)

+$V_{CC}$

R (150 Ω typical)

LED segment

(b) Equivalent circuit for an illuminated segment

**FIG. 14–2**

drive the indicator, and the four inputs to the 7447 are the four flip-flop outputs of the 7490 decade counter. Remember that the 7447 has active low outputs, so the equivalent circuit of an illuminated segment appears as in Fig. 14–1b. A 1-Hz square wave applied at the clock input of the 7490 will cause the counter to count upward, advancing one count each second, and the equivalent decimal number will appear on the display.

A similar single decimal digit display using a common-cathode-type LED indicator is shown in Fig. 14–2a. The seven-segment decoder used here is the 7448; its outputs are active high, and they are intended to drive buffer amplifiers since their output current capabilities are too small to drive LEDs directly. The seven *npn* transistors simply act as switches to connect $+V_{CC}$ to a segment. When an output of the 7448 is high, a transistor is on, and current is supplied to a LED segment. The equivalent circuit for an illuminated segment is shown in Fig. 14–2b. When an output of the 7448 is low, the transistor is off, and there is no segment current and thus no illumination.

Let's take a look at the power required for the single-digit display in Fig. 14–1a. A segment is illuminated whenever an output of the 7447 goes low (essentially to ground). If we assume a 2-Vdc drop across an illuminated segment (LED), a current $I = (5 - 2)/150 = 20$ mA is required to illuminate each segment. The largest current is required when the number 8 is displayed, since this requires all segments to be illuminated. Under this condition, the indicator will require $7 \times 20 = 140$ mA. The 7447 will also require about 64 mA, so a maximum of around 200 mA is required for this single digit display. An analysis of the display circuit in Fig. 14–2 will yield similar results.

A digital instrument that has a four-digit decimal display will require four of the circuits in Fig. 14–1 and thus has a current requirement of $4 \times 200 = 800$ mA. A six-digit instrument would require 1200 mA, or 1.2 A, just for the displays! Clearly these current requirements are much too large for small instruments, but they can be greatly reduced using a technique known as *multiplexing*.

Basically, multiplexing is accomplished by applying current to each display digit in short, repeated pulses rather than continuously. If the pulse repetition rate is sufficiently high, your eye will perceive a steady illumination without any flicker. (For instance, hardly any flicker is noticeable with indicators illuminated using 60 Hz.) The single-digit display in Fig. 14–3a has +5 V dc (and thus current) applied through a *pnp* transistor that acts as a switch. When DIGIT is high, the transistor (switch) is off, and the indicator current is zero. When DIGIT is low, the transistor is on, and a number is displayed. If the waveform in Fig. 14–3b is used as DIGIT, the transistor will be on and the segment will display a number for only 1 out of every 4 ms. Even though the display is not illuminated for 3 out of 4 ms, the illumination will appear to your eye as if it were continuous. Since the display is illuminated with a pulse that occurs once every 4 ms, the repetition rate (*RR*) is given as $RR = 1/0.004 = 250$ Hz. As a guideline, any *RR* greater than around 50 or 60 Hz will provide steady illumination without any perceptible flicker. The great advantage here is that this single-digit display requires only one-fourth the current of a continuously illuminated display. This then is the great advantage of multiplexing!

Let's see how to multiplex the four-digit display in Fig. 14–4a. Assume that the four BCD inputs to each digit are unchanging. If the four waveforms in Fig. 14–4b are used as the four DIGIT inputs, each digit will be illuminated for one-fourth of the time and extinguished for three-fourths of the time. Looking at the time line, we see that digit 1 is illuminated during time $t_1$, digit 2 during time $t_2$, and so on. Clearly, $t_1 = t_2 = t_3 = t_4 = T/4$. The repetition rate is given as $RR = 1/T$, and if the rate is sufficient, no flicker will appear. For instance, if $t_1 = 1$ ms, then $T = 4$ ms, and $RR = 1/0.004 = 250$ Hz.

Now, here is an important concept; an illuminated digit requires 200 mA, and since only one digit is illuminated at a time, the current required from the $+V_{CC}$ supply is always 200 mA. Therefore, we are illuminating four indicators but using the current required of only a single indicator. In fact, in multiplexing displays in this way, the power supply current is simply the current required of a single display, no matter how many displays are being multiplexed!

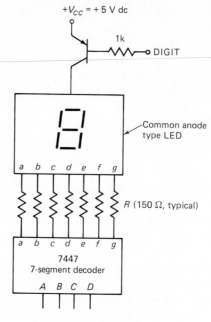

(a) Multiplexed display

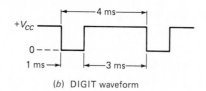

(b) DIGIT waveform

**FIG. 14–3**

☐ EXAMPLE 14—1

Explain the timing for a six-digit display that has a repetition rate of 125 Hz.

☐ SOLUTION

An *RR* of 125 Hz means that all digits must be serviced once every $\frac{1}{125} = 8$ ms. Dividing the time equally among the six digits means that each digit will be on for $\frac{8}{6} = 1.33$ ms and off for 6.67 ms. Note that as the pulse width is decreased, the display brightness will also decrease. It may thus become necessary to increase the peak current through each segment by reducing the size of the resistors *R* in Figs. 14–1 and 14–2.

☐ EXAMPLE 14—2

The circuit in Fig. 14–3 shows how to multiplex a common-anode-type display. Show how to multiplex a common-cathode-type display.

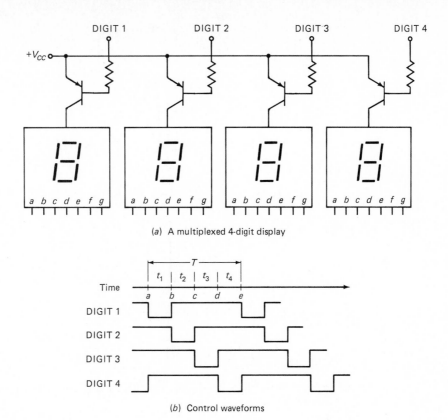

(a) A multiplexed 4-digit display

(b) Control waveforms

**FIG. 14–4**

☐ SOLUTION

The *npn* transistor in Fig. 14–5 is used as a switch between the cathode of the display and ground. When the transistor is on, current is allowed to pass through a segment for illumination. When the transistor is off, no current is allowed, and the segment cannot illuminate. The DIGIT waveform is shown in Fig. 14–5b. Notice that a positive pulse is required to turn the transistor on, and the display will be illuminated for 1 out of every 4 ms.

The flip-flop outputs of a 7490 decade counter are used to drive the seven-segment decoder-driver in Figs. 14–1 and 14–2, and as long as the counter is counting, the displays will be changing states. It is often more desirable to periodically strobe the contents of the counter into a four-flip-flop latch and use these latches to drive the seven-segment decoder-driver. Then the four BCD inputs to the decoder-driver as well as the display will be steady all of the time except when new data is being strobed in. The circuit shown in Fig. 14–6 is a complete single-digit display that will indicate the decimal equivalent of the binary number stored in the 7475 quad D-type latch by the positive STROBE pulse. It is also capable of being multiplexed by use of the DIGIT input.

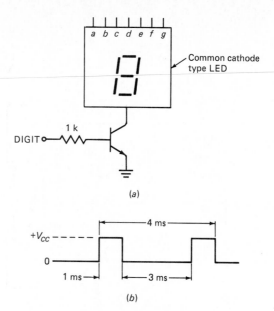

(a)

(b)

**FIG. 14–5**

☐ EXAMPLE 14–3

What are some possible methods for generating the DIGIT control waveforms shown in Fig. 14–4*b*?

☐ SOLUTION

Reflecting back on topics covered in previous chapters, a number of different methods come to mind—for instance:

1. A two-flip-flop counter with four decoding gates
2. A four-flip-flop ring counter
3. A two-flip-flop shift counter with four decoding gates
4. A 1-of-4, low-output multiplexer.

Can you think of any others?

The four-digit display in Fig. 14–7*a* uses four of the decimal digit displays in Fig. 14–6, and they are multiplexed to reduce power supply requirements. Notice that DIGIT 1 controls the LED on the left, and this is the most-significant digit (MSD). The right display is controlled by DIGIT 4, and this is the least-significant digit (LSD). If we assume that the decimal point for this display to be at the right, the LSD is the units digit, and the MSD is the thousands digit. This circuit is capable of displaying decimal numbers from 0000 up to 9999. The 54/74155 is a dual 2-line-to-4-line decoder-demultiplexer, and it is driven by a two-flip-flop binary counter called the *multiplexing counter*. As this multiplexing counter progresses through its four states, one and only one of the 54/74155 output lines will go low for each counter state. As a result, the DIGIT control waveforms

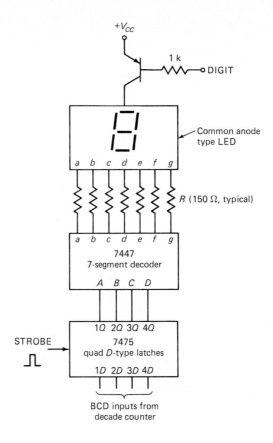

FIG. 14–6

exactly like those shown in Fig. 14–4*b* will be developed. You might like to take time to review the operation of decoder-demultiplexes as discussed in Chap. 2.

A savings in components as well as power can often be realized if the four inputs (*ABCD*) to the seven-segment decoder in Fig. 14–6 are multiplexed along with the DIGIT control. The four-digit display in Fig. 14–8 uses two 54/74153 dual 4- to-1-line multiplexers to apply the four outputs of each 7475 sequentially to a single seven-segment decoder. Here's how it works.

The BCD input data is stored in four 7475 *D*-type latches labeled 1, 2, 3, and 4. Latch 1 stores the MSD, and latch 4 stores the LSD. The 4-bit binary number representing the MSD is labeled $1_A 1_B 1_C 1_D$. For instance, if the MSD = 7, then $1_A 1_B 1_C 1_D = 0111$.

Each 74153 contains two multiplexers, and the four multiplexers are labeled $A, B, C,$ and $D$. The $A$ and $B$ SELECT lines of the two multiplexers are connected in parallel and are driven by the multiplexing counter (exactly as in Fig. 14–7). When the SELECT inputs are $AB = 00$, the number 1 line of each multiplexer will be connected to its output. So, the multiplexer outputs connected to the 7447 decoder will be $1_A 1_B 1_C 1_D$, which is the binary number for the MSD. This binary number is decoded by the 7447 and applied to all the LED displays in parallel. However, at this same time, DIGIT 1 is selected by the

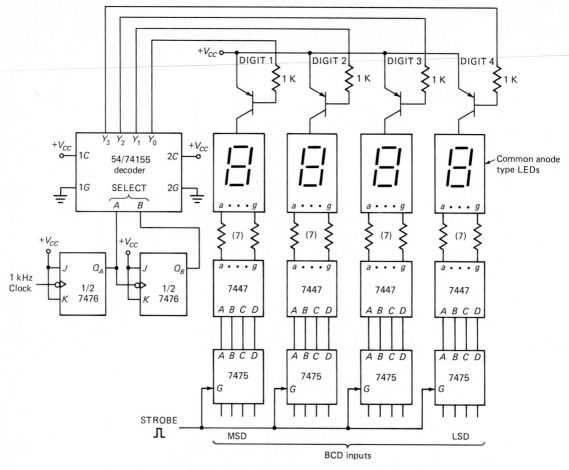

(a) Four-decimal-digit multiplexed display

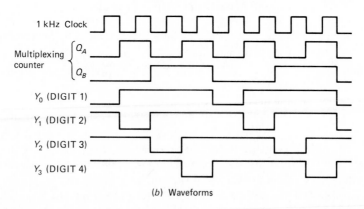

(b) Waveforms

**FIG. 14–7**

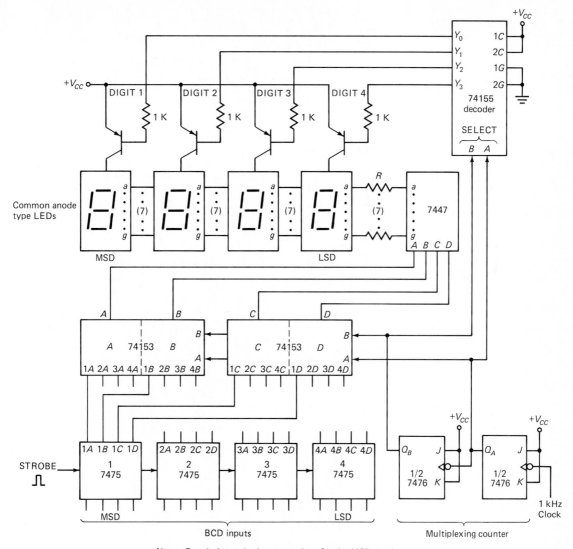

Note: For clarity, only the connections for the *MSD* are shown
(1A, 1B, 1C, 1D), but the other 12 connections must be made
between the input latches and the 74153s.

**FIG. 14–8**

74155 decoder, so the MSD will be displayed in the leftmost LED display. All the other
displays will be turned off.

Now, when the multiplexing counter advances to count $AB = 01$, the number 2 line
of each multiplexer will be selected, and the binary number applied to the 7447 will be
$2_A 2_B 2_C 2_D$, which is the next MSD (the hundreds digit). The decoded output of the 7447 is
again applied to all the displays in parallel, but DIGIT 2 is the only LOW DIGIT line, so

the "hundreds" digit is now displayed. (Again, all other displays are turned off during this time.)

In a similar fashion, the tens digit will be displayed when the SELECT inputs are $AB = 10$, and the units digit will be displayed when the SELECT inputs are $AB = 11$. Notice that only one digit is displayed at a time, and the $RR = 250$ Hz, so no flicker will be apparent. Again, we are illuminating four digits, but the power supply current is the same as for a single, continuously illuminated digit. At the same time, there is a modest

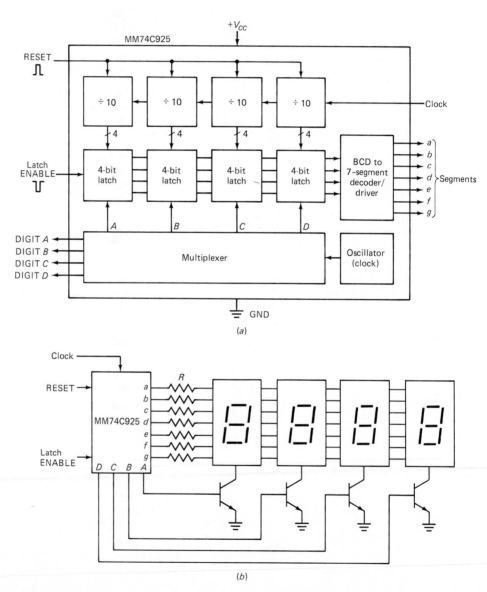

**FIG. 14–9**

saving of two chips. The savings in components increases as the number of decimal digits in the display increases.

The techniques used to multiplex the four-digit display in Fig. 14–8 are easily expanded to displays that have more than four decimal digits. It is necessary only to increase the size of the multiplexing counter and to replace the 74153 multiplexer with one that has a greater number of inputs. It is also a simple matter to alter the design to accommodate common-cathode-type LEDs instead of the common-anode types used here. (See the problems at the end of this chapter.)

All the display circuits discussed here are frequently constructed and used, but you should be aware that there are LSI chips available that have all the multiplexing accomplished on a single chip; examples of this are the National Semiconductor MM74C925, 926, 927, and 928. The MM74C925 shown in Fig. 14–9 is a four-digit counter with multiplexed seven-segment output drivers. The only external components needed are the seven-segment indicators and seven current-limiting resistors. In fact, a four-digit counter is even included on the chip! A positive pulse on the RESET input will reset the 4-bit counter, and then the counter will advance once with each negative transition of CLOCK. A negative pulse on LATCH ENABLE will then latch the contents of the counter into the four 4-bit latches. The four numbers stored are then multiplexed, decoded, and displayed on the four external seven-segment indicators. A simplified diagram is given in Fig. 14–9b. Notice that this is a common-cathode-type display.

## 14-2  A FREQUENCY COUNTER

A frequency counter is a digital instrument that can be used to measure the frequency of any periodic waveform. The fundamental concepts involved are illustrated in the block diagram in Fig. 14–10. The counter and display unit are exactly as described in Sec. 14–1. A GATE ENABLE signal that has a known period $t$ is generated with a clock oscillator and a divider circuit and is applied to one leg of an AND gate. The unknown signal is applied to the other leg of the AND gate and acts as the clock for the counter. The counter will advance one count for each transition of the unknown signal, and at the end of the known time period, the contents of the counter will equal the number of periods of the unknown signal that have occurred during $t$. In other words, the counter contents will be proportional to the frequency of the unknown signal. For instance, suppose that the gate signal is exactly 1 s and the unknown input signal is a 750-Hz square wave. At the end of

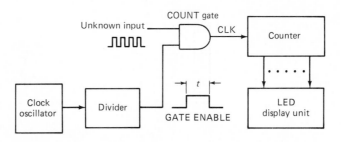

**FIG. 14–10**  Basic frequency counter

1 s, the counter will have counted up to 750, which is exactly the frequency of the input signal.

☐ EXAMPLE 14–4

Suppose that the unknown input signal in Fig. 14–10 is a 7.50-kHz square wave. What will the display indicate if the GATE ENABLE time is $t = 0.1$ s? What if $t = 1$, and then 10 s?

☐ SOLUTION

When $t = 0.1$ s, the counter will count up to 7500 (transitions per second) $\times$ 0.1 (second) = 750. When $t = 1$ s, the counter will display 7500 (transitions per second) $\times$ 1 (second) = 7500. When $t = 10$ s, the counter will display 7500 (transitions per second) $\times$ 10 (seconds) = 75,000. For this last case, we would have to have a five-decimal-digit display.

In Example 14–4, the contents of the counter are always a number that is proportional to the unknown input frequency. In this case, the proportionality constant is either 10, 1, or $\frac{1}{10}$. So, it is a simple matter to insert a decimal point between the indicators such that the unknown frequency is displayed directly. Figure 14–11 shows how the decimal point moves in a five-decimal-digit display as the gate width is changed. In the top display, the unknown frequency is the display contents multiplied by 10, so the decimal point is moved one place to the right. The middle display provides the actual unknown frequency directly. In the bottom display, the contents must be divided by 10 to obtain the unknown frequency, so the decimal point is moved one place to the left.

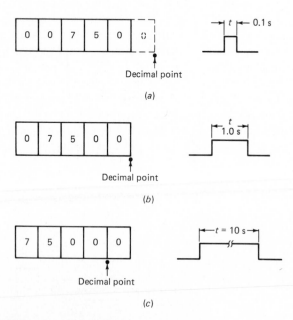

**FIG. 14–11** Decimal point movement for Example 14–4

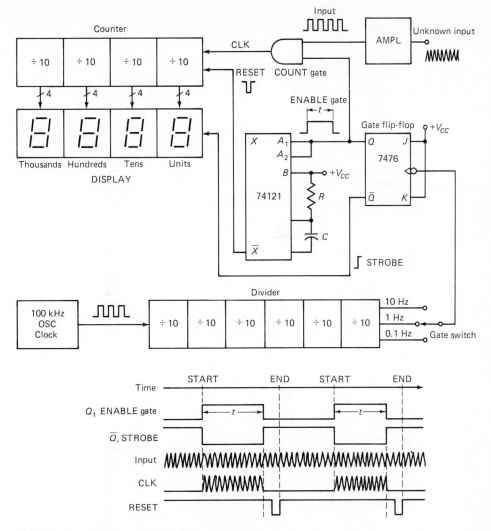

**FIG. 14–12** Four decimal digit frequency counter

The logic diagram in Fig. 14–12 shows one way to construct a four-decimal-digit frequency counter. The AMPLIFIER block is intended to condition the unknown input signal such that INPUT is a TTL-compatible signal—a series of positive pulses going from 0 to +5 V dc. When allowed to pass through the COUNT gate, INPUT will act as the clock for the COUNTER. The COUNTER can be constructed from four decade counters such as 54/74160s, and it can then be connected to a multiplexed LED DISPLAY such as the one shown in Fig. 14–8. Or, COUNTER and DISPLAY can be combined in a single chip such as the MM74C925 shown in Fig. 14–9.

The DIVIDER is composed of six decade counters (such as 54/74160s) connected in series. Its input is a 100-kHz square wave from OSC CLOCK, and it provides 10-, 1-, and 0.1-Hz square wave outputs that are used to generate the ENABLE–gate signal.

When the 1-Hz square wave is used to drive the GATE FLIP-FLOP, its output, $Q$, is a 0.5-Hz square wave. Output $Q$ will be high for exactly 1s and low for 1s, and it will thus be used for the ENABLE–gate signal. Notice that the 10-Hz signal will generate a 0.1-s gate and the 0.1-Hz signal will generate a 10-s gate. Let's use the waveforms in Fig. 14–12 to see exactly how the circuit functions.

A measurement period begins when the GATE flip-flop is toggled high—labeled *START* on the time line. INPUT now passes through the COUNT gate and advances the COUNTER. (Let's assume that the counter is initially at 0000.) At the end of the EN-ABLE–gate time $t$, the GATE flip-flop toggles low, the COUNTER ceases to advance, and this negative transition of $Q$ triggers the 74121 one-shot. Simultaneously, $\overline{Q}$ goes high, and this will strobe the contents of COUNTER into the DISPLAY latches. There is a propagation delay time of 30 ns minimum through the 74121, and then a negative RESET pulse appears at its output, $\overline{X}$. This propagation delay assures that the contents of COUNTER are strobed into DISPLAY before COUNTER is reset. The RESET pulse from the '121 has an arbitrary width of 1 $\mu$s, as set by its $R$ and $C$ timing components. The end of the RESET pulse is the end of one measurement period, labeled END on the time line.

For a 1.0-s gate, the decimal point will be at the right of the units digit, and the counter will be capable of counting up to 9999 full scale, with an accuracy of plus or minus one count (i.e., 1 part in $10^4$). With a 10-s gate, the decimal point is between the units and the tens digits, and with a 0.1-s gate, the decimal point is one place to the right of the units digit.

The CLOCK oscillator is set at 100 kHz, and this provides an accuracy on the ENABLE–gate time of 1 part in $10^4$ with the 0.1-s gate. Thus the accuracy here is compatible with that of the COUNTER.

☐ EXAMPLE 14–5

Explain what would happen if the instrument in Fig. 14–12 were set on a 1-s gate time and the input signal were 12 kHz.

☐ SOLUTION

Assuming that the counter began at 0000, the display would read 200 at the end of the first measurement period. It would then read 400, then 600, and so on at the end of succeeding periods. This is because the counter capacity is exceeded each time, and it simply recycles through 0000.

The design in Fig. 14–12 shows one method for constructing a frequency counter using readily available TTL chips, but you should be aware that there are numerous chips available that have all, or nearly all, of this design on a single chip, for instance, the Intersil ICM7226A. You will be asked to do a complete design of a frequency counter based on Fig. 14–12 in one of the problems at the end of this chapter.

## 14-3 TIME MEASUREMENT

With only slight modifications, the frequency counter in Fig. 14–10 can be converted into an instrument for measuring time. The logic block diagram in Fig. 14–13 illustrates the

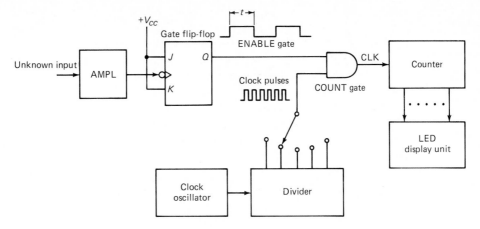

**FIG. 14–13**  Instrument to measure time period

fundamental ideas used to construct an instrument that can be used to measure the period of any periodic waveform. The unknown voltage is passed through a conditioning amplifier to produce a periodic waveform that is compatible with TTL circuits and is then applied to a *JK* flip-flop. The output of this flip-flop is used as the ENABLE–gate signal, since it is high for a time *t* that is exactly equal to the time period of the unknown input voltage. The oscillator and divider provide a series of pulses that are passed through the count gate and serve as the clock for the counter. The contents of the counter and display unit will then be proportional to the time period of the unknown input signal.

For instance, if the unknown input signal is a 5-kHz sine wave and the clock pulses from the divider are 0.1 $\mu$s in width and are spaced every 1.0 $\mu$s, the counter and display will read 200. Clearly this means 200 $\mu$s, since 200 of these 0.1-$\mu$s pulses will pass through the COUNT gate during the 200 $\mu$s that ENABLE–gate signal is high. Naturally the counter and the display have an accuracy of plus or minus one count.

☐ EXAMPLE 14–6

Suppose that the counter and the display unit in Fig. 14–13 have five-decimal-digit capacity and the divider switch is set to provide a 100-kHz square wave that will be used as clock pulses. What will the display read after one ENABLE–gate time *t*, if the unknown input is a 200-Hz square wave?

☐ SOLUTION

Assume that the counter and the display are initially at 00000. A 200-Hz input signal will produce an ENABLE–gate time of $t = \frac{1}{200} = 5000$ $\mu$s. The 100-kHz square wave used as the clock is essentially a series of positive pulses spaced by 10 $\mu$s. Therefore, during the gate time *t*, the counter will advance by, $\frac{5000}{10} = 500$ counts, and this is what will be viewed in the displays. Since each clock pulse represents 10 $\mu$s, the display should be read as $500 \times 10 = 5000$ $\mu$s—this is the time period of the unknown input.

□ EXAMPLE 14—7

Explain the meaning of an accuracy of plus or minus one count applied to the measurement in Example 14–6.

□ SOLUTION

An accuracy of plus or minus one count means that the display could read 499, 500, or 501 after the measurement period. This means that the period as measured could be 4990, 5000, or 5010 $\mu s$—in other words, 5000 plus or minus 10 $\mu s$. Since a single count represents a clock period of 10 $\mu s$, this instrument can be used for measurement only within this limit. For more precise measurement, say, to within 1 $\mu s$, the clock pulses would have to be changed from 10- to 1-$\mu s$ spacing.

The circuit in Fig. 14–14 is a four-decimal-digit instrument for measuring the time period of a periodic waveform. It is essentially the same as the frequency instrument in Fig. 14–12 with only slight modifications. First, the CLOCK has been increased to 1 MHz, and DIVIDER is composed of a buffer amplifier and three decade counters. This will provide clock pulses for COUNTER with 1-, 10-, and 100-$\mu s$ as well as 1-ms spacing. The unknown input is conditioned by AMPLIFIER and is then applied to the

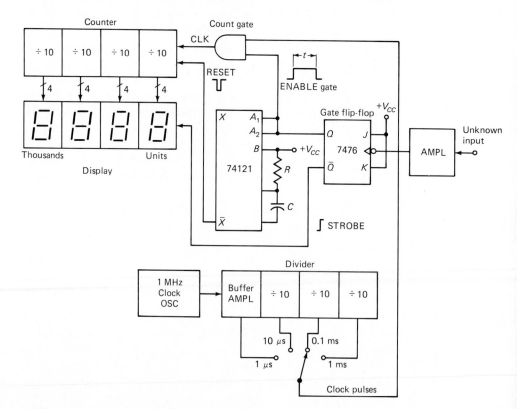

FIG. 14–14  Four decimal digit period measurement instrument

GATE flip-flop to generate the ENABLE–gate signal. STROBE and RESET are generated and applied as before. Notice that a single instrument for measuring both frequency and period could be easily designed by using a 1-MHz clock with a divider that has seven decade counters and some simple mechanical switches.

☐ EXAMPLE 14—8

Explain the DISPLAY ranges for the four-decimal-digit period measurement instrument in Fig. 14–14.

☐ SOLUTION

With CLOCK pulses switched to the 1-$\mu$s position, each count of COUNTER represents 1 $\mu$s. Therefore, it has a full scale of 9999 $\pm$ 1 $\mu$s. On 10 $\mu$s, it has a fullscale of 9999 $\times$ 10 = 99,990 $\pm$ 10 $\mu$s. Full scale on the 0.1-ms position is 9999 $\times$ 0.1 = 999.9 $\pm$ 0.1 ms. Full scale on the 1-ms position is 9999 $\pm$ 1 ms.

An interesting variation on the instrument in Fig. 14–14 is to use it to measure the time elapsed between two events. There will be two input signals, the first of which sets the ENABLE gate high and begins the count period. The second signal (or event) resets the ENABLE gate low and completes the time period. One method for handling this problem is to use the first event to set a flip-flop and then use the second event to reset it. Of course, both input signals must be first conditioned such that they are TTL-compatible. You are given the opportunity to design such an instrument in one of the problems at the end of the chapter.

# 14-4 MICROPROCESSOR-COMPATIBLE ADC

A fundamental requirement in many digital data acquisition systems is an A/D converter that is simple, reliable, accurate, inexpensive, and readily usable with a minicomputer or microprocessor. The National Semiconductor ADC3511 is a single-chip, A/D converter constructed with CMOS technology, that has $3\frac{1}{2}$ digit BCD outputs designed specifically for use with a microprocessor, and it is available for less than $9! The 3511 uses an ''integrating-type'' conversion technique and is considerably slower than ''flash''-type or SAR-type A/D converters. It is quite useful in digitizing quantities such as temperature, pressure, or displacement, where fewer than five conversions per second are adequate. The pinout and logic block diagram for an ADC3511 are shown in Fig. 14–15.

Only a single +5-V dc power supply is required, and the 3511 is completely TTL-compatible. This A/D converter is a very high precision analog device, and great care must be taken to ensure good grounding, power supply regulation, and decoupling. It is important that a single GROUND point be established at pin 13, to eliminate any ground loop currents. Voltage $V_{CC}$ on pin 1 is used to apply +5-V dc power. A 10-$\mu$F, 10-V dc capacitor is connected between pin 2 and GROUND; this capacitor, and the internal 100 $\Omega$ resistor shown on the logic block diagram are used to decouple the dc power used for the analog and digital circuits. Voltage $V_{ss}$ on pin 22 should also be connected directly to GROUND.

The conversion rate of the chip is established by a resistor $R$ connected between pins 17 and 18 and a capacitor $C$ connected between pin 17 and GROUND. The clock frequency developed by these two components is given by $f = 0.6/(RC)$ and should be set

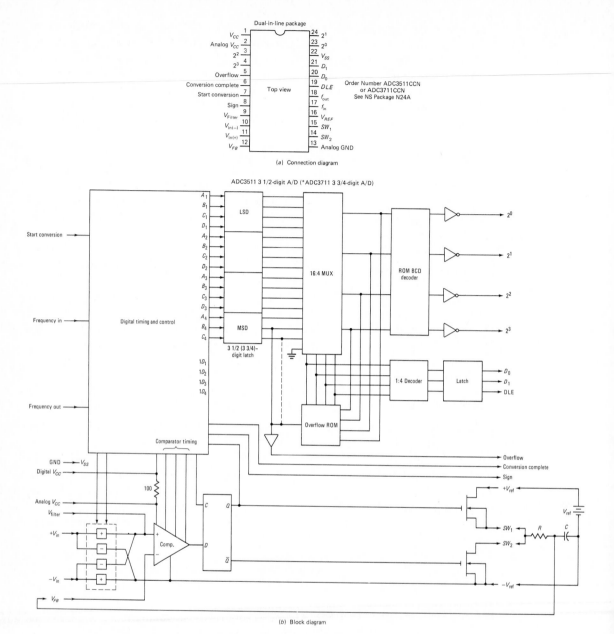

**FIG. 14–15** National Semiconductor ADC 3511 (3711)

between 100 and 640 kHz. This is the clock signal used to advance the internal counters that finally store the digitized value of the analog input voltage.

The analog signal to be digitized is applied between $+V_{in}$ and $-V_{in}$, pins 11 and 10, respectively. Negative signals are handled automatically by the converter through the switching network at the input to the comparator. A conversion is initiated with a low-to-high transition of START CONVERSION on pin 7. The waveform, CONVERSION COMPLETE, on pin 6 will go low at the beginning of a conversion cycle and then return high at the end of a conversion cycle. Connecting pin 7 to $+V_{CC}$ will cause the chip to continuously convert the analog input signal. The rising edge of the waveform on pin 6 indicates that new digital information has been transferred to the digit latches and is available for output.

The digitized analog signal is contained in the converter as four BCD digits. The LSD, or units digit, is $D_1C_1B_1A_1$, the tens digit is $D_2C_2B_2A_2$, the hundreds digit is $D_3C_3B_3A_3$, and the MSD is $C_4B_4A_4$. All digits can store the BCD equivalent of decimal 0, 1, 2, . . . 9, except the MSD. The MSD can have values of only decimal 0 or decimal 1. It is for this reason that the 3511 is called a $3\frac{1}{2}$ digit device—the MSD is referred to as a *half-digit*. The $16 \times 4$ MUX is used to multiplex one digit (4 bits) at a time to the outputs according to the input signals $D_0$ and $D_1$ as give in Fig. 14–16. For instance, when $D_1D_0 = 00$, the LSD appears on the four output lines 23, 22, 21, and 20. A low-to-high transition on DIGIT LATCH ENABLE (DLE), pin 19, will latch the inputs $D_1D_0$, and the selected digit will remain on the four output pins until DLE returns low. The polarity of the digitized input analog signal will also appear on pin 8, SIGN. The 3511 has a full-scale count of 1999, and if this count is exceeded, an overflow condition occurs and the four digit outputs will indicate *EEEE*.

The heart of the analog-to-digital conversion consists of the comparator, the $D$-type flip-flop, and an $RC$ network that is periodically switched between a reference voltage $V_{\mathrm{ref}}$ and ground. When the output of the $D$-type flip-flop, $Q$, is high, the transistor designated as $SW_1$ is on, and the other transistor designated as $SW_2$ is off. Under this condition, the capacitor $C$ charges through $R$ toward the reference voltage (usually $+2.00$ V dc), and the capacitor voltage $V_{\mathrm{fb}}$, is fed back to the negative input terminal of the comparator. When $V_{\mathrm{fb}}$ exceeds the analog input voltage, the comparator output switches low, and the next clock pulse will set $Q$ low. When $Q$ is low, $SW_1$ is off and $SW_2$ is on. The capacitor now discharges through resistor $R$ toward 0.0 V dc. As soon as $V_{\mathrm{fb}}$ discharges below the analog input voltage, the comparator output will switch back to a high state, and this process will repeat.

These components form a closed-loop system that will oscillate—that is, a rectangular waveform as shown in Fig. 14–17 will be produced at $SW_1$ and $SW_2$ (pins 15 and 14).

| DIGIT SELECT Inputs | | | |
|---|---|---|---|
| DLE | $D_1$ | $D_0$ | Selected DIGIT |
| L | L | L | DIGIT 0 (LSD) |
| L | L | H | DIGIT 1 |
| L | H | L | DIGIT 2 |
| L | H | H | DIGIT 3 (MSD) |
| H | X | X | No change |

L = Low logic level
H = High logic level
X = Irrelevent-logic level
The value of the selected digit is presented at the $2^3$, $2^2$, $2^1$, and $2^0$ outputs in BCD format.

**FIG. 14–16**  ADC 3511 (3711) control levels

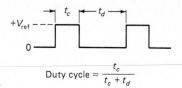

$$\text{Duty cycle} = \frac{t_c}{t_c + t_d}$$

**FIG. 14–17**  Waveforms at $SW_1$ and $SW_2$ (pins 15 and 14 respectively) for the ADC3511

The duty cycle of this waveform is given as

$$\text{Duty cycle} = \frac{t_c}{t_c + t_d}$$

and its dc value is given as

$$V_{dc} = V_{ref} \times (\text{duty cycle})$$

This dc voltage will appear at $V_{fb}$, and the closed-loop system will adjust itself such that

$$V_{in} = V_{dc} = V_{ref} \times (\text{duty cycle})$$

or

$$\frac{V_{in}}{V_{ref}} = \text{duty cycle} = \frac{t_c}{t_c + t_d}$$

The maximum allowable value for the analog input voltage is $V_{ref}$. When the input is equal to $V_{ref}$, the duty cycle must be equal to 1.0 ($t_d = 0$) and $Q$ is always high. If the input analog signal is 0.0, the duty cycle must be zero ($t_c = 0$), and $Q$ is always low. For an analog input voltage between 0.0 and $+V_{ref}$, the duty cycle is some value between 0.0 and 1.0.

The waveform $Q$ at the output of the $D$-type flip-flop has exactly the same duty cycle as $V_{fb}$, and it is used to gate a counter in the converter. The counter can only advance when $Q$ is high, and the gating is arranged such that for a duty cycle of 1.0, the counter will count full scale (1999), and for a duty cycle of 0, the counter will count 0000. For any duty cycle between 0.0 and 1.0, the counter will count a proportional amount between 0000 and 1999. In fact, the exact COUNT relationship is given as

$$\text{COUNT} = N \times \frac{V_{in}}{V_{ref}}$$

where is $N$ is the full-scale count of 2000.

## ☐ EXAMPLE 14–9

An ADC3511 is connected with a reference voltage of $+2.0$ V dc. What will be the count held in the counter for an analog input voltage of 1.25 V dc? What must be the duty cycle?

Using the expression given above, we obtain

$$\text{COUNT} = 2000 \times \frac{1.25}{2.00} = 1250$$

The duty cycle must be

$$\text{Duty cycle} = \frac{V_{\text{in}}}{V_{\text{ref}}} = \frac{1.25}{2.00} = 0.625$$

The circuit in Fig. 14–18a shows an ADC3511 (or an ADC3711) connected to convert 0.0 to +2.00 V dc into an equivalent digital signal in BCD form. The 3511 converts to 1999 counts full scale and thus has a 1-bit resolution of 1 mV. The 3711 converts to 3999 counts full scale and has a 1-bit resolution of 0.5 mV. The circuit in Fig. 14–18b utilizes an isolated power supply such that the converter can automatically handle input voltages of both polarities—from +2.0 to −2.0 V dc.

For both circuits, the reference voltage is derived from a National Semiconductor LM336, indicated by dotted lines. This is an active circuit that will provide 2.000 V dc with a very low thermal drift of around 20 ppm/° C.

A complete circuit used to interface the ADC3511 with an 8080A microprocessor is shown in Fig. 14–19. Tristate bus drivers (DM80LS95) are used between the 3511 digital outputs and the microprocessor data bus, and the OR–gate–NOR–gate combination is used for control. The analog input is balanced with 51-kΩ resistors, and the 200-Ω resistor connected to $SW_2$ is chosen to equal the source resistance of the voltage reference; this will provide equal time constants for charging or discharging the 0.47-μF capacitor.

In this application, the 3511 is a *peripheral mapped* device, which means that it is selected by an address placed on the address bus by the 8080A. The *unified bus comparator* is used to decode the proper address bits and select the ADC3511 with a low level at the $\overline{AD}$ input of the two control gates.

The CONVERSION COMPLETE output from the 3511 is used as an INTERRUPT signal to the 8080A, telling it that a digitized value is available to be read into the microprocessor. The receipt of an INTERRUPT signal causes the 8080A to read in the MSD (4 bits), the overflow (OFL), and the SIGN. If an overflow condition exists (OFL is high), an error signal is generated and the 8080A returns to its prior duties. Otherwise, the SIGN bit is examined and stored in the MSB of digit 4 (the LSD); a negative value is denoted by a 1 in this position. The 4 bits of the LSD, that now contains the sign bit, are shifted into the upper half of the 8080A data byte. (Note that the 8080A works with 1 byte, i.e., 8 bits, of data at a time on the data bus.) The 4 bits of digit 3 are then shifted into the lower half of this byte. In a similar fashion, digits 2 and 1 are shifted into the second byte, and the four digits are now stored in the 8080A memory.

It is beyond the intent and scope of this text to include the programming required on the 8080A to interface with the ADC3511, but the flowchart and service routine given in the National Semiconductor *Data Acquisition Handbook* are included in Fig. 14–20 for the convenience of those who might presently utilize the circuit. Additional information is available in the National Semiconductor Handbook.

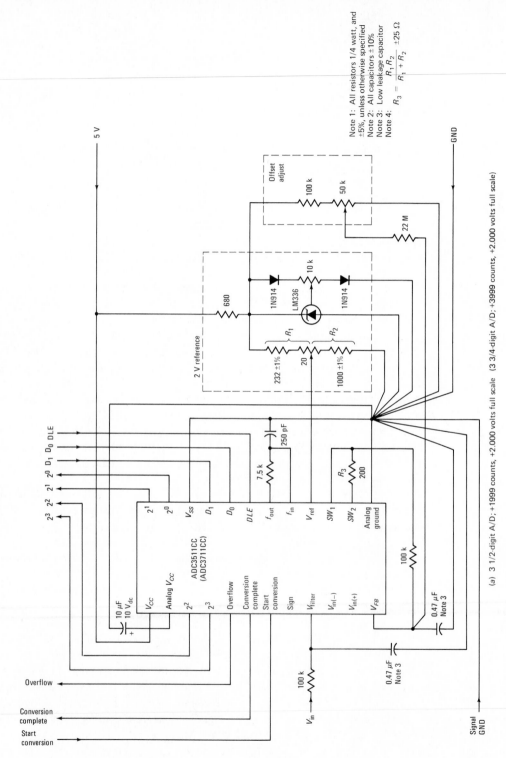

(a) 3 1/2-digit A/D: ±1999 counts, +2.000 volts full scale    (3 3/4-digit A/D: ±3999 counts, +2.000 volts full scale)

**FIG. 14–18**    (From National Semiconductor *Data Aquisition Handbook*)

Note 1:  All resistors 1/4 watt, and
±5%, unless otherwise specified
Note 2:  All capacitors ±10%
Note 3:  Low leakage capacitor
Note 4:  $R_3 = \dfrac{R_1 R_2}{R_1 + R_2} \pm 25 \ \Omega$

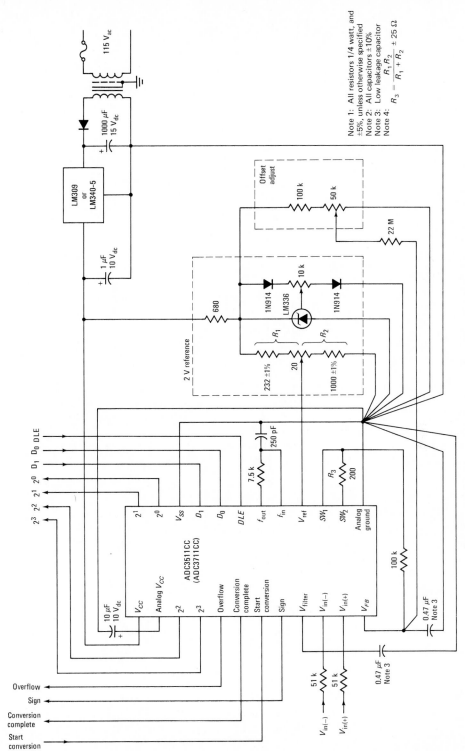

FIG. 14–18 (continued)

(b) 3 1/2-digit A/D; ±1999 counts, ±2.000 volts full scale  (3 3/4-digit A/D; ±3999 counts, ±2.000 volts full scale)

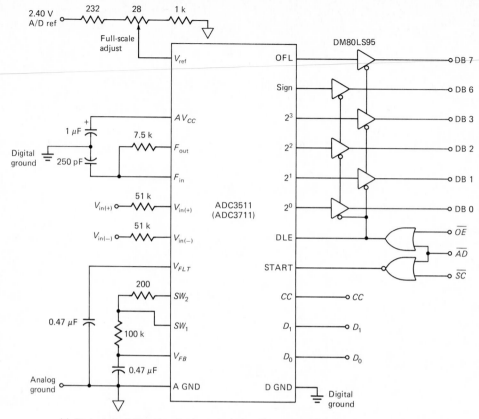

(a) Dual polarity A/D requires that inputs are isolated from the supply. Input range is ±1.999 V.

**FIG. 14–19**  (From National Semiconductor *Data Aquisition Handbook*)

☐ **EXAMPLE 14—10**

Explain why it is acceptable to place the sign bit of a voltage digitized by the ADC3511 (or 3711) in the MSB of the MSD.

☐ **SOLUTION**

The full-scale count for the 3511 is 1999 and for the 3711, is 3999. So, the largest value possible for the MSD in either case is 3 = 0011. Clearly the MSB is not needed for the magnitude of the MSD. It is thus convenient to specify a positive number when this bit is a 0 and a negative number when this bit is a 1.

## 14-5  DIGITAL VOLTMETER

The ADC3511 (or 3711) discussed in the previous section can be used as a digital voltmeter, but it is usually more convenient to have a circuit that will drive seven-segment LED displays directly. The National Semiconductor ADD3501 is a $3\frac{1}{2}$-digit DVM constructed

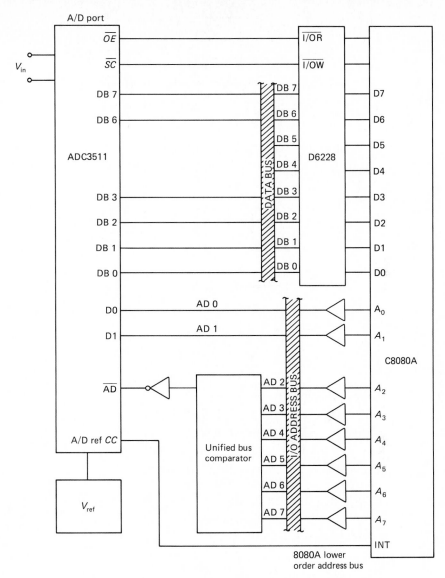

(b) Single channel A/D interface with peripheral mapped I/O

**FIG. 14–19** (continued)

using CMOS technology and available in a single dual in-line package (DIP). It operates from a single +5-V-dc power supply and will drive seven-segment indicators directly. The ADD3501 is widely used as a digital panel meter (DPM) as well as the basis for constructing a digital multimeter (DMM) capable of measuring voltage, current, and resistance, and it is readily available for around $9.00.

The connection and logic block diagrams for an ADD3501 are shown in Fig. 14–21. The only difference between this device and the ADC3511 are the outputs. There are seven segment outputs, $S_a, S_b, \ldots, S_g$, and the four digit outputs, DIGIT 1, . . . , DIGIT

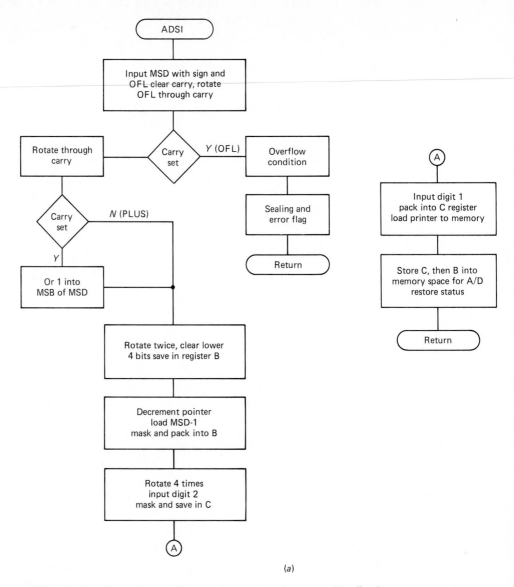

(a)

**FIG. 14–20**  (From National Semiconductor *Data Aquisition Handbook*)

4. These outputs are fully multiplexed and are designed to drive a common-cathode-type LED display directly. All the other inputs and controls are identical to the previously discussed ADC3511. The 3501 has a full-scale count of 1999 for a full-scale analog input voltage of +2.00 V dc. A resolution of 1 bit thus corresponds to 1 mV of input voltage.

The circuit shown in Fig. 14–22 shows how to use an ADD3501 as a digital voltmeter that has a full-scale analog input voltage of +2.00 V dc. The LM309 is a voltage regulator used to reduce jitter problems caused by switching. The NSB5388 is a $3\frac{1}{2}$ digit,

| Label | Opcode | Operand | Comment | Label | Opcode | Operand | Comment |
|-------|--------|---------|---------|-------|--------|---------|---------|
| ADIS: | PUSH | PSW | ; A/D interrupt service | | IN | ADD 2 | ; delay |
| | PUSH | H | ; save | | RAL | | ; rotate |
| | PUSH | B | ; current status | | RAL | | ; into |
| | IN | ADD 4 | ; input A/D digit 4 | | RAL | | ; upper |
| | IN | ADD 4 | ; delay | | RAL | | ; 4 bits |
| | ORA | | ; RESET carry | | ANI | FO | ; mask lower bits |
| | RAL | | ; rotate OFL through carry | | MOV | C, A | ; save in C |
| | JC | OFL | ; overflow condition | | IN | ADD 1 | ; in digit 1 |
| | RAL | | ; rotate sign through carry | | IN | ADD 1 | ; delay |
| | JC | PLUS | ; positive input | | ANI | OF | ; mask upper bits |
| | ORI | 20H | ; OR 1 into MSB negative input | | OR | C | ; pack |
| | | | | | MOV | C, A | ; save in C |
| PLUS: | RAL | | ; shift | | LXI | H, ADMS | ; load printer to A/D memory space |
| | RAL | | ; into position | | MOV | M, C | ; save C in memory |
| | ANI | FO | ; mask lower bits | | INX | H | ; point next |
| | MOV | BA | ; save in B | | MOV | M, B | ; save B in memory |
| | IN | ADD 3 | ; input digit 3 | | OUT | ADD 1 | ; start new conversion |
| | IN | ADD 3 | ; delay | | POP | B | ; restore |
| | ANI | OF | ; mask higher bits | | POP | H | ; previous |
| | OR | B | ; pack into B | | POP | PSW | ; status |
| | MOV | B, A | ; save in B | | EI | | ; ENABLE interrupts |
| | IN | ADD 2 | ; input digit 2 | | RET | | ; return to main program |

(b) Routine 1, single channel interrupt service routine

**FIG. 14–20** (continued)

0.5-in, common-cathode LED display. The LM336 is an active circuit used to provide the 2.00-V-dc reference voltage. When using this configuration, it is important to keep all ground leads connected to a single, central point as shown in Fig. 14–22, and care must be taken to prevent high currents from flowing in the analog $V_{CC}$ and ground wires. National Semiconductor has carefully designed the circuit to synchronize the multiplexing and the A/D conversion operations in an effort to eliminate switching noise due to power supply transients.

☐ EXAMPLE 14–11

What is the purpose of the 7.5-kΩ resistor and the 250-pF capacitor connected to pins 19 and 20 of the ADD3501 in Fig. 14–22?

☐ SOLUTION

These two components establish the internal oscillator frequency used as the clock frequency in the converter according to the relationship $f_{in} = 0.6/RC$. In this case, $f_{in} = 320$ kHz.

The DVM in Fig. 14–23 is modified slightly in order to accommodate analog input voltages of either polarity, and also of different magnitudes. Power for the circuit is obtained from the 115-V-ac power line through an isolation transformer, and the analog input is now applied at $V_{in}(+)$ and $V_{in}(-)$.

Scaling the analog input voltage for different ranges is accomplished by changing the feedback resistor between $SW_1$ on pin 17 and $V_{FB}$ on pin 14, or using a simple resistance divider across the analog input. First look at the 2.00-V-dc range, since this is the normal full-scale range for the 3501. In this position, the range switch connects 100 kΩ as the

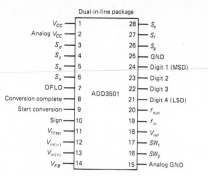

Order Number ADD3501CCN See NS Package N28A

Block diagram

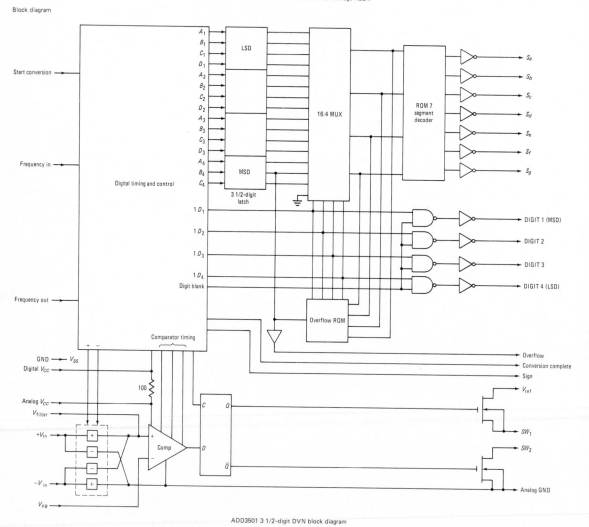

ADD3501 3 1/2-digit DVN block diagram

**FIG. 14–21**  National Semiconductor ADC 3501

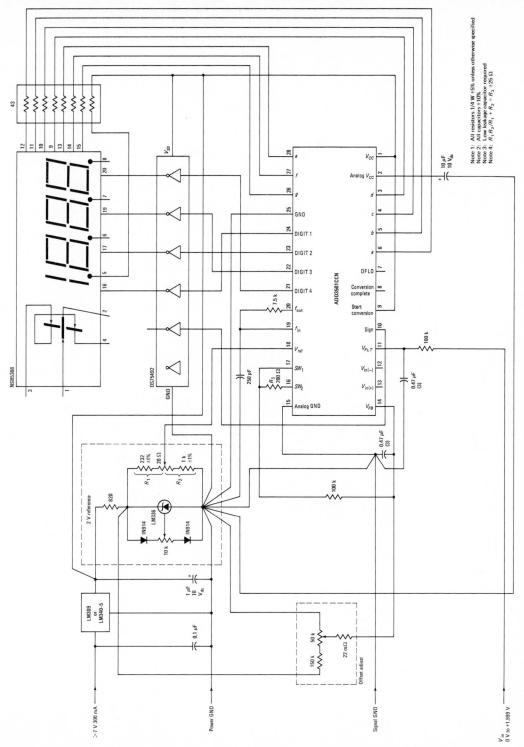

**FIG. 14–22** $3\frac{1}{2}$ digit DPM. $+1.999$ VDC full scale. (National Semiconductor)

Note 1: All resistors 1/4 W ±5% unless otherwise specified
Note 2: All capacitors ±10%
Note 3: Low leakage capacitor required
Note 4: $R_1 R_2/R_1 + R_2 = R_3 \pm 25\,\Omega$

**493**

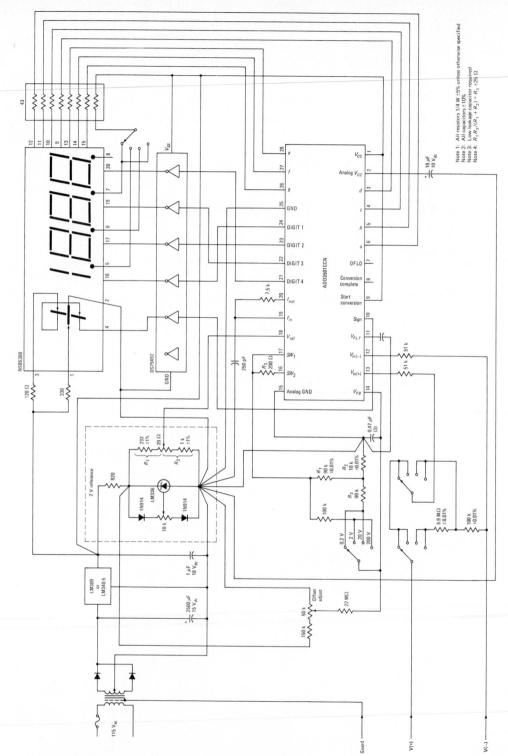

**FIG. 14-23** 3½ digit DVM. Four decade; ±0.2, +2.0, ±20, and ±200 VDC. (National Semiconductor)

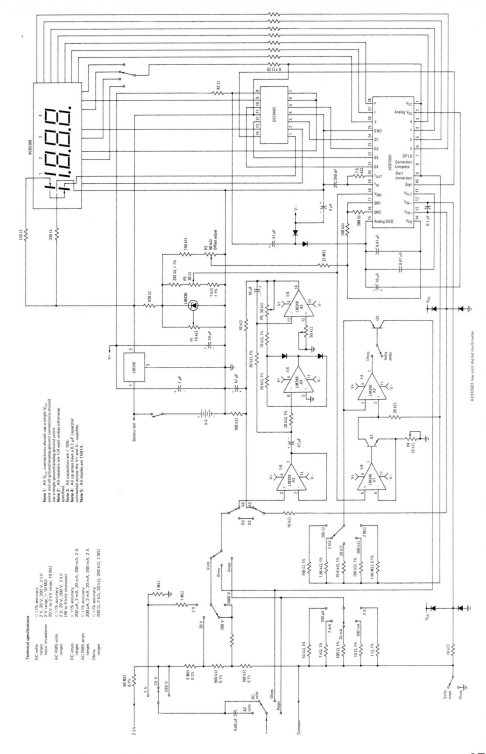

**FIG. 14–24** A low cost DMM using the ADD 3501. (National Semiconductor *Data Aquisition Handbook*)

feedback resistor, and the analog input goes directly to pin 13 $[V_{in}(+)]$. Also notice that the decimal point is between the 1 and the 8, giving 1.999 V dc as a full-scale reading.

On the 0.2-V-dc scale, the range switch still applies the analog input voltage directly to pin 13, but the reference voltage at $SW_1$ is reduced by a factor of 10 by a resistive voltage divider before being used as a feedback voltage. The resistive divider is composed of a 90-kΩ resistor $R_1$ and a 10-kΩ resistor $R_2$. The voltage developed at the node connecting these two resistors is 0.1 $V_{ref}$, and so the full-scale voltage is also reduced by a factor of 10. The 90-kΩ resistor $R_3$ is used to keep the charging time constant essentially the same on all ranges. The time constant is given as $RC = 100$ kΩ $\times$ 0.47 $\mu$F. Notice that the decimal-point position has moved to pin 7 on the NSB5388 to give a full-scale reading of 199.9 mV.

On the 200-V-dc position, the range switch puts back the original feedback resistor, but the analog input voltage is reduced by a factor of 100 with a resistive voltage divider composed of a 9.9-MΩ and a 100-kΩ resistor. The analog input to the 3501 is thus still 2.00 V dc full scale even though the actual input signal is 200 V dc full scale. The decimal point will be placed on pin 7 of the NSB5388 to give a full-scale reading of 199.9 V dc.

On the 20-V-dc full-scale position, the range switch still uses the input voltage divider to reduce the input signal by a factor of 100, but the feedback resistor is also used to effectively increase the full scale by a factor of 10. The net result is that the 3501 will count full scale when the analog input voltage is 20 V dc. Notice that the decimal point is now applied to pin 6 of the NSB5388 to give a full-scale reading of 19.99 V dc.

The circuit shown in Fig. 14–24 is a complete DMM taken from the National Semiconductor *Data Acquisition Handbook*. It utilizes the ADD3501 and is capable of measuring both dc and ac currents and voltages as well as resistances. The ranges and accuracies of the instrument are given in Fig. 14–25.

The different dc and ac voltage ranges are accommodated by a resistive voltage divider at the analog input. Alternating-current voltages are measured by using the three operational amplifiers $A_3$, $A_4$, and $A_5$ to develop a dc voltage that is proportional to the root-mean-square (RMS) value of the ac input voltage.

A series of current-sensing resistors are used to measure either dc or ac current. The current to be measured is passed through one of the sensing resistors, and the DMM digitizes the voltage developed across the resistor.

The DMM measures resistance by applying a known current from an internal current source (operational amplifiers $A_1$ and $A_2$) to the unknown resistance and then digitizing the resulting voltage developed.

| Measurement mode | Range | | | | | Frequency response | Accuracy | Overrange display |
|---|---|---|---|---|---|---|---|---|
| | 0.2 | 2.0 | 20 | 200 | 2000 | | | |
| DC volts | — | V | V | V | V | — | ≤ 1% FS | ± OFLO |
| AC volts | — | $V_{RMS}$ | $V_{RMS}$ | $V_{RMS}$ | $V_{RMS}$ | 40 Hz to 5 kHz | ≤ 1% FS | + OFLO |
| DC amps | mA | mA | mA | mA | mA | — | ≤ 1% FS | ± OFLO |
| AC amps | mARMS | mARMS | mARMS | mARMS | mARMS | 40 Hz to 5 kHz | ≤ 1% FS | + OFLO |
| Ohms | kΩ | kΩ | kΩ | kΩ | kΩ | — | ≤ 1% FS | + OFLO |

**FIG. 14–25** Performance of the DMM in Fig. 14–24

For those interested in pursuing this subject, complete details for the construction and calibration of this DMM are given in the National Semiconductor *Data Acquisition Handbook*.

## SUMMARY

The primary objective of this chapter is to demonstrate the use of many of the most fundamental principles discussed throughout the text by considering some of the more common digital circuit configurations encountered in industry. Multiplexing of LED displays, time and frequency measurement, and use of digital voltmeters of all types are widely used throughout industry. Although our coverage is by no means comprehensive, it will serve as an excellent introduction to industrial practices.

The problems at the end of this chapter will also provide a good transition into industry. They are in general longer than previously assigned problems. All the necessary information required to work a given problem may not be given—this is intentional since it will require you to seek information from industrial data sheets. However, the problems are more of a design nature, and usually deal with a practical, functional circuit that can be used to accomplish a given task; as such, they are much more interesting and satisfying to solve.

## PROBLEMS

In order to solve some of these problems, you may have to consult product data sheets that are not included in this text. It is intended that you discover a source for such information.

**14–1.** Pick one of the solutions suggested in Example 14–3 and do a detailed design, including part numbers and pin numbers.

**14–2.** Design a four-decimal-digit multiplexed display like the one in Fig. 14–7, but use common-cathode-type LEDs. Use a basic circuit like the one in Fig. 14–2, but you will now need to generate DIGIT waveforms that have positive pulses.

**14–3.** How often is each digit in Fig. 14–7 serviced, and for what period of time is it illuminated? Extinguished?

**14–4.** Design a multiplexed display like the one in Fig. 14–8 having eight decimal digits. Use a three-flip-flop multiplexing counter and four 74151 multiplexers.

**14–5.** Specify a ROM that could be used in place of the 7447 in Fig. 14–8. Draw a circuit, showing exactly how to connect it.

**14–6.** Design a four-decimal-digit display using 54/74143 and common-anode LEDs.

**14–7.** Using Fig. 14–12 as a pattern, design a four-digit frequency counter using 54/74143s and 54/74160s. Use a 1.0-MHz clock, and provide 0.1-, 1.0-, and 10.0-s gates. Specify the frequency range for each gate.

**14–8.** Following Fig. 14–12 as a guide, design a four-digit frequency counter using National Semiconductor MM74C925.

**14–9.** Design a circuit to measure "elapsed time" between two events in time—for instance, the time difference between a pulse occurring on one signal followed by a pulse occurring on another signal. Use as much of Fig. 14–14 as possible, but consider using a set-reset flip-flop in conjunction with the two input signals.

**14–10.** Combine the circuits in Figs. 14–12 and 14–14 into a single instrument. Use a 1.0-MHz clock and seven decade counters. Define the scales and readouts carefully.

**14–11.** Design a resistive voltage divider to use with the ADC3511 such that it will digitize an analog input voltage of 20 V dc as full-scale voltage input. What is the resolution in millivolts for this design?

**14–12.** Design a voltage divider such that the DVM in Fig. 14–23 will measure full-scale voltages of 2.0, 20.0, and 200.0 V dc without changing the feedback resistor. Leave the feedback resistor at 100 k$\Omega$. Draw the complete design. Is it possible to achieve a full scale of 0.2 V dc for this circuit without changing the feedback resistor?

# APPENDIX 1.  BINARY-HEXADECIMAL-DECIMAL EQUIVALENTS

| Binary | Hexadecimal | Upper Byte | Lower Byte |
|---|---|---|---|
| 0000 0000 | 00 | 0 | 0 |
| 0000 0001 | 01 | 256 | 1 |
| 0000 0010 | 02 | 512 | 2 |
| 0000 0011 | 03 | 768 | 3 |
| 0000 0100 | 04 | 1,024 | 4 |
| 0000 0101 | 05 | 1,280 | 5 |
| 0000 0110 | 06 | 1,536 | 6 |
| 0000 0111 | 07 | 1,792 | 7 |
| 0000 1000 | 08 | 2,048 | 8 |
| 0000 1001 | 09 | 2,304 | 9 |
| 0000 1010 | 0A | 2,560 | 10 |
| 0000 1011 | 0B | 2,816 | 11 |
| 0000 1100 | 0C | 3,072 | 12 |
| 0000 1101 | 0D | 3,328 | 13 |
| 0000 1110 | 0E | 3,584 | 14 |
| 0000 1111 | 0F | 3,840 | 15 |
| 0001 0000 | 10 | 4,096 | 16 |
| 0001 0001 | 11 | 4,352 | 17 |
| 0001 0010 | 12 | 4,608 | 18 |
| 0001 0011 | 13 | 4,864 | 19 |
| 0001 0100 | 14 | 5,120 | 20 |
| 0001 0101 | 15 | 5,376 | 21 |
| 0001 0110 | 16 | 5,632 | 22 |
| 0001 0111 | 17 | 5,888 | 23 |
| 0001 1000 | 18 | 6,144 | 24 |
| 0001 1001 | 19 | 6,400 | 25 |
| 0001 1010 | 1A | 6,656 | 26 |
| 0001 1011 | 1B | 6,912 | 27 |
| 0001 1100 | 1C | 7,168 | 28 |
| 0001 1101 | 1D | 7,424 | 29 |
| 0001 1110 | 1E | 7,680 | 30 |
| 0001 1111 | 1F | 7,936 | 31 |
| 0010 0000 | 20 | 8,192 | 32 |
| 0010 0001 | 21 | 8,448 | 33 |
| 0010 0010 | 22 | 8,704 | 34 |
| 0010 0011 | 23 | 8,960 | 35 |
| 0010 0100 | 24 | 9,216 | 36 |
| 0010 0101 | 25 | 9,472 | 37 |
| 0010 0110 | 26 | 9,728 | 38 |
| 0010 0111 | 27 | 9,984 | 39 |
| 0010 1000 | 28 | 10,240 | 40 |
| 0010 1001 | 29 | 10,496 | 41 |
| 0010 1010 | 2A | 10,752 | 42 |
| 0010 1011 | 2B | 11,008 | 43 |

| Binary | Hexadecimal | Upper Byte | Lower Byte |
|---|---|---|---|
| 0010 1100 | 2C | 11,264 | 44 |
| 0010 1101 | 2D | 11,520 | 45 |
| 0010 1110 | 2E | 11,776 | 46 |
| 0010 1111 | 2F | 12,032 | 47 |
| 0011 0000 | 30 | 12,288 | 48 |
| 0011 0001 | 31 | 12,544 | 49 |
| 0011 0010 | 32 | 12,800 | 50 |
| | | | |
| 0011 0011 | 33 | 13,056 | 51 |
| 0011 0100 | 34 | 13,312 | 52 |
| 0011 0101 | 35 | 13,568 | 53 |
| 0011 0110 | 36 | 13,824 | 54 |
| 0011 0111 | 37 | 14,080 | 55 |
| 0011 1000 | 38 | 14,336 | 56 |
| 0011 1001 | 39 | 14,592 | 57 |
| 0011 1010 | 3A | 14,848 | 58 |
| 0011 1011 | 3B | 15,104 | 59 |
| 0011 1100 | 3C | 15,360 | 60 |
| | | | |
| 0011 1101 | 3D | 15,616 | 61 |
| 0011 1110 | 3E | 15,872 | 62 |
| 0011 1111 | 3F | 16,128 | 63 |
| 0100 0000 | 40 | 16,384 | 64 |
| 0100 0001 | 41 | 16,640 | 65 |
| 0100 0010 | 42 | 16,896 | 66 |
| 0100 0011 | 43 | 17,152 | 67 |
| 0100 0100 | 44 | 17,408 | 68 |
| 0100 0101 | 45 | 17,664 | 69 |
| 0100 0110 | 46 | 17,920 | 70 |
| | | | |
| 0100 0111 | 47 | 18,176 | 71 |
| 0100 1000 | 48 | 18,432 | 72 |
| 0100 1001 | 49 | 18,688 | 73 |
| 0100 1010 | 4A | 18,944 | 74 |
| 0100 1011 | 4B | 19,200 | 75 |
| 0100 1100 | 4C | 19,456 | 76 |
| 0100 1101 | 4D | 19,712 | 77 |
| 0100 1110 | 4E | 19,968 | 78 |
| 0100 1111 | 4F | 20,224 | 79 |
| 0101 0000 | 50 | 20,480 | 80 |
| | | | |
| 0101 0001 | 51 | 20,736 | 81 |
| 0101 0010 | 52 | 20,992 | 82 |
| 0101 0011 | 53 | 21,248 | 83 |
| 0101 0100 | 54 | 21,504 | 84 |
| 0101 0101 | 55 | 21,760 | 85 |
| 0101 0110 | 56 | 22,016 | 86 |
| 0101 0111 | 57 | 22,272 | 87 |
| 0101 1000 | 58 | 22,528 | 88 |
| 0101 1001 | 59 | 22,784 | 89 |
| 0101 1010 | 5A | 23,040 | 90 |

| Binary | Hexadecimal | Upper Byte | Lower Byte |
|--------|-------------|------------|------------|
| 0101 1011 | 5B | 23,296 | 91 |
| 0101 1100 | 5C | 23,552 | 92 |
| 0101 1101 | 5D | 23,808 | 93 |
| 0101 1110 | 5E | 24,064 | 94 |
| 0101 1111 | 5F | 24,320 | 95 |
| 0110 0000 | 60 | 24,576 | 96 |
| 0110 0001 | 61 | 24,832 | 97 |
| 0110 0010 | 62 | 25,088 | 98 |
| 0110 0011 | 63 | 25,344 | 99 |
| 0110 0100 | 64 | 25,600 | 100 |
| 0110 0101 | 65 | 25,856 | 101 |
| 0110 0110 | 66 | 26,112 | 102 |
| 0110 0111 | 67 | 26,368 | 103 |
| 0110 1000 | 68 | 26,624 | 104 |
| 0110 1001 | 69 | 26,880 | 105 |
| 0110 1010 | 6A | 27,136 | 106 |
| 0110 1011 | 6B | 27,392 | 107 |
| 0110 1100 | 6C | 27,648 | 108 |
| 0110 1101 | 6D | 27,904 | 109 |
| 0110 1110 | 6E | 28,160 | 110 |
| 0110 1111 | 6F | 28,416 | 111 |
| 0111 0000 | 70 | 28,672 | 112 |
| 0111 0001 | 71 | 28,928 | 113 |
| 0111 0010 | 72 | 29,184 | 114 |
| 0111 0011 | 73 | 29,440 | 115 |
| 0111 0100 | 74 | 29,696 | 116 |
| 0111 0101 | 75 | 29,952 | 117 |
| 0111 0110 | 76 | 30,208 | 118 |
| 0111 0111 | 77 | 30,464 | 119 |
| 0111 1000 | 78 | 30,720 | 120 |
| 0111 1001 | 79 | 30,976 | 121 |
| 0111 1010 | 7A | 31,232 | 122 |
| 0111 1011 | 7B | 31,488 | 123 |
| 0111 1100 | 7C | 31,744 | 124 |
| 0111 1101 | 7D | 32,000 | 125 |
| 0111 1110 | 7E | 32,256 | 126 |
| 0111 1111 | 7F | 32,512 | 127 |
| 1000 0000 | 80 | 32,768 | 128 |
| 1000 0001 | 81 | 33,024 | 129 |
| 1000 0010 | 82 | 33,280 | 130 |
| 1000 0011 | 83 | 33,536 | 131 |
| 1000 0100 | 84 | 33,792 | 132 |
| 1000 0101 | 85 | 34,048 | 133 |
| 1000 0110 | 86 | 34,304 | 134 |
| 1000 0111 | 87 | 34,560 | 135 |
| 1000 1000 | 88 | 34,816 | 136 |
| 1000 1001 | 89 | 35,072 | 137 |
| 1000 1010 | 8A | 35,328 | 138 |
| 1000 1011 | 8B | 35,584 | 139 |
| 1000 1100 | 8C | 35,840 | 140 |

| Binary | Hexadecimal | Upper Byte | Lower Byte |
|--------|-------------|------------|------------|
| 1000 1101 | 8D | 36,096 | 141 |
| 1000 1110 | 8E | 36,352 | 142 |
| 1000 1111 | 8F | 36,608 | 143 |
| 1001 0000 | 90 | 36,864 | 144 |
| 1001 0001 | 91 | 37,120 | 145 |
| 1001 0010 | 92 | 37,376 | 146 |
| 1001 0011 | 93 | 37,632 | 147 |
| 1001 0100 | 94 | 37,888 | 148 |
| 1001 0101 | 95 | 38,144 | 149 |
| 1001 0110 | 96 | 38,400 | 150 |
| | | | |
| 1001 0111 | 97 | 38,656 | 151 |
| 1001 1000 | 98 | 38,912 | 152 |
| 1001 1001 | 99 | 39,168 | 153 |
| 1001 1010 | 9A | 39,424 | 154 |
| 1001 1011 | 9B | 39,680 | 155 |
| 1001 1100 | 9C | 39,936 | 156 |
| 1001 1101 | 9D | 40,192 | 157 |
| 1001 1110 | 9E | 40,448 | 158 |
| 1001 1111 | 9F | 40,704 | 159 |
| 1010 0000 | A0 | 40,960 | 160 |
| | | | |
| 1010 0001 | A1 | 41,216 | 161 |
| 1010 0010 | A2 | 41,472 | 162 |
| 1010 0011 | A3 | 41,728 | 163 |
| 1010 0100 | A4 | 41,984 | 164 |
| 1010 0101 | A5 | 42,240 | 165 |
| 1010 0110 | A6 | 42,496 | 166 |
| 1010 0111 | A7 | 42,752 | 167 |
| 1010 1000 | A8 | 43,008 | 168 |
| 1010 1001 | A9 | 43,264 | 169 |
| 1010 1010 | AA | 43,520 | 170 |
| | | | |
| 1010 1011 | AB | 43,776 | 171 |
| 1010 1100 | AC | 44,032 | 172 |
| 1010 1101 | AD | 44,288 | 173 |
| 1010 1110 | AE | 44,544 | 174 |
| 1010 1111 | AF | 44,800 | 175 |
| 1011 0000 | B0 | 45,056 | 176 |
| 1011 0001 | B1 | 45,312 | 177 |
| 1011 0010 | B2 | 45,568 | 178 |
| 1011 0011 | B3 | 45,824 | 179 |
| 1011 0100 | B4 | 46,080 | 180 |
| | | | |
| 1011 0101 | B5 | 46,336 | 181 |
| 1011 0110 | B6 | 46,592 | 182 |
| 1011 0111 | B7 | 46,848 | 183 |
| 1011 1000 | B8 | 47,104 | 184 |
| 1011 1001 | B9 | 47,360 | 185 |
| 1011 1010 | BA | 47,616 | 186 |
| 1011 1011 | BB | 47,872 | 187 |
| 1011 1100 | BC | 48,128 | 188 |
| 1011 1101 | BD | 48,384 | 189 |
| 1011 1110 | BE | 48,640 | 190 |

| Binary | Hexadecimal | Upper Byte | Lower Byte |
|--------|-------------|------------|------------|
| 1011 1111 | BF | 48,896 | 191 |
| 1100 0000 | C0 | 49,152 | 192 |
| 1100 0001 | C1 | 49,408 | 193 |
| 1100 0010 | C2 | 49,664 | 194 |
| 1100 0011 | C3 | 49,920 | 195 |
| 1100 0100 | C4 | 50,176 | 196 |
| 1100 0101 | C5 | 50,432 | 197 |
| 1100 0110 | C6 | 50,688 | 198 |
| 1100 0111 | C7 | 50,944 | 199 |
| 1100 1000 | C8 | 51,200 | 200 |
| 1100 1001 | C9 | 51,456 | 201 |
| 1100 1010 | CA | 51,712 | 202 |
| 1100 1011 | CB | 51,968 | 203 |
| 1100 1100 | CC | 52,224 | 204 |
| 1100 1101 | CD | 52,480 | 205 |
| 1100 1110 | CE | 52,736 | 206 |
| 1100 1111 | CF | 52,992 | 207 |
| 1101 0000 | D0 | 53,248 | 208 |
| 1101 0001 | D1 | 53,504 | 209 |
| 1101 0010 | D2 | 53,760 | 210 |
| 1101 0011 | D3 | 54,016 | 211 |
| 1101 0100 | D4 | 54,272 | 212 |
| 1101 0101 | D5 | 54,528 | 213 |
| 1101 0110 | D6 | 54,784 | 214 |
| 1101 0111 | D7 | 55,040 | 215 |
| 1101 1000 | D8 | 55,296 | 216 |
| 1101 1001 | D9 | 55,552 | 217 |
| 1101 1010 | DA | 55,808 | 218 |
| 1101 1011 | DB | 56,064 | 219 |
| 1101 1100 | DC | 56,320 | 220 |
| 1101 1101 | DD | 56,576 | 221 |
| 1101 1110 | DE | 56,832 | 222 |
| 1101 1111 | DF | 57,088 | 223 |
| 1110 0000 | E0 | 57,344 | 224 |
| 1110 0001 | E1 | 57,600 | 225 |
| 1110 0010 | E2 | 57,856 | 226 |
| 1110 0011 | E3 | 58,112 | 227 |
| 1110 0100 | E4 | 58,368 | 228 |
| 1110 0101 | E5 | 58,624 | 229 |
| 1110 0110 | E6 | 58,880 | 230 |
| 1110 0111 | E7 | 59,136 | 231 |
| 1110 1000 | E8 | 59,392 | 232 |
| 1110 1001 | E9 | 59,648 | 233 |
| 1110 1010 | EA | 59,904 | 234 |
| 1110 1011 | EB | 60,160 | 235 |
| 1110 1100 | EC | 60,416 | 236 |
| 1110 1101 | ED | 60,672 | 237 |
| 1110 1110 | EE | 60,928 | 238 |
| 1110 1111 | EF | 61,184 | 239 |
| 1111 0000 | F0 | 61,440 | 240 |

| Binary | Hexadecimal | Upper Byte | Lower Byte |
|---|---|---|---|
| 1111 0001 | F1 | 61,696 | 241 |
| 1111 0010 | F2 | 61,952 | 242 |
| 1111 0011 | F3 | 62,208 | 243 |
| 1111 0100 | F4 | 62,464 | 244 |
| 1111 0101 | F5 | 62,720 | 245 |
| 1111 0110 | F6 | 62,976 | 246 |
| 1111 0111 | F7 | 63,232 | 247 |
| 1111 1000 | F8 | 63,488 | 248 |
| 1111 1001 | F9 | 63,744 | 249 |
| 1111 1010 | FA | 64,000 | 250 |
| 1111 1011 | FB | 64,256 | 251 |
| 1111 1100 | FC | 64,512 | 252 |
| 1111 1101 | FD | 64,768 | 253 |
| 1111 1110 | FE | 65,024 | 254 |
| 1111 1111 | FF | 65,280 | 255 |

# APPENDIX 2.   2'S COMPLEMENT REPRESENTATION

| POSITIVE | | | NEGATIVE | | |
|---|---|---|---|---|---|
| Decimal | Hexadecimal | Binary | Binary | Hexadecimal | Decimal |
| 0 | 00H | 0000 0000 | 0000 0000 | 00H | 0 |
| 1 | 01H | 0000 0001 | 1111 1111 | FFH | −1 |
| 2 | 02H | 0000 0010 | 1111 1110 | FEH | −2 |
| 3 | 03H | 0000 0011 | 1111 1101 | FDH | −3 |
| 4 | 04H | 0000 0100 | 1111 1100 | FCH | −4 |
| 5 | 05H | 0000 0101 | 1111 1011 | FBH | −5 |
| 6 | 06H | 0000 0110 | 1111 1010 | FAH | −6 |
| 7 | 07H | 0000 0111 | 1111 1001 | F9H | −7 |
| 8 | 08H | 0000 1000 | 1111 1000 | F8H | −8 |
| 9 | 09H | 0000 1001 | 1111 0111 | F7H | −9 |
| 10 | 0AH | 0000 1010 | 1111 0110 | F6H | −10 |
| 11 | 0BH | 0000 1011 | 1111 0101 | F5H | −11 |
| 12 | 0CH | 0000 1100 | 1111 0100 | F4H | −12 |
| 13 | 0DH | 0000 1101 | 1111 0011 | F3H | −13 |
| 14 | 0EH | 0000 1110 | 1111 0010 | F2H | −14 |
| 15 | 0FH | 0000 1111 | 1111 0001 | F1H | −15 |
| 16 | 10H | 0001 0000 | 1111 0000 | F0H | −16 |
| 17 | 11H | 0001 0001 | 1110 1111 | EFH | −17 |
| 18 | 12H | 0001 0010 | 1110 1110 | EEH | −18 |
| 19 | 13H | 0001 0011 | 1110 1101 | EDH | −19 |
| 20 | 14H | 0001 0100 | 1110 1100 | ECH | −20 |
| 21 | 15H | 0001 0101 | 1110 1011 | EBH | −21 |
| 22 | 16H | 0001 0110 | 1110 1010 | EAH | −22 |
| 23 | 17H | 0001 0111 | 1110 1001 | E9H | −23 |
| 24 | 18H | 0001 1000 | 1110 1000 | E8H | −24 |
| 25 | 19H | 0001 1001 | 1110 0111 | E7H | −25 |
| 26 | 1AH | 0001 1010 | 1110 0110 | E6H | −26 |
| 27 | 1BH | 0001 1011 | 1110 0101 | E5H | −27 |
| 28 | 1CH | 0001 1100 | 1110 0100 | E4H | −28 |

| POSITIVE | | | NEGATIVE | | |
|---|---|---|---|---|---|
| Decimal | Hexadecimal | Binary | Binary | Hexadecimal | Decimal |
| 29 | 1DH | 0001 1101 | 1110 0011 | E3H | −29 |
| 30 | 1EH | 0001 1110 | 1110 0010 | E2H | −30 |
| 31 | 1FH | 0001 1111 | 1110 0001 | E1H | −31 |
| 32 | 20H | 0010 0000 | 1110 0000 | E0H | −32 |
| 33 | 21H | 0010 0001 | 1101 1111 | DFH | −33 |
| 34 | 22H | 0010 0010 | 1101 1110 | DEH | −34 |
| 35 | 23H | 0010 0011 | 1101 1101 | DDH | −35 |
| 36 | 24H | 0010 0100 | 1101 1100 | DCH | −36 |
| 37 | 25H | 0010 0101 | 1101 1011 | DBH | −37 |
| 38 | 26H | 0010 0110 | 1101 1010 | DAH | −38 |
| 39 | 27H | 0010 0111 | 1101 1001 | D9H | −39 |
| 40 | 28H | 0010 1000 | 1101 1000 | D8H | −40 |
| 41 | 29H | 0010 1001 | 1101 0111 | D7H | −41 |
| 42 | 2AH | 0010 1010 | 1101 0110 | D6H | −42 |
| 43 | 2BH | 0010 1011 | 1101 0101 | D5H | −43 |
| 44 | 2CH | 0010 1100 | 1101 0100 | D4H | −44 |
| 45 | 2DH | 0010 1101 | 1101 0011 | D3H | −45 |
| 46 | 2EH | 0010 1110 | 1101 0010 | D2H | −46 |
| 47 | 2FH | 0010 1111 | 1101 0001 | D1H | −47 |
| 48 | 30H | 0011 0000 | 1101 0000 | D0H | −48 |
| 49 | 31H | 0011 0001 | 1100 1111 | CFH | −49 |
| 50 | 32H | 0011 0010 | 1100 1110 | CEH | −50 |
| 51 | 33H | 0011 0011 | 1100 1101 | CDH | −51 |
| 52 | 34H | 0011 0100 | 1100 1100 | CCH | −52 |
| 53 | 35H | 0011 0101 | 1100 1011 | CBH | −53 |
| 54 | 36H | 0011 0110 | 1100 1010 | CAH | −54 |
| 55 | 37H | 0011 0111 | 1100 1001 | C9H | −55 |
| 56 | 38H | 0011 1000 | 1100 1000 | C8H | −56 |
| 57 | 39H | 0011 1001 | 1100 0111 | C7H | −57 |
| 58 | 3AH | 0011 1010 | 1100 0110 | C6H | −58 |
| 59 | 3BH | 0011 1011 | 1100 0101 | C5H | −59 |
| 60 | 3CH | 0011 1100 | 1100 0100 | C4H | −60 |
| 61 | 3DH | 0011 1101 | 1100 0011 | C3H | −61 |
| 62 | 3EH | 0011 1110 | 1100 0010 | C2H | −62 |
| 63 | 3FH | 0011 1111 | 1100 0001 | C1H | −63 |
| 64 | 40H | 0100 0000 | 1100 0000 | C0H | −64 |
| 65 | 41H | 0100 0001 | 1011 1111 | BFH | −65 |
| 66 | 42H | 0100 0010 | 1011 1110 | BEH | −66 |
| 67 | 43H | 0100 0011 | 1011 1101 | BDH | −67 |
| 68 | 44H | 0100 0100 | 1011 1100 | BCH | −68 |
| 69 | 45H | 0100 0101 | 1011 1011 | BBH | −69 |
| 70 | 46H | 0100 0110 | 1011 1010 | BAH | −70 |
| 71 | 47H | 0100 0111 | 1011 1001 | B9H | −71 |
| 72 | 48H | 0100 1000 | 1011 1000 | B8H | −72 |
| 73 | 49H | 0100 1001 | 1011 0111 | B7H | −73 |
| 74 | 4AH | 0100 1010 | 1011 0110 | B6H | −74 |
| 75 | 4BH | 0100 1011 | 1011 0101 | B5H | −75 |
| 76 | 4CH | 0100 1100 | 1011 0100 | B4H | −76 |
| 77 | 4DH | 0100 1101 | 1011 0011 | B3H | −77 |
| 78 | 4EH | 0100 1110 | 1011 0010 | B2H | −78 |
| 79 | 4FH | 0100 1111 | 1011 0001 | B1H | −79 |
| 80 | 50H | 0101 0000 | 1011 0000 | B0H | −80 |

| POSITIVE | | | NEGATIVE | | |
|---|---|---|---|---|---|
| Decimal | Hexadecimal | Binary | Binary | Hexadecimal | Decimal |
| 81 | 51H | 0101 0001 | 1010 1111 | AFH | −81 |
| 82 | 52H | 0101 0010 | 1010 1110 | AEH | −82 |
| 83 | 53H | 0101 0011 | 1010 1101 | ADH | −83 |
| 84 | 54H | 0101 0100 | 1010 1100 | ACH | −84 |
| 85 | 55H | 0101 0101 | 1010 1011 | ABH | −85 |
| 86 | 56H | 0101 0110 | 1010 1010 | AAH | −86 |
| 87 | 57H | 0101 0111 | 1010 1001 | A9H | −87 |
| 88 | 58H | 0101 1000 | 1010 1000 | A8H | −88 |
| 89 | 59H | 0101 1001 | 1010 0111 | A7H | −89 |
| 90 | 5AH | 0101 1010 | 1010 0110 | A6H | −90 |
| 91 | 5BH | 0101 1011 | 1010 0101 | A5H | −91 |
| 92 | 5CH | 0101 1100 | 1010 0100 | A4H | −92 |
| 93 | 5DH | 0101 1101 | 1010 0011 | A3H | −93 |
| 94 | 5EH | 0101 1110 | 1010 0010 | A2H | −94 |
| 95 | 5FH | 0101 1111 | 1010 0001 | A1H | −95 |
| 96 | 60H | 0110 0000 | 1010 0000 | A0H | −96 |
| 97 | 61H | 0110 0001 | 1001 1111 | 9FH | −97 |
| 98 | 62H | 0110 0010 | 1001 1110 | 9EH | −98 |
| 99 | 63H | 0110 0011 | 1001 1101 | 9DH | −99 |
| 100 | 64H | 0110 0100 | 1001 1100 | 9CH | −100 |
| 101 | 65H | 0110 0101 | 1001 1011 | 9BH | −101 |
| 102 | 66H | 0110 0110 | 1001 1010 | 9AH | −102 |
| 103 | 67H | 0110 0111 | 1001 1001 | 99H | −103 |
| 104 | 68H | 0110 1000 | 1001 1000 | 98H | −104 |
| 105 | 69H | 0110 1001 | 1001 0111 | 97H | −105 |
| 106 | 6AH | 0110 1010 | 1001 0110 | 96H | −106 |
| 107 | 6BH | 0110 1011 | 1001 0101 | 95H | −107 |
| 108 | 6CH | 0110 1100 | 1001 0100 | 94H | −108 |
| 109 | 6DH | 0110 1101 | 1001 0011 | 93H | −109 |
| 110 | 6EH | 0110 1110 | 1001 0010 | 92H | −110 |
| 111 | 6FH | 0110 1111 | 1001 0001 | 91H | −111 |
| 112 | 70H | 0111 0000 | 1001 0000 | 90H | −112 |
| 113 | 71H | 0111 0001 | 1000 1111 | 8FH | −113 |
| 114 | 72H | 0111 0010 | 1000 1110 | 8EH | −114 |
| 115 | 73H | 0111 0011 | 1000 1101 | 8DH | −115 |
| 116 | 74H | 0111 0100 | 1000 1100 | 8CH | −116 |
| 117 | 75H | 0111 0101 | 1000 1011 | 8BH | −117 |
| 118 | 76H | 0111 0110 | 1000 1010 | 8AH | −118 |
| 119 | 77H | 0111 0111 | 1000 1001 | 89H | −119 |
| 120 | 78H | 0111 1000 | 1000 1000 | 88H | −120 |
| 121 | 79H | 0111 1001 | 1000 0111 | 87H | −121 |
| 122 | 7AH | 0111 1010 | 1000 0110 | 86H | −122 |
| 123 | 7BH | 0111 1011 | 1000 0101 | 85H | −123 |
| 124 | 7CH | 0111 1100 | 1000 0100 | 84H | −124 |
| 125 | 7DH | 0111 1101 | 1000 0011 | 83H | −125 |
| 126 | 7EH | 0111 1110 | 1000 0010 | 82H | −126 |
| 127 | 7FH | 0111 1111 | 1000 0001 | 81H | −127 |
| 128 | — | — | 1000 0000 | 80H | −128 |

# APPENDIX 3.  TTL DEVICES

| Number | Function | Number | Function |
|---|---|---|---|
| 7400 | Quad 2-input NAND gates | 7455 | Expandable 4-input 2-wide AND-OR-INVERT gates |
| 7401 | Quad 2-input NAND gates (open collector) | | |
| 7402 | Quad 2-input NOR gates | 7459 | Dual 2-3 input 2-wide AND-OR-INVERT gates |
| 7403 | Quad 2-input NOR gates (open collector) | 7460 | Dual 4-input expanders |
| 7404 | Hex inverters | 7461 | Triple 3-input expanders |
| 7405 | Hex inverters (open collector) | 7462 | 2-2-3-3 input 4-wide expanders |
| 7406 | Hex inverter buffer-driver | 7464 | 2-2-3-4 input 4-wide AND-OR-INVERT gates |
| 7407 | Hex buffer-drivers | 7465 | 4-wide AND-OR-INVERT gates (open collector) |
| 7408 | Quad 2-input AND gates | | |
| 7409 | Quad 2-input AND gates (open collector) | 7470 | Edge-triggered JK flip-flop |
| 7410 | Triple 3-input NAND gates | 7472 | JK master-slave flip-flop |
| 7411 | Triple 3-input AND gates | 7473 | Dual JK master-slave flip-flop |
| 7412 | Triple 3-input NAND gates (open collector) | 7474 | Dual D flip-flop |
| 7413 | Dual Schmitt triggers | 7475 | Quad latch |
| 7414 | Hex Schmitt triggers | 7476 | Dual JK master-slave flip-flop |
| 7416 | Hex inverter buffer-drivers | 7480 | Gates full adder |
| 7417 | Hex buffer-drivers | 7482 | 2-bit binary full adder |
| 7420 | Dual 4-input NAND gates | 7483 | 4-bit binary full adder |
| 7421 | Dual 4-input AND gates | 7485 | 4-bit magnitude comparator |
| 7422 | Dual 4-input NAND gates (open collector) | 7486 | Quad EXCLUSIVE-OR gate |
| 7423 | Expandable dual 4-input NOR gates | 7489 | 64-bit random-access read-write memory |
| 7425 | Dual 4-input NOR gates | 7490 | Decade counter |
| 7426 | Quad 2-input TTL-MOS interface NAND gates | 7491 | 8-bit shift register |
| | | 7492 | Divide-by-12 counter |
| 7427 | Triple 3-input NOR gates | 7493 | 4-bit binary counter |
| 7428 | Quad 2-input NOR buffer | 7494 | 4-bit shift register |
| 7430 | 8-input NAND gate | 7495 | 4-bit right-shift–left-shift register |
| 7432 | Quad 2-input OR gates | 7496 | 5-bit parallel-in–parallel-out shift register |
| 7437 | Quad 2-input NAND buffers | 74100 | 4-bit bistable latch |
| 7438 | Quad 2-input NAND buffers (open collector) | 74104 | JK master-slave flip-flop |
| 7439 | Quad 2-input NAND buffers (open collector) | 74105 | JK master-slave flip-flop |
| 7440 | Dual 4-input NAND buffers | 74107 | Dual JK master-slave flip-flop |
| 7441 | BCD-to-decimal decoder-Nixie driver | 74109 | Dual JK positive-edge-triggered flip-flop |
| 7442 | BCD-to-decimal decoder | 74116 | Dual 4-bit latches with clear |
| 7443 | Excess 3-to-decimal decoder | 74121 | Monostable multivibrator |
| 7444 | Excess Gray-to-decimal | 74122 | Monostable multivibrator with clear |
| 7445 | BCD-to-decimal decoder-driver | 74123 | Monostable multivibrator |
| 7446 | BCD-to-seven segment decoder-drivers (30-V output) | 74125 | Three-state quad bus buffer |
| | | 74126 | Three-state quad bus buffer |
| 7447 | BCD-to-seven segment decoder-drivers (15-V output) | 74132 | Quad Schmitt trigger |
| | | 74136 | Quad 2-input EXCLUSIVE-OR gate |
| 7448 | BCD-to-seven segment decoder-drivers | 74141 | BCD-to-decimal decoder-driver |
| 7450 | Expandable dual 2-input 2-wide AND-OR-INVERT gates | 74142 | BCD counter-latch-driver |
| | | 74145 | BCD-to-decimal decoder-driver |
| 7451 | Dual 2-input 2-wide AND-OR-INVERT gates | 74147 | 10/4 priority encoder |
| 7452 | Expandable 2-input 4-wide AND-OR gates | 74148 | Priority encoder |
| 7453 | Expandable 2-input 4-wide AND-OR-INVERT gates | 74150 | 16-line–to–1-line multiplexer |
| | | 74151 | 8-channel digital multiplexer |
| 7454 | 2-input 4-wide AND-OR-INVERT gates | 74152 | 8-channel data selector-multiplexer |
| 74153 | Dual 4/1 multiplexer | 74190 | Up-down decade counter |

| Number | Function | Number | Function |
|--------|----------|--------|----------|
| 74154 | 4-line–to–16-line decoder-demultiplexer | 74191 | Synchronous binary up-down counter |
| 74155 | Dual 2/4 demultiplexer | 74192 | Binary up-down counter |
| 74156 | Dual 2/4 demultiplexer | 74193 | Binary up-down counter |
| 74157 | Quad 2/1 data selector | 74194 | 4-bit directional shift register |
| 74160 | Decade counter with asynchronous clear | 74195 | 4-bit parallel-access shift register |
| 74161 | Synchronous 4-bit counter | 74196 | Presettable decade counter |
| 74162 | Synchronous 4-bit counter | 74197 | Presettable binary counter |
| 74163 | Synchronous 4-bit counter | 74198 | 8-bit shift register |
| 74164 | 8-bit serial shift register | 74199 | 8-bit shift register |
| 74165 | Parallel-load 8-bit serial shift register | 74221 | Dual one-shot Schmitt trigger |
| 74166 | 8-bit shift register | 74251 | Three-state 8-channel multiplexer |
| 74173 | 4-bit three-state register | 74259 | 8-bit addressable latch |
| 74174 | Hex $F$ flip-flop with clear | 74276 | Quad $JK$ flip-flop |
| 74175 | Quad $D$ flip-flop with clear | 74279 | Quad debouncer |
| 74176 | 35-MHz presettable decade counter | 74283 | 4-bit binary full adder with fast carry |
| 74177 | 35-MHz presettable binary counter | 74284 | Three-state 4-bit multiplexer |
| 74179 | 4-bit parallel-access shift register | 74285 | Three-state 4-bit multiplexer |
| 74180 | 8-bit odd-even parity generator-checker | 74365 | Three-state hex buffers |
| 74181 | Arithmetic-logic unit | 74366 | Three-state hex buffers |
| 74182 | Look-ahead carry generator | 74367 | Three-state hex buffers |
| 74184 | BCD-to-binary converter | 74368 | Three-state hex buffers |
| 74185 | Binary-to-BCD converter | 74390 | Individual clocks with flip-flops |
| 74189 | Three-state 64-bit random-access memory | 74393 | Dual 4-bit binary counter |

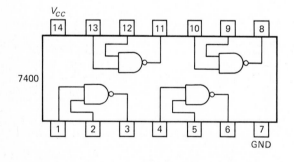

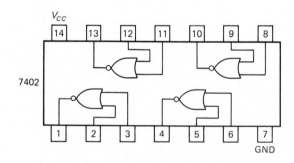

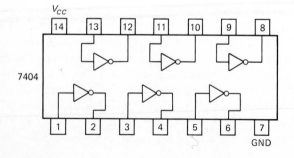

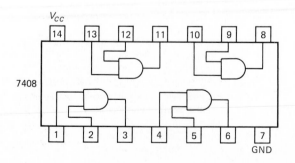

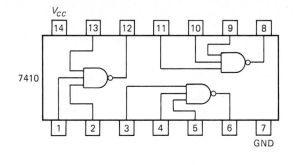

7410

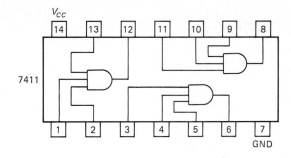

7411

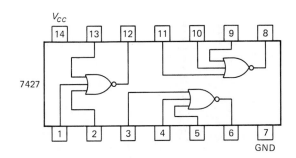

7420

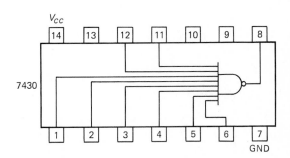

7421

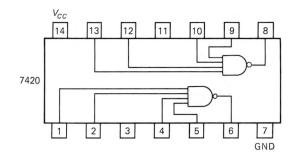

7427

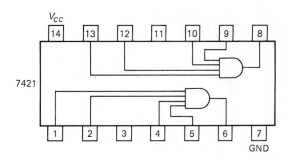

7430

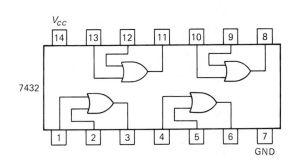

7432

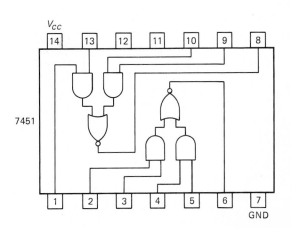

7451

# APPENDIX 4.   CODES

TABLE 1

| Decimal | BCD | Binary |
|---|---|---|
| 0 | 0000 | 0000 |
| 1 | 0001 | 0001 |
| 2 | 0010 | 0010 |
| 3 | 0011 | 0011 |
| 4 | 0100 | 0100 |
| 5 | 0101 | 0101 |
| 6 | 0110 | 0110 |
| 7 | 0111 | 0111 |
| 8 | 1000 | 1000 |
| 9 | 1001 | 1001 |
| 10 | 0001 0000 | 1010 |
| 11 | 0001 0001 | 1011 |
| 12 | 0001 0010 | 1100 |
| 13 | 0001 0011 | 1101 |
| ... | ...... | ... |
| 98 | 1001 1000 | 1100010 |
| 99 | 1001 1001 | 1100011 |
| 100 | 0001 0000 0000 | 1100100 |
| 101 | 0001 0000 0001 | 1100101 |
| 102 | 0001 0000 0010 | 1100110 |
| ... | ......... | ... |
| 578 | 0101 0111 1000 | 1001000010 |
| ... | ......... | ...... |

TABLE 2 4-BIT BCD CODES

| Decimal | 7421 | 6311 | 5421 | 5311 | 5211 |
|---|---|---|---|---|---|
| 0 | 0000 | 0000 | 0000 | 0000 | 0000 |
| 1 | 0001 | 0001 | 0001 | 0001 | 0001 |
| 2 | 0010 | 0011 | 0010 | 0011 | 0011 |
| 3 | 0011 | 0100 | 0011 | 0100 | 0101 |
| 4 | 0100 | 0101 | 0100 | 0101 | 0111 |
| 5 | 0101 | 0111 | 1000 | 1000 | 1000 |
| 6 | 0110 | 1000 | 1001 | 1001 | 1001 |
| 7 | 1000 | 1001 | 1010 | 1011 | 1011 |
| 8 | 1001 | 1011 | 1011 | 1100 | 1101 |
| 9 | 1010 | 1100 | 1100 | 1101 | 1111 |

TABLE 3 *MORE 4-BIT BCD CODES*

| Decimal | 4221 | 3321 | 2421 | $84\overline{2}\overline{1}$ | $74\overline{2}\overline{1}$ |
|---|---|---|---|---|---|
| 0 | 0000 | 0000 | 0000 | 0000 | 0000 |
| 1 | 0001 | 0001 | 0001 | 0111 | 0111 |
| 2 | 0010 | 0010 | 0010 | 0110 | 0110 |
| 3 | 0011 | 0011 | 0011 | 0101 | 0101 |
| 4 | 1000 | 0101 | 0100 | 0100 | 0100 |
| 5 | 0111 | 1010 | 1011 | 1011 | 1010 |
| 6 | 1100 | 1100 | 1100 | 1010 | 1001 |
| 7 | 1101 | 1101 | 1101 | 1001 | 1000 |
| 8 | 1110 | 1110 | 1110 | 1000 | 1111 |
| 9 | 1111 | 1111 | 1111 | 1111 | 1110 |

TABLE 4 *5-BIT BCD CODES*

| Decimal | 2-out-of-5 | 63210 | Shift-Counter | 86421 | 51111 |
|---|---|---|---|---|---|
| 0 | 00011 | 00110 | 00000 | 00000 | 00000 |
| 1 | 00101 | 00011 | 00001 | 00001 | 00001 |
| 2 | 00110 | 00101 | 00011 | 00010 | 00011 |
| 3 | 01001 | 01001 | 00111 | 00011 | 00111 |
| 4 | 01010 | 01010 | 01111 | 00100 | 01111 |
| 5 | 01100 | 01100 | 11111 | 00101 | 10000 |
| 6 | 10001 | 10001 | 11110 | 01000 | 11000 |
| 7 | 10010 | 10010 | 11100 | 01001 | 11100 |
| 8 | 10100 | 10100 | 11000 | 10000 | 11110 |
| 9 | 11000 | 11000 | 10000 | 10001 | 11111 |

TABLE 5 *MORE THAN-5-BIT CODES*

| Decimal | 50 43210 | 543210 | 9876543210 |
|---|---|---|---|
| 0 | 01 00001 | 000001 | 0000000001 |
| 1 | 01 00010 | 000010 | 0000000010 |
| 2 | 01 00100 | 000100 | 0000000100 |
| 3 | 01 01000 | 001000 | 0000001000 |
| 4 | 01 10000 | 010000 | 0000010000 |
| 5 | 10 00001 | 100001 | 0000100000 |
| 6 | 10 00010 | 100010 | 0001000000 |
| 7 | 10 00100 | 100100 | 0010000000 |
| 8 | 10 01000 | 101000 | 0100000000 |
| 9 | 10 10000 | 110000 | 1000000000 |

**TABLE 6** *EXCESS-3 CODE*

| Decimal | BCD | Excess-3 |
|---------|------|----------|
| 0 | 0000 | 0011 |
| 1 | 0001 | 0100 |
| 2 | 0010 | 0101 |
| 3 | 0011 | 0110 |
| 4 | 0100 | 0111 |
| 5 | 0101 | 1000 |
| 6 | 0110 | 1001 |
| 7 | 0111 | 1010 |
| 8 | 1000 | 1011 |
| 9 | 1001 | 1100 |

**TABLE 7** *GRAY CODE*

| Decimal | Gray Code | Binary |
|---------|-----------|--------|
| 0 | 0000 | 0000 |
| 1 | 0001 | 0001 |
| 2 | 0011 | 0010 |
| 3 | 0010 | 0011 |
| 4 | 0110 | 0100 |
| 5 | 0111 | 0101 |
| 6 | 0101 | 0110 |
| 7 | 0100 | 0111 |
| 8 | 1100 | 1000 |
| 9 | 1101 | 1001 |
| 10 | 1111 | 1010 |
| 11 | 1110 | 1011 |
| 12 | 1010 | 1100 |
| 13 | 1011 | 1101 |
| 14 | 1001 | 1110 |
| 15 | 1000 | 1111 |
| ... | ... | ... |

# APPENDIX 5.   BCD CODES

## Hollerith Code

Probably the most widely used system for recording information on a punched card is the *Hollerith code*. In this code the numbers 0 through 9 are represented by a single punch in a vertical column. For example, a hole punched in the fifth row of column 12 represents a 5 in that column. The letters of the alphabet are represented by two punches in any one column. The letters A and I are represented by a zone punch in row 12 and a punch in rows 1 through 9. The letters J through R are represented by a zone punch in row 11 and a punch in rows 1 through 9. The letters S through Z are represented by a zone punch in row 0 and a punch in rows 2 through 9. Thus, any of the 10 decimal digits and any of the 26 letters of the alphabet can be represented in a binary fashion by punching the proper holes in the card. In addition, a number of special characters can be represented by punching combinations of holes in a column which are not used for the numbers or letters of the alphabet.

An easy device for remembering the alphabetic characters is the phrase "JR. is 11." Notice that the letters J through R have an 11 punch, those before have a 12 punch, and those after have a 0 punch. It is also necessary to remember that S begins on a 2 and not a 1.

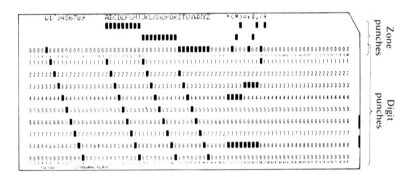

Standard punched card using Hollerith code.

## Eight-Hole Code

There are a number of codes for punching data in paper tape, but one of the most widely used is the *eight-hole code*. Holes, representing data, are punched in eight parallel channels which run the length of the tape. (The channels are labeled 1, 2, 4, 8, parity, 0, *X*, and end of line.) Each character—numeric, alphabetic, or special—occupies one column of eight positions across the width of the tape.

Numbers are represented by punches in one or more channels labeled 0, 1, 2, 4, and 8, and each number is the sum of the punch positions. For example, 0 is represented by a single punch in the 0 channel; 1 is represented by a single punch in the 1 channel; 2 is a single punch in channel 2; 3 is a punch in channel 1 and a punch in channel 2, etc.

## The American Standard Code for Information Exchange*

| | 000 | 001 | 010 | 011 | 100 | 101 | 110 | 111 |
|---|---|---|---|---|---|---|---|---|
| 0000 | NULL | ① $DC_0$ | b | 0 | @ | P | | |
| 0001 | SOM | $DC_1$ | ! | 1 | A | Q | | |
| 0010 | EOA | $DC_2$ | '' | 2 | B | R | | |
| 0011 | EOM | $DC_3$ | # | 3 | C | S | | |
| 0100 | EOT | $DC_4$ (Stop) | $ | 4 | D | T | | |
| 0101 | WRU | ERR | % | 5 | E | U | | |
| 0110 | RU | SYNC | & | 6 | F | V | | |
| 0111 | BELL | LEM | ' | 7 | G | W | | |
| 1000 | $FE_0$ | $S_0$ | ( | 8 | H | X | Unassigned | |
| 1001 | HT / SK | $S_1$ | ) | 9 | I | Y | | |
| 1010 | LF | $S_2$ | * | : | J | Z | | |
| 1011 | $V_{TAB}$ | $S_3$ | + | ; | K | [ | | |
| 1100 | FF | $S_4$ | , | < | L | \ | | ACK |
| 1101 | CR | $S_5$ | − | = | M | ] | | ② |
| 1110 | SO | $S_6$ | * | > | N | ↑ | | ESC |
| 1111 | SI | $S_7$ | / | ? | O | ← | | DEL |

Example

| 100 | 0001 | = | A |
|---|---|---|---|

$b_1$ - - - - - - - - $b_1$

The abbreviations used in the figure mean:

| | | | |
|---|---|---|---|
| NULL | Null idle | CR | Carriage return |
| SOM | Start of message | SO | Shift out |
| EOA | End of address | SI | Shift in |
| EOM | End of message | $DC_0$ | Device control   1 |
| | | | Reserved for data |
| | | | Link escape |
| EOT | End of transmission | $DC_1$—$DC_2$ | Device control |
| WRU | "Who are you?" | ERR | Error |
| RU | "Are you . . . ?" | SYNC | Synchronous idle |
| BELL | Audible signal | LEM | Logical end of media |
| FE | Format effector | $SO_0$—$SO_7$ | Separator (information) |
| HT | Horizontal tabulation | | Word separator (blank, normally non-printing) |
| SK | Skip (punched card) | ACK | Acknowledge |
| LF | Line feed | 2 | Unassigned control |
| V/TAB | Vertical tabulation | ESC | Escape |
| FF | Form feed | DEL | Delete idle |

*Reprinted from *Digital Computer Fundamentals* by Thomas C. Bartee. Copyright 1960, 1966 by McGraw-Hill, Inc. Used with permission of McGraw-Hill Book Company.

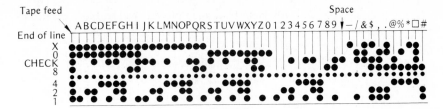

Alphabetic characters are represented by a combination of punches in channels $X$, 0, 1, 2, 4, and 8. Channels $X$ and 0 are used much as the zone punches in punched cards. For example, the letter $A$ is designated by punches in channels $X$, 0, and 1. The special characters are represented by combinations of punches in all channels which are not used to designate either numbers or letters. A punch in the end-of-line channel signifies the end of a block of information, or the end of record. This is the only time a punch appears in this channel.

As a means of checking the validity of the information punched on the tape, the parity channel is used to ensure that each character is represented by an *odd* number of holes. For example, the letter C is represented by punches in channels $X$, 0, 1, and 2. Since an odd number of holes is required for each character, the code for the letter C also has a punch in the parity channel, and thus a total of five punches is used for this letter.

## Universal Product Code (UPC)

The Universal Product Code (UPC) symbol in Fig. A is an example of a machine-readable label that appears on virtually every kind of retail grocery product. It is the result of an

**FIG. A.** UPC symbol from box of Kleenex tissues. Registered trademark of Kimberly-Clark Corp., Neenah, Wis.

industry-wide attempt to improve productivity through the use of automatic checkstand equipment. The standard symbol consists of a number of parallel light and dark bars of variable widths.

The symbol is designed around a 10-digit numbering system, 5 digits being assigned as an identification number for each manufacturer and the remaining 5 digits being used to identify a specific product, e.g., creamed corn, pea soup, or catsup. Each symbol can be read by a fixed-position scanner, as on a conveyor belt, or by a hand-held wand. The code numbers are printed on each symbol under the bars as a convenience in the event of equipment failure.

## Specifications

Each 10-digit symbol is rectangular in shape and consists of exactly 30 dark and 29 light vertical bars, as seen in Fig. B. Each digit is represented by two dark bars and two light

**FIG. B.** UPC standard symbol.

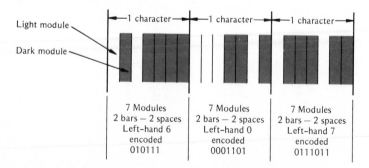

**FIG. C.** UPC character construction.

spaces. To account for the variable widths of the bars, each digit or character is broken down into seven modules. A module can be either dark or light, and each dark bar is made up of 1, 2, 3, or 4 dark modules. An example is shown in Fig. C. Notice that each digit or character has exactly seven modules, and each digit has two dark bars and two light spaces. Dark modules are 1s while light modules are 0s.

Characters are encoded differently at the left and at the right of center. A left-side character begins with a light space and ends with a dark bar and always consists of either three or five dark modules (odd parity). A right-side character begins with a dark bar and ends with a light space and always has either two or four dark modules (even parity). The encoding for each character is summarized in Fig. D.

| Decimal number | Left characters (odd parity) | Right characters (even parity) |
|---|---|---|
| 0 | 0001101 | 1110010 |
| 1 | 0011001 | 1100110 |
| 2 | 0010011 | 1101100 |
| 3 | 0111101 | 1000010 |
| 4 | 0100011 | 0011100 |
| 5 | 0110001 | 0001110 |
| 6 | 0101111 | 1010000 |
| 7 | 0111011 | 1000100 |
| 8 | 0110111 | 1001000 |
| 9 | 0001011 | 1110100 |

**FIG. D.** UPC character encoding.

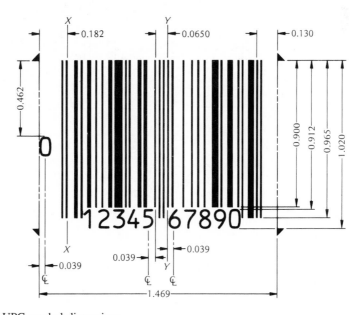

**FIG. E.** UPC symbol dimensions.

## CHAPTER 1

**1–1.** 01110000

**1–3.** a. 1  b. 2  c. 3  d. 4

**1–5.** 2 mA for either $h_{FE}$

**1–7.** Low; high

**1–9.** The truth table is:

| A | B | C | Y |
|---|---|---|---|
| 0 | 0 | 0 | 0 |
| 0 | 0 | 1 | 1 |
| 0 | 1 | 0 | 1 |
| 0 | 1 | 1 | 1 |
| 1 | 0 | 0 | 1 |
| 1 | 0 | 1 | 1 |
| 1 | 1 | 0 | 1 |
| 1 | 1 | 1 | 1 |

This is the truth table of a 3-input OR gate. Therefore, a cascade of two 2-input OR gates is equivalent to a 3-input OR gate.

**1–11.** The truth table is

| A | B | C | Y |
|---|---|---|---|
| 0 | 0 | 0 | 1 |
| 0 | 0 | 1 | 0 |
| 0 | 1 | 0 | 0 |
| 0 | 1 | 1 | 0 |
| 1 | 0 | 0 | 0 |
| 1 | 0 | 1 | 0 |
| 1 | 1 | 0 | 0 |
| 1 | 1 | 1 | 0 |

**1–13.** The truth table is

| A | B | C | Y |
|---|---|---|---|
| 0 | 0 | 0 | 0 |
| 0 | 0 | 1 | 0 |
| 0 | 1 | 0 | 0 |
| 0 | 1 | 1 | 0 |
| 1 | 0 | 0 | 0 |
| 1 | 0 | 1 | 0 |
| 1 | 1 | 0 | 0 |
| 1 | 1 | 1 | 1 |

**1–15.** The truth table is

| A | B | C | Y |
|---|---|---|---|
| 0 | 0 | 0 | 1 |
| 0 | 0 | 1 | 1 |
| 0 | 1 | 0 | 1 |
| 0 | 1 | 1 | 1 |
| 1 | 0 | 0 | 1 |
| 1 | 0 | 1 | 1 |
| 1 | 1 | 0 | 1 |
| 1 | 1 | 1 | 0 |

**1–17.** The Boolean equations are

$$Y = A + B + C$$
$$Y = \overline{(A + B)}$$
$$Y = \overline{(A + B + C)}$$

**1–19.** The Boolean equations are

$$Y = ABC$$
$$Y = \overline{AB}$$
$$Y = \overline{ABC}$$

**1–21.** Fig. 1 shows the logic circuit.

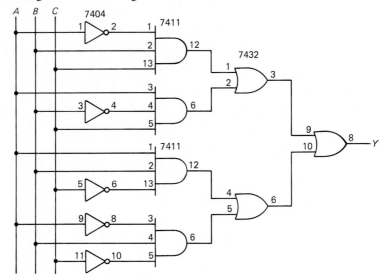

**1–23.** Here is a summary of the truth table. $Y$ equals 1 when $ABCD = 0000$; $Y$ equals 0 for all other $ABCD$ inputs. There are 16 $ABCD$ inputs, starting with 0000 and ending with 1111.

**1–25.** d

**1–27.** $Y = \overline{A_0 A_1 A_2 A_3 A_4 A_5 A_6 A_7}$

**1–29.** a

# CHAPTER 2

**2–1.** Draw an AND–OR circuit with two AND gates and one OR gate. The upper AND gate has inputs of $A$, $\overline{B}$, and $C$. The lower AND gate has inputs of $A$, $B$, and $C$. The simplified logic circuit is an AND gate with inputs of $A$ and $C$.

**2–3.** The lower input gate

**2–5.** d

**2–7.** $Y = \overline{A}CD + A\overline{B}C + AB\overline{C}$ which means an AND–OR circuit that ORs the foregoing logical products

**2–9.** $Y = \overline{A}B + A\overline{B}$, which implies an AND–OR circuit that ORs the foregoing products

**2–11.** Figure 2a shows the Karnaugh map.

|  | $\overline{C}\overline{D}$ | $\overline{C}D$ | $CD$ | $C\overline{D}$ |
|---|---|---|---|---|
| $\overline{A}\overline{B}$ | 0 | 1 | 1 | 1 |
| $\overline{A}B$ | 0 | 0 | 1 | 0 |
| $AB$ | 0 | 1 | 0 | 0 |
| $A\overline{B}$ | 1 | 1 | 0 | 1 |

|  | $\overline{C}\overline{D}$ | $\overline{C}D$ | $CD$ | $C\overline{D}$ |
|---|---|---|---|---|
| $\overline{A}\overline{B}$ | 0 | 1 | 1 | 1 |
| $\overline{A}B$ | 0 | 0 | 1 | 0 |
| $AB$ | 0 | 1 | 0 | 0 |
| $A\overline{B}$ | 1 | 1 | 0 | 1 |

(a)  (b)

**2–13.** Figure 2b shows the Karnaugh map.

**2–15.** The simplified equation is

$$Y = \overline{A}BD + \overline{A}CD + A\overline{B}\,\overline{C} + A\overline{C}D + \overline{B}C\overline{D}$$

The corresponding AND–OR circuit has five AND gates driving an OR gate.

**2–17.** The simplified logic circuit is an AND gate with inputs of $\overline{A}$, $\overline{C}$, and $D$.

**2–19.** $Y = AB + AC$; use an AND–OR circuit to produce this equation

**2–21.** Figure 3 shows the unsimplified logic circuit.

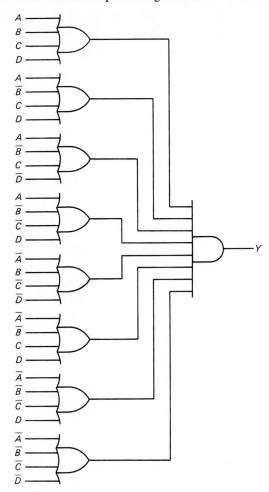

**2–23.** Figure 4 shows the map and the circuit.

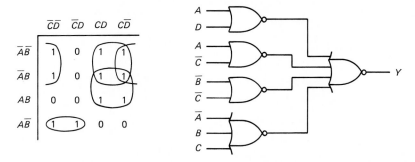

**2–25.** The $Y$ waveform is low between 0 and 7. Then, it is high between 7 and 16.

**2–27.** Figure 5 shows the simplified NAND–NAND circuits.

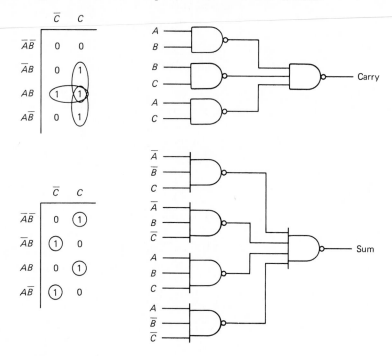

# CHAPTER 3

**3–1.** $Y$ equals $D_9$.

**3–3.** Connect the data inputs as follows: +5 V—$D_0$, $D_4$, $D_5$, $D_6$, $D_{11}$, $D_{12}$, $D_{14}$, and $D_{15}$; Ground—$D_1$, $D_2$, $D_3$, $D_7$, $D_8$, $D_9$, $D_{10}$, and $D_{13}$.

**3–5.** $Y_3$ multiplexer: ground $D_1$, $D_6$, $D_7$, and $D_{14}$; all other data inputs high. $Y_2$ multiplexer: ground $D_3$, $D_8$, and $D_{13}$; all other data inputs high. $Y_1$ multiplexer: ground $D_0$, $D_1$, $D_{14}$, and $D_{15}$; all other data inputs high. $Y_0$ multiplexer: ground $D_8$, $D_9$, and $D_{13}$; all other data inputs high.

**3–7.** None; $Y_5$

**3–9.** c

**3–11.** The chip on the right; $Y_6$

**3–13.** a. 67   b. 813   c. 7259

**3–15.** $Y_7$

**3–17.** c

**3–19.** Approximately 3 mA

**3–21.** Pin 5; 0111

**3–23.** a. 0   b. 1   c. 0   d. 1

**3–25.** a. 0  b. 1  c. 0

**3–27.** Ground pin 4 (after disconnecting from +5 V) and connect pin 3 to +5 V (after disconnecting from ground).

**3–29.** 256

**3–31.** 1100

**3–33.** Figure 6 shows the PROM.

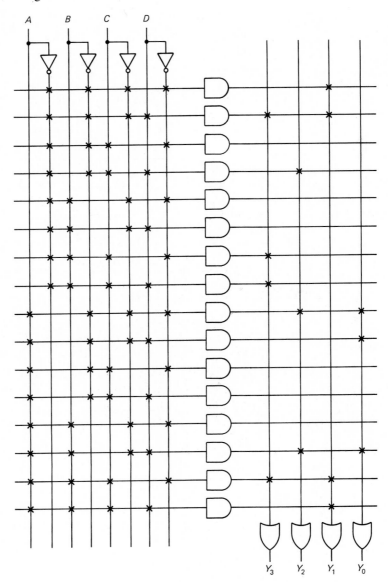

**3–35.** Figure 7 shows the PAL circuit.

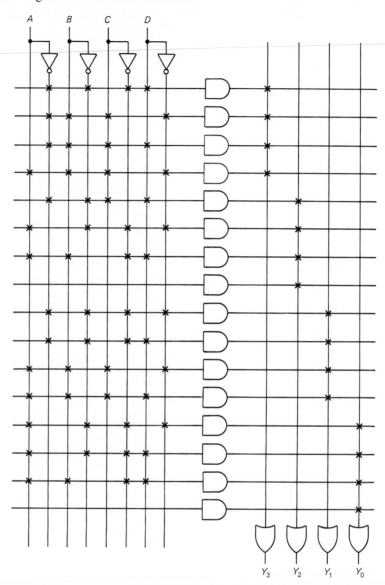

## CHAPTER 4

**4–1.** 1000011, 1011011, 0011000, 1110011, 0100100, 1100110

**4–3.** a. 188  b. 255

**4–5.** 131,072

**4–7.** 100001100

**4–9.** 10100$^1$0100000

**4–11.** 011 010 101.111 011 110

**4–13.** 504.771

**4–15.** a. 257   b. 15.331   c. 123.55

**4–17.** a. 1110 0101   b. 1011 0100 1101   c. 0111 1010 1111 0100

**4–19.** 12,121

**4–21.** a. 0000   b. 0100   c. 1010   d. 1111

**4–23.** a. 011 0111   b. 101 0111   c. 110 0110   d. 111 1001

**4–25.**

| Address | Alphanumeric | Hex contents |
|---------|--------------|--------------|
| 2000 | G | C7 |
| 2001 | O | 4F |
| 2002 | O | 4F |
| 2003 | D | C4 |
| 2004 | B | C2 |
| 2005 | Y | D9 |
| 2006 | E | 45 |

**4–27.** 3694

**4–29.** 0001 0001

**4–31.** a, b, and d

**4–33.** e

# CHAPTER 5

**5–1.** a. $12_8$   b. $13_8$   c. $10_{16}$   d. $17_{16}$

**5–3.** 0000 0101 0000 1000

**5–5.** 0001 1000

**5–7.** a. 0001 0111   b. 0111 1011   c. 1011 1000   d. 1110 1011

**5–9.** a. DCH   b. BAH   c. 36H   d. O2H

**5–11.** a. 0100 1110   b. 1110 1001   c. 1010 0110   d. 1000 0111

**5–13.**
a.     0010 1101      b.     0101 1001      c.     0100 0011
     +0011 1000            +1101 1110            +1001 1110
     ─────────            ─────────            ─────────
      0110 0101             0011 0111             1110 0001

**5–15.** 02CBH, 0000 0010 1100 1011

**5–17.** Binary 0101 0011, or decimal 83

# CHAPTER 6

**6–1.** a. LSI   b. MSI   c. SSI   d. LSI

**6–3.** Low-power Schottky

**6–5.** 100

**6–7.** 20

**6–9.** a. 1   b. 1   c. 0   d. 0

**6–11.** High; low

**6–13.** $312\ \Omega$

**6–15.** Use a low DISABLE with register $B$; all other DISABLES are high.

**6–17.** $3.03$ mA

**6–19.** $14.4$ mA

**6–21.** a. Active-low   b. Active-low   c. Active-high   d. Active-low

**6–23.** On; negative true

**6–25.** Use the circuit of Fig. 8a, except include three hex inverters as shown in Fig. 8b.

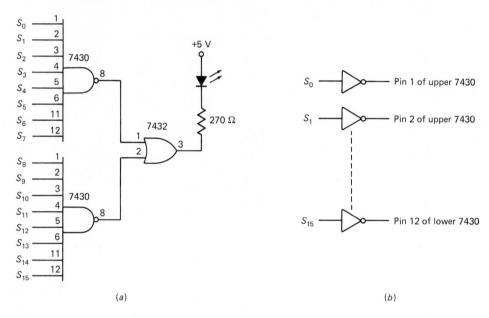

(a)                                                    (b)

# CHAPTER 7

**7–1.**   a. Non   b. Non   c. Non   d. Conducting

**7–3.**   12 V; 1.2 mA; conducting

**7–5.**   300 ns

**7–7.**   Low; high

**7–9.**   Connecting pin 13 to the supply voltage will produce a permanent high input. This will force the output to stick in the low state, regardless of what values $A$ and $B$ have.

**7–11.**  Grounding pin 1 will force the output to remain permanently in the high state, no matter what the values of $A$ and $B$.

**7–13.**  3.6 mA

**7–15.**  2.27 mA; 88 mV

**7–17.**  d

## CHAPTER 8

**8–3.**

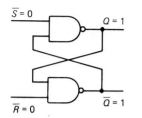 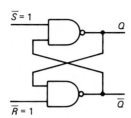 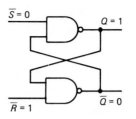 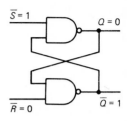

**8–5.** a. $C$ b. $G$

**8–7.** When the clock is low, the flip-flop is insensitive to levels on either $R$ or $S$ input. (Only first case is shown here.)

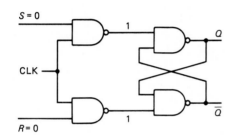 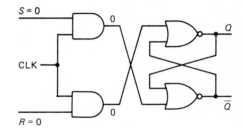

**8–9.** $\phi$

**8–11.** a. 5 ns b. 10 ns c. 15 ns

**8–13.**

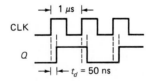

**8–15.** Clock period $= 1$ $\mu$s. Period of $Q = 2$ $\mu$s ($f = 500$ kHz)

**8–17.**

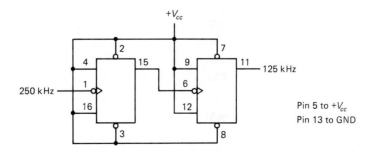

**8–19.**

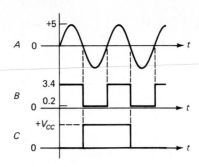

# CHAPTER 9

**9–1.** a. 100 ns b. 167 ns c. 1.33 $\mu$s

**9–3.** 13.3 MHz

**9–5.** 0.45/4.05

**9–7.** 3.5 MHz plus or minus 28 Hz

**9–9.**

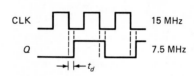

**9–11.** 48 kHz. $t_1 = 13$ $\mu$s, $t_2 = 7.8$ $\mu$s

**9–13.** 33.3 percent, 37.5 percent

**9–15.** $R_A + R_B = 15$ k$\Omega$. $R_A = 3.75$ k$\Omega$,
$R_B = 11.25$ k$\Omega$.

**9–17.** 3.88 ms

**9–19.** 0.136 $\mu$F

**9–21.**

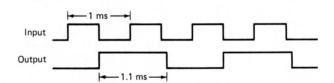

**9–23.** Connect as in Example 9–6. 21.3 nF.

**9–25.** a.

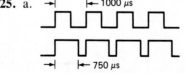

b.

**9–27.** Connect $\overline{A}_1$ to GND and apply input to $B_2$. $C = 44.6$ nF.

**9–29.** Same as Problem 9–25.

**9–31.** $Q$ would time out and go low resulting in multiple pulses at $A'$.

**9–33.**

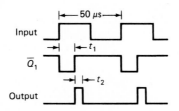

**9–35.** Let $R_1 = R_2 = 1$ k$\Omega$.   a. $t_1 = 2.5$ $\mu$s, $t_2 = 7.5$ $\mu$s, $C_1 = 7500$ pF, $C_2 = 22,500$ pF.   b. $t_1 = t_2 = 1$ $\mu$s, $C_1 = C_2 = 3000$ pF.

# CHAPTER 10

**10–1.**   a. 6  b. 6  c. 4

**10–3.**   See Fig. 10–1

**10–5.**

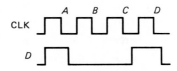

**10–7.**

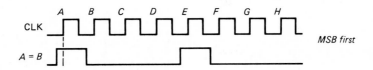

**10–9.**   a. 8 $\mu$s   b. 1.6 $\mu$s

**10–11.** 16.7 MHz

**10–13.** a. $R = 1, S = 0, Q = 0$   b. $R = 0, S = 1, Q = 1$

**10–15.**

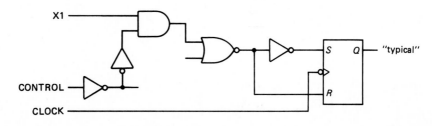

**10–17.** MSB first shift/load is low, *ABCD EFGH* = 1011 1110.

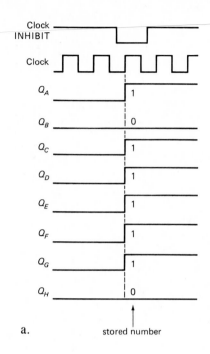

a.                  stored number

**b.** MSB first, shift/load is high.

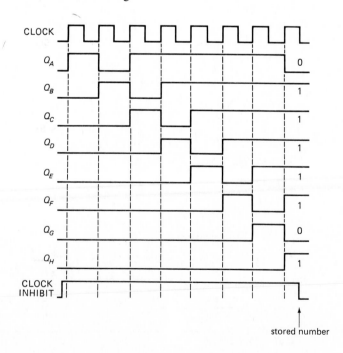

stored number

**10–19.** Same as 10–15.

**10–21.**

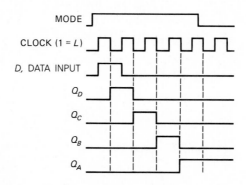

**10–23.** b. For alternate 1s and 0s, replace feedback with:

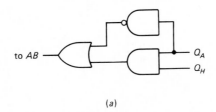

(a)

Include a ''power-on-CLEAR'' circuit like:

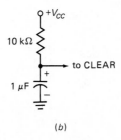

(b)

This will CLEAR all flip-flops to zeros when power is first applied.

# CHAPTER 11

**11–1.**

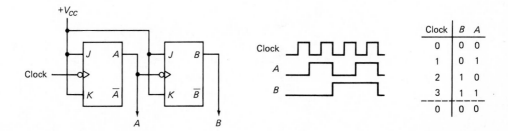

**11–3.** 4 MHz

**11–5.** Difficult to get clock and first few flip-flop outputs on the same page with the last few flip-flop outputs.

**11–7.** Same as Fig. 11–3 where period of $QB$ is 2 $\mu$s.

**11–9.** Sixteen, 4-input NAND–gates with inputs, $\overline{A}\overline{B}\overline{C}\overline{D}$, $A\overline{B}\overline{C}\overline{D}$, $\overline{A}B\overline{C}\overline{D}$, $AB\overline{C}\overline{D}$, $\overline{A}\overline{B}C\overline{D}$, $A\overline{B}C\overline{D}$, $\overline{A}BC\overline{D}$, $ABC\overline{D}$, $\overline{A}\overline{B}\overline{C}D$, $A\overline{B}\overline{C}D$, $\overline{A}B\overline{C}D$, $AB\overline{C}D$, $\overline{A}\overline{B}CD$, $A\overline{B}CD$, $\overline{A}BCD$, and $ABCD$.

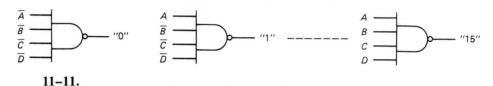

**11–11.**

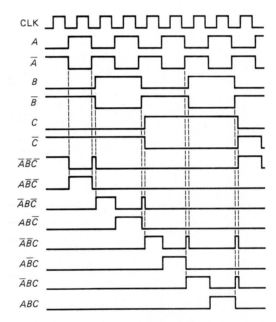

**11–13.** Same as Fig. 11–12c, except transitions occur on low-to-high clock.

**11–15.** As in Fig. 11–15.

**11–17.** a. 3 b. 4 c. 4 d. 5 e. 5

**11–19.**

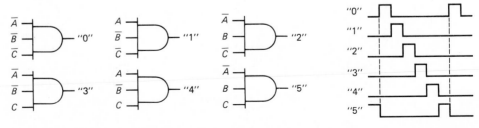

**11–21.**

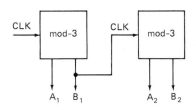

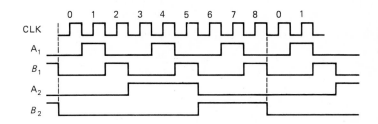

**11–23.**

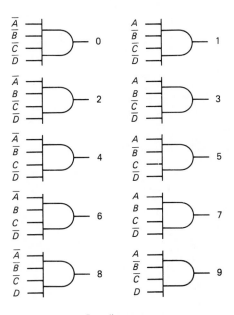

Decoding gates

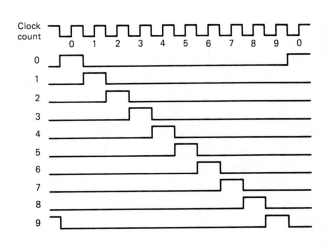

Decade-counter outputs

**11–25.**

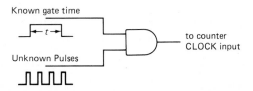

**11–27.**

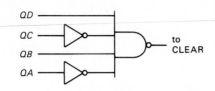

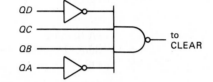

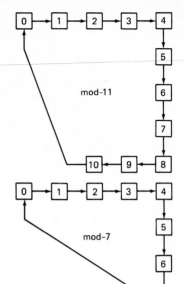

mod-11

mod-7

**11–29.** Like problem 11–27.

**11–31.** a. 3  b. 4  c. 5  d. 5  e. 11

**11–33.**

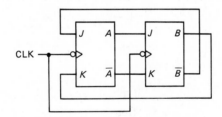

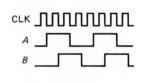

| B | A | Count |
|---|---|-------|
| 0 | 1 | 1 |
| 1 | 1 | 2 |
| 1 | 0 | 3 |
| 0 | 0 | 4 |

**11–35.** Five clock periods.

**11–39.** Reconnect the $J$ input on flip-flop $A$ to $\overline{D}$.

**11–41.** Mod-5 illegal states are 2(010), 5(101), and 7(111).
AND–gate will detect $\overline{A}B$ and force counter to $CB\overline{A}$.
Count 7 will progress to count 6(110). Mod-3 illegal state is 3(11), which
will progress naturally to count 2(10).

**11–43.**

# CHAPTER 12

**12–1.** a. RAM  b. ROM  c. EPROM

**12–3.** Loss of power results in loss of memory.

**12–5.** 1x32, 2x16, 4x8

**12–7.** 1001

**12–9.** 16,384 bits of memory.   16 address bits.

**12–11.** $EDCBA$ = 11011 = 1B.

**12–13.** "15" = $(\overline{EDCBA})\overline{G}$

**12–15.** Exactly like Fig. 12–8 with $t_p$ = 35 ns.

**12–17.** $\overline{CS}/WE$ is low selects chip for reading data out and must be high when chip is programmed.

**12–19.** Two identical chips connected together as follows:
a) Select inputs ($A$ to $A$, $B$ to $B$, $C$ to $C$, $D$ to $D$)
b) Data inputs ($D_1$ to $D_1$, $D_2$ to $D_2$, $D_3$ to $D_3$, $D_4$ to $D_4$)
c) Sense outputs ($S_1$ to $S_1$, $S_2$ to $S_2$, $S_3$ to $S_3$, $S_4$ to $S_4$)
Now, $ME$ and $WE$ are used to select one chip or the other.

**12–21.** You must accomplish the following:
READ: a. $ME$ goes low for a given time period.
       b. The SELECT inputs (address) must be stable while $ME$ is low. Then DATA is valid at the outputs for the time shown in Fig. 12–13$d$.
WRITE: a. $ME$ must be low for a given time period.
       b. The Select inputs (address) and the Data inputs must be stable while $ME$ is low.
       c. $WE$ must go low for a time $t_w$ as in Fig. 12–13c.

**12–23.** Draw two circuits exactly like Fig. 12–16, one below the other. Now, connect:
a. The ADDRESS lines in parallel.
b. the two $ME$ lines together.
c. the two $WE$ lines together.
The four DATA IN and DATA OUT lines from the upper circuit can be considered the 4 LSBs, and those from the lower circuit are the 4 MSBs.

**12–25.**

**12–27.**

**12–29.** $Q_1$ is ON and $Q_2$ is OFF. DATA OUT = $\overline{\text{DATA OUT}}$ = high.

**12–31.** $Q_1$ is ON and $Q_2$ is OFF. DATA OUT = 1, $\overline{\text{DATA OUT}}$ = 0.

**12–33.** $V_o$ is: a. high   b. low   c. Isolated from the input. With a resistive load to ground, $V_o$ is low. With a resistive load to $+V_{CC}$, the output is high.

**12–35.** Select cell by taking ROW and COLUMN both high. Data bit stored then appears at DATA OUT.

**12–37.** Same as Problem 12–35.

# CHAPTER 13

**13–1.** $\frac{1}{63}$, $\frac{2}{63}$, $\frac{4}{63}$, $\frac{8}{63}$, $\frac{16}{63}$, $\frac{32}{63}$

**13–5.** 51.2 mA

**13–7.** a. 0.641 V   b. 0.923 V   c. 0.766 V

**13–9.** 1 part in 4096; 2.44 mV

**13–11.** 31

**13–13.** $A_2A_1A_0 =$ a. 001   b. 010   c. 110

**13–15.** 7 MHz

**13–17.** 12 $\mu$s, not counting delay and control times.

**13–19.** $i_R = i_C$, $V_i/R = V_oC/t$, $\therefore V_o = \dfrac{V_i}{RC} \times t$

**13–21.** 0.01 $\mu$F

**13–25.** They should all be comparable.

# CHAPTER 14

Solutions for the problems at the end of this chapter are not unique—that is, there are many different designs that will satisfy each requirement. Nevertheless, typical designs for each problem can be found in the literature and are therefore not included here. It is intended that you search application notes and other publications supplied by manufacturers in order to satisfy a particular design requirement. This will provide the opportunity to see numerous different applications, and at the same time challenge you to improve on existing logic configurations. Here are some suggested references:

"Data Sheets and Application Notes," Intersil, Inc. Cupertino, CA;

"Linear Integrated Circuits Data Book," Technical Information Center, Motorola Semiconductor Products, Inc. Phoenix, AZ, 1979;

"Linear Applications Handbook," National Semiconductor Corporation, Santa Clara, CA, 1980;

"Data Acquisition Handbook," National Semiconductor Corporation, Santa Clara, CA, 1978;

"TTL Data Book for Design Engineers," "Linear Circuits Data Book," "Interface Circuits Data Book," "Optoelectronics Data Book," Texas Instruments, Information Publishing Center, P.O. Box 225012 MS-54, Dallas, TX 75265

# Index

# Index

N-channel MOSFET, 233
NAND gate, 28–34, 202, 236–237
NAND–NAND circuit, 33
Natural count, 375
Negative false, 220
Negative gate, 218
Negative logic, 188, 218
Negative true, 220, 225
Nibble, 3, 35
Nibble multiplexers, 83
90 percent point, 297
Noise immunity, 198–199, 252
Nonvolatile storage, 421
NOR gates, 23–28, 202, 237
NOR–NOR network, 27
NOT gate, 1
NOT operation, 17–18

Octal dabble, 133
Octal numbers, 123, 132–135, 149
Octal odometer, 132–133
Octal-to-binary conversion, 134
Octal-to-decimal conversion, 133
Octets, 57, 72
Odd parity, 102
Odd-parity generator, 104, 116
On-chip decoding, 107–108
One-shot, 297
1's complement, 167
1-of-10 decoder, 93
1-of-16 decoder, 89
1-to-16 demultiplexer, 85
Op-amp drive, 215
Open-collector gate, 209–210
Open-drain interface, 248
OR gate, 1, 8–13, 35, 203
OR operation, 18–19, 43–44
Output, 128
Overflow, 164–165, 170–171, 175–177
Overlapping groups, 58, 72

P-channel (PMOS), 189
Pairs, 54, 72
Parallel counter, 375
Parallel shift, 323

Parity bit, 144–145
Parity checker, 102–103
Parity generation, 103–104, 116
Parity generators-checkers, 102–106
Passive load, 233
Passive pull-up, 210
Positional notation, 125, 134–135
Positive false, 219
Positive gate, 218
Positive logic, 218
Positive true, 219, 225
Power dissipation, 238
Power-on-reset, 321
Preset, 266
Presettable counter, 356, 375
Priority encoder, 99
Product-of-sums equations, 22, 35, 72
Product-of-sums method, 65–67
Profile, 196
Programmable array logic (PAL), 110–111, 114, 116
Programmable ROM (PROM), 110, 380, 386, 421
Propagation delay, 267, 275
Propagation delay time, 191, 235, 238
Pull-up resistance, 214

Quads, 56, 72
Quantization error, 462

Random-access memory (RAM), 127, 380, 394, 421
Read-only memory (ROM), 106–110, 116, 380, 386
Redundant group, 60, 72
Refresh, 405
Reset-and-carry, 2
Ring counter, 319, 323
Ripple counter, 328, 375
Rise time, 297

Sample and hold, 438
Saturation delay time, 192, 225
Schmitt trigger, 204, 225, 272, 275

Schottky TTL, 192
Semiconductor memory, 379
Serial shift, 324
Setup time, 268, 275
74C00 devices, 236–239
7400 devices, 190–193
Seven-segment decoder, 95–97
Seven-segment indicator, 95–96
  multiplexing, 465
Shift counter, 364, 375
Shift register, 301, 324
  capacity, 323
  parallel in-parallel out, 315
  parallel in-serial out, 311
  serial in-parallel out, 308
  serial in-serial out, 303
  types, 302
Sign-magnitude numbers, 165–167
Sink, 225
16-to-1 multiplexer, 38
Small-scale integration (SSI), 189
Soft saturation, 4, 35
Speed-up capacitor, 245, 252
Standard loading, 199–200
State diagram, 362
Static memory, 394, 421
Static power dissipation, 238
Steering logic, 337
Strobe, 71–72, 80, 116
Stuck nodes, 148
Subscripts, 158
Substrate ($p$ region), 232
Sum-of-products equation, 21, 35, 49, 72
Sum-of-products method, 48–52
Summation, 105
Switch, 4
Switch drive, 213–214
Synchronous, 275

10 percent point, 297
Thévenin circuit, 147
Three inputs, 9, 14
Three-state buffer, 212
Three-state TTL, 210–213, 225
Threshold voltage, 232–233
Time measurement, 478
Timers, 278
Timing diagram, 8, 11–13, 35